C0-AVZ-692

FEATURES AND BENEFITS

Elementary Algebra, Part 1, New Edition

The **content of a standard Algebra 1 course is spread over a 2-year period** so that students who need to proceed at a slower pace can experience success, pp. v–xi (contents).

Arithmetic skills are reviewed and reinforced as students learn algebra, pp. 2–4, 28, 56–58, 65–67, 96–98.

Problem solving lessons and special word-problem sets give students the instruction and practice that they need to become good problem solvers, pp. 134–136, 198–200, 337–338.

The **helpful and attractive format,** with **many worked-out examples** and a **minimum of reading,** helps students grasp concepts easily, pp. 8, 17, 30, 127, 175.

Abundant practice and review are provided by oral and written exercise sets (graded by difficulty level) and by Self-Tests, Chapter Tests, Cumulative Reviews, and Extra Practice exercises, pp. 296–297, 304, 313, 406, 410.

Motivating special features include Calculator Corner, Consumer Notes, Career Capsule, Challenge Topics, and an appendix on Programming in BASIC, pp. 13, 33, 161, 261, 441.

This **Teacher's Edition** provides Diagnostic Tests in Arithmetic, Alternate Chapter Tests, and a Guide to Individualized Assignments, as well as Lesson Commentary that gives teaching suggestions for each lesson in the book. The following supplementary materials are available:

> **Solution Key** includes complete solutions to all written exercises.
> **Progress Tests** booklets (with annotated Teacher's Edition) are a useful evaluation tool.

Teacher's Edition

Elementary
Algebra Part 1

New Edition

Richard A. Denholm

Robert G. Underhill

Mary P. Dolciani

EDITORIAL ADVISERS

Andrew M. Gleason
Albert E. Meder, Jr.

HOUGHTON MIFFLIN COMPANY/BOSTON
Atlanta   Dallas   Geneva, Ill.   Lawrenceville, N.J.   Palo Alto   Toronto

About the Authors

RICHARD A. DENHOLM, Supervisor, Office of Teacher Education and Lecturer, Department of Mathematics, University of California at Irvine.

ROBERT G. UNDERHILL, Associate Professor and Coordinator of Mathematics Education, Department of Curriculum and Instruction, University of Houston.

MARY P. DOLCIANI, formerly Professor of Mathematics, Hunter College, City University of New York.

About the Editorial Advisers

ANDREW M. GLEASON, Hollis Professor of Mathematics and Natural Philosophy, Harvard University.

ALBERT E. MEDER, JR., Dean and Vice Provost and Professor of Mathematics, Emeritus, Rutgers, The State University of New Jersey.

The authors of *Elementary Algebra, Part 1* wish to express appreciation to PERSIS O. REDGRAVE, formerly head of the Mathematics Department at the Norwich Free Academy in Norwich, Connecticut, for her valuable contribution to this Teacher's Edition.

Copyright © 1988, 1985, 1983, 1980, 1977 by Houghton Mifflin Company
All rights reserved. No part of this work may be reproduced or transmitted in any form or by any means, electronic or mechanical, including photocopying and recording, or by any information storage or retrieval system, without permission in writing from the publisher.

Printed in the United States of America
ISBN: 0-395-43441-6

ABCDEFGHIJ-RM-8987

CONTENTS

INTRODUCTION

This textbook has been specially designed for those students who wish to take a first-year algebra course but would be unable to maintain the pace of a standard one year course. Together, *Elementary Algebra, Parts 1* and *2* cover the equivalent of a first-year course in algebra, as well as offering review and practice of topics such as integers, fractional numbers, factors, and multiples, usually covered in previous courses. The emphasis in this course is on a gradual, thorough approach to first year algebra without unnecessary stress on structure.

We invite you to browse through the next few pages which illustrate many of the features of this new edition by showing reduced sample pages from the text.

Organization of the Text

The text consists of fourteen chapters, each divided into at least two major sections. The main sections are in turn divided into numbered lessons, most of which are one page or less. Each lesson begins with one or more stated objectives and ends with oral exercises, written exercises and/or word problems.

Self-Tests at the end of each major section allow students to evaluate their progress. Answers to Self-Tests are provided at the end of the text. In addition, the Self-Tests review new vocabulary and symbols when they occur, with convenient references to the appropriate page in the text.

5-6 Using Formulas to Solve Problems

OBJECTIVE
Use perimeter and area formulas to solve problems.

Formulas, which are special kinds of equations, can often be used in solving problems.

EXAMPLE The mainsail on a boat has the shape of a triangle, with a base 6 meters in length. The height of the sail is 8.5 meters. What is the area of the sail in square meters (m^2)?

The area formula for triangles is $A = \dfrac{bh}{2}$.

Replacements for variables ▶ A is unknown.
$$b = 6$$
$$h = 8.5$$

Equation ▶ $A = \dfrac{bh}{2} = \dfrac{6(8.5)}{2}$

$$= \dfrac{51}{2} = 25.5$$

The area of the sail is $25.5\,m^2$.

Match each description in Column 1 with the correct formula in Column 2.

Oral EXERCISES

COLUMN 1	COLUMN 2
1. Perimeter of a square	**A.** $A = lw$
2. Area of a square	**B.** $P = a + b + c$
3. Perimeter of a rectangle	**C.** $P = 4s$
4. Area of a rectangle	**D.** $A = \frac{1}{2}bh$
5. Perimeter of a triangle	**E.** $C = \pi d$
6. Area of a triangle	**F.** $A = s^2$
7. Circumference of a circle	**G.** $A = \pi r^2$
8. Area of a circle	**H.** $P = 2l + 2w$

EQUATIONS AND INEQUALITIES / 137

SELF-TEST 1

Be sure that you understand these terms and symbols.

simplify (p. 2)
equation (p. 5)

= (p. 5)
variable (p. 8)

Simplify:

Section 1-1, p. 2

1. $2 \times (3 + 4)$ 2. $\dfrac{24}{2^2 \times 3}$

True or false?

Section 1-2, p. 5

3. $10 - 6 = 2 \times 3$ 4. $14 - \dfrac{8}{2} = 2 \times 5$

chapter summary

1. Number sentences that contain the symbol = are called **equations**. Those that contain the symbols ≠, <, >, ≤, or ≥ are called **inequalities**.

2. A number sentence that contains a variable is neither true nor false until the variable is replaced by a member of a specified **replacement set**.

3. To solve an equation or inequality, the variable in the equation or in

chapter test

Name the coordinate.

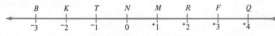

1. B
2. R
3. Q
4. M
5. K
6. The point half the distance from *T* to *M*
7. T
8. F
9. The point half the distance from *K* to *N*

Each chapter is followed by a Chapter Summary, a Chapter Test, and a Review of Skills, which provides practice in skills needed in the next chapter. Alternate Chapter Tests are provided in this Teacher's Edition, beginning on page T12.

review of skills

Simplify.

1. $6 \cdot (10 + 2)$
2. $8 \cdot (9 - 2)$
3. $6(7 \cdot 5)$
4. $2\frac{1}{2} + 3\frac{1}{4}$
5. $\frac{2}{5} + \left(\frac{1}{4} + \frac{1}{2}\right)$
6. $8.6 + 9.3$
7. $\frac{30 - 18}{2}$
8. $\frac{1.8}{6} + 4.2$
9. $\frac{1}{2}(15 + 7)$
10. $2 \cdot 5\frac{3}{8} \cdot 1$
11. $25 + \frac{300}{3}$
12. $0 + \frac{5}{8}$

There are two Cumulative Reviews, one following Chapter 7, which covers Chapters 1–7, and one following Chapter 14, which covers Chapters 8–14.

cumulative review

Simplify.

1. $\frac{15 + 5}{4}$
2. $3^2 - 3$
3. $16 - (4 \times 4)$

True or false?

4. $\frac{30}{5} = 2 \times 3$
5. $58 = 13 + 35$
6. $5^2 - 5 = 4 \times 5$

Name the value. Let $b = 2$, $l = 6$, $u = 10$.

Teaching Features

The text contains minimal reading. Illustrative examples take the place of lengthy discussion. Each lesson is brief—most are one page or less—and is enclosed in a box rule. Important definitions and properties appear in color for maximum visibility. In addition, annotations to the student, which give brief explanations or solving hints, also appear in color. A design that is both attractive and functional makes this text easy for students to use.

Basic Properties

7-1 *Combining Similar Terms*

OBJECTIVE
Simplify expressions by combining similar terms.

We can simplify an expression like $4t + 3t$. We justify the idea that $4t + 3t = 7t$ by using the distributive property: $4t + 3t = (4 + 3)t = 7t$.

Terms such as $4t$ and $3t$ which contain the same variables are called similar terms or like terms. Terms which contain no variables are also called similar terms.

$$5x + 7x \qquad 8rs + 7 - 1 \qquad a^2 + 3a - a$$

similar terms   similar terms   similar terms

These examples further illustrate how we use the distributive property to combine similar terms.

EXAMPLE 1 $\begin{aligned} 8mn + 6mn &= (8 + 6)mn \quad \text{◀ The Distributive Property} \\ &= 14mn \end{aligned}$

EXAMPLE 2 $\begin{aligned} 5a - a &= 5a - 1a \quad \text{◀} a = 1 \cdot a \\ &= (5 - 1)a \quad \text{◀ The Distributive Property} \\ &= 4a \end{aligned}$

EXAMPLE 3 $5t + 9s$ ◀ The terms are unlike. The expression cannot be simplified any further.

EXAMPLE 4 $\begin{aligned} 5x^2 + 4x + 2x &= 5x^2 + (4 + 2)x \\ &= 5x^2 + 6x \end{aligned}$

Oral EXERCISES

Simplify.

Sample: $6m - 2m$ *What you say:* $(6 - 2)m = 4m$

1. $5a + 2a$	2. $3b + 5b$	3. $10k - 4k$
4. $2m + 3m + 4m$	5. $5r + r$	6. $10s + 10s$
7. $k + k$	8. $6xy + 4xy$	9. $6y - y - y$

Simplify. Use exponents.

EXERCISES

1. $m \cdot m \cdot m \cdot m \cdot m$
2. $6 \cdot x \cdot x^5 \cdot x$
3. $n \cdot n \cdot n^4$
4. $2^3 \cdot 2$
5. $(ab)^5 \cdot (ab)^4$
6. $w \cdot w \cdot z \cdot z \cdot z$
7. $5 \cdot r \cdot r \cdot s \cdot s \cdot s$
8. $(ef)^3 \cdot (ef)^4$
9. $(x \cdot x)(y \cdot y \cdot y)$
10. $p \cdot p^3 \cdot p$

Give the value. Use $x = 3$ and $z = 2$.

Written
EXERCISES

A

1. z^2
2. $10z^2$
3. $(x \cdot x)(z \cdot z \cdot z)$
4. x^3
5. $x \cdot x \cdot x \cdot z$
6. $3 \cdot z \cdot z$
7. $4 \cdot x \cdot x$
8. $(x \cdot z)^2$
9. $(3z)(xz)$
10. $5z^4$
11. $8x^3$
12. $x^3 x^2$

Solve. Use the formula $A = lw$ (Area = length × width).

Problems

1. The length of a rectangular swimming pool is 19 meters. The area of the pool is 133 square meters. Find the width.
2. The dimensions of a rectangular picture are 12 cm by 18 cm. What is the area of the picture?

Solve. Use the formula $I = prt$ (Interest = principal × rate × time).

3. Sheila borrowed $2700.00 from the bank at a rate of $9\frac{1}{4}\%$. She paid $499.50 in interest. How long did she keep the loan?
4. Rudy borrowed $1000 for a period of 9 months $\left(\frac{3}{4} \text{ year}\right)$. He paid $30.00 interest. What was the rate of interest?
5. A homeowner borrowed money at the rate of $8\frac{1}{2}\%$ for a period of 2 years. The interest charged was $255. What was the amount of the loan?
6. A couple borrowed $4000 for 2 years at 7% interest. How much interest did they pay?

Solve. Use the formula Area $= \dfrac{1}{2} \times$ altitude × sum of bases,

$A = \dfrac{1}{2}as$.

7. The area of a trapezoid is 640 square centimeters. The sum of the bases is 80 cm. Find the altitude.

8. What is the area of this trapezoid?

9. The area of this trapezoid is 18 square centimeters. What is the altitude?

10. The area of this trapezoid is 540 square meters. What is the sum of the lengths of the bases?

Each section is followed by a substantial number of Oral and Written Exercises. Written Exercises are graded A, B, and C, in order of increasing difficulty. Answers to odd-numbered written exercises and problems appear at the end of the text. In addition, there are twenty pages of Extra Practice Exercises geared toward computation and equation solving. References to Extra Practice exercises are annotated on the appropriate page of the student text.

Word problems using relevant applications test understanding of material and make algebraic concepts more meaningful. The metric system of measurement is used throughout.

The following references provide students with quick access to specific information and page references to appropriate sections of the text:

A list of symbols (page xii)
Glossary (pages 448–451)
Index (pages 452–456)
Answers to Odd-Numbered Exercises (pages 457–476)
Answers to Self-Tests (pages 477–480)

Special Features

Striking chapter opener pages which show "then and now" photographs of a particular career area or process are designed to spark the students' interest.

Left: Telephone switchboard, 1908.

Right: Modern electronic switching system.

5 Equations and Inequalities

consumer notes *Electricity*

Do you turn off the radio when you're not listening to it? If you do, you are saving electricity. Electricity usage is measured in kilowatt-hours (KWH) by an electric meter. To read a meter, note the position of the pointer on each of the four dials. If the pointer is between numbers, read the smaller number. This meter reads 4726 KWH.

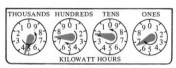

Find the electric meter in your house or apartment. Make readings at 9 A.M., 1 P.M., 6 P.M., and 10 P.M. Calculate the amount of electricity used during each period. Over which of these periods was the most electricity used? Have your family think of and use ways of saving electricity. After a week make another set of readings. Do you notice a difference?

Numerous features throughout the text provide a break from purely academic work. *Consumer Notes* offer practical information about everyday situations. *Challenge Topics* enrich the mathematical content of the text and provide supplemental activities of varying difficulty. *Historical Notes* give background information about distinguished representatives of many career fields.

challenge topics *Probability*

Suppose you toss a penny and a nickel at the same time. What combinations of heads and tails might come up? Study the chart.

Nickel

		H	T
Penny	h	h,H	h,T
	t	t,H	t,T

The chart shows that there are four possible **outcomes,** that is, four ways for the coins to come up heads and tails. A coin is **honest** if after many tosses it comes up heads very nearly the same number of times it comes up tails. Assuming the two coins are honest, each of the four outcomes is said to be **equally likely.** From the chart we see that the probability of an outcome of two heads (h,H) is 1 out of 4 or $\frac{1}{4}$. We write: $P(h,H) = \frac{1}{4}$.

What are the possible outcomes when two dice are tossed? The sum of the number of dots on the upper faces is considered the outcome.

Kotaro Honda   *1870–1954*

Kotaro Honda was one of Japan's leading metallurgists. (A metallurgist is a scientist who experiments with metals.) In 1916 he found that the addition of cobalt to tungsten steel produced a more powerful magnet than steel. Nothing more advanced in the field of magnetics was discovered until the mid 20th century. In 1937 Honda was awarded the Cultural Order of the Rising Sun, Japan's equivalent of the Nobel Prize.

Calculator Corners offer enjoyable, clever exercises that the student can do on a simple electronic calculator.

calculator corner

How can an arctic scientist go from one experiment site to another? You can find out by using your calculator. Begin with the number of hours in one half-day. Multiply by the number of degrees in two right angles. Divide by the number of months in a year. Multiply by the number of centimeters in a meter. Add the number of grams in 0.455 kilograms. Multiply by the number of huskies in the scientist's kennel (25). To see the answer, turn your calculator upside down.

Time Out activities break the pace of the text and provide intriguing problems, puzzles, and games.

Time out

A horse trader sold a horse that had four shoes and six nails in each shoe. The price was set in this way: the buyer was to pay 1¢ for the first nail, 2¢ for the second, 4¢ for the third, and so on, doubling the amount for each nail until all were paid for. What was the price of the horse?

Career Capsules lead students to think about various occupations and their educational requirements.

career capsule

Television and Radio Service Technician

Television and radio service technicians repair electronic products, television sets, radios, stereo components, tape recorders, intercoms and public address systems. Using voltmeters and signal generators they check suspected circuits for loose or broken connections and other probable causes of trouble. Technicians refer to

Supplementary Materials

This Teacher's Edition contains section-by-section commentary on the student text: mathematical background material, teaching suggestions, chalkboard examples, and extensions. Annotated alternate chapter tests, an individualized assignment guide, and a table of related references are also provided. Answers to oral and written exercises, problems, and puzzles are included, most appearing right on the full-sized facsimiles of the student pages.

Progress Tests are a convenient way to measure achievement and to keep track of each student's performance. Each test is keyed to the student text. Answers to all tests appear in the Teacher's Annotated Edition of the Progress Tests. The Solution Key provides step-by-step solutions for every written exercise and problem.

Diagnostic Tests in Arithmetic*

1. Whole Numbers—Addition

1.	2.	3.	4.	5.
7 +9 **16**	2 3 +3 **8**	2 5 +4 **11**	44 +25 **69**	14 +16 **30**

6.	7.	8.	9.
714 +191 **905**	758 +174 **932**	604 755 +469 **1828**	7701 9819 5626 +4213 **27,359**

2. Whole Numbers—Subtraction

1.	2.	3.	4.	5.
5 −3 **2**	14 − 7 **7**	86 −23 **63**	777 −206 **571**	656 −309 **347**

6.	7.	8.	9.	10.
859 −377 **482**	9505 −2454 **7051**	844 −466 **378**	8550 −7797 **753**	40,505 −27,886 **12,619**

3. Whole Numbers—Multiplication

1.	2.	3.	4.	5.
9 ×7 **63**	22 ×4 **88**	37 ×3 **111**	64 ×8 **512**	278 ×3 **834**

6.	7.	8.	9.	10.
34 ×20 **680**	137 ×200 **27,400**	154 ×300 **46,200**	6751 ×2000 **13,502,000**	3911 ×3000 **11,733,000**

4. Whole Numbers—Division

1. $8\overline{)72}$ **9**   2. $4\overline{)44}$ **11**   3. $9\overline{)648}$ **72**   4. $5\overline{)2130}$ **426**   5. $24\overline{)792}$ **33**   6. $19\overline{)629}$ **33 R2**

7. $675\overline{)509886}$ **755 R261**

8. Express the remainder in Exercise 7 as a fraction. $\dfrac{261}{675}$

9. Express the quotient to the nearest hundredth: $7\overline{)57}$ **8.14**

5. Fractions—Basic Skills

1. Identify the figure that is divided into thirds.   B

 A   B   C

* Used by permission from INDIVIDUALIZED COMPUTATIONAL SKILLS PROGRAM, COMPUTER VERSION by Bryce R. Shaw, Miriam M. Schaefer, and Petronella M. W. Hiehle. Copyright © 1973 by Houghton Mifflin Company.

2. What is the denominator of $\frac{2}{4}$?　4

3. The area shaded in Figure A is represented by which fraction in Row B?

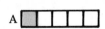

A 　B $\frac{3}{5}, \frac{2}{5}, \frac{1}{5}, \frac{5}{1}$

4. Which of the figures represents the fraction $\frac{1}{2}$?　D

C　　　　D　　　E

5. Write the fraction represented by the set diagram.　$\frac{1}{4}$

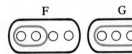

6. Which diagram represents the fraction $\frac{3}{4}$?　G

　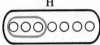

F　　　　G　　　　H

7. Which fraction at right represents the number 1?　$\frac{2}{1}, \frac{6}{6}, \frac{2}{4}, \frac{1}{5}$　$\frac{6}{6}$

8. Which fraction at right represents B on the number line?　$\frac{3}{2}, \frac{3}{3}, \frac{1}{2}, \frac{1}{4}$　$\frac{3}{3}$

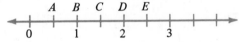

A　B　C　D　E

0　　1　　2　　3

9. Write the consecutive multiples of 4: 4, __?__, __?__, __?__　8, 12, 16

10. Find the least common multiple of 5, 3, and 15.　15

11. Write 18 as the product of prime factors.　2 × 3 × 3

12. Find the greatest common factor of 18 and 24.　6

13. Find the fraction in Row Y that is equal to each fraction or mixed numeral in Row X.

X: (a) $\frac{1}{3}$　(b) $\frac{6}{8}$　(c) $3\frac{1}{5}$

(a) $\frac{3}{9}$　(b) $\frac{3}{4}$　(c) $\frac{16}{5}$

Y: $\frac{3}{4}, \frac{8}{6}, \frac{16}{5}, \frac{3}{9}, \frac{5}{16}, \frac{1}{4}$

14. Find the lowest common denominator for the fractions $\frac{2}{3}, \frac{1}{4}$, and $\frac{1}{6}$.　12

15. Find the fraction or mixed numeral in Row Y that is equal to each fraction or mixed numeral in Row X.

X: (a) $\frac{7}{6}$　(b) $2\frac{1}{3}$　(a) $1\frac{1}{6}$　(b) $\frac{7}{3}$

Y: $\frac{1}{6}, 1\frac{1}{6}, \frac{7}{3}, \frac{3}{7}$

6. Fractions—Addition

1. $\frac{1}{5} + \frac{2}{5} = \underline{\ ?\ }$ $\frac{3}{5}$ **2.** $3\frac{1}{8} + 4\frac{1}{8} = \underline{\ ?\ }$ $7\frac{1}{4}$ **3.** $\begin{array}{r} 3\frac{7}{10} \\ +2\frac{9}{10} \\ \hline 6\frac{3}{5} \end{array}$ **4.** $\frac{1}{4} + \frac{1}{3} = \underline{\ ?\ }$ $\frac{7}{12}$ **5.** $\begin{array}{r} 4\frac{8}{20} \\ +4\frac{4}{5} \\ \hline 9\frac{1}{5} \end{array}$

7. Fractions—Subtraction

1. $\begin{array}{r} \frac{4}{4} \\ -\frac{1}{4} \\ \hline \frac{3}{4} \end{array}$ **2.** $\begin{array}{r} 3\frac{6}{8} \\ -1\frac{4}{8} \\ \hline 2\frac{1}{4} \end{array}$ **3.** $\begin{array}{r} 3 \\ -2\frac{8}{9} \\ \hline \frac{1}{9} \end{array}$ **4.** $\begin{array}{r} 7\frac{3}{8} \\ -1\frac{7}{8} \\ \hline 5\frac{1}{2} \end{array}$ **5.** $\begin{array}{r} \frac{1}{2} \\ -\frac{1}{5} \\ \hline \frac{3}{10} \end{array}$ **6.** $\begin{array}{r} 12\frac{1}{3} \\ -3\frac{3}{6} \\ \hline 8\frac{5}{6} \end{array}$

8. Fractions—Multiplication

1. $\frac{1}{8} \times \frac{1}{3} = \underline{\ ?\ }$ $\frac{1}{24}$ **2.** $\frac{6}{7} \times \frac{5}{9} = \underline{\ ?\ }$ $\frac{10}{21}$ **3.** $\frac{2}{3} \times 6 = \underline{\ ?\ }$ 4

4. $1\frac{7}{8} \times \frac{1}{2} = \underline{\ ?\ }$ $\frac{15}{16}$ **5.** $2\frac{5}{8} \times 4\frac{2}{3} = \underline{\ ?\ }$ $12\frac{1}{4}$

9. Fractions—Division

1. $1 \div \frac{1}{3} = \underline{\ ?\ }$ 3 **2.** $3 \div \frac{5}{8} = \underline{\ ?\ }$ $4\frac{4}{5}$ **3.** $\frac{7}{9} \div \frac{1}{6} = \underline{\ ?\ }$ $4\frac{2}{3}$

4. $3\frac{4}{5} \div \frac{1}{3} = \underline{\ ?\ }$ $11\frac{2}{5}$ **5.** $4\frac{1}{3} \div 10 = \underline{\ ?\ }$ $\frac{13}{30}$ **6.** $5\frac{5}{6} \div 4\frac{1}{4} = \underline{\ ?\ }$ $1\frac{19}{51}$

10. Decimals

1. Write the decimal which represents "four and four tenths." 4.4

2. $\frac{5}{1000} = \underline{\ ?\ }$ (Decimal) 0.005 **3.** $0.201 = \underline{\ ?\ }$ (Fraction) $\frac{201}{1000}$ **4.** $2\frac{4}{5} = \underline{\ ?\ }$ (Decimal) 2.8

5. Round 61.6505 to the nearest tenth. 61.7 **6.** $1.051 + 0.8 + 40.012 = \underline{\ ?\ }$ 41.863

7. $153.199 - 30.03 = \underline{\ ?\ }$ 123.169 **8.** $\begin{array}{r} 71.2 \\ 500.35 \\ +54.84 \\ \hline 626.39 \end{array}$ **9.** $\begin{array}{r} 825.01 \\ -8.109 \\ \hline 816.901 \end{array}$ **10.** $\begin{array}{r} 5519 \\ \times0.032 \\ \hline 176.608 \end{array}$ **11.** $0.8\overline{)2.5608}$ 3.201

11. Percents

1. $\frac{5}{4} = \frac{?}{100}$ 125 **2.** $0.57 = \underline{\ ?\ }\%$ 57 **3.** $\frac{11}{12} = \underline{\ ?\ }\%$ $91\frac{2}{3}$ **4.** $6\frac{1}{2} = \underline{\ ?\ }\%$ 650

5. $2.404 = \underline{\ ?\ }\%$ 240.4 **6.** $413\% = \underline{\ ?\ }$ (Mixed numeral) $4\frac{13}{100}$

7. $34\frac{1}{2}\% = \underline{\ ?\ }$ (Decimal) 0.345 **8.** 25% of $27 = \underline{\ ?\ }$ 6.75

9. $\underline{\ ?\ }\%$ of $20 = 5$ 25 **10.** $66\frac{2}{3}\%$ of $\underline{\ ?\ } = 4$ 6

Alternate Chapter Tests

TEST FOR CHAPTER 1
Simplify.

1. $\dfrac{5 \times 4}{10}$ 2

2. $8^2 + 3$ 67

3. $3 \times 3 \times 3$ 27

4. $5 \times 17 \times 0$ 0

True or False?

5. $18 + 4 = 8 \times 3$ False

6. $\dfrac{8}{8} = 6 - 5$ True

7. $\dfrac{10}{2} - 5 = \dfrac{10}{2} \times 0$ True

Name the value. Let $b = 3$, $t = 4$, and $m = 8$.

8. $t + 5$ 9

9. $t + m + 1$ 13

10. $t^2 + 43$ 59

11. $b \times t$ 12

12. $4 + m^2$ 68

13. $\dfrac{45}{b}$ 15

Write the equation in symbols.

14. The sum of thirty and fifteen equals forty-five.
$30 + 15 = 45$

15. The product of five and some number is equal to thirty-five. $5x = 35$

Solve each equation.

16. $5 + n = 12$ $n = 7$

17. $18 - t = 13$ $t = 5$

18. $\dfrac{40}{k} = 8$ $k = 5$

19. $s \times 6 = 60$ $s = 10$

20. $50 = c + 40$ $c = 10$

21. $11 = w - 5$ $w = 16$

22. $7 = \dfrac{49}{n}$ $n = 7$

23. $39 = b \times 3$ $b = 13$

Find the value.

24. 6^2 36

25. 4^3 64

26. 7^2 49

27. 20^3 8000

TEST FOR CHAPTER 2
Name the coordinate.

1. F $^-2$

2. Q $^+3$

3. T $^+2$

4. P 0

5. N $^-1$

6. The point half the distance from F to T 0

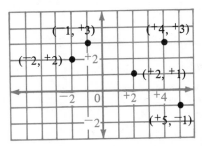

7. A $^-3$

8. K $^+4$

9. The point half the distance from P to F $^-1$

Name the opposite.

10. $^+7$ $^-7$

11. $^-5$ $^+5$

12. 0   0

13. $\dfrac{^+2}{3}$ $\dfrac{^-2}{3}$

14. $^-15$ $^+15$

15. $^+12$ $^-12$

16. $\dfrac{^-1}{3}$ $\dfrac{^+1}{3}$

17. $^+2.3$ $^-2.3$

Complete each pattern.

18. $^-4$, $^-3$, $\underline{^-2}$, $\underline{^-1}$, $\underline{0}$, $^+1$, $^+2$

19. $\underline{^-4}$, $\underline{^-3}$, $^-2$, $^-1$, $\underline{0}$, $^+1$

20. $^-2$, $^-1$, 0, $\underline{^+1}$, $\underline{^+2}$, $\underline{^+3}$, $\underline{^+4}$

21. $^-30$, $^-29$, $\underline{^-28}$, $\underline{^-27}$, $\underline{^-26}$, $^-25$

Graph. Use the number line.

22. The integers between $^+2$ and $^-3$.

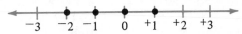

23. The integers between $^-1$ and $^+2$, inclusive.

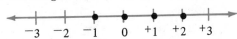

24. The numbers greater than $^-2$.

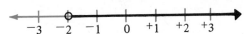

25. The numbers between $^-3$ and $^+3$.

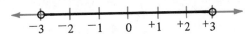

Complete. Use < or >.

26. $^+7 \underline{>} ^+4$

27. $^+2 \underline{>} ^-3$

28. $0 \underline{>} ^-12$

29. $0 \underline{<} ^+5$

30. $^-6 \underline{<} ^-3$

31. $^-5 \underline{>} ^-8\tfrac{1}{3}$

Draw axes and graph.

32. $(^-2, ^+2)$, $(^-1, ^+3)$, $(^+4, ^+3)$, $(^+5, ^-1)$, and $(^+2, ^+1)$

33. $(^-2, ^-3)$, $(0, ^+2)$, $(0, ^-2)$, $(^+4, 0)$, and $(^-4, 0)$

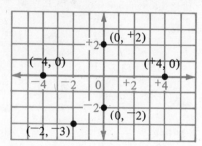

TEST FOR CHAPTER 3
Name the sets of multiples and the LCM.
1. 4 and 5
$\{4, 8, 12, 16, 20, \ldots\}$
$\{5, 10, 15, 20, 25, \ldots\}$
LCM: 20

2. 3 and 6
$\{3, 6, 9, 12, \ldots\}$
$\{6, 12, 18, 24, \ldots\}$
LCM: 6

3. 8 and 10
$\{8, 16, 24, 32, 40, \ldots\}$
$\{10, 20, 30, 40, \ldots\}$
LCM: 40

4. 2 and 7
$\{2, 4, 6, 8, 10, 12, 14, \ldots\}$
$\{7, 14, 21, 28, \ldots\}$
LCM: 14

Name the sets of factors and the GCF.
5. 32 and 24
$\{1, 2, 4, 8, 16, 32\}$
$\{1, 2, 3, 4, 6, 8, 12, 24\}$
GCF: 8

6. 15 and 20
$\{1, 3, 5, 15\}$
$\{1, 2, 4, 5, 10, 20\}$
GCF: 5

7. 3 and 8
$\{1, 3\}$
$\{1, 2, 4, 8\}$
GCF: 1

8. 7 and 14
$\{1, 7\}$
$\{1, 2, 7, 14\}$
GCF: 7

Write in the form $2 \times n$ or $(2 \times n) + 1$.
9. 13 $(2 \times 6) + 1$ 10. 18 2×9
11. 26 2×13 12. 49 $(2 \times 24) + 1$

Complete.
13. $\frac{1}{3} = \frac{4}{12}$ 14. $\frac{12}{16} = \frac{3}{4}$

15. $\frac{5}{6} = \frac{15}{18}$ 16. $\frac{12}{20} = \frac{6}{10}$

Write as a fraction in lowest terms.
17. 8 out of 12 $\frac{2}{3}$ 18. 4 out of 20 $\frac{1}{5}$

19. 5 out of 6 $\frac{5}{6}$ 20. 30 out of 100 $\frac{3}{10}$

Write the decimal equivalent.
21. $\frac{4}{5}$ 0.8   22. $1\frac{1}{4}$ 1.25

23. $\frac{7}{10}$ 0.7   24. $\frac{35}{100}$ 0.35

Write as a percent.
25. $\frac{25}{100}$ 25%   26. $\frac{85}{100}$ 85%

27. 0.06 6%   28. 0.23 23%

Complete.
29. 2 m $\leftrightarrow$ _200_ cm   30. 60 cm $\leftrightarrow$ _0.6_ m

31. 2 cm $\leftrightarrow$ _20_ mm   32. 20 cm $\leftrightarrow$ _200_ mm

33. 2 m $\leftrightarrow$ _2000_ mm   34. 350 mm $\leftrightarrow$ _0.35_ m

TEST FOR CHAPTER 4
Add or subtract.
1. $\frac{4}{9} + \frac{2}{9}$ $\frac{2}{3}$ 2. $\frac{7}{10} - \frac{6}{10}$ $\frac{1}{10}$

3. $\frac{1}{3} + \frac{4}{9}$ $\frac{7}{9}$ 4. $3\frac{2}{7} + 2\frac{4}{7}$ $5\frac{6}{7}$

5. $5\frac{5}{8} - 2\frac{1}{8}$ $3\frac{1}{2}$ 6. $\frac{5}{6} - \frac{2}{5}$ $\frac{13}{30}$

Multiply or divide.
7. $\frac{5}{6} \times \frac{1}{2}$ $\frac{5}{12}$ 8. $2\frac{1}{4} \div \frac{1}{2}$ $4\frac{1}{2}$

9. $3\frac{1}{3} \div 3$ $1\frac{1}{9}$ 10. $\frac{3}{4} \div \frac{4}{5}$ $\frac{15}{16}$

11. $\frac{1}{3} \times 3\frac{1}{2}$ $1\frac{1}{6}$ 12. $8 \times 1\frac{1}{6}$ $9\frac{1}{3}$

Find the value when $a = 3$, $b = 1$, and $c = 2$.
13. $\frac{b}{8} + \frac{b}{c}$ $\frac{5}{8}$ 14. $\frac{a}{c} \times \frac{b}{c}$ $\frac{3}{4}$

15. $\frac{b}{c} \times \frac{b}{a} \times \frac{a}{c}$ $\frac{1}{4}$

Solve.
16. $n = 26.2 + 5.7$ $n = 31.9$
17. $t = 63.07 - 21.35$ $t = 41.72$
18. $x = 15.08 \times 3.6$ $x = 54.288$
19. $s = 113.9 \div 0.85$ $s = 134$

Find the percent.
20. $35\% \times 64$ 22.4   21. $6\% \times \$135$ $8.10

Simplify.

22. $(10 - 7)(2 + 8)$ 30   **23.** $[(4)(5)]$ 20

24. $\dfrac{9 + 7}{4}$ 4   **25.** $3x(y)(z)$ $3xyz$

Find the values of the expression. The replacement set for n is $\{1, 3, 5\}$.

26. $2n + 11$ 13, 17, 21   **27.** $3(n + 5)$ 18, 24, 30

Complete each factorization of the given product.

28. Product: $10ab$; $10(\underline{ab})$, $5(\underline{2ab})$, $2a(\underline{5b})$

29. Product: $\dfrac{4s}{7}$; $\dfrac{4}{7}(\underline{s})$, $4\left(\dfrac{\underline{s}}{7}\right)$, $\dfrac{1}{7}(\underline{4s})$

Complete the table and graph the function.

Rule: $2m - 1$	Input Number	Output Number	Ordered Pair
Sample	$\dfrac{1}{2}$	0	$\left(\dfrac{1}{2}, 0\right)$
	1	1	$(1, 1)$
	$1\frac{1}{2}$	2	$(1\frac{1}{2}, 2)$
	2	3	$(2, 3)$
	$2\frac{1}{2}$	4	$(2\frac{1}{2}, 4)$

TEST FOR CHAPTER 5

Write a sentence. Do not determine the solution set.

1. If some number is increased by 20, the result is 32. $n + 20 = 32$

2. If 19 is decreased by some number, the difference is 12. $19 - n = 12$

3. If some number is multiplied by 5, the product is less than 23. $5n < 23$

Find the solution set. The replacement set is given.

4. $x + 6 = 2x + 3$; $\{1, 3, 5, 7\}$ $\{3\}$

5. $t^2 = 3t$; $\{0, 1, 2, 3\}$ $\{0, 3\}$

6. $3n < 12$; $\{0, 2, 4, 6\}$ $\{0, 2\}$

7. $3a - 2 \leq 4$; $\{1, 2, 3, 4\}$ $\{1, 2\}$

To solve, perform the operation as indicated.

8. $5m = 15$; divide by 5 $m = 3$

9. $k + 5 = 9$; subtract 5 $k = 4$

10. $b - 6 = 2$; add 6 $b = 8$

11. $t + 10 = 12$; subtract 10 $t = 2$

12. $\dfrac{s}{2} = 5$; multiply by 2 $s = 10$

13. $r - \dfrac{1}{5} = 3$; add $\dfrac{1}{5}$ $r = 3\frac{1}{5}$

Graph the solution set on the number line. The replacement set is {the numbers of arithmetic}.

14. $3y > 12$

15. $h - 2 < 4$

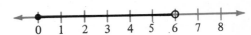

Write an algebraic expression.

16. The sum of a number and 5, multiplied by 4 $4(n + 5)$

17. 3 times a given number, increased by 2 $3n + 2$

18. The perimeter of a rectangle with width 6 cm and length n cm $12 + 2n$ or $2(6 + n)$

Write a number sentence. Then solve it to answer the question.

19. If the length of a rectangle is 12 m and the width is 7 m, what is the perimeter? $P = 2(7) + 2(12)$; 38 meters

20. Jean's father is more than twice as old as Jean. If Jean is 16, what numbers represent how old his father might be? $F > 2(16)$; Jean's father is more than 32 years old.

TEST FOR CHAPTER 6

Match each statement or equation in **1-20** with the appropriate property **A–N** below. You may use a property more than once.

1. $3(2 + n) = 3 \cdot 2 + 3 \cdot n$ K

2. $a(x + 3) = a(x + 3)$ C

3. $15 = 1 \cdot 15$ L

4. $8(xy) = 8x(y)$ H

5. $n + 3 = 3 + n$ F

6. $5x + 0 = 5x$ N

7. $m \cdot \dfrac{5}{5} = m \cdot 1$ J

8. $3 + (a + 5) = (3 + a) + 5$ I

9. If $9 = 3y$, then $3y = 9$.   A

10. If a and b are whole numbers, ab is a whole number.   D

11. $5(ab) = ab(5)$ G

12. $(5 - n)y = 5y - ny$ K

13. $(m + 3)(m - 1) = (m - 1)(m + 3)$ G

14. $\dfrac{2}{6} = \dfrac{1}{3}$ and $\dfrac{1}{3} = \dfrac{3}{9}$, so $\dfrac{2}{6} = \dfrac{3}{9}$.   B

15. $3(5 + n) = 3(n + 5)$ F
16. $x + (5 - 5) = x + 0$ J
17. If x and y are whole numbers, $x + y$ is a whole number.   E
18. If $x + 5 = 8 + 3$, and $8 + 3 = 11$, then $x + 5 = 11$.   B
19. $7ab \cdot 0 = 0$ M
20. $x + (y + 10) = (x + y) + 10$ I

A. Symmetric property of equality
B. Transitive property of equality
C. Reflexive property of equality
D. Closure under multiplication
E. Closure under addition
F. Commutative property of addition
G. Commutative property of multiplication
H. Associative property of multiplication
I. Associative property of addition
J. Substitution principle
K. Distributive property
L. Multiplicative property of one
M. Multiplicative property of zero
N. Additive property of zero

Complete.
21. $f(p) = p^2 + 3$; replacement set: $\{1, 3, 7, 10\}$

p	$f(p) = p^2 + 3$	$(p, f(p))$
1	4	(1, 4)
3	12	(3, 12)
7	52	(7, 52)
10	103	(10, 103)

TEST FOR CHAPTER 7
Solve. Check your answer.
1. $y + 11 = 35$ $y = 24$
2. $x - 3 = 14$ $x = 17$
3. $7r = 70$ $r = 10$
4. $\frac{1}{5}s = 3$ $s = 15$
5. $6b - 3 = 15$ $b = 3$
6. $n + \frac{1}{3} = 10$ $n = 9\frac{2}{3}$
7. $160 = 8x + 40$ $x = 15$
8. $36 - \frac{1}{4}a = 30$ $a = 24$
9. $14b - 1 = 0$ $b = \frac{1}{14}$
10. $\frac{m}{3} = \frac{1}{9}$ $m = \frac{1}{3}$

11. $\frac{3}{5}z = 18$ $z = 30$
12. $18 - w = 8w$ $w = 2$
13. $17 - 4c = 3c - 4$ $c = 3$
14. $\frac{p}{3} + 6 = 10$ $p = 12$

Write an equation. Explain what number the variable represents. Do not solve the equation.
15. The sum of 17 and a number is 35. Find the number. $17 + n = 35$
16. If a number is increased by ten, the result is the same as 13 less than twice the number. Find the number. $m + 10 = 2m - 13$
17. The sum of an even number and the next smaller even number is 54. Find the numbers. $s + (s - 2) = 54$
18. Each of two sides of a triangle is three-fourths as long as the third side. The perimeter is 150 cm. Find the length of the third side.
$$l + \frac{3}{4}l + \frac{3}{4}l = 150$$

TEST FOR CHAPTER 8
Name the directed number described.

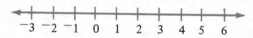

1. 2 units to the right of 0   2
2. 3 units to the left of 0   −3
3. the negative number of magnitude 2   −2
4. the positive number 6 units from 0   6

The statement refers to moves on the number line. Make a sketch and tell where you finish.
5. Start at 0. Move 3 units in the positive direction. Then move 1 unit in the negative direction.   2

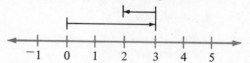

6. Start at 0. Move 4 units in the negative direction. Then move 2 units in the positive direction.   −2

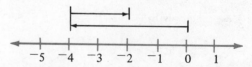

7. Start at ⁻2 and move 5 units in the positive direction.   3

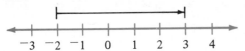

Complete. Use right or left and $>$ or $<$.

8. ⁻1 is to the __left__ of 0, so ⁻1 __$\leq$__ 0.

9. ⁻1 is to the __right__ of ⁻5, so ⁻1 __$>$__ ⁻5.

Show whether a true or a false statement results when the variable is replaced by each member of the replacement set. Then state the solution set.

10. $x > $ ⁻1; {⁻3, ⁻2, ⁻1, 0}   F, F, F, T; {0}

11. $y < 3$; {1, 2, 3, 4}   T, T, F, F; {1, 2}

Graph.

12. {the directed numbers greater than ⁻3}

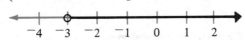

13. {the directed numbers between ⁻3 and 3}

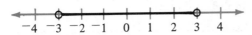

Name the solution set. The replacement set is {the directed numbers}.

14. $m < $ ⁻2
{the directed numbers less than ⁻2}

15. $n > $ ⁻3
{the directed numbers greater than ⁻3}

16. $p \geq 1$
{1 and the directed numbers greater than 1}

17. ⁻4 $< q <$ 0
{the directed numbers between ⁻4 and 0}

Name the property illustrated.

10. $3 + (-8) = -8 + 3$
Commutative property

11. $3 + [5 + (-3)] = (3 + 5) + (-3)$
Associative property

12. $15 + (-15) = 0$
Additive property of inverses

13. $-5 + [5 + (-10)] = (-5 + 5) + (-10)$
Associative property

Solve. Begin by writing an equivalent addition equation.

14. $y = 6 - 13$ $y = 6 + (-13) = -7$

15. $w = -5 - 3$ $w = -5 + (-3) = -8$

16. $5 - (-1) = m$ $m = 5 + 1 = 6$

17. $-4 - (-2) = z$ $z = -4 + 2 = -2$

18. $n = -6 - 5$ $n = -6 + (-5) = -11$

19. $0 - (-8) = x$ $x = 0 + 8 = 8$

Complete according to the function equation.

20. $f(x) = -2 - x$; {(5, __⁻7__), (⁻3, __1__)}

21. $f(y) = y + (-1)$; {(0, __⁻1__), (⁻2, __⁻3__)}

22. $f(a) = a - (-5)$; {(⁻3, __2__), (3, __8__)}

Complete the table according to the given function rule. Then write the set of ordered pairs and graph the function.

23. $f(n) = -n + 2$

	n	$f(n)$	$(n, f(n))$
Sample:	−1	3	(−1, 3)
	−2	4	(−2, 4)
	−3	5	(−3, 5)
	0	2	(0, 2)
	1	1	(1, 1)
	2	0	(2, 0)
	3	−1	(3, −1)

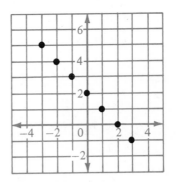

TEST FOR CHAPTER 9

Name the opposite.

1. 7 -7 **2.** -5 5   **3.** ⁻3   3

Solve.

4. $y + $ ⁻3 $= 0$ $y = 3$

5. $-6 + 6 = x$ $x = 0$

6. $-11 + n = 0$ $n = 11$

7. $8 + m = 5$ $m = -3$

8. $5 = -1 + b$ $b = 6$

9. $3 + c = -1$ $c = -4$

TEST FOR CHAPTER 10
Simplify.

1. $5(-7)$ -35

2. $-7 \cdot 8$ -56

3. $\dfrac{45}{-9}$ -5

4. $\dfrac{-63}{7}$ -9

5. $-1(-1)(-1)$ -1

6. $-4(-4)(-4)$ -64

7. $-1 \cdot 1(-10)$ 10

8. $-2 \cdot 3(-6)$ 36

9. $\dfrac{3 + (-23)}{5}$ -4

10. $\dfrac{-33}{-3}$ 11

11. $8a + 7b - 6a$ $2a + 7b$

12. $7w - 5c + 2c - 8w$ $-w - 3c$

Name the reciprocal.

13. $-\dfrac{1}{3}$ -3

14. $1\frac{3}{5}$ $\dfrac{5}{8}$

15. -0.6 $-\dfrac{5}{3}$

Solve.

16. $-5a = -6 \cdot 10$ $a = 12$

17. $6b + (-b) = 30$ $b = 6$

18. $3h + (-1) = 26$ $h = 9$

19. $7d - 3 = 11$ $d = 2$

20. $\dfrac{3}{5}y = -6$ $y = -10$

21. $\dfrac{n}{4} = -3$ $n = -12$

Complete using the given function equation.

22. $f(c) = 4c - 3$; $\{(0, \underline{-3}), (2, \underline{5}), (5, \underline{17})\}$

23. $f(b) = 3b + 5$;

$\{(0, \underline{5}), (-1, \underline{2}), (-3, \underline{-4}), (3, \underline{14})\}$

TEST FOR CHAPTER 11
Solve.

1. $y - 3 = -8$ $y = -5$

2. $3n + 7n = 5$ $n = \dfrac{1}{2}$

3. $-7m = 63$ $m = -9$

4. $21 = -10 + x$ $x = 31$

5. $6n = 42$ $n = 7$

6. $3m - 7m = 28$ $m = -7$

7. $\dfrac{t}{8} - 4 = 6$ $t = 80$

8. $-\dfrac{3}{5}s + \dfrac{2}{5}s = 10$ $s = -50$

9. $-\dfrac{4x}{5} = 8$ $x = -10$

10. $-2.3x - 0.3 = 2$ $x = -1$

Solve for n. Assume that no divisor has the value 0.

11. $n - a = b$ $n = a + b$

12. $an - b = -c$ $n = \dfrac{b - c}{a}$

13. $na = b$ $n = \dfrac{b}{a}$

14. $bn - cn = a$ $n = \dfrac{a}{b - c}$

Solve. Graph the solution set.

15. $\dfrac{1}{3}x > -1$ $x > -3$

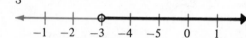

16. $-3x > -6$ $x < 2$

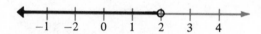

17. $3x + 5x \leq 24$ $x \leq 3$

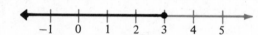

18. $x + 2 > -2$ $x > -4$

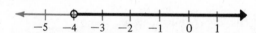

19. $-5x + 10 \geq 0$ $x \leq 2$

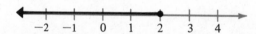

20. $-3 + x > -5$ $x > -2$

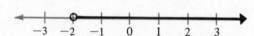

21. The figure has volume 72 cm³. Use the formula $V = lwh$ and solve for h. Then substitute and find the value of h. $h = 2$ cm

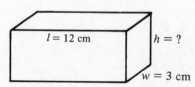

22. A rectangular garden plot is to be 13 m long. How wide will you make it if you want its total area to be 65 m²?   5 m

TEST FOR CHAPTER 12
Tell whether the polynomial is a monomial, a binomial, or a trinomial.

1. $9mn$ monomial

2. $3a^3 - 3a$ binomial

3. $\frac{1}{3}a^3 - a - 3$ trinomial

4. $\frac{2}{5}t$ monomial

5. -5 monomial
6. $7 - a^2$ binomial

Write in standard form.
7. $5 - 3x^2y$ $-3x^2y + 5$
8. $3a^2 - 4a^3 + a$ $-4a^3 + 3a^2 + a$
9. $r^3 - 7 + 3r^4$ $3r^4 + r^3 - 7$

Add.

10. $\begin{array}{r} 5n^2 - 3n + 18 \\ 11n^2 - 6n + 21 \\ \hline 16n^2 - 9n + 39 \end{array}$

11. $\begin{array}{r} 12x^2 + 8 \\ -\ 3x^2 - 2 \\ \hline 9x^2 + 6 \end{array}$

12. $(3c + 5) + (6c - 2)$ $9c + 3$
13. $(-7y - 5) + (7y + 2)$ -3

Subtract the second polynomial from the first.

14. $\begin{array}{r} 7y^2 + 3y - 2 \\ 3y^2 -\ y + 7 \\ \hline 4y^2 + 4y - 9 \end{array}$

15. $\begin{array}{r} 3m^2 - 12m + 5 \\ 3m - 8 \\ \hline 3m^2 - 15m + 13 \end{array}$

Simplify.
16. $(3b^2 - 7) - (b^2 - 2)$ $2b^2 - 5$
17. $(2z + 2) - (-3z - 5)$ $5z + 7$

Simplify. Remove grouping symbols and combine similar terms.
18. $(3w - 7) + (2w - 1)$ $5w - 8$
19. $(6t + 5) - (t - 3)$ $5t + 8$
20. $(c^2 - 3c) + (3c - 5)$ $c^2 - 5$
21. $(8a - 5b) - (5b - 8a)$ $16a - 10b$

Write as a polynomial in standard form.
22. The perimeter of a square with sides of length $5n - 2$. $20n - 8$

23. The perimeter of a rectangle with length $2s + 1$ and width $\frac{1}{2}s - 2$. $5s - 2$

Solve and check.
24. $(3x - 5) + 4x = 16$ $x = 3$
25. $h - (5 - 3h) = 3$ $h = 2$
26. $2n + (n - 9) = 0$ $n = 3$
27. $(-n + 2) - (3n - 2) = 12$ $n = -2$

TEST FOR CHAPTER 13
Simplify. Use exponents.

1. $y^3 \cdot y^5$ y^8
2. $(5z \cdot z)(3 \cdot w \cdot w \cdot w)$ $15z^2w^3$
3. $n^3(-3np^3)(4p)$ $-12n^4p^4$
4. $(3c^2d)^2$ $9c^4d^2$
5. $(3ab)^3$ $27a^3b^3$

Multiply.

6. $-p(p^2 - 3p - 2)$ $-p^3 + 3p^2 + 2p$
7. $3t(tw^2 + t^2w)$ $3t^2w^2 + 3t^3w$
8. $7s(s + 5)$ $7s^2 + 35s$
9. $rs(r^2 - 3rt + 4)$ $r^3s - 3r^2st + 4rs$
10. $(a - 3)^2$ $a^2 - 6a + 9$
11. $(4y + 3)(4y - 3)$ $16y^2 - 9$
12. $3n(n + 5) + 2n(2n + 1)$ $7n^2 + 17n$
13. $(h - 3)(2h + 5)$ $2h^2 - h - 15$
14. $-4d(d^2 - 3d + 5)$ $-4d^3 + 12d^2 - 20d$
15. $(k - 3)(2k^2 - 7k + 2)$ $2k^3 - 13k^2 + 23k - 6$

Express the answer as a polynomial in simplest form.
16. A rectangle has length $3n - 2$ and width $n + 3$. Find the area. $3n^2 + 7n - 6$
17. A square has sides of length $3z - 5$. Find the area. $9z^2 - 30z + 25$

Divide. Assume no denominator is 0.

18. $\frac{n^7}{n^3}$ n^4

19. $\frac{-3a^3b}{ab}$ $-3a^2$

20. $\frac{12m^2 + 16m}{4m}$ $3m + 4$

21. $\frac{15x^2 - 20x + 7}{5x}$ $3x - 4 + \frac{7}{5x}$

22. $\frac{c^2 - 10c + 25}{c - 5}$ $c - 5$

23. $\frac{d^2 - 12d + 20}{d - 2}$ $d - 10$

TEST FOR CHAPTER 14

Write the prime factorization.

 1. 40 $\ 2^3 \cdot 5$ 2. 42 $\ 2 \cdot 3 \cdot 7$ 3. 16 $\ 2^4$

Name the GCF.

 4. $15a^2$, $10ab$ $\quad 5a$ 5. 12 and $3x$ $\quad 3$
 6. $18c^2d$, $27cd^2$ $\quad 9cd$

Factor.

 7. $9 + 6s^2$ $\quad 3(3 + 2s^2)$ 8. $9b^2 + 45c$ $\quad 9(b^2 + 5c)$
 9. $4a^2 + 6a - 16$ $\quad 2(2a^2 + 3a - 8)$
10. $7m^2 - 14n$ $\quad 7(m^2 - 2n)$ 11. $p^2r + r$ $\quad r(p^2 + 1)$
12. $3w^2x - 7wx + 6x$ $\quad x(3w^2 - 7w + 6)$

Group terms. Then factor.

13. $st + 2s + 5t + 10$ $\quad (s + 5)(t + 2)$
14. $ab + 4b - 12 - 3a$ $\quad (b - 3)(a + 4)$

Factor.

15. $n^2 - 49$ $\quad (n + 7)(n - 7)$
16. $h^2 + 12h + 36$ $\quad (h + 6)^2$
17. $z^2 - 4z + 4$ $\quad (z - 2)^2$
18. $p^2 - 16$ $\quad (p + 4)(p - 4)$
19. $m^2 + 18m + 81$ $\quad (m + 9)^2$
20. $b^2 - 10b + 9$ $\quad (b - 9)(b - 1)$
21. $9a^2 - c^2$ $\quad (3a + c)(3a - c)$
22. $d^2 - 16d + 64$ $\quad (d - 8)^2$
23. $r^2 + 9r + 20$ $\quad (r + 4)(r + 5)$

Factor. Check by multiplication.

24. $a^2 + 3a - 28$ $\quad (a + 7)(a - 4)$
25. $k^2 - 5k - 24$ $\quad (k - 8)(k + 3)$
26. $f^2 + 7f - 18$ $\quad (f + 9)(f - 2)$
27. $s^2 - 2s - 15$ $\quad (s - 5)(s + 3)$

Guide to Individualized Assignments

The following guide outlines a suggested sequence of assignments for an essential and a comprehensive course. Naturally, the needs of your students will dictate the amount of time you spend on particular topics. The guide merely provides an overview which we hope will make your planning easier.

The essential course covers all the prerequisite material for *Elementary Algebra, Part 2*. The comprehensive course treats the text material in more depth, providing more challenging work and somewhat less drill than the essential course. All assignments refer either to written exercises or to problems, designated by a *P*. For your convenience, the following **Suggested Time Schedule** is divided into trimesters by red vertical rules and into semesters by a black vertical rule.

Suggested Time Schedule

Chapter	1	2	3	4	5	6	7	8	9	10	11	12	13	14	Total
Lessons	12	9	10	11	13	12	13	9	11	11	10	12	14	13	160

Semester 1 — Semester 2

Trimester 1 — Trimester 2 — Trimester 3

Essential Course

Lesson	Section	*Pages*/Exercises
		Chapter 1
1	1-1	**3**/1–24
2	1-2	**6**/1–10; **7**/11–22
3	1-2 1-3	**7**/23–32, odd **9**/1–13, odd
4	1-3	**9**/1–14, even; 15–22
5	1-4	**11**/1–18; **12**/*P*:1–3
6	 1-5	**13**/Self-Test 1 **15**/1–20
7	1-5 1-6	**15**/21–30 **18**/1–17, odd
8	1-6 1-7	**18**/2–18, even **19**/19–30 **21**/1–5

Comprehensive Course

Section	*Pages*/Exercises
	Chapter 1
1-1	**3**/1–13, odd **4**/25–34
1-2	**6**/1–9, odd **7**/11–32
1-2 1-3	**7**/33–44 **9**/1–14
1-3	**9**/15–34
1-4	**11**/1–21, odd **12**/22–24; *P*:1–6
 1-5	**13**/Self-Test 1 **15**/13–26
1-5 1-6	**15**/27–31 **18**/1–17, odd **19**/19–24
1-6 1-7	**18**/2–18, even **19**/25–36 **21**/1–5; **22**/6–11

Essential Course

Lesson	Section	Pages/Exercises
9	1-7 1-8	**22**/6–20 **24**/1–10
10	1-8	**25**/11–25; Self-Test 2
11		**26–27**/Chapter Test
12		**28**/Review of Skills

Chapter 2

Lesson	Section	Pages/Exercises
13	2-1	**32**/1–28
14	2-2	**35**/1–15; **36**/16–23
15	2-2 2-3	**36**/24–37 **38**/1–11
16	2-3	**38**/12–17 **39**/Self-Test 1
17	2-4	**42**/1–20; **43**/21–28
18	2-5	**45**/1–8; **46**/9–20
19	2-6	**48**/1–15; **49**/16–32, even
20	2-6	**49**/17–33, odd **50**/Self-Test 2
21		**52**/Chapter Test **54**/Review of Skills

Chapter 3

Lesson	Section	Pages/Exercises
22	3-1	**57**/1–4; **58**/5–26
23	3-2	**60**/1–5; **61**/6–26
24	3-3	**63**/1–29; **64**/30–38
25	 3-4	**64**/Self-Test 1 **66**/1–12; **67**/13–30
26	3-5	**69**/1–17; *P.*1–3

Comprehensive Course

Section	Pages/Exercises
1-7 1-8	**22**/12–29 **24**/1–10
1-8	**25**/11–33; Self-Test 2
	26–27/Chapter Test Challenge Topic
	28/Review of Skills

Chapter 2

Section	Pages/Exercises
2-1	**32**/1–20, odd; 21–36 **33**/37–47
2-2	**35**/1–15; **36**/16–33
2-2 2-3	**36**/34–42 **38**/1–17
2-3	**39**/18–31 **39**/Self-Test 1
2-4	**42**/2–20, even **43**/21–42
2-5	**45**/1–8; **46**/9–15, odd; 17–29
2-6	**48**/1–5; 6–14, even; **49**/16–25
2-6	**49**/26–35; **50**/36–37 **50**/Self-Test 2
	52/Chapter Test **54**/Review of Skills

Chapter 3

Section	Pages/Exercises
3-1	**58**/5–8; 9–25, odd; 27–41
3-2	**60**/1–5 **61**/6–26, even; 27–36
3-3	**63**/1–29, odd; **64**/30–50
3-4	**64**/Self-Test 1 **67**/13–36
3-5	**69**/1–17; *P.*1–3 **70**/*P.*4–8

Essential Course

Lesson	Section	*Pages*/*Exercises*
27	3-6	**72**/1–12; **73**/13–35
28	3-7	**75**/1–21; **76**/22–30; *P.*1–3
29	3-8	**77**/Self-Test 2 **79**/1–17
30	3-8	**80**/18–29 **80**/Self-Test 3
31		**82**/Chapter Test **84**/Review of Skills

Comprehensive Course

Section	*Pages*/*Exercises*
3-6	**73**/13–40
3-7	**75**/1–21 **76**/22–36; *P*/1–3; **77**/*P.*4
3-8	**77**/Self-Test 2 **79**/1–17; **80**/18–25
3-8	**80**/26–41 **80**/Self-Test 3
	82/Chapter Test **84**/Review of Skills

Chapter 4

Lesson	Section	*Pages*/*Exercises*
32	4-1	**87**/1–12; **88**/13–30
33	4-2	**91**/1–15; **92**/16–42, even
34	4-2 4-3	**92**/17–41, odd **94**/1–22
35	4-3 4-4	**95**/23–26; *P.*1–5 **97**/1–18
36	4-4	**98**/19–32; *P.*1–5
37	4-5	**99**/Self-Test 1 **101**/1–10; **102**/11–22
38	4-5 4-6	**102**/23–32 **105**/1–20
39	4-7	**107**/1–8; **108**/9–28 **109**/*P.*1–4
40	4-8	**112**/1–6 **113**/7–14
41	4-8	**113**/15–25 **114**/Self-Test 2
42		**115–116**/Chapter Test **118**/Review of Skills

Chapter 4

Section	*Pages*/*Exercises*
4-1	**87**/2–12, even; **88**/13–39
4-2	**91**/1–15; **92**/16–42
4-2 4-3	**92**/43–48 **94**/1–22; **95**/23–32
4-3 4-4	**95**/*P.*1–5 **97**/1–18; **98**/19–28
4-4	**98**/29–36; *P.*1–5
4-5	**99**/Self-Test 1 **101**/1–10; **102**/11–32
4-5 4-6	**102**/33–44 **105**/1–26
4-6 4-7	**105**/27–32 **108**/9–28; **109**/*P.*1–4
4-7 4-8	**109**/29–39, odd **112**/1–6; **113**/7–25
4-8	**114**/26–31 **114**/Self-Test 2
	115–116/Chapter Test **118**/Review of Skills

Chapter 5

Lesson	Section	*Pages*/*Exercises*
43	5-1	**121**/1–10; **122**/11–18
44	5-2	**125**/1–3; **126**/4–15

Chapter 5

Section	*Pages*/*Exercises*
5-1	**121**/1–10; **122**/11–26
5-2	**126**/4–20

Essential Course

Lesson	Section	Pages/Exercises
45	5-2 5-3	**126**/16–20 **128**/1–20; **129**/21–30
46	5-4	**131**/1–14 **132**/15–24
47	5-4	**132**/25–30 **133**/Self-Test 1
48	5-5	**135**/1–14; **136**/15–23
49	5-6	**138**/1–11 **139**/Self-Test 2
50	5-7	**141**/1–8; **142**/9–14
51	5-7	**142**/15–29
52	5-8	**144**/1–16
53	5-8	**145**/17–27
54	5-9	**147**/P.1–4; **148**/P.5–7 **148–149**/Self-Test 3
55		**150**/Chapter Test **152**/Review of Skills

Chapter 6

Lesson	Section	Pages/Exercises
56	6-1	**155**/1–12; **156**/13–20
57	6-2	**158**/1–10 **159**/11–20
58	6-3	**160**/Self-Test 1 **164**/1–7
59	6-3	**164**/8–20; **165**/21–30
60	6-4	**167**/1–16
61	6-4 6-5	**167**/17–22; **168**/23–26 **170**/1–7
62	6-5	**171**/8–10; 11–25, odd; 26–30
63	6-6	**172**/Self-Test 2 **174**/1–16

Comprehensive Course

Section	Pages/Exercises
5-2 5-3	**126**/21–26 **128**/2–20, even **129**/21–34
5-3 5-4	**129**/35–40 **131**/1–14; **132**/15–24
5-4	**132**/25–30; **133**/31–35 **133**/Self-Test 1
5-5	**135**/1–14; **136**/15–27
5-6	**138**/1–13; **139**/14 **139**/Self-Test 2
5-7	**141**/1–8; **142**/9–20
5-7	**142**/21–41
5-8	**144**/1–16; **145**/17–21
5-8	**145**/22–39
5-9	**147**/P.1–4; **148**/P.5–10 **148–149**/Self-Test 3
	150/Chapter Test **152**/Review of Skills

Chapter 6

Section	Pages/Exercises
6-1	**155**/2–12, even **156**/13–24
6-1 6-2	**156**/25–28 **158–160**/1–22
6-3	**160**/Self-Test 1 **164**/1–15
6-3	**164**/16–20; **165**/21–37
6-4	**167**/1–22
6-4 6-5	**168**/23–30 **170**/1–7
6-5	**171**/8–30
6-6	**172**/Self-Test 2 **174**/1–24

Essential Course

Lesson	Section	*Pages*/*Exercises*
64	6-6 6-7	**174**/17–30 **176**/1–12
65	6-7	**176**/13–24; **177**/25–32
66	6-7	**177**/33–38 **178**/Self-Test 3
67		**180–181**/Chapter Test **182**/Review of Skills

Comprehensive Course

Section	*Pages*/*Exercises*
6-6 6-7	**174**/25–38 **176**/1–18
6-7	**176**/19–24; **177**/25–38
6-7	**178**/39–44 **178**/Self-Test 3
	180–181/Chapter Test **182**/Review of Skills

Chapter 7

Chapter 7

Lesson	Section	*Pages*/*Exercises*
68	7-1	**185**/1–26
69	7-1 7-2	**185**/27–34 **187**/1–8; **188**/9–18
70	7-2 73	**188**/19–30 **190**/1–12
71	7-3 7-4	**190**/13–24 **192**/1–15
72	7-4	**192**/16–21; **193**/22–27 **193**/Self-Test 1
73	7-5	**195**/1–12; **196**/13–24
74	7-5 7-6	**196–197**/*P*:1–10 **199**/1, 2; **200**/3–8
75	7-6 7-7	**200**/9–12 **202**/1–22
76	7-7	**202**/23–28; *P*:1–3; **203**/*P*:4–6 **203**/Self-Test 2
77	Cum. Review	**204–205**/Chapter Test **206**/1–24
78	Cum. Review	**206–207**/25–45
79	Cum. Review	**207–209**/46–76
80	Cum. Review	**209**/77–101 **210**/Review of Skills

Section	*Pages*/*Exercises*
7-1	**185**/1–15, odd; 17–36
7-1 7-2	**185**/37–42 **187**/1–8; **188**/9–24
7-2 7-3	**188**/25–40 **190**/11–24
7-3 7-4	**190**/25–34 **192**/1–21
7-4	**193**/22–39 **193**/Self-Test 1
7-5	**196**/13–36
7-5 7-6	**196–197**/*P*1–10 **199**/1, 2; **200**/3–12
7-6 7-7	**200**/13–19 **202**/1–28
7-7	**202–203**/29–34; *P*:1–8 **203**/Self-Test 2
Cum. Review	**204–205**/Chapter Test **206**/1–24
Cum. Review	**206–207**/25–45
Cum. Review	**207–209**/46–76
Cum. Review	**209**/77–101 **210**/Review of Skills

Essential Course

Lesson	Section	Pages/Exercises
		Chapter 8
81	8-1	**214**/1–28
82	8-1 8-2	**214**/29–32 **218**/1–9; **219**/10–12
83	8-2	**219**/13–26
84	8-3	**222**/1–14; **223**/15–29
85	8-4	**224**/Self-Test 1 **226**/1–12
86	8-4	**226**/13–15; **227**/16–24
87	8-5	**229**/1–3; **230**/4–14
88	8-5	**230**/15–22 **231**/Self-Test 2
89		**232**/Chapter Test **234**/Review of Skills
		Chapter 9
90	9-1	**237**/1–6; **238**/7–20
91	9-1	**238**/21–30; **239**/P.1–9
92	9-2	**241**/1–10; **242**/11–25
93	9-2 9-3	**242**/26–40 **244**/1–9
94	9-3	**245**/10–35
95	9-4	**247**/1–20; **248**/21–30
96	9-4	**248**/31–40 **249**/Self-Test 1
97	9-5	**251**/1–12; **252**/13–38
98	9-6	**256**/1–18
99	9-6	**257**/19–28 **258**/Self-Test 2
100		**260**/Chapter Test **262**/Review of Skills

Comprehensive Course

Section	Pages/Exercises
	Chapter 8
8-1	**214**/1–32
8-1 8-2	**214**/33–36 **218**/1–9; **219**/10–20
8-2	**219**/21–26; **220**/27–36
8-3	**222**/1–14; **223**/15–42
8-4	**224**/Self-Test 1 **226**/1–15
8-4	**227**/16–32
8-5	**229**/1–3; **230**/4–22
8-5	**230**/23–32 **231**/Self-Test 2
	232/Chapter Test **234**/Review of Skills
	Chapter 9
9-1	**237**/1–6; **238**/7–19, odd; 21–30
9-1	**238**/31–40; **239**/P.1–9
9-2	**241**/1–10; **242**/11–34
9-2 9-3	**242**/35–52 **244**/1–9; **245**/10–15
9-3	**245**/16–42
9-4	**247**/1–20; **248**/21–40
9-4	**248**/41–43; **249**/44–46 **249**/Self-Test 1
9-5	**251**/2–12, even; **252**/14–20, even; 22–50
9-6	**256**/1–18; **257**/19–23
9-6	**257**/24–34 **258**/Self-Test 2
	260/Chapter Test **262**/Review of Skills

Essential Course

Lesson	Section	*Pages*/Exercises
		Chapter 10
101	10-1	**265**/1–17; **266**/18–30
102	10-2	**268**/1–12; **269**/13–20
103	10-2 10-3	**269**/21–30 **271**/1–12
104	10-3	**272**/13–33
105	10-4	**274**/1–20
106	 10-5	**275**/Self-Test 1 **277**/1–30
107	10-6	**280**/1–27; **281**/28–30
108	10-6 10-7	**281**/31–42; *P*:1–7 **284**/1–4
109	10-7	**285**/5–20
110	10-7	**286**/21–25 **286–287**/Self-Test 2
111		**289**/Chapter Test **290**/Review of Skills
		Chapter 11
112	11-1	**293**/1–14; **294**/15–20
113	11-1 11-2	**294**/21–29 **296**/1–10
114	11-2 11-3	**297**/11–33 **300**/1–6
115	11-3	**301**/7–28
116	11-3 11-4	**301**/29–36 **303**/*P*:1–10
117	 11-5	**304**/Self-Test 1 **306**/1–6; **307**/7–12
118	11-5	**307**/13–34
119	11-6	**309**/1–10; **310**/11–22

Comprehensive Course

Section	*Pages*/Exercises
	Chapter 10
10-1	**265**/16, 17 **266**/18–24, even; 26–43
10-2	**268**/1–12; **269**/13–24
10-2 10-3	**269**/25–43 **271**/1–12
10-3	**272**/13–33, odd; 34–43
10-4	**274**/2–12, even; 13–26 **275**/27–36
10-5	**275**/Self-Test 1 **277**/1–36
10-6	**280**/1–27; **281**/28–36
10-6 10-7	**281**/37–42; *P*:1–7 **284**/1–4; **285**/5–14
10-7	**285**/15–20; **286**/21–25
10-7	**286**/26–30 **286–287**/Self-Test 2
	289/Chapter Test **290**/Review of Skills
	Chapter 11
11-1	**293**/1–14; **294**/15–26
11-1 11-2	**294**/27–38 **296**/1–10; **297**/11–19
11-2 11-3	**297**/20–42 **300**/1–6
11-3	**301**/7–34
11-3 11-4	**301**/35–42 **303**/*P*:1–10
11-5	**304**/Self-Test 1 **306**/1–6; **307**/7–20
11-5	**307**/21–40
11-6	**309**/1–10; **310**/11–28

Essential Course

Lesson	Section	Pages/Exercises
120	11-6	**310**/23–34 **311**/Self-Test 2
121		**313**/Chapter Test **314**/Review of Skills

Chapter 12

Lesson	Section	Pages/Exercises
122	12-1 12-2	**317**/1–15 **319**/1–18
123	12-2 12-3	**319**/19–26 **321**/1–8; **322**/9–14
124	12-3 12-4	**322**/15–26 **324**/1–21
125	12-4	**325**/22–36
126	12-5	**327**/1–13; **328**/14–20
127	12-5	**328**/21–30 **329**/Self-Test 1
128	12-6	**331**/1–14; **332**/15–28
129	12-7	**334**/1–14; **335**/15–30
130	12-7	**335**/31–36; **336**/37–39 **336**/Self-Test 2
131	12-8 12-9	**337–338**/P:1–9 **339**/1–14
132	12-9	**340**/15–25 **340**/Self-Test 3
133		**341–342**/Chapter Test **344**/Review of Skills

Chapter 13

Lesson	Section	Pages/Exercises
134	13-1	**347**/1–43
135	13-2	**349**/1–24
136	13-3	**351**/1–40
137	13-4	**353**/1–33

Comprehensive Course

Section	Pages/Exercises
11-6	**310**/29–42 **311**/Self-Test 2
	313/Chapter Test **314**/Review of Skills

Chapter 12

Section	Pages/Exercises
12-1 12-2	**317**/1–15 **319**/1–26
12-3	**321**/1–8; **322**/9–25
12-3 12-4	**322**/26–32 **324**/1–21
12-4	**325**/22–46
12-5	**327**/1–13; **328**/14–27
12-5	**328**/28–35 **329**/Self-Test 1
12-6	**331**/2–14, even **332**/15–38
12-7	**334**/2–14, even **335**/16, 18, 19–36
12-7	**336**/37–44 **336**/Self-Test 2
12-8 12-9	**337–338**/P:1–9 **339**/1–14; **340**/15–20
12-9	**340**/21–30 **340**/Self-Test 3
	341–342/Chapter Test **344**/Review of Skills

Chapter 13

Section	Pages/Exercises
13-1 13-2	**347**/1–43 **349**/1–10
13-2	**349**/11–35
13-3	**351**/1–44
13-4	**353**/1–33; **354**/34–38

Essential Course

Lesson	Section	*Pages*/*Exercises*
138	13-4	**354**/34–38; *P*:1–7 **355**/Self-Test 1
139	13-5	**357–358**/1–30
140	13-5 13-6	**358**/31–40; **359**/*P*:1–5 **361**/1–12
141	13-6	**362**/13–30; *P*:1–7
142	13-7	**366**/1–16 **367**/Self-Test 2
143	13-8	**370**/1–30
144	13-9	**372–373**/1–26
145	13-9 13-10	**373**/27–32 **375**/1–22
146	13-10	**375**/23–30 **376**/Self-Test 3
147		**379**/Chapter Test **380**/Review of Skills

Comprehensive Course

Section	*Pages*/*Exercises*
13-4	**354–355**/39–45; *P*:1–10 **355**/Self-Test 1
13-5	**357–358**/1–40
13-5 13-6	**358**/41–48; **359**/*P*:1–8 **361**/1–12
13-6	**362**/13–39; *P*:1–7
13-7	**366–367**/1–8; 9–29, odd **367**/Self-Test 2
13-8	**370**/1–37
13-9	**372–373**/1–32
13-10	**375**/1–30
13-10	**376**/31–38 **376**/Self-Test 3
	379/Chapter Test **380**/Review of Skills

Chapter 14

148	14-1	**383**/1–32
149	14-2	**385**/1–19 **386**/20–35
150	14-3	**388**/1–39
151	14-4	**390**/1–18 **391**/19–24
152	14-5	**391**/Self-Test 1 **393**/1–27
153	14-5 14-6	**393**/28–36 **395**/1–34
154	14-7	**397**/1–32; **398**/33–50
155	14-8	**400**/1–30
156	14-9	**402**/1–36

Chapter 14

14-1 14-2	**383**/1–32 **385**/1–19
14-2 14-3	**386**/20–41 **388**/1–8
14-3 14-4	**388**/9–39 **390**/1–10
14-4	**390**/11–18; **391**/19–24 **391**/Self-Test 1
14-5	**393**/1–42
14-6	**395**/1–46
14-7	**397**/1–32; **398**/33–56
14-8	**400**/1–34
14-9	**402**/1–46

Essential Course

Lesson	Section	*Pages*/Exercises
157		**403**/Self-Test 2 **404**/Chapter Test
158	Cum. Review	**406–407**/1–41
159	Cum. Review	**407–408**/42–83
160	Cum. Review	**409**/84–125

Comprehensive Course

Section	*Pages*/Exercises
	403/Self-Test 2 **404**/Chapter Test
Cum. Review	**406–407**/1–41
Cum. Review	**407–408**/42–83
Cum. Review	**409**/84–125

Table of Related References

Chapter	Progress Tests	For use with Sections
1	Test 1 Test 2 Test 3	1–4 5,6 7,8
2	Test 4 Test 5 Test 6	1,2 3,4 5,6
3	Test 7 Test 8	1–3 4–8
4	Test 9 Test 10 Test 11	1,2 3,4 5–8
Cumulative Test	Test 12	
5	Test 13 Test 14 Test 15	1–4 5,6 7–9
6	Test 16 Test 17 Test 18	1,2 3–5 6,7
7	Test 19 Test 20 Test 21	1,2 3,4 5–7
Cumulative Test	Test 22	

Chapter	Progress Tests	For use with Sections
8	Test 23 Test 24	1–3 4,5
9	Test 25 Test 26	1–4 5,6
10	Test 27 Test 28	1–4 5–7
11	Test 29 Test 30	1–4 5,6
Cumulative Test	Test 31	
12	Test 32 Test 33 Test 34	1–5 6,7 8,9
13	Test 35 Test 36 Test 37	1–4 5–7 8–10
14	Test 38 Test 39 Test 40	1–4 5–7 8,9
Cumulative Test	Test 41	

1 Working with Integers

CHAPTER OVERVIEW

This opening chapter reviews basic skills needed in solving problems. The emphasis on the positive integers allows the student to review the fundamental operations of addition, subtraction, multiplication and division, with none of the distractions involved in working with fractions and decimals. Squares and cubes are introduced as a "shorthand" for repeated multiplication. The symbol "=" is stressed as meaning "names the same number as." Equations that involve variables are solved intuitively, and the student will then consider whether equations are true or false. The student begins to "translate" word phrases into algebraic expressions and word statements into equations.

1-1 *Numerical Expressions*

TEACHING THESE PAGES

You may wish to introduce this lesson with a class activity. On the chalkboard write the expression:

$$\frac{(15 - 3) + 8}{5}$$

Work through the simplification with the class, arriving at the value 4. Now get the class involved in either or both of the following ways:

(1) Have each student write another expression that names "four," using one or more of the operations of addition, subtraction, multiplication, and division. List the expressions offered on the board, while the class checks each expression to verify the value.

(2) Have each student write an expression for some other number. Select a volunteer to write his or her expression on the board, and another to simplify the expression. The student who simplifies the expression correctly then writes another expression to be simplified, and so on. The more students who participate, the better.

CHALKBOARD EXAMPLES

Simplify.

1. $17 - (2 \times 3) = 17 - 6 = 11$
2. $(17 - 2) \times 3 = 15 \times 3 = 45$
3. $\dfrac{3^2 + 1}{5 - 3} = \dfrac{(3 \times 3) + 1}{2} = \dfrac{9 + 1}{2} = \dfrac{10}{2} = 5$
4. $\dfrac{(7 \times 0) + 34}{2} = \dfrac{0 + 34}{2} = \dfrac{34}{2} = 17$

SUGGESTED EXTENSION

Write on the chalkboard a combination of integers and operation symbols, such as:

$$5 + 3 - 2 + 6 \times 5 \div 2$$

Now have the students simplify the expression. Right away the students should realize that there is more than one way to interpret the expression. For example:

$$5 + (3 - 2) + (6 \times 5) \div 2 = 5 + 1 + 30 \div 2$$
$$= 36 \div 2 = 18$$

Or does it equal: $6 + 15 = 21$
What about $(5 + 3) - (2 + 6) \times 5 \div 2$
or $5 + (3 - 2) + 6 \times (5 \div 2)$?

Discussion will lead the better students to see the need for agreement as to the order of operations when there is more than one way to simplify an expression. The rule is: *Do all multiplications and divisions first, in order from left to right. Then do all additions and subtractions, in order from left to right.* Using this rule:

$$5 + 3 - 2 + 6 \times 5 \div 2 = 5 + 3 - 2 + (6 \times 5) \div 2$$
$$= 5 + 3 - 2 + (30 \div 2)$$
$$= 5 + 3 - 2 + 15 = 21$$

1-2 *Expressions and Equations*

TEACHING THESE PAGES
You will want to stress the meaning of the symbol = as "is equal to" or "names the same number as." Then write on the chalkboard a true statement, such as $2 \times 8 = 19 - 3$, and a false statement, such as $7 \times 11 = 4 + 40 + 8$. Discuss true and false statements and point out that every *equation* (statement with the symbol =) is not necessarily true.

Ask each member of the class to write down two equations, one true and the other false. Then ask several students to write their equations on the board for general discussion. Next show the class a sentence in which one number is "missing," such as:

$$16 + \underline{\ ?\ } = 25 - 6$$

Discuss the fact that this sentence cannot be classified as either true or false unless we know what number is missing. Ask for possible replacements for $\underline{\ ?\ }$ that will make the statement true, and discuss whether there is more than one such number.

CHALKBOARD EXAMPLES
Show whether each statement is true or false.

1. $3 \times 8 = 26 - 4$ False; $24 \neq 22$

2. $\dfrac{17 + 3}{5} = 2^2 + 0$ True; $4 = 4$

3. $2 + 4 + 6 + 8 = 2^2 \times 10$ False; $20 \neq 40$
4. $300 + 60 + 9 = 3 \times 123$ True; $369 = 369$

Complete to make a true statement.

5. $\underline{\ ?\ } \times 21 \times 1 = 7 \times 3 \times 2$ 2

SUGGESTED EXTENSION
Ask students to find a replacement for $\underline{\ ?\ }$ in an identity for which there will be many replacements.

Examples: $19 + \underline{\ ?\ } = \underline{\ ?\ } + 19$
$5 + \underline{\ ?\ } + 30 = 35 + \underline{\ ?\ }$
$\underline{\ ?\ } \times 0 = (9 - 3^2) \times \underline{\ ?\ }$

1-3 *Expressions and Variables*

TEACHING THESE PAGES
In this section we introduce the use of variables instead of $\underline{\ ?\ }$ to represent unknown numbers. Encourage students to use a variety of letters as variables. Point out that no variable should have different values in any one expression or equation, although it is possible for two *different* variables to represent the same number.

You might begin by asking each student to pick a whole number from 0 to 10, inclusive. Then write an expression on the board, such as:

$$47 - n + 13$$

Ask the students to name the variable in the expression. Then have each student evaluate the expression for the number he or she has chosen. Ask several volunteers to state their numbers and the corresponding values of the expression. Have them show their work on the board. This activity can be continued either by changing the expression on the board, or by having students select replacements greater than 10.

CHALKBOARD EXAMPLES
Find the value.

1. $3 + a^2$, when $a = 3$ 12

2. $\dfrac{34}{8 + x}$, when $x = 9$ 2

3. $\dfrac{18 + k - 2}{3}$, when $k = 5$ 7

4. $7 \times (r + 4)$, when $r = 0$ 28

SUGGESTED EXTENSIONS
Fractional and negative numbers are considered in detail in later chapters. You may preview these numbers by having students evaluate expressions like these:

Fractional Numbers

$$\dfrac{16 - e}{7}, \text{ when } e = 8$$

$$r \div 6, \text{ when } r = 15$$

$$(m + 5) \div 4, \text{ when } m = 0$$

Negative Numbers

$3 + 4 - m$, when $m = 9$

$x - 12$, when $x = 5$

$\dfrac{4 - x}{3}$, when $x = 10$

1-4 *Special Equations: Formulas*

TEACHING THESE PAGES

This section serves to bring together what we have done so far, and should help students to see a use for the material in the previous sections. Since the students have probably worked with formulas before, you might introduce the lesson by asking for examples of formulas that they have used in the past. List each formula on the chalkboard and discuss its meaning. Point out that in each instance the value of any one variable in a formula depends upon the values assigned to the other variables. As an aside, you might ask students why they think we call these letters *variables*.

From the list of formulas, select several formulas which will always produce a positive integer in the left member, such as $A = l \times w$. Ask students to suggest values for the variables in the right member, then compute the value of the left member. Repeat this for several other values of each variable.

CHALKBOARD EXAMPLES

Complete.

1. $A = P + I$ When $P = 500$ and $I = 30$, $A = \underline{\ ?\ }$
   530

2. $C = \dfrac{22}{7} \times d$ When $d = 70$, $C = \underline{\ ?\ }$ 220

3. $P = 2 \times l + 2 \times w$ When $l = 15$ and $w = 10$, $P = \underline{\ ?\ }$ 50

4. $A = \dfrac{(B + b) \times h}{2}$ When $B = 12$, $b = 8$, and $h = 5$, $A = \underline{\ ?\ }$ 50

SUGGESTED EXTENSIONS

You might wish to discuss informally some restrictions on the values of variables in certain formulas. For instance, in $A = l \times w$, ask the class whether such values as $l = 0$, $w = 5$ or $l = {}^-4$, $w = 6$ are possible. Even though we can multiply 5×0 to find $A = 0$, an area of 0 simply makes no sense. Likewise, a negative value for A has no meaning. The idea is to get the students thinking about the replacement set for a variable in an informal manner.

1-5 *Using Symbols for Words*

TEACHING THESE PAGES

The ability to translate expressions from words into symbols is an essential skill for problem solving, which is a basic reason for learning mathematics at all levels. You might introduce the lesson by writing expressions using symbols on the chalkboard and translating them into words. List as many translations as possible for each. For example:

Numerical Expression: $32 - 5$

Word Expression

The difference between thirty-two and five.
Thirty-two minus five.
Five less than thirty-two.
Five subtracted from thirty-two.
Thirty-two decreased by five.

Numerical Expression: $2x + 7$

Word Expression

The sum of twice some number and seven.
Twice some number plus seven.
Twice some number increased by seven.
Seven more than twice some number.
Seven added to twice some number.

Discourage the tendency to say "Twice some number more than seven" for the second example. The student should think of "$+7$" as "seven more than" to parallel "-5" as "five less than." For variety, you might write a few word expressions on the board and have students translate them into numerical expressions.

Students may enjoy the following game: Write a numerical expression or an equation on the chalkboard. Then have each student write down one possible word translation. The translations are compared. Correct responses which are unique score 5 points. Responses given only two or three times score 4 points. All other correct responses score 3 points.

CHALKBOARD EXAMPLES

Write the meaning in words.

1. 18×7
   The product of eighteen and seven.

2. $\dfrac{n}{36}$
   Some number divided by thirty-six.

3. $4a = 20$
   Four times some number equals twenty.
4. $6y - 3 = 21$
   Three less than six times some number equals twenty-one.

1-6 *Solving Equations: Addition and Subtraction*

TEACHING THESE PAGES

We emphasize again that an equation makes a true statement when both members *name the same number*. At this point we do not want to set up rules for solving equations, but rather to have the students find the appropriate replacement for a variable intuitively or by experimentation. This sort of solution is aided by the previous practice in "translating" symbols into words.

The lesson might be introduced by writing the following equation on the chalkboard:

$$15 + t = 27$$

Solving this equation can be thought of as answering the question "What number can be added to 15 to give 27?" Let values of t be suggested, and test each value to see whether the resulting equation is true or false. If you ask students how they arrived at a certain value, their replies might include:

"I know that $15 + 12 = 27$."
"Because $27 - 15 = 12$."
"I subtracted 15 from 27."
"I tried some numbers and 12 was the only one that worked."

Accept all replies, and do not attempt to formulate a rule at this point.

In this course, formal notation for solution sets is avoided except where needed for clarity. Thus, for the equation $15 + t = 27$, students may say, "The solution is $t = 12$," rather than "The solution set is $\{12\}$."

CHALKBOARD EXAMPLES

Solve.

1. $m - 8 = 17$
   $25 - 8 = 17$, so $m = 25$
2. $17 - a = 8$
   $17 - 9 = 8$, so $a = 9$
3. $d = 3 + 7$
   $10 = 3 + 7$, so $d = 10$
4. $h + 3 = 15 - 4$
   $8 + 3 = 15 - 4$, so $h = 8$

SUGGESTED EXTENSIONS

For $15 + t = 27$, ask why it seems obvious that $t = 12$ is the only solution. Ask:

(1) Can you think of any other solution?
(2) What happens to the value of $15 + t$ if numbers greater than 12 are used for t?
(3) What happens when numbers less than 12 are used?

1-7 *Solving Equations: Multiplication and Division*

TEACHING THESE PAGES

This lesson follows so naturally after Section 1-6 that it might be introduced by comparison:

(1) In $n + 7 = 21$, we decide that $n = 14$ because $14 + 7 = 21$. So to find what number n represents in

$$n \times 3 = 21,$$

we note that $7 \times 3 = 21$, and decide that:

$$n = 7$$

(2) In $t - 5 = 4$, we decide that $t = 9$ because $9 - 5 = 4$. So to find what number t represents in

$$t \div 5 = 4,$$

we note that $20 \div 5 = 4$, and decide that:

$$t = 20$$

Try to foster intuitive recognition of the uniqueness of sums, differences, products, and quotients of two numbers by discussing the solutions to the two examples above. Ask the students why they feel that $n = 7$ is the *only* solution for $n \times 3 = 21$, and $t = 20$ is the *only* solution for $t \div 5 = 4$. Be careful *not* to formulate rules for solving equations at this point.

CHALKBOARD EXAMPLES

Write an equation. Then solve the equation.

1. What number multiplied by 3 equals 18?
   $n \times 3 = 18; n = 6$
2. What number divided by 3 equals 18? $m \div 3 = 18;$
   $m = 54$

Use the true statement to solve each equation.

3. $5 \times 6 = 30$

   $5 \times a = 30, a = \underline{}$ 6
   $b \times 6 = 30, b = \underline{}$ 5
   $5 \times 6 = c, c = \underline{}$ 30

4. $\dfrac{24}{8} = 3$

$\dfrac{24}{x} = 3, \; x = \underline{\;?\;}$ 8

$\dfrac{y}{8} = 3, \; y = \underline{\;?\;}$ 24

$\dfrac{24}{8} = z, \; z = \underline{\;?\;}$ 3

SUGGESTED EXTENSIONS

At this point, you may wish to discuss inverse operations such as addition and subtraction, multiplication and division. Since $7 + 9 = 16$, we also know that $7 = 16 - 9$, and $9 = 16 - 7$. Since $5 \times 9 = 45$, we also know that $5 = 45 \div 9$, and $9 = 45 \div 5$.

Do not set up rules for solving equations, but show how the appropriate inverse operation can help us to select the solution for an equation. For example:

$x + 23 = 52$ (Think: $52 - 23 = ?$)
$n - 81 = 106$ (Think: $106 + 81 = ?$)
$p \times 9 = 144$ (Think: $144 \div 9 = ?$)
$q \div 17 = 21$ (Think: $21 \times 17 = ?$)

1-8 *Equations and Exponents*

TEACHING THESE PAGES

In this lesson we discuss the use of an exponent as a short way of showing a product in which the same factor is used more than once. Right now we will consider only squares and cubes.

You might begin by writing several expressions on the board, such as:

8×8 $6 \times 6 \times 6$ 4×4

Ask students if they know a shorter way to write these expressions. If no one offers 8^2, 6^3, and so forth, show them the exponential notation for the expressions and explain its meaning. Point out that 8^2 can be read as "8 to the second power" as well as "8 squared." Similarly, 6^3 may be read both as "6 cubed" and "6 to the third power."

In solving equations such as $n^2 = 25$, students will consider only the positive root 5. They will not encounter the negative roots of quadratic equations in this course, so $n = 5$ is an acceptable solution to the equation $n^2 = 25$.

CHALKBOARD EXAMPLES

Complete.

1. $5^2 = \underline{\;?\;} \times \underline{\;?\;}$ 5×5
   $5^2 = \underline{\;?\;}$ 25

2. $7^3 = \underline{\;?\;} \times \underline{\;?\;} \times \underline{\;?\;}$ $7 \times 7 \times 7$
   $7^3 = \underline{\;?\;}$ 343

3. $\underline{\;?\;} \times \underline{\;?\;} = 31^2$ 31×31
   $\underline{\;?\;} = 31^2$ 961

4. $\underline{\;?\;} \times \underline{\;?\;} \times \underline{\;?\;} = 20^3$ $20 \times 20 \times 20$
   $\underline{\;?\;} = 20^3$ 8000

SUGGESTED EXTENSION

Discuss the meaning of other positive integral exponents, such as:

$5^4 = 5 \times 5 \times 5 \times 5 = 625$

$4^5 = 4 \times 4 \times 4 \times 4 \times 4 = 1024$

$1^8 = 1 \times 1 \times 1 \times 1 \times 1 \times 1 \times 1 \times 1 = 1$

Explain that 5^4 is read "five to the fourth power," 4^5 is read "four to the fifth power," and so forth.

2 Positive and Negative Numbers

CHAPTER OVERVIEW

Thus far we have been dealing exclusively with whole numbers. We now introduce negative and fractional numbers. We use the number line as a convenient device for illustrating ideas about numbers. The whole numbers and their opposites make up the set of numbers called the integers, while the numbers between the integers on the number line (except for the irrationals) are fractional numbers. The irrational π is mentioned in passing, but irrational numbers as a distinct set are not discussed until *Elementary Algebra, Part 2.*

Set terminology is introduced only for the sake of clarity, and should not be taught as an end in itself. Symbols and terminology are kept to a minimum: we enclose the set in braces ({ }) to avoid repeating the words "the set of," and we use the symbol Ø for the empty set.

In this chapter, the student will graph both discrete and continuous sets of numbers on the number line. Discrete sets such as $\{^+1, 0, ^-1\}$ are graphed as points. Continuous sets such as {the numbers between $^-1$ and $^+1$} are graphed by shading a portion of the number line. In the final lesson of the chapter, the student will graph ordered pairs of numbers as points in the coordinate plane.

2-1 *The Whole Numbers*

TEACHING THESE PAGES

In this lesson we introduce both the set concept and the number line very casually. We shall use these ideas when they are helpful, but they should not be over-emphasized. The intent of the lesson is to familiarize students with the "tools" and terminology they will need later on, so be sure the students understand these conventions:

(1) We use braces to designate a set: $\{0, 3, 6, 9\}$

(2) We use three dots to show that a set continues indefinitely: $\{0, 2, 4, 6, 8, \ldots\}$

(3) We use Ø as the symbol for the empty set—the set with no members. Some students may have seen the symbol { } used for the empty set, but in this course, we will use the symbol Ø. Remind students that the empty set symbol never appears in braces.

(4) We use the number line to represent numbers. The arrowheads on both ends of a number line mean

that the number line extends in both directions, even though for the moment we do not mention numbers less than zero.

(5) The tic marks on the number line that stand for consecutive whole numbers are equally spaced.

(6) The point marked 0 is called the origin.

CHALKBOARD EXAMPLES

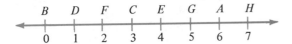

Name the coordinate.

1. Point G 5   **2.** Point H 7   **3.** Point B 0

4. A point between F and E 3
5. The point halfway between B and F 1
6. The point half the distance from D to G 3

Name the set of whole numbers:

7. Greater than 3   The set 4, 5, 6, 7, . . .
8. Less than 7   The set 0, 1, 2, 3, 4, 5, 6
9. Between 0 and 3   The set 1, 2
10. Between 4 and 5   ∅

2-2 *The Integers*

TEACHING THESE PAGES

At this point we introduce the negative integers and extend the number line to the left of zero. The symbols + and − may be used in three ways: (1) to indicate the operations of addition and subtraction; (2) to indicate a directed number, such as ⁻6; and (3) to indicate the opposite of a number, such as −(2) or −a. To avoid confusion, for the time being, we will use raised plus and minus signs to indicate directed numbers, for example, ⁺12, ⁻3. Be careful that the student reads "⁺3" as "positive three," "⁻5" as "negative five," "6 + 3" as "six plus three," and "8 − 5" as "eight minus five."

Point out that when we choose to name {the integers} by listing some of its elements, we should use three dots *at each end:*

$$\{. . . , ^-3, ^-2, ^-1, 0, ^+1, ^+2, ^+3, . . .\}$$

The concept of the opposite of a number may be new to students. In this lesson, it is introduced intuitively, by example. There is no need for a formal definition of opposites here. Later, in Chapter 8, opposites are defined as two different numbers that are the same distance from zero on the number line.

CHALKBOARD EXAMPLES

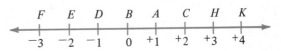

Name the coordinate.

1. Point F ⁻3   **2.** Point B 0   **3.** Point K ⁺4

4. A point between E and B ⁻1
5. The point halfway between E and C 0

Complete the pattern.

6. ⁻10, ⁻9, ⁻8, __?__ , __?__ , __?__   ⁻7, ⁻6, ⁻5
7. ⁻2, __?__ , __?__ , __?__ , ⁺2, ⁺3   ⁻1, 0, ⁺1

8. Name the set of integers between ⁻3 and ⁺7.
{⁻2, ⁻1, 0, ⁺1, ⁺2, ⁺3, ⁺4, ⁺5, ⁺6}

SUGGESTED EXTENSIONS

Some students may be ready to appreciate the fact that the integers are *directed* numbers. That is, the notation "⁻7" tells us two things about the point that corresponds to ⁻7 on the number line: (1) the side of zero on which it lies, and (2) its distance from zero. Thus, an integer indicates a direction as well as a distance from zero. Get the students thinking about direction on the number line by discussing the following exercises:

1. Name the coordinate of a point on the number line that lies five units to the left of zero.   (⁻5)
2. Name two different points on the number line that are nine units away from zero.   (⁺9, ⁻9)
3. Name the coordinate of the point that lies four units to the right of ⁻1 on the number line.   (⁺3)

2-3 *Graphing Integers on the Number Line*

TEACHING THESE PAGES

We now introduce the idea that the graph of a number is the point on the number line that corresponds to the number. In this section, we are working only with integers, and most of the sets discussed are finite. The graph of a set of integers is a set of discrete (separate) points on the number line. The important ideas in this lesson are as follows:

(1) A set may be named by listing its members, or by stating a rule or description. In either case, braces are used, and the notation is read: "the set. . . ."

(2) A set of numbers described as *between* two numbers does not include the two numbers unless the description ends with the word "inclusive."

This lesson is suitable for a student-participation exercise. Have each student write a set of numbers, either describing the set or listing its members. Remind them to use braces. Then have a volunteer write his or her set on the board, and a second volunteer come to the board and graph the set on the number line. The whole class should check the graph. When everyone agrees that the graph is correct, another student may write a set on the board, and the activity may be repeated.

Alternatively, you may have a student draw a graph on the board and have the class name the corresponding set of numbers. Discuss the fact that a set may be described in more than one way. Consider the following graph:

Students may describe this set correctly in any of the following ways:

(1) $\{^-2, ^-1, 0\}$
(2) {the integers between $^-3$ and $^+1$}
(3) {the integers between $^-2$ and 0, inclusive}

CHALKBOARD EXAMPLES

Name the set graphed by listing the members.

1.
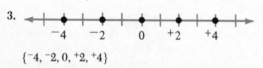
$\{^-3, ^-2, ^-1, 0\}$

2.
$\{^-1, 0, ^+1, ^+2\}$

3.
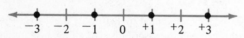
$\{^-4, ^-2, 0, ^+2, ^+4\}$

Graph the set of numbers on the number line.

4. $\{^-3, ^-1, ^+1, ^+3\}$

5. {the integers between $^-2$ and $^+2$}

6. {the integers between $^-2$ and $^+2$, inclusive}

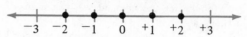

SUGGESTED EXTENSIONS

You might preview the vertical axis of the coordinate plane by drawing the number line in a vertical position. Then discuss the graphing of sets of numbers on such a line. Ask students which direction should be positive and which should be negative.

2-4 *Numbers on the Number Line*

TEACHING THESE PAGES

In this lesson we consider coordinates of points that lie between the integers on the number line. We work chiefly with rational numbers, although we do mention the familiar irrational π. There is no formal discussion of irrationals in this course. The density of the number line is only implied here. You might hint at it by having students name many fractional numbers that lie between two given integers. Do not formulate a general rule, however.

Point out that the number line is made up of points, so when we say "a number on the number line," we actually mean "a number that is the coordinate of a point on the number line." Thinking of each number as related to a point on the number line helps us to "order" the numbers. That is, a given number on the number line is greater than any number which lies to its left and less than any number which lies to its right. Be sure students understand the *greater than* and *less than* symbols.

CHALKBOARD EXAMPLES

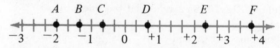

Name the coordinate of each point.

1. Point A $^-2$ 2. Point C $\dfrac{^-2}{3}$ 3. Point E $^+2\frac{1}{3}$

4. Point F $^+3\frac{2}{3}$ 5. Point B $^-1\frac{1}{3}$ 6. Point D $\dfrac{^+2}{3}$

Complete. Use **left** or **right** and $<$ or $>$.

7. $^+4$ is to the __?__ of 0, so $^+4$ __?__ 0.   right, $>$
8. $^-3$ is to the __?__ of $^-1$, so $^-3$ __?__ $^-1$.   left, $<$

Arrange in order, least to greatest.

9. $^-10$, 0, $^+3\frac{1}{7}$, $^-5\frac{1}{4}$, $^+6$ $^-10$, $^-5\frac{1}{4}$, 0, $^+3\frac{1}{7}$, $^+6$

10. $\dfrac{^-2}{9}$, $^+1$, $\dfrac{^-5}{9}$, 0, $^+2$, $^+1\frac{3}{4}$ $\dfrac{^-5}{9}$, $\dfrac{^-2}{9}$, 0, $^+1$, $^+1\frac{3}{4}$, $^+2$

SUGGESTED EXTENSIONS

A single point on the number line may have many equivalent coordinates. For instance, the point that corresponds to $\dfrac{^+1}{2}$ also corresponds to $^+0.5$, $\dfrac{^+5}{10}$, $\dfrac{^+2}{4}$, and so on.

Draw a number line on the board, similar to the one shown.

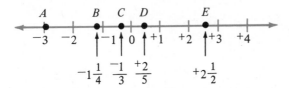

Have the students name five other numbers that have the same graph as each of the points A through E $\Big($For example: A: $\dfrac{^-3}{1}$, $\dfrac{^-6}{2}$, $^-3.0$, $^-3.00$, $\dfrac{^-9}{3}\Big)$.

2-5 *More About Graphing Numbers*

TEACHING THESE PAGES

In this lesson we again use the number line for graphing numbers. In the last lesson we graphed finite sets of integers as discrete points. Now we shall be graphing continuous, infinite sets. The students will need to understand the following techniques to graph these sets:

(1) To graph a set of numbers greater than (or less than) a given number, we darken the entire portion of the number line to the right (or to the left) of the given number. We use a heavy arrow to show the graph goes on and on. The number itself is represented by a hollow dot to indicate it is not included in the graph.

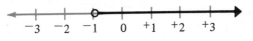

{the numbers greater than $^-1$}

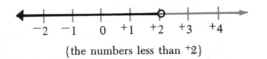

{the numbers less than $^+2$}

(2) When the set includes the given number, a solid dot is used to represent the number.

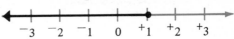

We shall call this set {$^+1$ and the numbers less than $^+1$}, since we have not yet introduced the "less than or equal to" notation.

(3) To graph a set between two given numbers, we darken the portion of the number line between the numbers, and use either hollow or solid dots for the endpoints, depending on whether they are included.

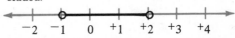

{the numbers between $^-1$ and $^+2$}

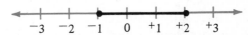

{the numbers between $^-1$ and $^+2$, inclusive}

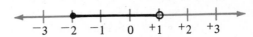

{$^-2$ and the numbers between $^-2$ and $^+1$}

CHALKBOARD EXAMPLES

Graph.

1. {the numbers between $^-4$ and $^-1$}

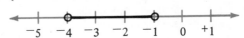

2. {the numbers between 0 and $^+5$, inclusive}

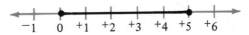

3. {the numbers greater than $^-3$}

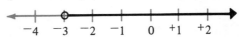

4. {the numbers less than $^+2\frac{1}{2}$}

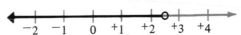

5. {$^+1$ and the numbers less than $^+1$}

6. {⁻1 and the numbers between ⁻1 and ⁺3}

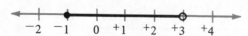

SUGGESTED EXTENSIONS

Your more capable students might try graphing the following sets, which are slightly different from anything they have previously encountered.

1. {the numbers greater than ⁺2 and less than ⁻2}
2. {all numbers except the numbers between 0 and ⁺4}
3. {all numbers except the even integers}
4. {all numbers except 0}
5. {the numbers greater than ⁺3 and the numbers between ⁻1 and ⁺1}

2-6 *Graphing Pairs of Integers*

TEACHING THESE PAGES

In this lesson we graph ordered pairs of integers on the coordinate plane. We use two number lines, one horizontal, the other vertical. The lesson material is a logical extension of the one-dimensional graphing in the previous five sections. Students have already become familiar with the terms *origin* and *coordinate* in graphing points on the number line. In order to make the transition to two-dimensional graphing, students should understand the following:

(1) The horizontal line is the number line we have been using all along. Positive numbers lie to the right of zero, and negative numbers to the left. On the vertical number line, points *above* zero represent positive numbers, while those *below* represent negative numbers. We call these number lines *coordinate axes*.

(2) The two number lines cross at zero on both lines. We call this point of intersection the *origin*.

(3) Each point in the plane can be named by an ordered pair of numbers, called the coordinates of the point. The first coordinate tells us the location of the point with respect to the horizontal number line. The second coordinate tells us the location of the point with respect to the vertical number line. Stress the importance of the order of the coordinates. You might, for instance, show that (⁺4, ⁺2) and (⁺2, ⁺4) name different points.

(4) If the first coordinate of a point is zero, then the point lies on the vertical axis. If the second coordinate of a point is zero, then the point lies on the horizontal axis.

CHALKBOARD EXAMPLES

Name the coordinates of each point.

1. *A* (⁺2, ⁺1)
2. *B* (⁻1, ⁺3)
3. *C* (⁻2, 0)
4. *D* (0, ⁻2)
5. *E* (⁻1, ⁻3)
6. *F* (⁺3, ⁻2)

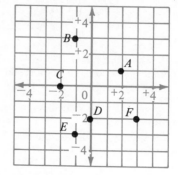

Graph the ordered pairs.

7. (⁺4, 0)
8. (⁻3, ⁺1)
9. (⁺1, ⁻3)
10. (⁺2, ⁺4)

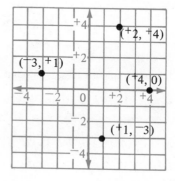

SUGGESTED EXTENSIONS

1. Have students graph points which do not have integral coordinates, such as (⁻5½, ⁺1¼) and (⁺3, ⁻2½). Some students might suggest pairs for others to graph.

2. To develop a feeling for the properties of the coordinate plane, you might ask students the following questions:

Tell where a point is located that:

a. Has two positive coordinates.

b. Has a positive first coordinate and a negative second coordinate.

c. Has a negative first coordinate and a positive second coordinate.

d. Has two negative coordinates.

You may introduce the term *quadrant* to aid in this discussion, or you may prefer to have students use descriptions like "above the horizontal axis and to the right of the vertical axis."

3 Factors and Multiples; Decimals and the Metric System

CHAPTER OVERVIEW

This chapter serves to review and strengthen students' understanding of concepts met in previous years. Topics include whole number factors and multiples, equivalent fractions, the relationship between fractions, decimals, and percents, and metric units of length. The metric system of measurement, which demonstrates an important application of decimals, is used exclusively throughout this text. Three common units of length—the meter, the centimeter, and the millimeter—are discussed. Students will also encounter word problems for the first time in this course in the sections on ratios and fractional numbers.

3-1 *Multiples and Common Multiples*

TEACHING THESE PAGES

In Chapter 2, we used a raised plus sign to stress that positive numbers are also directed numbers. From this point on, we will omit the plus sign for positive numbers, but students should understand it is implied. In order to permit students to concentrate on the skills reviewed in the chapter, they will work only with nonnegative numbers.

   This lesson introduces multiples and common multiples, leading up to the least common multiple, which is useful in finding the least common denominator of two fractions. Discuss the following points:

(1) Since the first counting number is 1, every number is a multiple of itself.

(2) Since 0 is not a counting number, the only number of which 0 is a multiple is the number 0 itself.

(3) The least common multiple (LCM) of two numbers is the smallest multiple of both numbers.

CHALKBOARD EXAMPLES

1. Solve each equation. Then write the set of multiples.

$$1 \times 6 = a \quad a = 6 \qquad 3 \times 6 = c \quad c = 18$$
$$2 \times 6 = b \quad b = 12 \qquad 4 \times 6 = d \quad d = 24$$

{the multiples of 6} = $\underline{\quad?\quad}$ {6, 12, 18, 24, ...}

Name the least common multiple (LCM).

2. 6 and 8   24
3. 15 and 6   30
4. 5 and 7   35
5. 20 and 5   20
6. 3 and 10   30
7. 2 and 3   6

SUGGESTED EXTENSIONS

Discuss the following questions:

1. Is every multiple of a whole number a whole number?
2. Name the set of multiples of $2\frac{1}{2}$; of $4\frac{1}{3}$; of $1\frac{1}{5}$.
3. Is a multiple of a fraction ever a whole number? If so, when?

3-2 *Common Factors*

TEACHING THESE PAGES

This lesson uses sets of factors and common factors to introduce the idea of greatest common factor (GCF). Later on, the GCF will be used to write fractions in lowest terms. Unless otherwise specified, when we talk about the factors of a whole number we mean the whole number factors. Stress the following ideas:

(1) The number 1 is a factor of every number.
(2) Every number is a factor of itself.
(3) The set of common factors of two whole numbers is made up of all the whole numbers that are factors of *both* given numbers.

You might begin by asking the class for a number. Then ask them to name the number as a product of two whole numbers in as many different ways as possible. For example:

$$60: \begin{array}{lll} 1 \times 60 & 2 \times 30 & 3 \times 20 \\ 5 \times 12 & 6 \times 10 & 4 \times 15 \end{array}$$

After several numbers have been named as products, select pairs of the numbers and write their sets of common factors. Then ask the students to name the greatest common factor for each set of common factors.

CHALKBOARD EXAMPLES

List the members of each set.

1. {the factors of 5} $\{1, 5\}$
2. {the factors of 12} $\{1, 2, 3, 4, 6, 12\}$
3. {the factors of 25} $\{1, 5, 25\}$
4. {the factors of 27} $\{1, 3, 9, 27\}$

Name the sets of factors and the greatest common factor (GCF).

5. 24 and 27
   {the factors of 24} = $\{1, 2, 3, 4, 6, 8, 12, 24\}$
   {the factors of 27} = $\{1, 3, 9, 27\}$
   The GCF is 3.
6. 9 and 19
   {the factors of 9} = $\{1, 3, 9\}$
   {the factors of 19} = $\{1, 19\}$ The GCF is 1.

7. 26 and 8
   {the factors of 26} = $\{1, 2, 13, 26\}$
   {the factors of 8} = $\{1, 2, 4, 8\}$
   The GCF is 2.

SUGGESTED EXTENSIONS

A discussion of tests for divisibility is appropriate at this point. Students may find the following tests helpful.

1. Simple inspection of the final digit
   a. A number ending in 2, 4, 6, 8, or 0 is divisible by 2.
   b. A number ending in 0 or 5 is divisible by 5.
   c. A number ending in 0, then, is divisible by both 2 and 5, so it is divisible by 10.
2. Addition of digits
   a. If the sum of the digits is divisible by 3, the number is divisible by 3; if the sum is divisible by 9, the number is divisible by 9.
   b. Number the digits of the numeral starting from the right. Subtract the sum of the even-numbered digits from the sum of the odd-numbered digits. If the difference is zero or a multiple of 11, the number is divisible by 11. (For example, to test 8723: $(7 + 3) - (8 + 2) = 0$; so 8723 is divisible by 11.)

Now have students do exercises like the following:

1. Which of the following numbers are divisible by 2? by 3? by 5? by 9? by 10? by 11?

876	2, 3; 6	3212	2, 11
595	5	6841	
710	2, 5, 10	3374	2
543	3	58603	
654	2, 3; 6	49782	2, 3; 6
1982	2	31590	2, 3, 5, 9, 10; 6, 18, 27
73165	5	2736	2, 3, 9; 6, 18, 27

2. Which of the numbers in Exercise 1 are divisible by 6? by 18? by 27? How did you decide?

3-3 *Special Sets of Whole Numbers*

TEACHING THESE PAGES

While the idea of odd and even numbers is probably familiar to the entire class, the expression of an even number as $2n$ and an odd number as $2n + 1$, where n is a whole number, may be new to some students. Note that the terms *odd* and *even* apply here to whole numbers. The concept of prime numbers may be less familiar to the class. Again, we are considering only primes that are whole numbers. Discuss the following points with the class:

(1) The final digit of an even number is one of the digits 0, 2, 4, 6, and 8; the final digit of an odd number is one of the digits 1, 3, 5, 7, and 9.

(2) The even numbers together with the odd numbers make up the whole numbers.

(3) Given two consecutive numbers, one is odd, the other is even.

(4) Every whole number is the product of itself and 1. A prime number has no other (whole number) factors except itself and 1.

(5) The number 1 is not a prime, because it does not have exactly two *different* factors, itself and 1.

CHALKBOARD EXAMPLES

Tell whether each number is odd or even. Then write the number in either the form $2 \times n$ or the form $(2 \times n) + 1$.

1. 16   even; 2×8 **2.** 25   odd; $(2 \times 12) + 1$
3. 13   odd; $(2 \times 6) + 1$ **4.** 42   even; 2×21
5. 36   even; 2×18 **6.** 37   odd; $(2 \times 18) + 1$

Add. Label each sum *odd* or *even*.

7. $11 + 6 = 17$
odd + even = odd

8. $11 + 9 = 20$
odd + odd = even

9. $12 + 6 = 18$
even + even = even

10. $12 + 9 = 21$
even + odd = odd

Write the set of factors. Tell whether the number is prime or not prime.

11. 13   {1, 13}; prime
12. 29   {1, 29}; prime
13. 9   {1, 3, 9}; not prime
14. 43   {1, 43}; prime

SUGGESTED EXTENSIONS

Have the students construct a "sieve of Eratosthenes" to determine the primes less than 50. This method for finding primes is based on eliminating numbers that are multiples of a lesser number.

1. Make an array like the one shown.
2. Cross out or circle each number as follows: Cross out 1, which is not prime. Circle 2, then cross out each second number. Circle 3, then cross out each third number. The next uncrossed number is 5; circle it, then cross out each fifth number. The next uncrossed number is 7; circle it, then cross out each seventh number. Continue in this manner until each number in the array is either circled or crossed out.

Point out that this method could be used to determine the set of prime numbers less than *any* given number, although the process becomes very tedious for very great numbers.

Having made the array, discuss the conjecture that there is an infinite number of pairs of primes that differ by 2, such as 3 and 5, 5 and 7, 11 and 13, 17 and 19, and so on. Point out that not every prime belongs to such a pair. For example, 23 is a prime, but neither $23 - 2$ nor $23 + 2$ is prime. (A *conjecture* is a statement that *seems* to be true, but has never been proven.)

You might also discuss Goldbach's Conjecture that any even number greater than 2 can be written as the sum of two prime numbers. Then do Written Exercises 45–53 on page 64 as a group.

3-4 *Fractions*

TEACHING THESE PAGES

In this lesson, our aim is to review and strengthen skills which students already possess for dealing with equivalent forms of a fraction. It is important that students understand the following concepts:

(1) We can write an equivalent fraction for any fraction by either multiplying or dividing the numerator and the denominator by the same nonzero whole number.

(2) When the greatest common factor (GCF) of the numerator and the denominator is 1, the fraction is in *lowest terms*. The shortest way to write a fraction in lowest terms is to divide the numerator and the denominator by their GCF. However, students may divide by any common factor, but remind them then to continue dividing until the only common factor of the numerator and denominator is 1. For example:

$$\frac{45}{75} = \frac{45 \div 15}{75 \div 15} = \frac{3}{5}, \quad \text{or}$$

$$\frac{45}{75} = \frac{45 \div 5}{75 \div 5} = \frac{9}{15} = \frac{9 \div 3}{15 \div 3} = \frac{3}{5}$$

CHALKBOARD EXAMPLES

Solve.

1. $\dfrac{5 \times 3}{8 \times 3} = a \quad a = \dfrac{15}{24}$ **2.** $\dfrac{18 \div 6}{30 \div 6} = b \quad b = \dfrac{3}{5}$

3. $\dfrac{7 \times 1}{9 \times 1} = c \quad c = \dfrac{7}{9}$ **4.** $\dfrac{14 \div 7}{49 \div 7} = d \quad d = \dfrac{2}{7}$

Write in lowest terms.

5. $\dfrac{16}{24} \quad \dfrac{2}{3}$ **6.** $\dfrac{8 + 7}{27} \quad \dfrac{5}{9}$

7. $\dfrac{8 \times 5}{100} \quad \dfrac{2}{5}$ **8.** $\dfrac{13}{78} \quad \dfrac{1}{6}$

Replace the variable to name the equivalent fraction.

9. $\dfrac{35}{100} = \dfrac{x}{20} \quad x = 7$ **10.** $\dfrac{63}{18} = \dfrac{y}{2} \quad y = 7$

11. $\dfrac{7}{10} = \dfrac{z}{80} \quad z = 56$

3-5 *Ratios*

TEACHING THESE PAGES

This lesson reviews the important concept of ratio. The ratio of 1 to 4 can be expressed in each of the following ways: 1 out of 4, 1 to 4, 1:4, and $\frac{1}{4}$. The fraction form of stating a ratio is especially convenient, because it is the easiest form to simplify. For example, $\dfrac{8}{10}$ is more easily simplified to $\dfrac{4}{5}$, than "8 to 10" to "4 to 5," or than "8:10" to "4:5."

Point out that the order of comparison is important. For example, suppose a changepurse contains 15 nickels and 20 pennies. The ratio of the number of nickels to the total number of coins is 15 to 35, or $\dfrac{15}{35}$, not 35 to 15. The fractional form $\dfrac{15}{35}$ can easily be simplified to show that the ratio in lowest terms is $\dfrac{3}{7}$. You might use coins to illustrate this ratio concretely.

CHALKBOARD EXAMPLES

Write each ratio as a fraction in lowest terms or as a whole number.

1. 6 out of 40 $\quad \dfrac{3}{20}$ **2.** 8 out of 100 $\quad \dfrac{2}{25}$

3. 17 out of 20 $\quad \dfrac{17}{20}$ **4.** 7 out of 10 $\quad \dfrac{7}{10}$

5. 15 out of 15 $\quad 1$ **6.** 85 out of 100 $\quad \dfrac{17}{20}$

Complete the statements about the shapes:

7. __?__ out of __?__ shapes are squares. 4, 6
8. In lowest terms, the ratio of the number of squares to the total number of shapes is __?__. $\dfrac{2}{3}$

SUGGESTED EXTENSIONS

Using the 15 nickels and 20 pennies mentioned in *Teaching These Pages*, have students name five more ratios:

1. Number of pennies (P) to total number of coins (T). $\frac{4}{7}$
2. Number of nickels (N) to number of pennies (P). $\frac{3}{4}$
3. Number of pennies (P) to number of nickels (N). $\frac{4}{3}$
4. Number of coins (T) to number of nickels (N). $\frac{7}{3}$
5. Number of coins (T) to number of pennies (P). $\frac{7}{4}$

Point out the following relationships among the six ratios:

Because $\dfrac{N}{P}$ and $\dfrac{P}{N}$ are reciprocals: $\dfrac{3}{4} \times \dfrac{4}{3} = 1$

Likewise, $\dfrac{3}{7} \times \dfrac{7}{3} = 1$ and $\dfrac{4}{7} \times \dfrac{7}{4} = 1$.

Because there are only nickels and pennies in the changepurse, it is not too surprising that:

$$\dfrac{3}{7} + \dfrac{4}{7} = 1$$

More surprising is the fact that:

$$\dfrac{7}{3} + \dfrac{7}{4} = \dfrac{49}{12} \quad \text{and} \quad \dfrac{7}{3} \times \dfrac{7}{4} = \dfrac{49}{12}$$

Have students try other numbers of nickels and pennies to see if these relationships still hold true. Amazingly, they do! If we have n nickels, and p pennies, we have $n + p$ coins. Then:

$$\dfrac{n + p}{n} + \dfrac{n + p}{p} = \dfrac{np + p^2 + n^2 + np}{np}$$

$$= \dfrac{(n + p)^2}{np}$$

and $\dfrac{n+p}{n} \times \dfrac{n+p}{p} = \dfrac{(n+p)^2}{np}$.

3-6 *Fractions and Decimals*

TEACHING THESE PAGES

The material in this lesson will be familiar to most students. The idea that a fraction indicates a division is the key to writing decimal equivalents of fractions. The denominator of a decimal fraction is a power of ten. Only the numerator of a decimal fraction is written— the denominator is indicated by the location of the decimal point.

Every decimal has an equivalent fraction form, but not every fraction has an equivalent terminating decimal form. For example, the division for the fraction $\dfrac{1}{3}$ produces the repeating decimal $0.3333\ldots$. Repeating decimals and the bar notation $0.3333\ldots = 0.\overline{3}$ are introduced in the C exercises of this section.

To point out the difference between significant and nonessential zeros in decimal fractions, you might write a decimal like 0.0250700 on the board and ask students which zeros are significant.

CHALKBOARD EXAMPLES

Write as a decimal.

1. $\dfrac{3}{5}$ 0.6
2. $\dfrac{7}{7}$ 1.0
3. $\dfrac{9}{10}$ 0.9
4. $\dfrac{23}{100}$ 0.23
5. $\dfrac{5}{8}$ 0.625
6. $\dfrac{5}{4}$ 1.25

Complete.

7. $\dfrac{9}{10} = 0.9 = \underline{0.90} = \underline{0.900}$

8. $0.06 = 0.060 = \underline{0.0600} = \underline{0.06000}$

9. $\dfrac{8}{1} = \dfrac{80}{10} = \dfrac{800}{100}$

10. $5.7 = 5.70 = \underline{5.700} = \underline{5.7000}$

Write as a fraction in lowest terms.

11. 0.8 $\dfrac{4}{5}$
12. 0.125 $\dfrac{1}{8}$
13. 3.2 $\dfrac{16}{5}$
14. 1.5 $\dfrac{3}{2}$

SUGGESTED EXTENSION

Writing a repeating decimal as an equivalent fraction requires a bit of simple equation solving. For example,

to show that $0.999\ldots = 1$:

Let $n = 0.999\ldots$
Then $10n = 9.999\ldots$

$$\begin{array}{r} 10n \\ -\ n \\ \hline 9n \end{array} \quad \blacktriangleright \quad \begin{array}{r} 9.999\ldots \\ -0.999\ldots \\ \hline 9 \end{array}$$

Therefore, $9n = 9$
$n = 1$

3-7 *Fractions, Decimals, and Percents*

TEACHING THESE PAGES

Students often have little difficulty with percent once they realize that *percent* means *per 100*. Thus the expression 5% is equivalent to the fraction $\dfrac{5}{100}$. Writing a decimal as an equivalent percent or a percent as an equivalent decimal is a simple process:

$$0.7 = \dfrac{7}{10} = \dfrac{70}{100} = 70\%$$

$$0.083 = \dfrac{83}{1000} = \dfrac{8.3}{100} = 8.3\%$$

$$23\% = 23\circlearrowleft\% = 0.23$$

$$5\% = \dfrac{5}{100} = 0.05$$

Explain that when we "move" the decimal point two places to the right (or left), we are actually multiplying (or dividing) the number by 100.

Writing a percent as an equivalent fraction in lowest terms is also a quite straightforward process.

$$56\% = \dfrac{56}{100} = \dfrac{14}{25}$$

However, writing a fraction as an equivalent percent requires a facility with the same division process needed for writing a fraction as a decimal:

$$\dfrac{5}{8} = 8\overline{)5.000}^{\,0.625} = 0.625 = 62.5\%$$

The ability to interpret charts and graphs is an important skill. You may wish to use the problems on page 76 for a group activity.

CHALKBOARD EXAMPLES

Write as a decimal and as a percent.

1. $\dfrac{53}{100}$ 0.53; 53%
2. $\dfrac{9}{10}$ 0.9; 90%

3. $\frac{4}{5}$ 0.8; 80%   **4.** $\frac{3}{8}$ 0.375; 37.5%

5. $\frac{3}{3}$ 1.0; 100%   **6.** $\frac{5}{4}$ 1.25; 125%

7. $\frac{200}{1000}$ 0.2; 20%   **8.** $\frac{430}{1000}$ 0.43; 43%

Write as a fraction in lowest terms.

9. 45% $\frac{9}{20}$ **10.** 36% $\frac{9}{25}$ **11.** 14% $\frac{7}{50}$

12. 52% $\frac{13}{25}$ **13.** 9% $\frac{9}{100}$ **14.** 24% $\frac{6}{25}$

SUGGESTED EXTENSION

This is an appropriate point to introduce the symbol $\doteq$ for "is approximately equal to." Students have already seen the bar notation for writing repeating decimals: $0.4545\ldots = 0.\overline{45}$. Now explain that sometimes it is more convenient to work with an approximate value for a repeating decimal. Thus, rounding $0.4545\ldots$ to thousandths, we may write:

$$0.4545\ldots \doteq 0.455$$

Rounding to hundredths: $0.4545\ldots \doteq 0.45$
Rounding to tenths: $0.4545\ldots \doteq 0.5$

Students may complete exercises like the following.

Use either $=$ or $\doteq$ to complete the sentence.

1. $0.666\ldots$ _?_ $0.\overline{6}$ $=$ **2.** $0.2727\ldots$ _?_ 0.273 $\doteq$

3. $0.\overline{1}$ _?_ 0.111 $\doteq$ **4.** $0.777\ldots$ _?_ $0.\overline{7}$ $=$

Complete the statement by rounding to the nearest hundredth.

5. $0.\overline{36} \doteq$ _?_   0.36   **6.** $0.\overline{142857} \doteq$ _?_   0.14

7. $0.1\overline{7} \doteq$ _?_   0.18   **8.** $0.\overline{129} \doteq$ _?_   0.13

3-8 *Metric Measurement*

TEACHING THESE PAGES

The big selling point for the use of the metric system is the ease of conversion from one metric unit to another. In this lesson we focus attention on the meter—a little over a yard in length—and two other frequently used units, the centimeter and the millimeter. We have used the symbol $\leftrightarrow$ to mean "is the same as" in order to reserve the equals symbol exclusively for numerical equality.

Point out that the prefix *centi* means *hundredth* and the prefix *milli* means *thousandth*.

If a meter stick is available, have students measure various distances in the classroom to develop a feeling for the length of a meter. They might enjoy measuring each other's heights in centimeters.

Students should see that they can convert from one metric unit to another by merely "moving" the decimal point. You might have the students compare the ease of conversion from one unit to another within the metric system and the U.S. Standard system. Ask them to complete statements like the following:

7 yd $\leftrightarrow$ _?_ in. 252	7 m $\leftrightarrow$ _?_ cm 700
19 ft $\leftrightarrow$ _?_ in. 228	19 m $\leftrightarrow$ _?_ cm 1900
33 ft $\leftrightarrow$ _?_ in. 396	3.3 m $\leftrightarrow$ _?_ cm 330
576 in. $\leftrightarrow$ _?_ ft 48	576 cm $\leftrightarrow$ _?_ m 5.76
576 in. $\leftrightarrow$ _?_ yd 16	576 mm $\leftrightarrow$ _?_ cm 57.6

CHALKBOARD EXAMPLES

Draw line segments. Use a metric ruler.

1. 25 cm   **2.** 480 mm   **3.** 8 cm
4. 1.5 m   **5.** 113 cm   **6.** 0.6 m

Complete.

7. 275 cm $\leftrightarrow$ _?_ m _?_ cm   2; 75
8. 73 cm $\leftrightarrow$ _?_ m _?_ cm   0; 73
9. 63 mm $\leftrightarrow$ _?_ cm _?_ mm   6; 3
10. 27 mm $\leftrightarrow$ _?_ cm _?_ mm   2; 7

SUGGESTED EXTENSION

Introduce the kilometer, the metric unit used for large distances. *Kilo* means *thousand*, so a kilometer is a thousand meters.

$$1 \text{ km} \leftrightarrow 1000 \text{ m}$$

Students may do exercises like the following:

1. 2000 m $\leftrightarrow$ _?_ km   2
2. 1500 m $\leftrightarrow$ _?_ km   1.5
3. 200 m $\leftrightarrow$ _?_ km   0.2
4. 5725 m $\leftrightarrow$ _?_ km   5.725
5. 372 m $\leftrightarrow$ _?_ km   0.372
6. 10,000 m $\leftrightarrow$ _?_ km   10

Then have the class discuss the most appropriate metric unit to use for common measurements of length, such as:

1. The height of a person   (cm)
2. The length of a room   (m)
3. The width of a piece of fabric   (cm)
4. The distance between two cities   (km)
5. The length of a football field   (m)
6. The thickness of a sheet of plastic   (mm)

4 Fractions, Decimals, Expressions, Functions

CHAPTER OVERVIEW

This chapter starts with a review of the fundamental operations of addition, subtraction, multiplication and division with fractions, and then with decimals. In the remainder of the chapter we introduce some basic algebraic concepts which students will need for the work in subsequent chapters. First, students will evaluate numerical expressions that contain grouping symbols. This leads naturally to the evaluation of expressions containing variables by replacing each variable by the members of its replacement set. We introduce the concepts of *coefficient* and *term* and discuss factorization of monomials. The final lesson introduces functions through the use of a "function machine." The machine uses a rule, stated as an expression, to generate a set of ordered pairs that is a function. Graphing sets of ordered pairs on the coordinate plane is extended from the initial treatment in Chapter 2.

4-1 *Fractions: Addition and Subtraction*

TEACHING THESE PAGES

Students should have little difficulty adding and subtracting with fractions with like denominators. However, to add or subtract with fractions with unlike denominators, we must first rewrite them as equivalent fractions with the same denominator. It is not necessary to use the least common denominator (LCD), although to do so will often save the extra step of simplifying the sum to lowest terms. In some cases, however, it may be easier to use a common denominator greater than the LCD.

CHALKBOARD EXAMPLES

Solve.

1. $\dfrac{2}{7} + \dfrac{4}{7} = a \quad a = \dfrac{6}{7}$ 2. $\dfrac{7}{8} - \dfrac{3}{8} = b \quad b = \dfrac{1}{2}$

3. $1\frac{1}{2} + 3\frac{1}{2} = c \quad c = 5$ 4. $5\frac{5}{8} - 2\frac{1}{2} = d \quad d = 3\frac{1}{8}$

5. $\dfrac{3}{5} + \dfrac{3}{4} = e \quad e = 1\frac{7}{20}$ 6. $6\frac{2}{3} - 3\frac{1}{4} = f \quad f = 3\frac{5}{12}$

SUGGESTED EXTENSION

One possible common denominator for two fractions is the product of the denominators of the fractions. This common denominator may or may not be the LCD. We can write the following general formulas, in which a, b, c, and d are integers, and b and d are not zero:

$$\frac{a}{b} + \frac{c}{d} = \frac{ad + bc}{bd}$$

and

$$\frac{a}{b} - \frac{c}{d} = \frac{ad - bc}{bd}$$

Work several examples using the formulas. Include examples for which bd is the LCD as well as examples for which it is not. For example:

$$\frac{3}{5} + \frac{1}{8} = \frac{(3 \times 8) + (5 \times 1)}{5 \times 8} = \frac{24 + 5}{40} = \frac{29}{40}$$

$$\frac{7}{8} + \frac{1}{4} = \frac{(7 \times 4) + (8 \times 1)}{8 \times 4} = \frac{28 + 8}{32}$$

$$= \frac{36}{32} = \frac{9}{8} = 1\frac{1}{8}$$

$$2\frac{5}{7} - \frac{3}{4} = \frac{19}{7} - \frac{3}{4} = \frac{(19 \times 4) - (7 \times 3)}{7 \times 4}$$

$$= \frac{76 - 21}{28} = \frac{55}{28} = 1\frac{27}{28}$$

4-2 *Fractions: Multiplication and Division*

TEACHING THESE PAGES

Multiplication is the easiest of the four operations with fractions, since it merely involves multiplying numerator by numerator and denominator by denominator. A few students may need to be reminded how to deal with a mixed numeral, such as $2\frac{5}{8}$. Many will remember this from past experience, but if a review is necessary, explain that $2\frac{5}{8}$ means $2 + \frac{5}{8}$. We know that another name for 2 is $\frac{16}{8}$, so:

$$2\frac{5}{8} = 2 + \frac{5}{8} = \frac{16}{8} + \frac{5}{8} = \frac{21}{8}$$

With practice, students should be able to do most of the steps mentally, and write directly $2\frac{5}{8} = \frac{21}{8}$. The product of two mixed numerals then becomes as simple as:

$$2\frac{5}{8} \times 1\frac{1}{3} = \frac{21}{8} \times \frac{4}{3}$$

$$= \frac{84}{24} = \frac{7}{2} = 3\frac{1}{2}$$

Once skill in multiplication is attained, division by a fraction follows easily, since it involves multiplication by the reciprocal of the divisor. Students must under-

stand that they are to take the reciprocal of the divisor (the second fraction), not of the dividend (the first fraction). They must also recognize that the divisor must be written as a fraction, not as a mixed numeral, before the reciprocal can be named. For example:

$$5\frac{1}{2} \div 2\frac{1}{5} = \frac{11}{2} \div \frac{11}{5}$$

$$= \frac{11}{2} \times \frac{5}{11} = \frac{5}{2} = 2\frac{1}{2}$$

CHALKBOARD EXAMPLES
Multiply.

1. $\frac{1}{6} \times \frac{3}{4}$ $\frac{1}{8}$ 2. $\frac{5}{4} \times \frac{1}{3} \times \frac{3}{5}$ $\frac{1}{4}$

3. $2\frac{2}{3} \times 3\frac{3}{4}$ 10

Name the reciprocal.

4. $\frac{1}{5}$ 5   5. $\frac{5}{6}$ $\frac{6}{5}$ 6. 8 $\frac{1}{8}$ 7. $3\frac{3}{5}$ $\frac{5}{18}$

Solve.

8. $x = \frac{2}{3} \div \frac{5}{6}$ $x = \frac{4}{5}$ 9. $\frac{4}{7} \div 8 = y$ $y = \frac{1}{14}$

10. $8 \div \frac{4}{7} = z$ $z = 14$

SUGGESTED EXTENSIONS

Develop formulas for multiplying and dividing the fractions $\frac{a}{b}$ and $\frac{c}{d}$:

$$\frac{a}{b} \times \frac{c}{d} = \frac{ac}{bd}$$

and

$$\frac{a}{b} \div \frac{c}{d} = \frac{ad}{bc}$$

Discuss the restrictions on the values of the variables in the preceding formulas. For both formulas, $b \neq 0$, $d \neq 0$. For division, it is also necessary that $c \neq 0$. Ask why there are more restrictions on the variables for division than for the other three operations.

4-3 *Decimals: Addition and Subtraction*

TEACHING THESE PAGES

Addition and subtraction with decimals is quite similar to addition and subtraction with whole numbers. Stress the need for lining up the decimal points in writing additions or subtractions vertically. Point out that adding zeros on the right is a way of expressing the two decimal fractions with a common denominator. This makes the addition or subtraction easier to perform. Note that the check by rounding simply confirms the reasonableness of the answer, but does not insure its exact accuracy. For example, for the addition $17.63 + 5.013$, both the incorrect answer 22.76 and the correct answer 22.643 check by estimation: $18 + 5 = 23$. However, the importance of the check by estimation is that it verifies the location of the decimal point.

CHALKBOARD EXAMPLES

Add or subtract. Estimate to check your work.

1. $15.78 + 6.23 = \underline{22.01}$; $16 + 6 = 22$
2. $15.78 - 6.23 = \underline{9.55}$; $16 - 6 = 10$
3. $2.1 + 9.82 + 5.203 = \underline{17.123}$; $2 + 10 + 5 = 17$
4. $30.2 - 8.753 = \underline{21.447}$; $30 - 9 = 21$
5. $13.09 - 5.005 = \underline{8.085}$; $13 - 5 = 8$
6. $43.1 + 4.31 = \underline{47.41}$; $43 + 4 = 47$

4-4 *Decimals: Multiplication and Division*

TEACHING THESE PAGES

Locating the decimal point in a product or a quotient is a bit more complicated than locating it in a sum or a difference. Its location in the product of two decimals can be found by counting the number of digits to the right of the decimal point in each factor. The number of decimal places in the product, then, is equal to the sum of the numbers of decimal places in the factors. For example:

$$
\begin{array}{r}
5.36 \leftarrow\text{2 digits} \\
\times\ 2.5 \leftarrow\text{1 digit} \\
\hline
2680 \qquad\text{3 digits} \\
10720 \\
\hline
13.400
\end{array}
$$

Estimate $5 \times 3 = 15$

$$
\begin{array}{r}
0.0296 \leftarrow\text{4 digits} \\
\times\ 3.2 \leftarrow\text{1 digit} \\
\hline
592 \qquad\text{5 digits} \\
8880 \\
\hline
0.09472
\end{array}
$$

Estimate $0.03 \times 3 = 0.09$

In the case of division, the decimal point is located before the division is performed by "moving" the decimal points in both dividend and divisor the same number of places to the right—enough places to make the divisor a whole number. Students should understand that they are actually multiplying each decimal by a power of ten.

$$
\begin{array}{r}
1\ \ 62.8 \\
0.25\,)\overline{40\,70.0}
\end{array}
$$

Estimate: $41 \div \dfrac{1}{4} = 41 \times 4 = 164$

Related to multiplying with a decimal is finding a percent of a number. The percent is first written as a decimal—a simple enough process—then the number is multiplied by the decimal.

CHALKBOARD EXAMPLES

Solve. Check by estimating the solution.

1. $35.7 \times 2.83 = m$ $m = 101.031$; $36 \times 3 = 108$
2. $5.01 \times 7.8 = n$ $n = 39.078$; $5 \times 8 = 40$
3. $q = 2.003 \times 1.7$ $q = 3.4051$; $2 \times 2 = 4$
4. $r = 5.9 \times 0.59$ $r = 3.481$; $6 \times 0.6 = 3.6$
5. $t = 53.16 \div 4$ $t = 13.29$; $53 \div 4$, about 13
6. $9.842 \div 3.8 = x$ $x = 2.59$; $10 \div 4 = 2.5$
7. $13.05 \div 9 = y$ $y = 1.45$; $13 \div 9$, about 1.5
8. $s = 40.872 \div 5.24$ $s = 7.8$; $41 \div 5$, about 8
9. $p = 6\% \times 94$ $p = 5.64$; $0.06 \times 100 = 6$
10. $z = 8\% \times 103$ $z = 8.24$; $0.1 \times 100 = 10$

4-5 *Order of Operations*

TEACHING THESE PAGES

To be able to simplify algebraic expressions, students must understand the meaning of grouping symbols. Point out that brackets are used in the same way as parentheses, and have the same meaning. The omission of an operation sign between a numeral and an expression within parentheses or brackets, or between a numeral and a variable indicates multiplication. Thus:

$$5(13 - 2) = 5 \times 11 = 55$$
$$2a = 2 \times a$$

When grouping symbols have been omitted, the convention is that multiplications and divisions are performed first, in order of occurrence, *before* additions and subtractions. Thus:

$$5 \times 3 + 12 \div 4 = 15 + 3 = 18$$

Now have students simplify the same expression grouped in the following ways:

$$5(3 + 12) \div 4 = (5 \times 15) \div 4 = 75 \div 4 = 18.75$$

$$5(3 + 12 \div 4) = 5(3 + 3) = 5 \times 6 = 30$$

$$(5 \times 3 + 12) \div 4 = (15 + 12) \div 4$$
$$= 27 \div 4 = 6.75$$

CHALKBOARD EXAMPLES
Simplify.

1. $(3 + 5)(1 + 8)$ 72
2. $(19 + 3)7$ 154
3. $(37 - 9)3$ 84
4. $23 + (15 - 7)$ 31
5. $\dfrac{8 + 1}{(12)(5)}$ $\dfrac{3}{20}$
6. $\dfrac{6 + 7 + 9}{2}$ 11
7. $56 - (3 + 5)$ 48
8. $(63 + 5) - 12$ 56

SUGGESTED EXTENSION
To preview the next lesson and at the same time emphasize the importance of grouping signs, have students consider the following numerals, variables, and operations:

$$4, a, +, 7, b, -, a, b$$

Let $a = 2$ and $b = 3$; then ask the students to evaluate the expression for each of the following groupings:

1. $4a + (7b - ab)$ 23
2. $4[(a + 7b) - ab]$ 68
3. $(4a + 7b) - ab$ 23
4. $4a + [(7b - a)b]$ 65
5. $4(a + 7b) - ab$ 86
6. $(4a + 7)(b - ab)$ -45
7. $4[a + (7b - ab)]$ 68
8. $[(4a + 7b) - a]b$ 81
9. $4a + 7(b - ab)$ -13
10. $4[(a + 7)(b - ab)]$ -108

Note that some answers are negative numbers. Operations with negative numbers will be presented later in the text (Chapters 9 and 10), but some students may have worked with them before. Others may be able to reason out the sign.

4-6 Evaluating Expressions
TEACHING THESE PAGES
When an expression contains variables, its numerical value depends, of course, on the values assigned to the variables. The set of values agreed upon for replacing a variable is called its *replacement set*. If the replacement set for each variable in an expression is a finite set, the set of values of the expression is also finite.

The idea of a replacement set is important because we cannot determine the solution set of an equation or inequality without considering the replacement sets for the variables. In the extensions of Sections 4-1 and 4-2, we discussed some restrictions on the replacement sets of several variables. Although often the replacement set for a variable is {the directed numbers}, sometimes the replacement set is more restricted. For instance, suppose we solve the equation $x + 1 = -1$ where the replacement set for x is {the positive numbers}. Although we find that $x = -2$, the *solution set* for the equation is $\emptyset$, since -2 is not a member of the replacement set for the variable.

CHALKBOARD EXAMPLES
Find all values of the expression. The replacement set for n is $\{1, 2, 3\}$.

1. $n + 5$ $6, 7, 8$
2. $\dfrac{n}{3}$ $\dfrac{1}{3}, \dfrac{2}{3}, 1$
3. $3n - 1$ $2, 5, 8$

Find the value of the expression. Let $x = 3$, $y = 2$, and $z = 7$.

4. $xy + xz$ 27
5. $x + z - y$ 8
6. $y(z - 3)$ 8

4-7 Factors, Coefficients, and Terms
TEACHING THESE PAGES
This lesson introduces some basic algebraic definitions. Point out that the term *coefficient* may be used to refer to any factor of a product. For example, in the product $5xy$:

(1) 5 is the coefficient of xy, xy is the coefficient of 5.
(2) $5x$ is the coefficient of y, y is the coefficient of $5x$.
(3) $5y$ is the coefficient of x, x is the coefficient of $5y$.

However, the word *coefficient* most often refers to the numerical factor, as in the first case given above. If there is a possibility of confusion, the word *numerical* can be inserted: 5 is the *numerical coefficient* of the term $5xy$.

When the numerical coefficient is 1, the "1" is usually omitted: The (numerical) coefficient of ab is 1, or $ab = 1ab$.

Terms of an algebraic expression are separated by plus and minus signs.

(1) $7 + n$ has two terms, 7 and n.
(2) $5a - 6c$ has two terms, $5a$ and $6c$.
(3) $3x^2yz^3$ is all one term.
(4) $4m^2 + 3m + 1$ has three terms, $4m^2$, $3m$, and 1.

CHALKBOARD EXAMPLES

Complete each factorization for the given product.

1. Product: $8amx$; $8(\underline{amx})$, $2a(\underline{4mx})$, $mx(\underline{8a})$

2. Product: $\dfrac{bc}{3}$; $\dfrac{1}{3}(\underline{bc})$, $\dfrac{b}{3}(\underline{c})$, $(\underline{b})\left(\dfrac{c}{3}\right)$

Show that each factorization has the same value as the product $\dfrac{3m}{4s}$, when $m = 2$ and $s = 6$.

$$\frac{3m}{4s} = \frac{3 \times 2}{4 \times 6} = \frac{1}{4}$$

3. Factorization:

$$\frac{3}{4}m\left(\frac{1}{s}\right) = \frac{3}{4}(2)\left(\frac{1}{6}\right) = \frac{6}{4} \times \frac{1}{6} = \frac{1}{4}$$

4. Factorization: $3\left(\dfrac{m}{4s}\right) = 3\left(\dfrac{2}{4 \cdot 6}\right) = 3\left(\dfrac{1}{12}\right) = \dfrac{1}{4}$

Name the product.

5. $5(a)(r)$ $5ar$ 6. $(0.4)(4.0)(x)(y)$ $1.6xy$

4-8 *Functions*

TEACHING THESE PAGES

The concept of function is, of course, one of the fundamental concepts in higher mathematics. We introduce functions using a "function machine," which accepts a number, x, works on it, and turns out another number, $f(x)$. This gives us the ordered pair $(x, f(x))$. The "function machine" works on a number according to a stated rule, such as $2x + 1$. Stress the fact that for a rule to be a function rule, it must produce *only one* output for each input. More than one input may result in the same output, but a given input must have only a single output for the resulting set of ordered pairs to be a function. For example, $\{(1, 2), (2, 2), (3, 2)\}$ represents a function while $\{(1, 3), (1, 4), (2, 5), (2, 6)\}$ does not.

Students will graph sets of ordered pairs in the first quadrant of the coordinate plane. Most of the functions they will encounter in this lesson are *linear functions*, that is, functions that result in a straight line when graphed.

CHALKBOARD EXAMPLES

Complete the set of ordered pairs by using the given rule.

1. $2n - 3$: $\{(2, 1), (3, \underline{\ ?\ }), (4, \underline{\ ?\ }), (5, \underline{\ ?\ })\}$ $3; 5; 7$

2. $\dfrac{s}{3}$: $\{(0, 0), (2, \underline{\ ?\ }), (4, \underline{\ ?\ }), (6, \underline{\ ?\ })\}$ $\dfrac{2}{3}; \dfrac{4}{3}; 2$

Graph the function.

3. $\{(0, 2), (1, 4), (2, 6), (3, 8), (4, 10)\}$

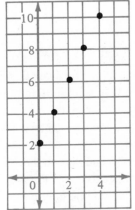

Match the set of ordered pairs with the correct rule.

4. $\left\{\left(\dfrac{1}{2}, 1\right), (1, 2), \left(1\tfrac{1}{2}, 3\right), (2, 4)\right\}$

    A. $\dfrac{n}{2}$ (B.) $2n$ C. $n + \dfrac{1}{2}$

5. $\{(0, 1), (3, 2), (6, 3), (9, 4)\}$

    A. $x + 1$ B. $\dfrac{1}{2}x$ (C.) $\dfrac{1}{3}x + 1$

SUGGESTED EXTENSIONS

1. Use the rule $2n - 3$, and the replacement set $\{2\tfrac{1}{2}, 3\tfrac{1}{3}, 4.2\}$ to write a set of ordered pairs. Now graph the ordered pairs. Do the three points seem to lie on a straight line?

2. Use the rule $2n - 3$ to write more ordered pairs using the replacement set $\{2, 3, 4, 5\}$. Graph the new pairs on the same axes that you used for Exercise 1. Do you find that all of these points also lie on the line that contained the first set of points?

3. Replace n with other fractional values that lie between 2 and 5. Do the new ordered pairs name points on the same line?

4. Replace n with values greater than 5. Where are these points? What do you think happens when n has a value less than 2?

5 Equations and Inequalities

CHAPTER OVERVIEW

In this chapter, students will develop intuitive equation solving skills. The emphasis here is on intuition rather than rules, so that when properties of equality and inequality are introduced later, students will have already met them in an informal context. We introduce the concept of an equation as a balanced scale. We continue to use the numbers of arithmetic, that is, the nonnegative numbers, in the development of ideas. Thus there will be times when the solution set for a given open sentence over {the numbers of arithmetic} will be only part of the solution set over {the directed numbers}. This may be the case for some quadratics, as well as for some linear equations for which the only solution is a negative number.

After the treatment of equation-solving methods, we turn to solving inequalities. Students will solve inequalities intuitively—no rules for manipulating an inequality will be introduced here. This avoids the problem of multiplication of an inequality by a negative number, which necessitates reversing the direction of the inequality symbol.

5-1 *Equations and Inequalities*

TEACHING THESE PAGES

Up to this point, we have only dealt with number sentences expressing equality. Here we introduce the symbols $\neq$, $<$, and $>$ as ways of expressing inequality. Point out that a number sentence which is false for the verb "=" is true for "$\neq$." If a number sentence contains a variable, its truth depends on which member of the replacement set is used for the variable.

Some group practice with the translation from a word expression to a symbolic expression might prove helpful. Encourage student participation by writing a sentence such as "The difference of 12 and 7 is not equal to 10" on the chalkboard and having a volunteer write the appropriate equation or inequality. The volunteer then gets to write another sentence on the board and to pick another student to write a translation into symbols. You might vary the activity by having students translate equations or inequalities into word sentences. Discuss which sentences are true, which are false, and which are neither true or false.

CHALKBOARD EXAMPLES

Read each sentence aloud, then tell whether the sentence is an equation or inequality.

1. $13 > 3n$ inequality
2. $13 = n^3 - 4$ equation
3. $13 + n < 4n$ inequality
4. $13 = 3n + 4$ equation
5. $n^2 + 4 \neq 13$ inequality

Write a number sentence.

6. 5 more than some number is 12. $n + 5 = 12$

7. 5 times the cube of some number is less than 50.
$5n^3 < 50$

8. The product of 6.3 and some number is greater than 75. $6.3n > 75$

5-2 *Equations and Solution Sets*

TEACHING THESE PAGES

In this section, students use tables to help them determine the solution sets of equations when the replacement sets are restricted to only a few values. In Example 2 on page 124, some students may notice that $2m = 7 + 1$ is true when $m = 4$. However, since the replacement set is $\{0, 2, 5\}$, and does *not* include 4, we say that the solution set is $\emptyset$. That is, every member of the solution set of a sentence must also be a member of its replacement set. Since $\emptyset$ names the set with no members, it is the answer here.

Graphing solution sets on the number line provides further practice in graphing and gives a visual representation of the numbers in the solution set.

Note that the word *solution* may be used in two senses. A solution of an equation is a specific set, such as $\{1, 2\}$. We also use the word solution for the method of solving a problem. This is a fine point, and you need not mention it to the class unless there is confusion.

CHALKBOARD EXAMPLES

Make a table to find the solution set. Use the given replacement set.

1. $y - 2 = 5 + 3$; $\{10, 12, 14\}$ Solution set: $\{10\}$

Replacement	$y - 2 = 5 + 3$	True/False
10	$10 - 2 = 5 + 3$	True
12	$12 - 2 = 5 + 3$	False
14	$14 - 2 = 5 + 3$	False

2. $12 - a = 3$; $\{7, 9, 11\}$ Solution set: $\{9\}$

Replacement	$12 - a = 3$	True/False
7	$12 - 7 = 3$	False
9	$12 - 9 = 3$	True
11	$12 - 11 = 3$	False

3. $2n^2 + 3 = 11$; $\{1, 3, 5\}$ Solution set: $\emptyset$

Replacement	$2n^2 + 3 = 11$	True/False
1	$2 + 3 = 11$	False
3	$18 + 3 = 11$	False
5	$50 + 3 = 11$	False

Find the solution set and graph it. Use the replacement set given.

4. $3t - 3 = 2t$; $\{1, 3, 5, 7\}$

Replacement	$3t - 3 = 2t$	True/False
1	$3 - 3 = 2$	False
3	$9 - 3 = 6$	True
5	$15 - 3 = 10$	False
7	$21 - 3 = 14$	False

Solution set: $\{3\}$

5-3 *Equation Solving Strategies*

TEACHING THESE PAGES

In this lesson we introduce informally the strategy of "undoing" what has been done to the variable by addition or subtraction of some number from both members of an equation. Since subtraction "undoes" addition, and addition "undoes" subtraction, we can sometimes isolate the variable by adding the same term to, or subtracting it from, both members of the equation. The illustration of a balanced scale helps students see the idea of keeping the two sides "in balance." We introduce {the numbers of arithmetic}, that is, {0 and all numbers greater than 0}, as our replacement set.

Students are expected to decide intuitively on the appropriate addition or subtraction to perform to solve an equation. To see if they are on the right track, you might ask them to explain how they arrived at a given addition or subtraction. However, at this stage you may expect some students to guess at random until they find the solution. This is acceptable for the present—just make sure they are adding to or subtracting from *both* sides of an equation.

CHALKBOARD EXAMPLES

Add or subtract as directed. Simplify the result.

1. $5 + 7 = 12$; add 2 to both members.
$$5 + 7 + 2 = 12 + 2$$
$$14 = 14$$

2. $5 + 7 = 12$; subtract 5 from both members.
$$5 + 7 - 5 = 12 - 5$$
$$7 = 7$$

3. $n + 7 = 12$; subtract 7 from both members.
$$n + 7 - 7 = 12 - 7$$
$$n = 5$$

4. $10 + m = 15$; subtract 10 from both members.
$$10 + m - 10 = 15 - 10$$
$$m = 5$$

5. $p - 8 = 2$; add 8 to both members.
$$p - 8 + 8 = 2 + 8$$
$$p = 10$$

6. $17 = s - 5$; add 5 to both members.
$$17 + 5 = s - 5 + 5$$
$$22 = s$$

Complete the solution. The replacement set is {the numbers of arithmetic}.

7. $b + 3 = 5$
$$b + 3 - 3 = 5 - \underline{\,?\,} \quad 3$$
$$b = \underline{\,?\,} \quad 2$$

8. $12 = g - 7$
$$12 + \underline{\,?\,} = g - 7 + 7 \quad 7$$
$$\underline{\,?\,} = g \quad 19$$

5-4 More Equation Solving Strategies

TEACHING THESE PAGES

In this section we use multiplication or division of both members of an equation to isolate the variable. Once again, the two operations are inverses and we restrict our consideration to {the numbers of arithmetic}. Instead of dividing by a number, we could, of course, have students multiply by the reciprocal of the number. However, since we will not discuss multiplication and division as inverse operations until Chapter 10, for the time being, students will use division rather than multiplication by a reciprocal.

CHALKBOARD EXAMPLES

Multiply or divide both members. Simplify the result.

1. $14 + 6 = 20$; divide by 5.
$$\frac{14}{5} + \frac{6}{5} = \frac{20}{5}; \frac{20}{5} = \frac{20}{5}$$

2. $\frac{15}{3} = \frac{20}{4}$; multiply by 4.
$$4 \cdot \frac{15}{3} = 4 \cdot \frac{20}{4}$$
$$\frac{60}{3} = \frac{80}{4}$$
$$20 = 20$$

3. $\frac{y}{5} = 7$; multiply by 5.
$$5 \cdot \frac{y}{5} = 5 \cdot 7$$
$$y = 35$$

4. $10m = 70$; divide by 10.
$$\frac{10m}{10} = \frac{70}{10}$$
$$m = 7$$

Complete the solution. The replacement set is {the numbers of arithmetic}.

5. $\frac{c}{5} = 4$
$$5 \cdot \frac{c}{5} = \underline{\,?\,} \cdot 4 \quad 5$$
$$c = \underline{\,?\,} \quad 20$$

6. $3x = 21$
$$\frac{3x}{3} = \frac{21}{\underline{\,?\,}} \quad 3$$
$$x = \underline{\,?\,} \quad 7$$

5-5 Writing Expressions Using Variables

TEACHING THESE PAGES

The focus of this lesson is translation. Before students can answer a question by solving an equation, they must first be able to *write* the equation. The ability to translate words into symbols is a fundamental skill.

Explain to the students that the letter chosen as a variable is unimportant, though sometimes we use a letter that reminds us of the word it represents, such as a for area, t for time, and so forth. The important thing is that students understand the relationship of the variable to the known numbers in a problem. Class practice might help students acquire this skill. For example, suppose we select the letter d to stand for a number of dollars. Read the class several statements involving dollars, and ask the class to write algebraic expressions for each statement. For example:

(1) Mary has $3 more than d dollars. $\quad d + 3$
(2) Tim had d dollars, then spent $5. $\quad d - 5$
(3) Bruno spent $1 more than half of d dollars.
$$\frac{d}{2} + 1$$

(4) Suppose you have d dollars and your sister has $2 more than you have. How much do you both have, in all? $d + (d + 2)$

(5) Paula has d dollars. She spends $1, then mows some lawns and earns $9. How much does she have then? $d - 1 + 9$

After the answers are read and discussed, select another variable with another meaning, such as number of hours worked, or number of fish caught. Ask students to make up statements using the new variable, which other students may then translate.

CHALKBOARD EXAMPLES

Write an algebraic expression to answer the question.

1. Given the number t, what is six more than that number? $t + 6$

2. Given the number q, what is three more than four-fifths that number? $\frac{4}{5}q + 3$

3. Given the number s, what is two less than twice the square of that number? $2s^2 - 2$

4. Given the number y, what is five less than twice that number? $2y - 5$

Write an algebraic expression.

5. 3 less than n $n - 3$
6. 2 less than one-fourth q $\frac{1}{4}q - 2$
7. $4k$ more than x $x + 4k$
8. two-fifths the product of 7 and y $\frac{2}{5}(7y)$

5-6 *Using Formulas to Solve Problems*

TEACHING THESE PAGES

When values are substituted for all except one of the variables in a formula, an equation in one unknown results. In this lesson we deal with some very common formulas, and students will substitute given quantities for all except one variable in a formula. Little, if any, manipulation of the equation is required to isolate the variable. The emphasis here, rather, is on (1) selection of the appropriate formula, and (2) correct substitution of information in word sentences in formulas. The *Oral Exercises* review the formulas students need to know to do the *Written Exercises*.

CHALKBOARD EXAMPLES

Tell what formula is required. Then complete the solution.

1. A rectangle is 16 cm long and 10.5 cm wide. Find (a) the perimeter, and (b) the area.

(a) $P = 2l + 2w$
$P = 2(16) + 2(10.5)$
$= 32 + 21 = 53$
The perimeter is 53 cm.

(b) $A = lw$
$A = 16 \times 10.5$
$= 168.0$
The area is 168 cm^2.

2. The perimeter of a square is 640 mm. Find (a) the length of a side, and (b) the area.

(a) $P = 4s$
$640 = 4s$
$160 = s$
The length of a side is 160 mm.

(b) $A = s^2$
$A = (160)^2$
$= 25,600$
The area is 25,600 mm^2

3. A circle has diameter 42 cm. Use $\pi = \frac{22}{7}$ to find (a) the circumference, and (b) the area.

(a) $C = \pi d$
$C = \frac{22}{7} \times 42$
$= 132$
The circumference is 132 cm.

(b) $A = \pi r^2$
$A = \frac{22}{7} \times 21 \times 21$
$= 1386$
The area is 1386 cm^2.

SUGGESTED EXTENSIONS

1. Using the formulas in the section, have students solve problems which involve manipulation of the equation to isolate the variable. For example:

1. A rectangle with an area of 96 m^2 is 15 m long. How wide is it? (6.4 m)
2. The perimeter of a rectangle is 70 cm. If it is 12 cm wide, how long is it? (23 cm)
3. The area of a square is 64 cm^2. Find the length of one side (8 cm), and the perimeter (32 cm).
4. The perimeter of a triangle is 106 cm. The lengths of two sides are 26 cm and 48 cm. How long is the third side? (32 cm)
5. The area of a triangle is 803 m^2. If the base is 73 m long, what is the height? (22 m)

2. In several formulas, have students find one variable in terms of the other variables. For instance, you might have them solve $P = 4s$ for s, obtaining the result $s = \dfrac{P}{4}$. Work through an example with the class to show them the method, then have them solve exercises like the following:

Solve each formula for the given variable.

1. $A = lw$; solve for w. $\left(w = \dfrac{A}{l}\right)$

2. $P = a + b + c$; solve for b. $(b = P - a - c)$

3. $A = \dfrac{1}{2}bh$; solve for h. $\left(h = \dfrac{2A}{b}\right)$

4. $C = \pi d$; solve for d. $\left(d = \dfrac{C}{\pi}\right)$

5-7 Solving Inequalities: The $<$ and $>$ Relationships

TEACHING THESE PAGES

In this lesson, students will solve inequalities intuitively, as they solved equations in Section 5-2. We do not introduce techniques for isolation of the variable yet. However, the students may see the similarity between solving equations and solving inequalities. The important thing here is that students recognize whether or not a given number satisfies the conditions of an inequality. Students had some experience in graphing inequalities in Chapter 2. Now they will graph solution sets of inequalities on the number line.

CHALKBOARD EXAMPLES

State the question that is suggested by the inequality. The replacement set is {the numbers of arithmetic}.

1. $2x < 10$
   Which numbers of arithmetic, when multiplied by 2, are less than 10?

2. $y + 5 > 6$
   Which numbers of arithmetic, when increased by 5, are greater than 6?

For each member of the replacement set, show whether a true or a false statement results when the variable in the sentence is replaced by the number.

3. $8 > n + 2$; $\{2, 4, 6\}$
   $8 > 2 + 2 \rightarrow 8 > 4$; True
   $8 > 4 + 2 \rightarrow 8 > 6$; True
   $8 > 6 + 2 \rightarrow 8 > 8$; False

4. $10 < p - 3$; $\{8, 12, 16\}$
   $10 < 8 - 3 \rightarrow 10 < 5$; False
   $10 < 12 - 3 \rightarrow 10 < 9$; False
   $10 < 16 - 3 \rightarrow 10 < 13$; True

5. $5a + 3 > 15$; $\{1, 3, 5\}$
   $5 \cdot 1 + 3 > 15 \rightarrow 8 > 15$; False
   $5 \cdot 3 + 3 > 15 \rightarrow 18 > 15$; True
   $5 \cdot 5 + 3 > 15 \rightarrow 28 > 15$; True

6. $2.5 + y < 10$; $\{5, 10, 15\}$
   $2.5 + 5 < 10 \rightarrow 7.5 < 10$; True
   $2.5 + 10 < 10 \rightarrow 12.5 < 10$; False
   $2.5 + 15 < 10 \rightarrow 17.5 < 10$; False

Using {the numbers of arithmetic} as the replacement set, graph the solution set on the number line.

7. $4x < 12$

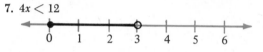

8. $c + 15 > 17$

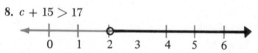

5-8 Solving Inequalities: The $\leq$ and $\geq$ Relationships

TEACHING THESE PAGES

Point out that a sentence which uses $\leq$ or $\geq$ is actually a contraction of two sentences—a compound sentence with the connective "or." For example, $a + 3 \leq 5$ is a contraction of $a + 3 < 5$ or $a + 3 = 5$. The statement $a + 3 \leq 5$ is true for any replacement of a that makes either part true.

$a + 3 < 5$ is true for $a < 2$
$a + 3 = 5$ is true for $a = 2$

If the replacement set is {the numbers of arithmetic}, the solution set is {the numbers of arithmetic less than or equal to 2}.

Graph: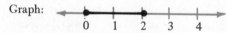

Compare this with the two graphs:

$a < 2$:

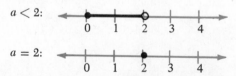

$a = 2$:

Explain that if we combine the two graphs above, we get the graph of $a + 3 \leq 5$.

Note that the replacement set for Example 3 on page 143 is {all the numbers on the number line}, which includes the negative numbers. Students are aware of the existence of negative numbers, but they have not yet been introduced to operations with negative numbers. Therefore, in the exercises, care is taken not to require students to operate with negative numbers.

CHALKBOARD EXAMPLES

State the question that is suggested by the inequality. The replacement set is {the numbers of arithmetic}.

1. $a + 5 \leq 12$
   Which numbers of arithmetic, when increased by 5, are less than or equal to 12?
2. $3b \geq 9$
   Which numbers of arithmetic, when multiplied by 3, are greater than or equal to 9?

For each member of the replacement set, show whether a true or a false statement results when the variable is replaced by the number.

3. $z - 8 \leq 5$; {11, 13, 15}
   $11 - 8 \leq 5 \rightarrow 3 \leq 5$; True
   $13 - 8 \leq 5 \rightarrow 5 \leq 5$; True
   $15 - 8 \leq 5 \rightarrow 7 \leq 5$; False

4. $7 \leq x + 1$; {6, 7, 8}
   $7 \leq 6 + 1 \rightarrow 7 \leq 7$; True
   $7 \leq 7 + 1 \rightarrow 7 \leq 8$; True
   $7 \leq 8 + 1 \rightarrow 7 \leq 9$; True

5. $4d \geq 14$; {3, 4, 5}
   $4 \cdot 3 \geq 14 \rightarrow 12 \geq 14$; False
   $4 \cdot 4 \geq 14 \rightarrow 16 \geq 14$; True
   $4 \cdot 5 \geq 14 \rightarrow 20 \geq 14$; True

6. $2n + 1 \neq 8$; {3, 3.5, 4}
   $2(3) + 1 \neq 8 \rightarrow 7 \neq 8$; True
   $2(3.5) + 1 \neq 8 \rightarrow 8 \neq 8$; False
   $2(4) + 1 \neq 8 \rightarrow 9 \neq 8$; True

Using the replacement set given, graph the solution set on the number line.

7. $3c + 2 \geq 5$; {the numbers of arithmetic}

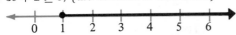

8. $5f \leq 15$; {all the numbers on the number line}

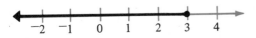

5-9 *Using Inequalities to Solve Problems*

TEACHING THESE PAGES

In this lesson we consider word problems which may be solved by writing inequalities rather than equations. You might begin by discussing some everyday situations which may lead to statements of inequality. Suppose, for example, you work a total of 20 hours at several different hourly rates, the lowest of which is $2.75 per hour. Then, your paycheck (before deductions) should be at least 20 × $2.75, or $55. (If it is not, you'd better ask why!) We may write: $p > 55$. Now consider another situation: a highway tunnel has a clearance of 6 m. Any vehicle that goes into the tunnel must be less than 6 m high, for obvious reasons. We may write: $h < 6$. Ask students to think up other situations that might be expressed by inequalities.

CHALKBOARD EXAMPLES

Write an inequality. Then write the solution set if the replacement set is {the numbers of arithmetic}.

1. Ruth works at a part time job that pays $2.50 per hour. In a certain week she earned more than $25. Represent the number of hours she worked that week. $2.5a > 25$; $a > 10$; She worked more than 10 hours. (Note: the solution set of the inequality is {the numbers of arithmetic >10}, but when we consider the situation of the word problem, we see that it imposes an upper limit on a also. If Ruth works "part time," we can assume she worked less than 40 hours. Therefore, $a > 10$ and $a < 40$.)
2. When 2 is increased by some number, the result is at least 5. What numbers meet this condition? $2 + n \geq 5$; $n \geq 3$; The solution set is {the numbers of arithmetic ≥ 3}.

SUGGESTED EXTENSIONS

Suppose a, b, c, and d are numbers of arithmetic and $a > b$, $c > d$. Try various replacements for a, b, c, and d to see whether the following relationships are true.

(1) $a + c > b + d$ (2) $ac > bd$

Do you think (1) and (2) are true for all numbers of arithmetic? (yes) What about these relationships:

(3) $a - c > b - d$ (4) $\dfrac{a}{c} > \dfrac{b}{d}$

Are they true for $a = 10$, $b = 7$, $c = 8$, and $d = 4$? (no) for $a = 10$, $b = 7$, $c = 5$, and $d = 4$? (yes) Can you say that (3) and (4) are true for *all* numbers of arithmetic?

6 Axioms and Properties

CHAPTER OVERVIEW

The time has arrived when we need to list the basic properties on which we base our reasoning. Some of these are so commonly accepted that the students' reaction may be, "Of course! Why mention it?" This may be true especially of the axioms of equality. Point out that we state these properties to give us a basis for less obvious reasoning that is to come. The general purpose of the chapter is to pull together the basic axioms and properties as a foundation for ideas that will be developed later in the course.

In this chapter we continue to direct attention to operations with non-negative numbers. Consideration of negative numbers and their corresponding properties will follow shortly.

6-1 *Basic Axioms of Equality*

TEACHING THESE PAGES

The equality axioms seem so obvious that students may wonder why we bother to state them. You might remind students that the development of algebra begins with these simple rules, which will be the building blocks for more complex concepts later on.

There are no reflexive and symmetric properties of inequality. Only the transitive property of equality has a corresponding property of inequality:

For any numbers r, s, and t, if $r > s$ and $s > t$, then $r > t$, and if $r < s$ and $s < t$, then $r < t$.

CHALKBOARD EXAMPLES

Tell which property is illustrated.

1. $n + 5 = n + 5$
   Reflexive Property of Equality
2. $12 \times 3 = 36$ and $36 = 4 \times 9$, so $12 \times 3 = 4 \times 9$.
   Transitive Property of Equality

3. $a^2 = 25$ so $25 = a^2$.
   Symmetric Property of Equality
4. $x + 2y = 10$ so $10 = x + 2y$.
   Symmetric Property of Equality

Assume the equation is true. Use the symmetric property of equality to write another equation.

5. $x + 3y = 18$ $18 = x + 3y$
6. $3(n + 5) = 3n + 15$ $3n + 15 = 3(n + 5)$

Use the transitive property of equality to complete the statement.

7. $21 \div 3 = 7$ and $7 = 5 + 2$, so __?__.
   $21 \div 3 = 5 + 2$
8. If $a + b = 50$ and $50 = 5c$, then __?__.
   $a + b = 5c$

SUGGESTED EXTENSION

Have the class decide whether the three basic properties of equality apply also to the relationship expressed by the symbol $\neq$. Ask: "Are the following properties true for all numbers?"

(1) Reflexive: For any number a, $a \neq a$

(2) Symmetric: For any numbers a and b, if $a \neq b$, then $b \neq a$

(3) Transitive: For any numbers a, b, and c, if $a \neq b$, and $b \neq c$, then $a \neq c$

(We can see that (1) is obviously false and (2) is obviously true. (3) is not true for *all* numbers, so we call it false, also.)

6-2 *The Closure Property and the Substitution Principle*

TEACHING THESE PAGES

In this lesson we add two more properties, still applying them only to the nonnegative numbers. The closure property, the most subtle property students will encounter in this chapter, perhaps is best illustrated with several concrete examples.

If we add any two whole numbers, we know that the sum will also be a whole number. Therefore, we can say that the set of whole numbers is *closed under addition*. Note that we must always specify an operation when we speak of closure. After considering a few examples, students should readily see that some sets are closed under one operation and not under another, and some sets are not closed under any operation.

The substitution principle is a readily obvious and yet important property. It is a property we will use again and again in simplifying expressions and solving equations and inequalities later on.

CHALKBOARD EXAMPLES

Tell whether or not the set is closed under **(a)** addition, **(b)** subtraction, **(c)** multiplication, and **(d)** division.

1. $\{3, 6, 9, 12, \ldots\}$
   (a) closed
   (b) not closed; for example, $3 - 6$ is not a member of $\{3, 6, 9, 12, \ldots\}$.
   (c) closed
   (d) not closed; for example, $3 \div 6$ is not a member of $\{3, 6, 9, 12, \ldots\}$.
2. $\{0, 2\}$
   (a) not closed; for example, $2 + 2 = 4$.
   (b) not closed; for example, $0 - 2$ does not have a result in $\{0, 2\}$.
   (c) not closed; since $2 \times 2 = 4$, and 4 is not a member of $\{0, 2\}$.
   (d) not closed; since $2 \div 2 = 1$, and 1 is not a member of $\{0, 2\}$.

Simplify. Use the substitution principle as indicated.

3. $1.2 + 5.7 + 3.4$

   $6.9 \underbrace{\underline{} + 3.4}$

   $\underbrace{\underline{}}\quad 10.3$

4. $\underbrace{\frac{1}{2} \cdot \frac{1}{3}} + 3 \cdot 2$

   $\frac{1}{6} \underbrace{\underline{} + \underline{}}\quad 6$

   $\underline{} \quad 6\frac{1}{6}$

6-3 *The Commutative and Associative Properties*

TEACHING THESE PAGES

In this lesson we discuss other properties that students have probably more or less taken for granted. The commutative and associative properties do not hold true for all four operations. Both properties are true for the operations of addition and multiplication, but not for the inverse operations, subtraction and division.

CHALKBOARD EXAMPLES

Show that the sentence is true. Name the property illustrated.

1. $3 + \left(\frac{1}{2} + \frac{1}{4}\right) = \left(3 + \frac{1}{2}\right) + \frac{1}{4}$ — Associative prop. of addition

$3 + \frac{3}{4}$	$3\frac{1}{2} + \frac{1}{4}$
$3\frac{3}{4}$	$3\frac{3}{4}$

2. $3\left(\frac{1}{2} + \frac{1}{4}\right) = 3\left(\frac{1}{4} + \frac{1}{2}\right)$ — Commutative prop. of addition

$3(\frac{3}{4})$	$3(\frac{3}{4})$
$\frac{9}{4}$	$\frac{9}{4}$

3. $0.5(1.4 + 3.6) = (1.4 + 3.6)0.5$ — Commutative prop. of mult.

$0.5(5.0)$	$(5.0)0.5$
2.5	2.5

4. $0.6\left(3 \cdot \frac{1}{3}\right) = (0.6 \cdot 3)\frac{1}{3}$ — Associative prop. of multiplication

$0.6(1)$	$(1.8)\frac{1}{3}$
0.6	0.6

5. Show that the statement $a(b + c) = a(c + b)$ is true when $a = 8$, $b = 3$, and $c = 5$.

   $a(b + c) = a(c + b)$

$8(3 + 5)$	$8(5 + 3)$
$8 \cdot 8$	$8 \cdot 8$
64	64

SUGGESTED EXTENSION

The following activity should give the students a more graphic idea of the meaning of commutativity. Discuss various nonmathematical "operations" to decide whether they are commutative. For example:

(1) Does putting on your hat, then your coat have the same result as putting on your coat, then your hat?
(2) Does putting on your socks, then your shoes have the same result as putting on your shoes, then your socks?
(3) Is washing your hands, then your face the same as washing your face, then your hands?
(4) Is opening the garage door, then driving the car in the same as driving the car in, then opening the door?

6-4 *The Distributive Property*

TEACHING THESE PAGES

In this lesson we consider the distributive property, in particular, the distribution of multiplication over addition. This is another property that students probably already take for granted. Suppose, for example, you buy 3 greeting cards at 25¢ each, and then 8 more greeting cards at the same price, 25¢ each. You automatically assume that the total cost is the same as the cost of 11 cards at 25¢ each. That is:

$$3(0.25) + 8(0.25) = (3 + 8)0.25 = 11(0.25)$$

To give another example, suppose your rate of pay is $3.25 per hour and you work 6 hours Monday and 4 hours Tuesday. You can figure your pay for the two days either as:

$$3.25(6) + 3.25(4) \text{ or } 3.25(6 + 4)$$

Ask students which is the easier computation. They should readily see that the second expression can be simplified mentally.

CHALKBOARD EXAMPLES

Show that the statement is true.

1. $7(0.8 + 9.3) = 7(0.8) + 7(9.3)$

$7(10.1)$	$5.6 + 65.1$
70.7	70.7

2. $4(8.3) + 5(8.3) = (4 + 5)(8.3)$

$33.2 + 41.5$	$9(8.3)$
74.7	74.7

3. $4\left(\dfrac{3}{5}\right) + 4\left(\dfrac{1}{5}\right) = 4\left(\dfrac{3}{5} + \dfrac{1}{5}\right)$

$\dfrac{12}{5} + \dfrac{4}{5}$	$4\left(\dfrac{4}{5}\right)$
$\dfrac{16}{5}$	$\dfrac{16}{5}$

Tell how to complete the sentence to illustrate the distributive property.

4. $7(300 + 55) = \underline{\ ?\ }$ $7(300) + 7(55)$
5. $2.5(8) + 2.5(6) = \underline{\ ?\ }$ $2.5(8 + 6)$

6-5 *More About the Distributive Property*

TEACHING THESE PAGES

Since the treatment of operations on directed numbers is postponed until Chapters 9 and 10, it is necessary at this point to extend the distributive property to deal with subtraction. Later (in Chapter 9) we shall see that to subtract a number we add its opposite. Until then, we show that, for the nonnegatives, multiplication is distributive over subtraction as well as over addition.

Since we know that dividing by a number is the same as multiplying by its reciprocal, we can demonstrate that division is distributive over both addition and subtraction (Examples 2 and 3 on page 169).

CHALKBOARD EXAMPLES

Show that the statement is true.

1. $(40 - 7)3 = 40 \cdot 3 - 7 \cdot 3$

$33 \cdot 3$	$120 - 21$
99	99

2. $\dfrac{3}{7}(14 - 5) = \dfrac{3}{7} \cdot 14 - \dfrac{3}{7} \cdot 5$

$\dfrac{3}{7}(9)$	$\dfrac{42}{7} - \dfrac{15}{7}$
$\dfrac{27}{7}$	$\dfrac{27}{7}$

3. $\dfrac{600}{4} + \dfrac{60}{4} - \dfrac{8}{4} = \dfrac{600 + 60 - 8}{4}$

$150 + 15 - 2$	$\dfrac{652}{4}$
163	163

4. $\dfrac{1}{3}\left(\dfrac{4}{5} - \dfrac{1}{4}\right) = \dfrac{1}{3}\cdot\dfrac{4}{5} - \dfrac{1}{3}\cdot\dfrac{1}{4}$

$\dfrac{1}{3}\left(\dfrac{11}{20}\right)$	$\dfrac{4}{15} - \dfrac{1}{12}$
$\dfrac{11}{60}$	$\dfrac{16}{60} - \dfrac{5}{60}$
$\dfrac{11}{60}$	$\dfrac{11}{60}$

Use the substitution principle and the distributive property to find the product.

5. $7(89) = 7(90 - \underline{})$

$7(89) = 7(90 - 1)$

$\qquad = 630 - 7$

$\qquad = 623$

6. $8630 \div 5 = \dfrac{8000 + 600 + ?}{5}$

$8630 \div 5 = \dfrac{8000}{5} + \dfrac{600}{5} + \dfrac{30}{5}$

$\qquad = 1600 + 120 + 6$

$\qquad = 1726$

6-6 *The Properties of Zero and One*

TEACHING THESE PAGES
In this lesson we introduce several more properties that, though they may seem obvious, have important applications in solving algebraic equations and inequalities. The additive property of zero follows readily from the meaning of the number zero. Note that there is no corresponding subtractive property of zero. That is, for every number r:

$$r + 0 = 0 + r = r, \text{ but } r - 0 \neq 0 - r$$

($r - 0 = 0 - r$ is true only when $r = 0$.)

Likewise, there are no corresponding division properties for the multiplicative properties of 0 and 1. For every number r:

$r \cdot 0 = 0 \cdot r = 0$, but we cannot write $r \div 0 = 0 \div r$, since $r \div 0$ has no meaning.

$r \cdot 1 = 1 \cdot r = r$, but $r \div 1 \neq 1 \div r$.

(The only value of r for which $r \div 1 = 1 \div r$ is true is $r = 1$.)

CHALKBOARD EXAMPLES
Solve.

1. $7 \cdot n = 7$ $\quad n = 1$ **2.** $7 + n = 7$ $\quad n = 0$

3. $\dfrac{7}{7} \cdot n = 7$ $\quad n = 7$

4. $5\left(\dfrac{4}{4} + n\right) = 5$ $\quad n = 0$

5. $3(n + 8) = 24$ $\quad n = 0$

6. $n + 0 = n$ $\quad n = $ any number

6-7 *Function Equations*

TEACHING THESE PAGES
We return here to the concept of function and introduce the $(x, f(x))$ notation. We again use the "function machine" to illustrate the idea of a function rule which generates sets of ordered pairs. We also provide further experience in the graphing of sets of ordered pairs in the coordinate plane. Now we will have students label the axes x and $f(x)$.

CHALKBOARD EXAMPLES
Use the function rule to complete the table.

Rule: $x^2 + 3$

	x	$f(x)$	$(x, f(x))$
1.	0	3	(0, 3)
2.	1	4	(1, 4)
3.	2	7	(2, 7)
4.	4	19	(4, 19)
5.	8	67	(8, 67)

Write the set of ordered pairs that represents the function. Draw axes and graph the set of ordered pairs.

6. Rule: $\dfrac{4x}{x} - 2$; replacement set: $\{1, 2, 3, 4\}$

$\{(1, 2), (2, 2), (3, 2), (4, 2)\}$

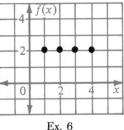

Ex. 6

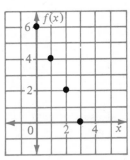

Ex. 7

7. Rule: $6 - 2x$; replacement set: $\{0, 1, 2, 3\}$

$\{(0, 6), (1, 4), (2, 2), (3, 0)\}$

7 Equations and Problem Solving

CHAPTER OVERVIEW

Up to now we have used intuitive methods for finding solutions for equations. In this chapter we begin to formalize procedures for solving equations. We start, however, with a lesson on combining similar terms. This necessary process in simplifying many equations is actually an application of the distributive property. Consideration of the basic addition, subtraction, multiplication, and division properties of equality is followed by a discussion of strategies for transforming equations using combinations of the basic properties.

7-1 *Combining Similar Terms*

TEACHING THESE PAGES

At first most students will be wise to write out the steps to show the use of the distributive property in combining terms. However, as soon as they understand the process, encourage them to shorten their work. For example:

$$5x + 7x = (5 + 7)x = 12x$$

As soon as students can do the middle step mentally, the work can be simplified to $5x + 7x = 12x$.
In the same way

$$8n - n = 8n - 1n = (8 - 1)n = 7n$$

may be shortened to $8n - n = 7n$.

When a student makes repeated errors, however, the student should return to the longer form until he or she becomes more proficient in combining terms.

CHALKBOARD EXAMPLES

Combine similar terms to simplify.

1. $3a + 8a$ $11a$
2. $6b + b + 3b$ $10b$
3. $15c - 4c$ $11c$
4. $am + 7am - 5am$ $3am$
5. $3(t + 5) + 4$ $3t + 15 + 4 = 3t + 19$
6. $3(t + 5) + 4t$ $3t + 15 + 4t = 7t + 15$
7. $7(5x + 1) + 3(x - 1)$ $35x + 7 + 3x - 3 = 38x + 4$
8. $5(n + 3) + 7(3n - 1)$ $5n + 15 + 21n - 7 = 26n + 8$

SUGGESTED EXTENSIONS

In mathematics, as in everyday life, there is often more than one way to do something. An important skill for students to develop is the ability to choose the most efficient method for a given task, thus avoiding a longer process in which computational errors are more likely to occur. The simplifications below are all correct, though one might seem easier than another.

1. Discuss two possible approaches to simplifying expressions like the one that follows, in which the expression in parentheses is the same:

   $$(1)\ 5(a + b) + 6(a + b) = 5a + 5b + 6a + 6b$$
   $$= (5 + 6)a + (5 + 6)b$$
   $$= 11a + 11b$$
   $$(2)\ 5(a + b) + 6(a + b) = (5 + 6)(a + b)$$
   $$= 11(a + b) = 11a + 11b$$

2. Consider the simplification of expressions in which the quantities in parentheses are different, but the monomial multiplier is the same. For example:

(1) $5(x + 3y) + 5(3x - y) = 5x + 15y + 15x - 5y$
$$= (5 + 15)x + (15 - 5)y$$
$$= 20x + 10y$$

(2) $5(x + 3y) + 5(3x - y) = 5(x + 3y + 3x - y)$
$$= 5[(1 + 3)x + (3 - 1)y]$$
$$= 5(4x + 2y) = 20x + 10y$$

7-2 Addition and Subtraction Properties of Equality

TEACHING THESE PAGES

The addition and subtraction properties of equality are the properties that were presented informally in Section 1-6, when we looked at an equation as a balanced scale. You may wish to use the balanced scale again as a convenient device to illustrate the addition and subtraction properties.

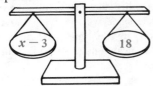

For the equation $x - 3 = 18$ to be "satisfied," the two pans of the scale must balance each other. In order to have the quantity x alone on the left pan, we add 3 to the left pan. This throws the scale off balance until 3 is also added to the other pan. To express this with equations, we write:

$$x - 3 = 18$$
$$x - 3 + 3 = 18 + 3$$
$$x = 21$$

Check:

$$x - 3 = 18$$

$21 - 3$	18
18	18 ✓

Similarly, to eliminate a number that is *added* to the variable, we *subtract* the number from both sides.

There is no need for students to memorize the properties presented in this chapter. They should, however, understand what they mean and be able to tell which property is used in a given step of a simplification.

CHALKBOARD EXAMPLES

Solve and check. Begin by stating what number should be added to or subtracted from both members.

1. $n + 7 = 10$
   Subtract 7.
   $$n + 7 - 7 = 10 - 7$$
   $$n = 3$$

$$n + 7 = 10$$

$3 + 7$	10
10	10 ✓

2. $s - 7 = 12$
   Add 7.
   $$s - 7 + 7 = 12 + 7$$
   $$s = 19$$

$$s - 7 = 12$$

$19 - 7$	12
12	12 ✓

3. $t - \dfrac{1}{8} = \dfrac{3}{8}$
   Add $\dfrac{1}{8}$.
   $$t - \frac{1}{8} + \frac{1}{8} = \frac{3}{8} + \frac{1}{8}$$
   $$t = \frac{4}{8} \text{ or } \frac{1}{2}$$

$$t - \frac{1}{8} = \frac{3}{8}$$

$\dfrac{4}{8} - \dfrac{1}{8}$	$\dfrac{3}{8}$
$\dfrac{3}{8}$	$\dfrac{3}{8}$ ✓

4. $\dfrac{5}{9} = r + \dfrac{3}{9}$
   Subtract $\dfrac{3}{9}$.
   $$\frac{5}{9} - \frac{3}{9} = r + \frac{3}{9} - \frac{3}{9}$$
   $$\frac{2}{9} = r$$

$$\frac{5}{9} = r + \frac{3}{9}$$

$\dfrac{5}{9}$	$\dfrac{2}{9} + \dfrac{3}{9}$
$\dfrac{5}{9}$	$\dfrac{5}{9}$ ✓

5. $1.43 = y - 0.34$
   Add 0.34.
   $$1.43 + 0.34 = y - 0.34 + 0.34$$
   $$1.77 = y$$

$$1.43 = y - 0.34$$

1.43	$1.77 - 0.34$
1.43	1.43 ✓

6. $0.85 = a + 0.07$
   Subtract 0.07.
   $$0.85 - 0.07 = a + 0.07 - 0.07$$
   $$0.78 = a$$

$$0.85 = a + 0.07$$

0.85	$0.78 + 0.07$
0.85	0.85 ✓

7-3 The Division Property of Equality

TEACHING THESE PAGES

Point out that in the statement of the division property of equality, it is necessary to make an exception of zero as a possible divisor, since division by zero has no meaning.

We are still aiming to "undo" operations in order to arrive at an equation that states the value of the variable. With three properties of equality at our disposal, we can now solve equations that need more than a single step. It is usually simpler to use the addition and subtraction properties first, then the division property.

CHALKBOARD EXAMPLES
Solve. Check your answer.

1. $9b = 63$
$b = 7$

$$9b = 63$$

$9 \cdot 7$	63
63	63 ✓

2. $34 = 2t$
$t = 17$

$$34 = 2t$$

34	$2 \cdot 17$
34	34 ✓

3. $3x + 5 = 20$
$x = 5$

$$3x + 5 = 20$$

$3(5) + 5$	20
$15 + 5$	20
	20 ‖ 20 ✓

4. $8c - 7 = 9$
$c = 2$

$$8c - 7 = 9$$

$8(2) - 7$	9
$16 - 7$	9
	9 ‖ 9 ✓

5. $7n - 4n = 15$
$n = 5$

$$7n - 4n = 15$$

$7(5) - 4(5)$	15
$35 - 20$	15
	15 ‖ 15 ✓

6. $15 + 3r = 35$
$r = 6\frac{2}{3}$

$$15 + 3r = 35$$

$15 + 3(6\frac{2}{3})$	35
$15 + 20$	35
	35 ‖ 35 ✓

SUGGESTED EXTENSIONS

1. Compare two ways of solving an equation that requires more than one transformation: (1) *first* using the division property, and (2) *first* using the addition or subtraction property. Note that using the division property involves dividing *each* term in the equation by the same number. For example:

(1)
$$4x + 3 = 17$$
$$\frac{4x}{4} + \frac{3}{4} = \frac{17}{4}$$
$$x + \frac{3}{4} = \frac{17}{4}$$
$$x + \frac{3}{4} - \frac{3}{4} = \frac{17}{4} - \frac{3}{4}$$
$$x = \frac{14}{4} = \frac{7}{2}$$

(2)
$$4x + 3 = 17$$
$$4x + 3 - 3 = 17 - 3$$
$$4x = 14$$
$$\frac{4x}{4} = \frac{14}{4}$$
$$x = \frac{7}{2}$$

Although the second solution has no fewer steps than the first, it involves less work with fractions, so it is probably the more desirable method.

2. Is it *always* better to do addition or subtraction before division? Consider this example:

(1)
$$3n - 15 = 54$$
$$\frac{3n}{3} - \frac{15}{3} = \frac{54}{3}$$
$$n - 5 = 18$$
$$n - 5 + 5 = 18 + 5$$
$$n = 23$$

(2)
$$3n - 15 = 54$$
$$3n - 15 + 15 = 54 + 15$$
$$3n = 69$$
$$\frac{3n}{3} = \frac{69}{3}$$
$$n = 23$$

Take your choice!

7-4 *The Multiplication Property of Equality*

TEACHING THESE PAGES

We now state the multiplication property of equality to complete the properties of equality for the four fundamental operations.

Point out that the statement of the multiplication property of equality does not require the exclusion of zero. However, since we are using the property in order to "undo" division, and since an expression divided by zero is meaningless, we shall not be using zero as a multiplier.

CHALKBOARD EXAMPLES
Solve. Check your answer.

1. $\frac{z}{3} = 5$

$z = 15$

$$\frac{z}{3} = 5$$

$\frac{15}{3}$	5
5	5 ✓

2. $\frac{b}{5} = 3.4$

$b = 17$

$$\frac{b}{5} = 3.4$$

$\frac{17}{5}$	3.4
3.4	3.4 ✓

3. $\frac{t}{4} = \frac{3}{4}$

$t = 3$

$$\frac{t}{4} = \frac{3}{4}$$

$\frac{3}{4}$	$\frac{3}{4}$ ✓

4. $\frac{1}{5} m = 0$

$m = 0$

$$\frac{1}{5} m = 0$$

$\frac{1}{5}(0)$	0
0	0 ✓

5. $\frac{1}{5} n = 1$

$n = 5$

$$\frac{1}{5} n = 1$$

$\frac{1}{5}(5)$	1
1	1 ✓

6. $0.6 = \frac{3}{8} w$

$w = 1.6$

$$0.6 = \frac{3}{8} w$$

0.6	$\frac{3}{8}(1.6)$
0.6	$3(0.2)$
0.6	0.6 ✓

7-5 *More about Equation Solving Strategy*

TEACHING THESE PAGES

Now that the four basic properties of equality have been developed, we come to where we can combine them to solve more complicated equations. The aim of the step-by-step procedure is to simplify the expressions on both sides of an equation until we get a simple statement of what number the variable equals. There will often be more than one acceptable approach to transforming an equation.

CHALKBOARD EXAMPLES

Solve and check.

1. $\dfrac{2}{3}a + 1\tfrac{1}{3}a = 19 - 5$

$a = 7$

$$\frac{2}{3}a + 1\tfrac{1}{3}a = 19 - 5$$

$\frac{2}{3}(7) + \frac{4}{3}(7)$	14
$\frac{6}{3}(7)$	14
$2(7)$	14
14	14 ✓

2. $3\tfrac{1}{2}d - 1\tfrac{1}{2}d + 5 = 31$

$d = 13$

$$3\tfrac{1}{2}d - 1\tfrac{1}{2}d + 5 = 31$$

$\frac{7}{2}(13) - \frac{3}{2}(13) + 5$	31
$\frac{4}{2}(13) + 5$	31
$2(13) + 5$	31
$26 + 5$	31
✓ 31	31

3. $\dfrac{y}{6} + 3 + \dfrac{5y}{6} = 8$

$y = 5$

$$\frac{y}{6} + 3 + \frac{5y}{6} = 8$$

$\frac{5}{6} + 3 + \frac{25}{6}$	8
$\frac{30}{6} + 3$	8
$5 + 3$	8
8	8 ✓

4. $\dfrac{k}{5} - 2 + \dfrac{4}{5} = 0$

$k = 6$

$$\frac{k}{5} - 2 + \frac{4}{5} = 0$$

$\frac{6}{5} - 2 + \frac{4}{5}$	0
$\frac{10}{5} - 2$	0
$2 - 2$	0
0	0 ✓

5. $10b + b - 9 = 70 - 2$

$b = 7$

$$10b + b - 9 = 70 - 2$$

$10(7) + 7 - 9$	68
$70 + 7 - 9$	68
68	68 ✓

6. $7(t + 3) + t = 85$

$t = 8$

$$7(t + 3) + t = 85$$

$7(8 + 3) + 8$	85
$7(11) + 8$	85
85	85 ✓

Write and solve an equation to answer the question.

7. If 17 is added to five times a number, the result is equal to 72. What is the number?

$5n + 17 = 72$

$n = 11$; The number is 11.

8. The length of a rectangle is four times its width. The perimeter of the rectangle is 150 cm. How long and how wide is the rectangle?

Let w = the width and $4w$ = the length.

$150 = 2(4w) + 2w$

$w = 15$

$4w = 60$

The rectangle is 60 cm long and 15 cm wide.

7-6 *Equations and Problem Solving*

TEACHING THESE PAGES

This lesson provides a purpose for the earlier work of the chapter—the use of equations for solving problems. Students will translate a real life situation into an equation, then solve the equation to answer a question. While there may be one obvious way to write an appropriate equation, sometimes there is more than one acceptable way.

The choice of the letter to use for the variable is less important than the choice of what the variable should represent. In general, when the problem has only a single missing number, the variable will simply represent that number. However, when the problem involves more than a single unknown quantity, students must consider which unknown the variable is to represent. For example, suppose we want to represent the lengths of two pieces of string, one of which is three times as long as the other. There are two possibilities using a single variable:

(1) Let s = the length of the shorter piece, and

$3s$ = the length of the longer piece.

(2) Let l = the length of the longer piece, and

$\dfrac{1}{3}l$ = the length of the shorter piece.

An equation based on (1) will not involve fractions, so the choice of variables in (1) is probably preferable.

CHALKBOARD EXAMPLES

Write an equation. Then solve it to answer the question.

1. In a ball game, the Red Sox won by 5 runs. The losing team scored 3 runs. How many runs did the Red Sox score?

   Let the Red Sox's score $= n$
   The losing team's score $= 3$
   Equation: $n = 3 + 5$
   $\qquad n = 8$; the Red Sox's score was 8.

2. Brand X costs $1.07, which is 19¢ more than the cost of Brand Y. How much does Brand Y cost?

   Let the cost in cents of Brand Y $= y$
   The cost in cents of Brand X $= 107$
   Equation: $107 = y + 19$
   $\qquad y = 88$; Brand Y costs 88¢.

3. Suppose you want to cut a piece of cloth 5 m long into two pieces so that one piece is 80 cm longer than the other. How many cm long is the shorter piece?

   Let the length in cm of the shorter piece $= s$
   The length in cm of the longer piece $= s + 80$
   Equation: $s + (s + 80) = 500$
   $\qquad\qquad\qquad s = 210$
   The shorter piece is 210 cm long.

4. The sum of three consecutive odd numbers is 153. What are the numbers?

   Let the smallest of the numbers $= n$
   The next consecutive odd number $= n + 2$
   The third consecutive odd number $= n + 4$
   Equation: $n + (n + 2) + (n + 4) = 153$
   $\qquad\qquad\qquad\qquad n = 49$
   $\qquad\qquad\qquad n + 2 = 51$ and
   $\qquad\qquad\qquad n + 4 = 53$
   The numbers are 49, 51, and 53.

5. The length of a rectangle is 5 m more than its width. If half the perimeter is 37 m, what are the length and the width?

   Let the width in meters $= w$
   Then, the length in meters $= w + 5$
   Half the perimeter $= w + (w + 5)$
   Equation: $w + (w + 5) = 37$
   $\qquad\qquad\qquad w = 16$
   $\qquad\qquad w + 5 = 21$
   The length is 21 m and the width is 16 m.

7-7 Equations with the Variable in Both Members

TEACHING THESE PAGES

When terms involving the variable occur in both members of an equation, the addition or the subtraction property of equality can be used for a transformation that removes the variable term or terms from one member.

CHALKBOARD EXAMPLES

Solve. Check your answer.

1. $5n = 14 + 3n$
   $\quad n = 7$

   $5n = 14 + 3n$

$5(7)$	$14 + 3(7)$
35	$14 + 21$
35	35 ✓

2. $c = 65 - 4c$
   $\quad c = 13$

   $c = 65 - 4c$

13	$65 - 4(13)$
13	$65 - 52$
13	13 ✓

3. $13b - 63 = 4b$
   $\qquad b = 7$

   $13b - 63 = 4b$

$13(7) - 63$	$4(7)$
$91 - 63$	28
28	28 ✓

4. $43 - x = 17 + x$
   $\qquad x = 13$

   $43 - x = 17 + x$

$43 - 13$	$17 + 13$
30	30 ✓

5. $\frac{1}{3}m + 17 = \frac{4}{3}m - 1$
   $\qquad m = 18$

   $\frac{1}{3}m + 17 = \frac{4}{3}m - 1$

$\frac{1}{3}(18) + 17$	$\frac{4}{3}(18) - 1$
$6 + 17$	$24 - 1$
23	23 ✓

Write an equation for the problem. Then solve the equation and answer the question.

6. If 8 is added to a number, the result is the same as three times the number. What is the number?
   $n + 8 = 3n$; $n = 4$; The number is 4.

7. If five times a number is decreased by 8, the result is the same as twice the number increased by 7. What is the number? $5m - 8 = 2m + 7$; $m = 5$; The number is 5.

8 Working with Directed Numbers

CHAPTER OVERVIEW

Although negative numbers have been treated briefly in an earlier chapter, it is at this point we choose to introduce the concept of directed numbers formally. We use the number line as a device to illustrate concepts about directed numbers. The position of a number on the number line with respect to 0 is as important as its distance from 0. Since directed numbers imply a move in either a positive or a negative direction, they can be represented by arrows on the number line.

The set of integers consists of the familiar set of whole numbers together with the set of opposites of the whole numbers—the negative integers. An inequality is a sentence that separates the set of directed numbers into two parts: (1) the set of numbers for which the sentence is true, and (2) the set of numbers for which it is false. Students will determine the solution sets of simple equalities and graph them on the number line.

8-1 *Directed Numbers and the Number Line*

TEACHING THESE PAGES

This lesson brings together many concepts presented earlier about directed numbers and adds a few new ideas. Opposites are now defined as two different numbers which are the same distance from 0. That is, $^-5$ is the opposite of 5. Its graph on the number line is just as far from 0 as the graph of 5, but it lies on the opposite side of 0. Likewise, 5 is the opposite of $^-5$. We say $^-5$ and 5 have the same *magnitude*.

The magnitude of a number is its distance from 0 on the number line. Point out that since magnitude does not depend on direction, the magnitude of a number is always positive. The magnitude of any number is the same as the magnitude of its opposite. The opposite of 0 is 0, and its magnitude is 0.

On the vertical number line, it is customary to graph positive numbers above the origin, or zero point, and negative numbers below it.

CHALKBOARD EXAMPLES

Name the coordinate of the point.

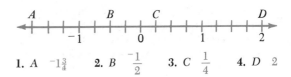

1. A $^-1\frac{3}{4}$ 2. B $\frac{^-1}{2}$ 3. C $\frac{1}{4}$ 4. D 2

Use a directed number to express the following.

5. 15 meters below sea level $^-15$

6. 32.5 degrees above zero 32.5

Name the directed number.

7. The positive number 5.3 units from 0 5.3

8. The negative number 15 units from 0 $^-15$

9. The number that lies the same distance from 0 as 3.5 $^-3.5$

10. Two directed numbers, each $3\frac{1}{3}$ units from 0 $3\frac{1}{3}$, $^-3\frac{1}{3}$

8-2 *Arrows to Represent Directed Numbers*

TEACHING THESE PAGES

Representing a directed number by an arrow on the number line can be helpful in visualizing sums and differences of directed numbers. The magnitude of the number determines the length of the arrow. The direction of the arrow indicates the sign of the number represented. For example, all of the arrows below represent the number 4.

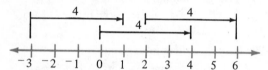

Each of the following arrows represents $^-3$.

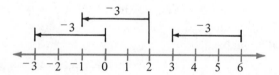

For classroom practice in representing directed numbers by arrows on the number line, draw on the chalkboard a number line that includes both positive and negative numbers.

(1) Ask a volunteer to draw an arrow above the line, marking the beginning and the end of the arrow clearly. The student who correctly identifies the number represented is next in turn to draw an arrow to be identified. This can go on as long as practice seems to be needed.

(2) Give directions for two successive moves on the number line. For example, you might say, "Start at 0 and move 6 units in the negative direction; then move $3\frac{1}{2}$ units in the positive direction." Ask a volunteer to sketch the arrows that represent the

moves, and then identify the point where the second arrow ends.

CHALKBOARD EXAMPLES

Make a number line sketch to show the moves. Tell where you finish.

1. Start at 0. Move 3 units in the positive direction. Then move 4 units in the negative direction.

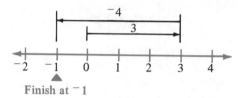

Finish at $^-1$

2. Start at 0. Move $1\frac{1}{3}$ units in the negative direction. Then move 3 units in the positive direction.

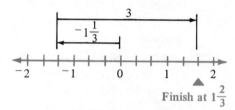

Finish at $1\frac{2}{3}$

Name the directed number suggested by the arrow.

3.

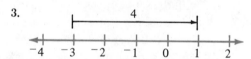

4.

The statement refers to moves on the number line. Make a number line sketch and tell where you finish.

5. Start at 5 and move 2 units in the negative direction.

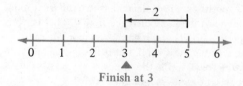

Finish at 3

6. Start at $^-3\frac{1}{2}$ and move $3\frac{1}{2}$ units in the positive direction. Then move 1 unit in the negative direction.

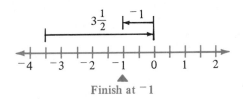

Finish at ⁻1

Complete. Use *positive* or *negative*.

7. A move from 3 to ⁻1 is in the ⎯?⎯ direction.   negative

8. A move from 1 to 5 is in the ⎯?⎯ direction.   positive

8-3 *Comparing Directed Numbers*

TEACHING THESE PAGES

The directed numbers are ordered. That is, of two directed numbers a and b, if $a \neq b$, then either $a < b$ or $a > b$. We can use the location of numbers on the number line to determine the order of two or more numbers. On the number line, "is to the right of" implies "is greater than" and "is to the left of" implies "is less than." Thus, if a is to the right of b, then $a > b$, and if a is to the left of b, then $a < b$.

Students should have no difficulty comparing two positive numbers. However, some practice in comparing two negative numbers may be necessary. For example, consider ⁻8 and ⁻5. If students think only of the magnitude and ignore the direction, they may reason that since $8 > 5$, it must also be true that ⁻8 > ⁻5. Reference to the location of the numbers on the number line, however, will show that ⁻8 is to the left of ⁻5, so it is true that ⁻8 < ⁻5.

The comparison of a negative number with a positive number should be quite simple, once students observe that *all* negative numbers lie to the left of 0 on the number line, and *all* positive numbers lie to the right of 0. So when we compare a negative number and a positive number, the positive number is always greater.

CHALKBOARD EXAMPLES

Complete to make a true statement. Use *right* or *left* and $>$ or $<$.

1. 4 is to the ⎯?⎯ of 1; so 4 ⎯?⎯ 1.   right; $>$

2. ⁻4 is to the ⎯?⎯ of ⁻3; so ⁻4 ⎯?⎯ ⁻3.   left; $<$

3. ⁻1 is to the ⎯?⎯ of $\frac{3}{5}$; so ⁻1 ⎯?⎯ $\frac{3}{5}$.   left; $<$

Complete to make a true statement. Use $>$ or $<$.

4. 5 ⎯?⎯ ⁻5 $>$ **5.** 0 ⎯?⎯ ⁻10 $>$

6. ⁻7 ⎯?⎯ 3 $<$ **7.** ⁻4 ⎯?⎯ ⁻7 $>$

8. $\frac{⁻4}{5}$ ⎯?⎯ 0 $<$ **9.** ⁻0.03 ⎯?⎯ ⁻0.004 $<$

True or false?

10. $9 > 13$ false   **11.** ⁻9 > ⁻13   true

12. $\frac{3}{5} < \frac{3}{10}$ false   **13.** ⁻3.01 ≥ 3.01   false

8-4 *Integers as Solutions of Inequalities*

TEACHING THESE PAGES

Although this lesson deals primarily with integers as solutions for inequalities, students should realize that inequalities may also have solutions that are not integers. Note that for most of the Written Exercises, the replacement set is a finite set of integers. In the next lesson students will solve inequalities for which the replacement set is {the directed numbers}.

CHALKBOARD EXAMPLES

Tell whether a true or a false statement results when the variable in the sentence is replaced by each member of the replacement set. Then state the solution set.

1. $n > ⁻3$; {⁻5, ⁻3, ⁻1, 1}
    false; false; true; true; {⁻1, 1}

2. $x < 2$; {⁻2, 0, 2, 4}
    true; true; false; false; {⁻2, 0}

3. $z \geq \frac{⁻1}{3}$; {⁻2, ⁻1, 0, 1, 2}

    false; false; true; true; true; {0, 1, 2}

4. $0 \geq m$; {⁻1, 0, 1, 2}
    true; true; false; false; {⁻1, 0}

Write and graph the solution set. The replacement set is {⁻3, ⁻2, ⁻1, 0, 1, 2, 3}

5. $t \geq 2$ {2, 3}

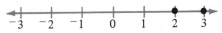

6. $y \leq 1$ {⁻3, ⁻2, ⁻1, 0, 1}

7. $\frac{1}{3} > c$ {⁻3, ⁻2, ⁻1, 0}

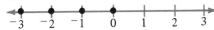

8-5 *Graphing Inequalities*

TEACHING THESE PAGES

In this lesson we review and extend the material on inequalities from Chapter 2. Be sure it is clear that when we speak of the graph of an inequality we mean the graph of its solution set on the number line. Students may prefer to graph solution sets of inequalities by means of a heavy black line (using the style of the graphs in the *Answers to Odd-Numbered Exercises*), rather than with the two-color method used in the text. Remind students that if an endpoint is included in the solution set, the point is indicated by a solid dot. To show that an endpoint (or any other single point) is *not* a solution, use an "open" dot at the point. For example:

(a) $x \geq 3$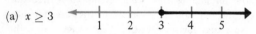

(b) $n < 3$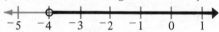

(c) $s \neq 3$

CHALKBOARD EXAMPLES

Graph.

1. {the directed numbers greater than ⁻4}

2. {the directed numbers less than or equal to 1}

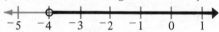

3. {the directed numbers between ⁻3 and 1}

Name the set of directed numbers graphed.

4.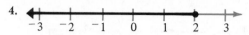

{the directed numbers less than or equal to 2}

5.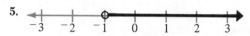

{the directed numbers greater than ⁻1}

6.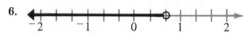

{the directed numbers less than $\frac{2}{3}$}

7.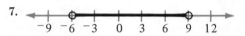

{the directed numbers between ⁻6 and 9}

Name and graph the solution set. The replacement set is {the directed numbers}.

8. $n > ^{-}1$
   {the directed numbers greater than ⁻1}

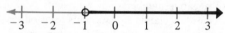

9. $m \leq 4$
   {the directed numbers less than or equal to 4}

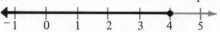

10. $^{-}1 < t \leq 3$
    {3 and the directed numbers between ⁻1 and 3}

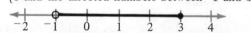

Addition and Subtraction of Directed Numbers

CHAPTER OVERVIEW

In this chapter we introduce another use of the minus sign, this time to indicate the opposite of a number. In the first two sections we continue to use raised minus signs to indicate negative numbers. We now use the standard minus sign—previously used only for subtraction—to indicate "the opposite of." Thus:

⁻6 means "negative six"

— 6 means "subtract six"

—6 means "the opposite of six"

After showing that ⁻6 and −6 name the same number (Section 9-3), we are able to simplify procedures by dropping the use of raised minus signs entirely. From here on, −6 may mean either "negative six" or "the opposite of six." Additive inverse, the more formal term for the opposite of a number, is mentioned to acquaint the students with the term. The terms *additive inverse* and *opposite* are used interchangeably in the text.

Four lessons are devoted to addition of directed numbers. A fifth lesson covers subtraction of directed numbers, which is shown to be the same as "adding the opposite." The inverse relationship of addition and subtraction is used to develop subtraction of directed numbers.

9-1 *Adding Directed Numbers on the Number Line*

TEACHING THESE PAGES

Diagrams of addition on the number line often help students understand what is being done when we add two directed numbers. However, using the number line for actually doing additions is only practical when the numbers being added are relatively small integers. Through practice in addition of lesser numbers, students should realize that additions of greater numbers and of non-integral numbers follow the same pattern:

(1) The sum of two positive numbers is positive and the sum of two negative numbers is negative.

(2) The sum of one positive number and one negative number is either positive or negative, depending on which number would be represented by the longer arrow.

The lengths of the arrows, of course, depend on the magnitudes of the numbers, which also determine the magnitude of each sum. The magnitude of the sum in (1) above is the sum of the magnitudes of the numbers to be added. The magnitude of the sum in (2) is the difference of the magnitudes of the two numbers.

CHALKBOARD EXAMPLES

Sketch a number line solution. Complete the equation to make a true statement.

1. $3 + {}^-6 = \underline{\ ?\ }$ $\quad {}^-3$

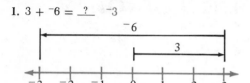

2. $\underline{\ ?\ } = {}^-3\frac{1}{2} + 1\frac{1}{2}$ $\quad {}^-2$

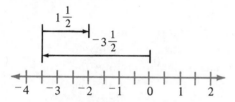

Tell whether the expression names a positive number, a negative number, or 0. Then simplify.

3. $3 + 5$ positive; 8
4. ${}^-4 + 1$ negative; ${}^-3$
5. $8 + {}^-8$ 0; 0
6. $15 + {}^-5$ positive; 10

7. Complete the addition table.

+	2	${}^-3$	5	${}^-4$
2	?4	?${}^-1$	?7	?${}^-2$
${}^-3$	?${}^-1$	?${}^-6$	?2	?${}^-7$

Find the sum.

8. $3 + {}^-3\frac{1}{3}$ $\quad \frac{{}^-1}{3}$
9. ${}^-301\frac{1}{4} + 301\frac{1}{4}$ $\quad 0$
10. $\begin{array}{r} 25 \\ {}^-17 \\ \hline 8 \end{array}$
11. $\begin{array}{r} {}^-30 \\ 11 \\ \hline {}^-19 \end{array}$

Express the problem as the sum of two directed numbers. Find the sum and answer the question.

13. Nancy had a package of Seal-All weather stripping that contained 12 m of weather stripping. She used 6.3 m of weather stripping around her back door. How much was left? $12 + {}^-6.3 = 5.7$; 5.7 m of weather stripping was left.

14. Two hikers walked 5 km south, then backtracked and walked 6.5 km north. Where were they then in relation to their starting point? ${}^-5 + 6.5 = 1.5$; 1.5 km north of their starting point.

9-2 Additive Inverses and the Identity Element for Addition

TEACHING THESE PAGES

Once again students meet the term "opposite." We use a centered minus sign $(-)$ as notation for "the opposite of." For example, "-7" means "the opposite of 7." We already have seen that the opposite of 7 is ${}^-7$. Since we may write the opposite of 7 as either -7 or ${}^-7$, we may now write:

$$-7 = {}^-7$$

Recall that we also know that the opposite of ${}^-7$ is 7. In symbols: $-({}^-7) = 7$. As we know, the opposite of 0 is 0, or $-0 = 0$.

The more formal term "additive inverse" is mentioned as another name for "opposite," and is used in some of the exercises. Students may use either term.

The number 0 is the identity element for addition. That is, when 0 is added to any number, the sum is the number to which 0 was added:

$$5 + 0 = 5 \qquad 0 + {}^-3 = {}^-3$$

CHALKBOARD EXAMPLES

Show two ways to express in symbols.

1. The opposite of 5.3 $\quad -5.3, {}^-5.3$
2. The opposite of $\frac{{}^-1}{4}$ $\quad -\left(\frac{{}^-1}{4}\right), \frac{1}{4}$

Name the two numbers described.

3. 3 units from zero on the number line $\quad 3, {}^-3$
4. 4 units from 4 on the number line $\quad 8, 0$

Tell whether or not the numbers are additive inverses.

5. $-({}^-5), 5$ no
6. ${}^-4.3, 4.3$ yes
7. ${}^-9, -9$ no
8. $-({}^-1.5), {}^-1.5$ yes

Solve.

9. $y = 15 + {}^-15$ $\quad y = 0$
10. ${}^-13 + m = 0$ $\quad m = 13$
11. $7.1 + c = 7.1$ $\quad c = 0$
12. $0 = a + 8$ $\quad a = {}^-8$

9-3 Simplifying Expressions

TEACHING THESE PAGES

It is in this lesson that we cease using raised minus signs to indicate negative numbers. We have seen, for example, that:

$$-3 = {}^-3$$

That is, the opposite of 3 is ⁻3. So we simplify our work by using -3 in place of ⁻3.

In this lesson, we show that the opposite of a sum is the same as the sum of the opposites. That is:

$$-(a + b) = -a + (-b)$$

We also use number line sketches to find solutions for simple equations that involve sums. However, students should be encouraged to drop the use of number line sketches in equation solving as soon as they feel able to proceed without a sketch.

CHALKBOARD EXAMPLES
Simplify.

1. $-(8 + 6)$ -14
2. $-7 + (-3)$ -10
3. $-(5.1 + 4.3)$ -9.4

Solve. Use a number line sketch.

4. $3 + n = 5$ $n = 2$

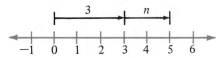

5. $1 = -3 + t$ $t = 4$

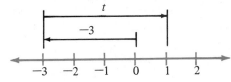

Show whether the statement is true or false.

6. $-(10 + 3) = -10 + 3$

$$-(10 + 3) \overset{?}{=} -10 + 3$$

$-(13)$	-7	
-13	-7	false

7. $3.3 - 1.4 = -(-3.3 + 1.4)$

$$3.3 - 1.4 \overset{?}{=} -(-3.3 + 1.4)$$

1.9	$-(-1.9)$	
1.9	1.9	true

Solve.

8. $-5 = 3 + b$ $b = -8$
9. $-(-10 + 7) = c$ $c = 3$
10. $8 = -1 + r$ $r = 9$

9-4 Addition Properties for Directed Numbers

TEACHING THESE PAGES
Three familiar properties of addition of non-negative numbers are the commutative property, the associative property, and the additive property of zero. These properties also apply to all directed numbers. In order to work with the negative numbers, we find we need to add two more properties to those above:

(1) The opposite of the sum of two numbers is the sum of the opposites of the numbers. (The Property of the Opposite of a Sum)

(2) The sum of a number and its opposite is zero. (The Additive Property of Inverses)

We can use these properties to add two numbers of opposite signs by renaming the number with the greater magnitude. For example:

$18 + (-3) = (15 + 3) + (-3)$ ◀ Rename 18 as $15 + 3$.

$$= 15 + [3 + (-3)]$$
$$= 15 + 0$$
$$= 15$$

Students who quickly become facile in finding sums of numbers with unlike signs should be encouraged to name the sum directly without writing out all the steps. Students will reach this stage at varying rates, but ultimately they should all be able to write simple sums directly. In any case, practice in going through all the steps at first helps to develop understanding and also provides a resource for reasoning through the addition process in more complicated expressions.

CHALKBOARD EXAMPLES
Name the property illustrated.

1. $-5 + (5 + 7) = (-5 + 5) + 7$ Assoc. Prop.
2. $-3 + (5 + 3) = -3 + (3 + 5)$ Comm. Prop.
3. $0 + (-8) = -8$ Additive Property of 0

Complete to make a true statement.

4. $\underline{\ ?\ } + (-37) = 0$ 37
5. $8 + \underline{\ ?\ } = 0$ -8
6. $(-1 + 1) + \underline{\ ?\ } = 11$ 11
7. $15 + (-15) = \underline{\ ?\ }$ 0

Add. Use the additive property of opposites.

8. $7 + (-7) + 18$
   $0 + 18 = 18$
9. $-13 + (-31) + 31$
   $-13 + 0 = -13$

10. $9 + (-9) + 37 + (-14) + 14$
$0 + 37 + 0 = 37$
11. $-4.5 + 17 + 4.5$
$0 + 17 = 17$

Show that the statement is true.

12. $16 + 33 + (-16) = -15 + 48$

$16 + 33 + (-16) \overset{?}{=} -15 + 48$

$16 + (-16) + 33$	$-15 + (15 + 33)$
$0 + 33$	$(-15 + 15) + 33$
33	$0 + 33$
33	33

Simplify.

13. $-5 + 13 + 7 + (-6) = [-5 + (-6)] + (13 + 7)$
$= -11 + 20$
$= -11 + (11 + 9)$
$= (-11 + 11) + 9$
$= 0 + 9 = 9$

14. $15 + (-8) + 31 = -8 + (15 + 31)$
$= -8 + 46$
$= -8 + (8 + 38)$
$= (-8 + 8) + 38$
$= 0 + 38 = 38$

9-5 *Subtracting Directed Numbers*

TEACHING THESE PAGES
Quite likely, students are already intuitively aware that subtracting a number has the same effect as adding the opposite of the number. It is in this lesson that we tie together the ideas of the first sections of the chapter and show the inverse relationship between addition and subtraction of directed numbers. If students recognize the opposite, or additive inverse, of a number, and can add directed numbers, then subtracting directed numbers should cause them little difficulty. If the class needs help at the start, some practice in writing subtraction expressions as equivalent addition expressions may start the thinking in the right direction. On the chalkboard, write several subtraction expressions and ask for volunteers to write equivalent addition expressions. You might write expressions like the following:

Subtraction	Equivalent Addition
1. $5 - 7$	$5 + (-7)$
2. $-5 - 7$	$-5 + (-7)$
3. $7 - 5$	$7 + (-5)$
4. $-7 - 5$	$-7 + (-5)$
5. $5 - (-7)$	$5 + 7$

CHALKBOARD EXAMPLES
Solve. Begin by writing the equivalent addition equation.

1. $17 - 3 = y$
$17 + (-3) = y$
$y = 14$

2. $5 - 11 = n$
$5 + (-11) = n$
$n = -6$

3. $d = 3 - (-10)$
$d = 3 + 10$
$d = 13$

4. $t = 9 - (-5)$
$t = 9 + 5$
$t = 14$

5. $-6 - (-13) = s$
$-6 + 13 = s$
$s = 7$

6. $c = -4 - 7$
$c = -4 + (-7)$
$c = -11$

Subtract. Add to check.

	7.		**8.**		**9.**	
	13	20	-17	-21	-30	-25
	-7	-7	4	4	-5	-5
	20	13	-21	-17	-25	-30

9-6 *Functions and Directed Numbers*

TEACHING THESE PAGES
We can now use directed numbers in the replacement sets for function equations. We again use the function "machine" as an illustrative device. Point out that, in a function, a negative input does not necessarily result in a negative output, nor does a positive input always result in a positive output.

CHALKBOARD EXAMPLES
Complete according to the given function equation.

1. $f(z) = z + 5$: $\{(-6, \overset{-1}{\underline{\ ?\ }}), (-4, \overset{1}{\underline{\ ?\ }}), (-2, \overset{3}{\underline{\ ?\ }}),$
$(0, \overset{5}{\underline{\ ?\ }}), (2, \overset{7}{\underline{\ ?\ }})\}$

2. $f(t) = t + (-3)$: $\{(11, \overset{8}{\underline{\ ?\ }}), (7, \overset{4}{\underline{\ ?\ }}), (3, \overset{0}{\underline{\ ?\ }}),$
$(-1, \overset{-4}{\underline{\ ?\ }}), (-5, \overset{-8}{\underline{\ ?\ }})\}$

3. $f(n) = n - (-4)$: $\{(-5, \overset{-1}{\underline{\ ?\ }}), (-3, \overset{1}{\underline{\ ?\ }}), (-1, \overset{3}{\underline{\ ?\ }}),$
$(0, \overset{4}{\underline{\ ?\ }}), (1, \overset{5}{\underline{\ ?\ }})\}$

Tell whether the set of number pairs is a function.

4. $\{(-1, 3), (0, 3), (1, 4), (3, 5), (5, 3)\}$ yes
5. $\{(-3, 5), (0, -1), (-3, 1), (0, -3), (7, 3)\}$ no
6. $\{(-3, -3), (2, 2), (-5, 5), (1, 6), (4, -3)\}$ yes

7. For the set of number pairs $\{(0, -4), (-1, -5), (-2, -6), (-3, -7)\}$ the function equation is $\underline{\ ?\ }$.
   A. $f(x) = x + 4$
   B. $f(y) = y - 4$
   C. $f(z) = z + (-6)$

10 Multiplication and Division of Directed Numbers

CHAPTER OVERVIEW

This chapter extends the properties of multiplication which hold for the numbers of arithmetic—that is, the positive numbers and zero—to multiplication of directed numbers. We find that all of the properties that were true for multiplication of the numbers of arithmetic are also true for multiplication of directed numbers.

We begin with the idea of multiplication by a positive integer as repeated addition. Working with a few specific examples helps us formulate generalizations about the signs of the products of two positives and of one negative and one positive.

To determine the result of multiplying two negative numbers, we use specific applications of the distributive property to demonstrate that the product of two negative numbers must be positive if the accepted properties of multiplication are to hold.

Once the facts about the signs for multiplication of directed numbers are well established, the treatment of division of directed numbers follows logically, since division is the inverse operation of multiplication.

10-1 *Multiplication by a Positive Number or by Zero*

TEACHING THESE PAGES

The interpretation of multiplication by a positive integer as a repeated addition is easily shown for small numbers on the number line. Such a demonstration helps to convince students of the logic of the basic statements about signs of products. For example:

$$3 \cdot 4 = 4 + 4 + 4 = 12$$

$$2 \cdot (-5) = -5 + (-5) = -10$$

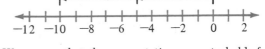

We assume that the commutative property holds for multiplication of directed numbers. Thus:

$$-6 \cdot 3 = 3 \cdot (-6) = -6 + (-6) + (-6) = -18$$

Two other properties of multiplication of non-negatives that we can assume to be true for all directed numbers are:

(1) The product of a number and 0 is 0.
(2) The product of 1 and a number is that number.

Some students may ask what happens when two negative numbers are multiplied. You might have them suggest answers for such a product as $(-3) \cdot (-2)$. Some students may already know that $(-3) \cdot (-2) = 6$, but discuss all answers that students offer. Tell them that 6 is the correct answer and assure them that such products will be discussed more fully after the basic multiplication properties have been summarized.

CHALKBOARD EXAMPLES

Simplify. Assume that the variable represents a positive number.

1. $3(-5)$ $\quad -15$
2. $-6 \cdot 7$ $\quad -42$
3. $13 \cdot 5$ $\quad 65$
4. $w\left(-\dfrac{3}{4}\right)$ $\quad -\dfrac{3w}{4}$
5. $-8\left(\dfrac{7}{8}\right)$ $\quad -7$
6. $\left(-\dfrac{3}{5}\right)\left(\dfrac{3}{8}\right)$ $\quad -\dfrac{9}{40}$

Simplify.

7. $(-3) \cdot 11 + 15$
$\quad -33 + 15 = -18$
8. $(-4 \cdot 7) + (-4 \cdot 3)$
$\quad -28 + (-12) = -40$

Solve.

9. $5n = -15$ $\quad n = -3$

10-2 *Multiplication Properties for Directed Numbers*

TEACHING THESE PAGES

We assume that multiplication of directed numbers is associative as well as commutative, and that the multiplicative properties of zero and one and the distributive property also hold for {the directed numbers}.

CHALKBOARD EXAMPLES

Write the expression as a sum of two products. Then simplify the expressions to show that they are equal.

1. $-5(7 + 8)$

$-5(7 + 8) = -5(7) + -5(8)$

| $-5(15)$ | $-35 + (-40)$ |
| -75 | -75 |

2. $3(-3 + 7)$

$3(-3 + 7) = 3(-3) + 3(7)$

| $3(4)$ | $-9 + 21$ |
| 12 | 12 |

3. $[-5 + (-4)] \cdot 6$

$[-5 + (-4)] \cdot 6 = (-5) \cdot 6 + (-4) \cdot 6$

| $(-9) \cdot 6$ | $-30 + (-24)$ |
| -54 | -54 |

Show that the statement is true.

4. $8(-5 \cdot 3) = [8(-5)] \cdot 3$

| $8(-15)$ | $(-40) \cdot 3$ |
| -120 | -120 |

5. $-3(9 + 7) = -3 \cdot 9 + (-3 \cdot 7)$

| $-3(16)$ | $-27 + (-21)$ |
| -48 | -48 |

6. $6[-3 + 5 + (-7)] = 6(-3) + 6(5) + 6(-7)$

| $6(-5)$ | $-18 + 30 + (-42)$ |
| -30 | -30 |

7. Show that the statement $x(y + w) = xy + xw$ is true when $x = -3$, $y = 1$, and $w = 4$.

$-3(1 + 4) = -3(1) + (-3) \cdot 4$

| $-3(5)$ | $-3 + (-12)$ |
| -15 | -15 |

10-3 *Multiplication of Negative Numbers*

TEACHING THESE PAGES

Now we are ready to complete our reasoning about the sign of the product of two directed numbers by considering the fourth possibility: negative × negative. We have seen that the distributive property holds for directed numbers, so we use it to show that the product of two negative numbers must be a positive number. On page 270, the student is shown several specific cases where this must be true. A more formal mathematical proof of the fact is rather involved, and is not needed here.

CHALKBOARD EXAMPLES

Tell whether the expression represented is positive, negative, or zero. Then simplify.

1. $-5 \cdot 7$ negative; -35
2. $-3 \cdot 1 \cdot (-6)$ positive; 18
3. $-7 \cdot 3 \cdot \dfrac{1}{3}$ negative; -7
4. $-1(-5)(6)$ positive; 30
5. $\dfrac{4}{7}(-7)$ negative; -4
6. $-1(-8)(-3)$ negative; -24

Rewrite the expression as a sum of two products. Then simplify both expressions to show that they are equal.

7. $-5[7 + (-5)]$

$$-5[7 + (-5)] = (-5)7 + (-5)(-5)$$

$-5 \cdot (2)$	$-35 + 25$
-10	-10

8. $-4[-7 + (-3)]$

$$-4[-7 + (-3)] = -4(-7) + (-4)(-3)$$

$-4(-10)$	$28 + 12$
40	40

9. $-\frac{1}{3}(0 - 6)$

$$-\frac{1}{3}(0 - 6) = -\frac{1}{3}(0) + \left(-\frac{1}{3}\right)(-6)$$

$-\frac{1}{3}(-6)$	$0 + 2$
2	2

10. $(-3 + 8)(-6)$

$$(-3 + 8)(-6) = -3(-6) + 8(-6)$$

$5(-6)$	$18 + (-48)$
-30	-30

Evaluate. Let $a = -3$, $b = \frac{1}{2}$, $c = 6$

11. $(-a) + bc \ -(-3) + \frac{1}{2}(6) = 3 + 3 = 6$

10-4 *The Distributive Property in Simplifying Expressions*

TEACHING THESE PAGES
We show the application of the distributive property to simplifying expressions for two reasons: (1) to justify the simplification process mathematically, and (2) to provide a step-by-step method for students to follow until they become sure of their ability to simplify mentally. In the text, we tell the student, "As soon as you understand this, you will probably do much of it in your head." By this, we do not mean to urge undue haste, but rather to eliminate unnecessary detail and consequently prevent boredom for students who have mastered the process. Some students may be able to shorten the simplification of $-5x + y + 3y - x$ to:

$$[-5 + (-1)]x + (1 + 3)y = -6x + 4y$$

Others may be able to write the final simplified form at once. This need not be discouraged as long as students

are simplifying correctly. However, students should continue, or return to, the more detailed approach when errors are made.

CHALKBOARD EXAMPLES
Simplify by combining similar terms.

1. $-n + 3p - 5p + 7n$
$(-1 + 7)n + (3 - 5)p = 6n - 2p$

2. $-4c^2 - 3c + 9c^2 + 8c$
$(-4 + 9)c^2 + (-3 + 8)c = 5c^2 + 5c$

3. $3a + 4ab - 7a + 10 - ab$
$(3 - 7)a + (4 - 1)ab + 10 = -4a + 3ab + 10$

4. $8(m - 5) + (-5m) + 7$
$8m - 40 + (-5m) + 7 =$
$(8 - 5)m + (-40 + 7) = 3m - 33$

5. $11w + [-6(w + 3)]$
$11w + (-6w) + (-18) = 5w - 18$

Find the value of the expression if $a = -3$, $b = 4$, $c = -5$, and $d = 6$.

6. $ac + (b + 3)^2$
$-3(-5) + (4 + 3)^2 = 15 + 49 = 64$

7. $-2(b + d) + c$
$-2(4 + 6) + (-5) = -20 + (-5) = -25$

10-5 *Division of Directed Numbers*

TEACHING THESE PAGES
Comparison of a division sentence with its related multiplication sentence (as in Examples 1–3 on page 276) reveals the same pattern of signs in division as in multiplication:

two signs alike $\rightarrow$ a positive result
two signs unlike $\rightarrow$ a negative result

Division by 0, of course, is impossible in {the directed numbers} as well as in {the numbers of arithmetic}.

As we have seen before, the decimal equivalent of a fraction can be found by performing the indicated division. The only new idea here is that we can now write decimal equivalents for negative fractions.

CHALKBOARD EXAMPLES
Simplify.

1. $-54 \div 6$ $\ -9$ **2.** $70 \div 5$ $\ 14$

3. $-40 \div 8$ $\ -5$ **4.** $\dfrac{-42}{-5 + (-2)}$ $\ 6$

5. $\dfrac{72}{3 + (-12)}$ $\ -8$ **6.** $\dfrac{-8 + (-10)}{-3}$ $\ 6$

Write the decimal equivalent.

7. $-\dfrac{27}{100}$ $\quad -0.27$ **8.** $\dfrac{-9}{-12}$ $\quad 0.75$ **9.** $\dfrac{5}{-8}$ $\quad -0.625$

True or false?

10. $\dfrac{5}{6} = \dfrac{-5}{-6}$ $\quad$ true $\qquad$ **11.** $\dfrac{-3}{5} = -0.6$ $\quad$ true

12. $-\dfrac{5}{9} = \dfrac{-5}{-9}$ $\quad$ false

10-6 *Reciprocals of Directed Numbers*

TEACHING THESE PAGES
In Chapter 4 we dealt with the reciprocals of positive numbers, and used multiplication by the reciprocal to divide with fractions. The same process can now be used to divide with directed numbers. Stress the following points:

(1) The product of a number and its reciprocal is 1.
(2) Since the product of any number and 0 is 0, 0 has no reciprocal.
(3) Positive numbers have positive reciprocals; negative numbers have negative reciprocals.
(4) We can solve some equations by multiplying both members by the reciprocal of the coefficient of the variable. That is, for every value of a except $a = 0$, we can solve the equation $ax = n$ by multiplying both members of the equation by $\dfrac{1}{a}$.

You might like to have the class practice solving equations of the form $ax = n$. Write several equations on the board and have volunteers come to the board and write the solutions, showing the use of reciprocals.

CHALKBOARD EXAMPLES
Complete to make a true statement.

1. $\dfrac{-8}{9}\left(-\dfrac{9}{8}\right) = \underline{\ ?\ }$ $\quad 1$ $\qquad$ **2.** $-\dfrac{3}{7} \cdot \underline{\ ?\ } = 1 -\dfrac{7}{3}$

Simplify. Use reciprocals.

3. $\dfrac{5}{6} \div 10$

$\quad \dfrac{5}{6} \cdot \dfrac{1}{10} = \dfrac{1}{12}$

4. $\dfrac{3}{-8} \div 6$

$\quad \dfrac{3}{-8} \cdot \dfrac{1}{6} = -\dfrac{1}{16}$

5. $-18 \div \dfrac{-9}{10}$

$\quad -18 \cdot \dfrac{10}{-9} = 20$

6. $-\dfrac{3}{5} \div \dfrac{3}{-5}$

$\quad -\dfrac{3}{5} \cdot \dfrac{-5}{3} = 1$

Solve.

7. $4x = -20$ $\quad x = -5$ $\qquad$ **8.** $-30 = -5n$ $\quad n = 6$

9. $-\dfrac{6}{n} = -9$ $\quad n = \dfrac{2}{3}$ $\qquad$ **10.** $\dfrac{-1}{12}b = -4$ $\quad b = 48$

Find the unknown number.

11. 35 divided by some number is equal to -5. Find the number. $\dfrac{35}{n} = -5; n = -7$

10-7 *Functions and Directed Numbers*

TEACHING THESE PAGES
Once again we return to the concept of function and extend it to include function equations which involve multiplication and division with directed numbers. Again the student uses the function machine with a stated finite replacement set to generate a function in the form of a set of ordered pairs. By this time, students should have little difficulty completing sets of ordered pairs.

Remind the students that the pairs are *ordered* pairs. That is, each pair is stated in the order $(x, f(x))$, where x is a value for the variable and $f(x)$ is the corresponding number that results from the function equation.

CHALKBOARD EXAMPLES
Complete the set of number pairs by using the given function equation. The replacement set is $\{-3, -1, 0, 4\}$.

1. $f(n) = -\dfrac{n}{5};$ $\quad \{(-3, \underline{\ ?\ }\ \overset{\frac{3}{5}}{}), (-1, \underline{\ ?\ }\ \overset{\frac{1}{5}}{}), (0, \underline{\ ?\ }\ \overset{0}{}), (4, \underline{\ ?\ })\} -\dfrac{4}{5}$

2. $f(s) = 3s - 2;$ $\quad \{(-3, \underline{\ ?\ }\ \overset{-11}{}), (-1, \underline{\ ?\ }\ \overset{-5}{}), (0, \underline{\ ?\ }\ \overset{-2}{}), (4, \underline{\ ?\ })\} 10$

3. $f(d) = \dfrac{-3d}{4};$ $\quad \{(-3, \underline{\ ?\ }\ \overset{\frac{9}{4}}{}), (-1, \underline{\ ?\ }\ \overset{\frac{3}{4}}{}), (0, \underline{\ ?\ }\ \overset{0}{}), (4, \underline{\ ?\ })\} -3$

Write the number pairs indicated by the given function equation and replacement set.

4. $f(y) = -5y + 3: \{-2, -1, 0, 1, 3\}$
$\{(-2, 13), (-1, 8), (0, 3), (1, -2), (3, -12)\}$

5. $f(b) = \dfrac{b}{3} - 3: \{-3, -1, 0, 3, 6\}$
$\{(-3, -4), (-1, -3\tfrac{1}{3}), (0, -3), (3, -2), (6, -1)\}$

11 Solving Equations and Inequalities

CHAPTER OVERVIEW

In the first sections of the chapter, methods for solving several basic types of linear equations are considered. In Section 11-4, equation-solving techniques are applied to solving problems using formulas. Sections 11-5 and 11-6 summarize the basic steps in solving linear inequalities, with special attention in 11-6 to the reversal of direction in an inequality when both members are multiplied by a negative number.

11-1 *Equations of Type* $x + a = b$

TEACHING THESE PAGES

Equations of the type $x + a = b$ include those which fit the pattern initially as well as those equations which can be transformed into the type by the application of a fundamental property. Examples:

$$x + 5 = 7$$
$$8 + n = 3 \rightarrow n + 8 = 3$$
$$9 = y + 4 \rightarrow y + 4 = 9$$
$$w - 3 = 6 \rightarrow w + (-3) = 6$$

Any equation that can be transformed into the pattern $x + a = b$ can be solved by adding the opposite of a to each member of the equation. For example:

In $8 + n = 3$, $a = 8$ and $b = 3$, so
$8 + (-8) + n = 3 + (-8)$; $n = -5$

In $9 = y + 4$, $a = -4$ and $b = 9$, so
$9 + (-4) = y + 4 + (-4)$; $5 = y$

In $w - 3 = 6$, $a = -3$ and $b = 6$, so
$w - 3 + 3 = 6 + 3$; $w = 9$

Urge students to form the habit of checking their answers—mentally, when possible. Encourage them to write out the steps of a solution in detail whenever they are not sure of their mental arithmetic.

CHALKBOARD EXAMPLES

Complete.

1. $n + 5 = 12$
   $n + 5 + (-5) = 12 + \underline{\ ?\ }\ (-5)$
   $n + \underline{\ ?\ } = 12 + (-5)$
   $0\ n = \underline{\ ?\ }\quad 7$

2. $-12 + b = 3$
   $-12 + (12) + b = 3 + \underline{\ ?\ }\ 12$
   $0\ \underline{\ ?\ } + b = 3 + 12$
   $b = \underline{\ ?\ }\quad 15$

Solve and check.

3. $m + (-5) = 7$ $m + (-5) \overset{?}{=} 7$
   $m = 12$

$12 + (-5)$	7
7	$7\ \checkmark$

4. $y + 18 = 0$ $y + 18 \overset{?}{=} 0$
   $y = -18$

$(-18) + 18$	0
0	$0\ \checkmark$

5. $z - 1.7 = 6.8$ $z - 1.7 \overset{?}{=} 6.8$
   $z = 8.5$

$8.5 - 1.7$	6.8
6.8	$6.8\ \checkmark$

6. $n - \left(-\frac{1}{4}\right) = \frac{1}{8}$

$\qquad n - \left(-\frac{1}{4}\right) \overset{?}{=} \frac{1}{8}$

$n = -\frac{1}{8}$

$$\begin{array}{c|c} -\frac{1}{8} - \left(-\frac{1}{4}\right) & \frac{1}{8} \\ \hline -\frac{1}{8} + \frac{1}{4} & \frac{1}{8} \\ \hline \frac{1}{8} & \frac{1}{8} \ \checkmark \end{array}$$

7. Solve for x:

$-a + x = -t$

$\qquad x = a - t$

11-2 *Equations of Type $ax = b$*

TEACHING THESE PAGES

The basic strategy for solving equations of the type $ax = b$ is to multiply both members by the reciprocal of a. The only value of a that has no reciprocal is 0. However, if a were 0, then the equation would have no meaning unless b were also 0. In that case, every possible value of x would make the sentence true, and there would not be a unique solution.

CHALKBOARD EXAMPLES

Solve and check.

1. $5n = 35$

$n = 7$

$5n \overset{?}{=} 35$

$$\begin{array}{c|c} 5(7) & 35 \\ \hline 35 & 35 \ \checkmark \end{array}$$

2. $5b = -8$

$b = -\frac{8}{5}$

$5b \overset{?}{=} -8$

$$\begin{array}{c|c} 5\left(-\frac{8}{5}\right) & -8 \\ \hline -8 & -8 \ \checkmark \end{array}$$

3. $-8y = 48$

$y = -6$

$-8y \overset{?}{=} 48$

$$\begin{array}{c|c} -8(-6) & 48 \\ \hline 48 & 48 \ \checkmark \end{array}$$

4. $q(-6) = -21$

$q = 3\frac{1}{2}$

$q(-6) \overset{?}{=} -21$

$$\begin{array}{c|c} -6(3\frac{1}{2}) & -21 \\ \hline -21 & -21 \ \checkmark \end{array}$$

Solve for the underlined variable. Assume that no divisor has the value 0.

5. $c = n\underline{p}$ $\quad p = \dfrac{c}{n}$

6. $6 = \pi\underline{d}$ $\quad d = \dfrac{6}{\pi}$

Solve.

7. $\dfrac{3n}{5} = 9$ $\quad n = 15$

8. $\dfrac{8}{-3}s = 16$ $\quad s = -6$

9. $35z = 7$ $\quad z = \dfrac{1}{5}$

10. $-1.7b = 34$ $\quad b = -20$

11-3 *Equations of Type $ax + bx = c$*

TEACHING THESE PAGES

Equations of this type, obviously, may be simplified to the form $ax = c$ of the previous lesson, once the similar terms are combined. Remind the students that the distributive property can be used in combining similar terms (as we saw in Section 7-1). For example:

$$\begin{array}{ll} (1) & 3n + 8n = 55 \\ (2) & (3 + 8)n = 55 \\ (3) & 11n = 55 \\ (4) & n = 5 \end{array}$$

By now most students will go from (1) to (3) mentally.

CHALKBOARD EXAMPLES

Complete.

1. $\dfrac{6}{7}n - \dfrac{1}{7}n = 10$

$\dfrac{5}{7}(\underline{\ ?\ })n = 10$

$(\underline{\ ?\ }) \cdot \dfrac{5}{7}n = 10 \cdot \dfrac{7}{5}$

$\dfrac{7}{5}$

$\qquad n = \underline{\ ?\ }$ $\quad 14$

2. $-5x = 3x - 40$

$-5x + (\underline{\ ?\ }) = 3x + \overset{-3x}{?} - 40$

$(\underline{\ ?\ })x = -40$

$\qquad x = \underline{\ ?\ }$ $\quad 5$

(annotations: $-3x$, -8)

3. Solve $an - bn = c$ for n. $\quad n = \dfrac{c}{a - b}$

Solve and check.

4. $5b = -3b + 40$

$8b = 40$

$b = 5$

$5b \overset{?}{=} -3b + 40$

$$\begin{array}{c|c} 5(5) & -3(5) + 40 \\ 25 & -15 + 40 \\ 25 & 25 \quad \checkmark \end{array}$$

5. $8x = 73 + x - 3$

$7x = 70$

$x = 10$

$8x \overset{?}{=} 73 + x - 3$

$$\begin{array}{c|c} 8(10) & 73 + 10 - 3 \\ 80 & 73 + 7 \\ 80 & 80 \quad \checkmark \end{array}$$

6. $-7m = 3m + 15$

$-10m = 15$

$m = -1.5$

$-7m \overset{?}{=} 3m + 15$

$$\begin{array}{c|c} -7(-1.5) & 3(-1.5) + 15 \\ 10.5 & -4.5 + 15 \\ 10.5 & 10.5 \quad \checkmark \end{array}$$

7. $d = 5.4 - 0.8d$

$1.8d = 5.4$

$d = 3$

$d \overset{?}{=} 5.4 - 0.8d$

$$\begin{array}{c|c} 3 & 5.4 - 0.8(3) \\ 3 & 5.4 - 2.4 \\ 3 & 3 \quad \checkmark \end{array}$$

11-4 *Applying Formulas*

TEACHING THESE PAGES

In this lesson we show students one possible method for solving problems using formulas. The method requires students first to solve the formula for the unknown variable in terms of the other variables. Then the transformed formula is used to find the value of the unknown variable when specific values for the other variables are substituted. This approach is most practical when several similar problems are presented. However, here, this method has the further advantage of providing practice in equation solving.

Consider the following problem: Suppose you borrow $1500 at the rate of 8% interest (using simple interest). To find out how long it would take for the interest to amount to $300, you could use either of the following solutions:

Solution A	Solution B
$I = prt$	$I = prt$
(1) $t = \dfrac{I}{pr}$	(1) $300 = 1500 \times 0.08t$
(2) $t = \dfrac{300}{1500(0.08)}$	(2) $\dfrac{300}{1500(0.08)} = t$
$t = 2\frac{1}{2}$	$t = 2\frac{1}{2}$

We find by both methods that it would take $2\frac{1}{2}$ years for the interest owed to amount to $300.

The advantage of using Solution A is evident when several problems of the same type are to be solved. Step (1) would need to be performed only once. Each problem after that could begin with Step (2), using the formula for t developed in the earlier solution. In solving a single problem of this type, however, Solution A has no real advantage over Solution B.

CHALKBOARD EXAMPLES

Solve.

1. Using the formula $A = lw$ (area = length × width), find the length of a rectangular mirror whose area is 1800 cm², if the width is 24 cm.

   $l = \dfrac{A}{w} = \dfrac{1800}{24} = 75$; length = 75 cm

2. Use the formula $I = prt$ (interest = principal × rate × time) to find the rate of interest on a loan if the interest was $90 on a principal of $800 for a period of $1\frac{1}{2}$ years. $r = \dfrac{I}{pt} = \dfrac{90}{1200} = 0.075$;

   rate = 7.5%

The area of a trapezoid may be found by the formula $A = \dfrac{1}{2}\,as$ (area = $\dfrac{1}{2}$ the altitude × the sum of the bases).

3. Find the area of the trapezoid.

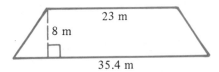

$A = \dfrac{1}{2}\,as = \dfrac{1}{2}(8)(23 + 35.4) = 233.6$

area = 233.6 m²

4. The area of the trapezoid is 74.5 cm². Find the altitude.

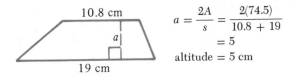

$a = \dfrac{2A}{s} = \dfrac{2(74.5)}{10.8 + 19}$
$= 5$

altitude = 5 cm

11-5 *The Addition Property of Inequality*

TEACHING THESE PAGES

We do not need to state a subtraction property of inequality since we have previously established that subtraction is the same as "adding the opposite." The addition property of inequality is similar to the corresponding property for equality and should be intuitively obvious.

CHALKBOARD EXAMPLES

Match each inequality in Column 1 with an equivalent inequality in Column 2.

Column 1	Column 2
1. $3 + y > 6$ B	A. $y \le 6$
2. $y - 7 < 3$ C	B. $y > 3$
3. $y - (-1) \le 7$ A	C. $y < 10$

Complete, then graph the solution set.

4. $\quad n + (-3) < 1$

   $n + (-3) + \underline{\ ?\ } < 1 + 3$

   $\quad\quad 3\ n < \underline{\ ?\ }\ \ 4$

   0 1 2 3 4 5

5. $(-5)\, 8 \geq t + 5$

$8 + \underline{\ ?\ } \geq t + 5 + \underline{\ ?\ }\ (-5)$

$3 \ \underline{\ ?\ } \geq t$

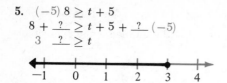

Solve the inequality and write its solution set.

6. $x - 9 > 4$

$x > 13$; {the directed numbers greater than 13}

7. $-3 + w \leq 11$

$w \leq 14$; {14 and the directed numbers less than 14}

8. $\frac{1}{4}(4m + 12) < 9$

$m < 6$; {the directed numbers less than 6}

9. $5\left(3 + \frac{b}{5}\right) \geq 13$

$b \geq -2$; {−2 and the directed numbers greater than −2}

11-6 *The Multiplication Property of Inequality*

TEACHING THESE PAGES

We now state the multiplication property of inequality. Point out that when a positive number is used as a multiplier, the result is similar to the result of multiplying the members of an equation by some number. That is, if $a < b$, and c is positive:

$$a \cdot c < b \cdot c$$

and if $a = b$, and c is any directed number:

$$a \cdot c = b \cdot c$$

However, the use of a negative number as a multiplier reverses the direction of the inequality. If $a > b$ and d is negative:

$$a \cdot d < b \cdot d$$

To convince students that the sign of the inequality must be reversed when the multiplier is negative, work several examples like the following on the board:

(1) $\quad 10 > 5$

$-2(10) < -2(5)$

$-20 < -10$

(2) $\quad 7 > -1$

$-5(7) < -5(-1)$

$-35 < 5$

(3) $\quad -5 < 0$

$-8(-5) > -8(0)$

$40 > 0$

(4) $\quad -12 < -4$

$-6(-12) > -6(-4)$

$72 > 24$

When both members are multiplied by 0, the products are 0, and the inequality becomes the rather dull equality $0 = 0$.

CHALKBOARD EXAMPLES

Use the multiplication property of inequality to change the first inequality into the second.

1. $-3 > -5$; $-9 > -15$

$3(-3) > 3(-5)$

$-9 > -15$

2. $12 \leq 18$; $6 \leq 9$

$\frac{1}{2}(12) \leq \frac{1}{2}(18)$

$6 \leq 9$

3. $-3 \leq 5$; $9 \geq -15$

$-3(-3) \geq -3(5)$

$9 \geq -15$

4. $12 > -18$; $6 > -9$

$\frac{1}{2}(12) > \frac{1}{2}(-18)$

$6 > -9$

Solve. Then graph the solution set.

5. $-11v \leq 22$

$v \geq -2$

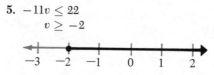

6. $5d > -15$

$d > -3$

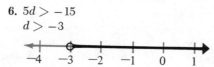

Solve each inequality and write the solution set.

7. $3y \geq 9$

$y \geq 3$;

{3 and the directed numbers greater than 3}

8. $-6n + 12 > n$

$n < 3$

{the directed numbers less than 3}

9. $2p + 3 < 6 + p$

$p < 3$

{the directed numbers less than 3}

10. $-10\left(\frac{1}{5} - \frac{x}{5}\right) \leq 3x$

$x \geq -2$

{−2 and the directed numbers greater than −2}

12 Addition and Subtraction of Polynomials

CHAPTER OVERVIEW

The chapter opens with lessons on what constitutes a polynomial and the standard form of a polynomial. After the development of addition of polynomials and a discussion of opposites of polynomials, subtraction is easily accomplished by adding the "opposite," or the additive inverse, of the polynomial to be subtracted.

12-1 *Polynomials*

TEACHING THESE PAGES

Make it clear that the name "polynomial" refers to any expression made up of *one or more* terms. The terms of a polynomial are separated by plus and/or minus signs. Special names—monomial, binomial, trinomial—are used for polynomials of one, two, and three terms, respectively. If a polynomial has more than three terms, we shall generally simply use the name "polynomial," or, if we wish to be more specific, "a polynomial with four terms" (or however many terms there may be).

CHALKBOARD EXAMPLES

Tell whether the polynomial is a monomial, a binomial, or a trinomial.

1. $5x^2 + y + z$ trinomial   2. $5x^2yz$ monomial
3. $5x^2 - yz$ binomial

Tell whether the right member is a monomial, a binomial, or a trinomial.

4. $P = 2l + 2w$ binomial
5. $N = x^2 - 3x - 28$ trinomial
6. $A = \frac{1}{2}lw$ monomial

Use the symbols 5, n^2, x, and x^3 to write a polynomial:

7. with four terms
   For example: $5n^2 + n^2x - 5x + 5x^3$
8. with six terms
   For example: $5x^3 - n^2x^3 + n^2x - 5x + 5n^2 + 5$

12-2 *Standard Form*

TEACHING THESE PAGES

Point out that the commutativity property of addition assures us that a polynomial is the same regardless of the order in which its terms are written. However, some operations with polynomials are performed more neatly when the terms are ordered in some way. This is especially true of division (discussed in Chapter 13). The standard form of a polynomial, then, is simply a convenient method for writing the polynomial. After a while, students may come to write polynomials in standard form automatically.

Note that, in rewriting an expression:

(1) When a term preceded by a plus sign is moved to become the first term, the plus sign is dropped.
(2) A positive first term which is moved to another position must be preceded by a plus sign.

CHALKBOARD EXAMPLES

Write in standard form.

1. $5y^2 - 3 - 7y^4$ $-7y^4 + 5y^2 - 3$
2. $4n^2 - 5 - n^5 + n^3$ $-n^5 + n^3 + 4n^2 - 5$
3. $m^2 + 15 - 6m$ $m^2 - 6m + 15$
4. $7c^6 - 3c^7 + c^4$ $-3c^7 + 7c^6 + c^4$
5. $x^2 - 6t^3 + 4xt^2 - 3$
    $x^2 + 4xt^2 - 6t^3 - 3$ (or $-6t^3 + 4xt^2 + x^2 - 3$)
6. $5r^2 - w^2 - 4rw$
    $5r^2 - 4rw - w^2$ (or $-w^2 - 4rw + 5r^2$)

Write in standard form. Insert any "missing" terms.

7. $z^4 + 5 - 3z^3$ $z^4 - 3z^3 + 0z^2 + 0z + 5$
8. $10 - 8a^3 - 5a^2$ $-8a^3 - 5a^2 + 0a + 10$

12-3 *Polynomials and Function Machines*

TEACHING THESE PAGES

Once again the function machine appears, this time to show that a function rule may be a polynomial. In addition to learning to recognize polynomials, the student also gains further practice in finding values of polynomials for given replacements of the variable.

CHALKBOARD EXAMPLES

Find the value of $f(a)$. Use the given replacement for a.

1. $f(a) = a^2 - 8a + 12$
   Let $a = -3$.
   $f(a) = 9 + 24 + 12$
       $= 45$

2. $33 - 15a + 9a^2 = f(a)$
   Let $a = \dfrac{1}{3}$.
   $f(a) = 33 - 5 + 1$
       $= 29$

3. $f(a) = (3a + 5) - (5a - 6)$
   Let $a = 0.3$.
   $f(a) = (0.9 + 5) - (1.5 - 6)$
       $= 5.9 - (-4.5) = 10.4$

Use the function equation and replacement set to write the function.

4. $f(n) = (3n^2 + 5) - n$; $\{-2, -1, 0, 1, 2\}$
   $\{(-2, 19), (-1, 9), (0, 5), (1, 7), (2, 15)\}$

5. $f(k) = 11 + 3k - k^3$; $\{-3, -1, 1, 3, 5\}$
   $\{(-3, 29), (-1, 9), (1, 13), (3, -7), (5, -99)\}$

12-4 *Addition of Polynomials*

TEACHING THESE PAGES

Students are given experience with both vertical and horizontal arrangements for adding polynomials. After they understand both forms, they may use the form they prefer.

An addition may be checked by substituting a value for the variable. Although the computation is simplest when the replacement for the variable is 1, this check is not as accurate as a check using other values for the variable, because every power of 1 is 1. However, in most cases, 1 may be used quite satisfactorily as a replacement, unless students have made the error of adding polynomials of different degrees together (as $3m^2 + 2m = 5m^2$, which checks correctly for $m = 1$).

CHALKBOARD EXAMPLES

Add.

1. $\begin{array}{r} 3a + 7b \\ 5a - 3b \\ \hline 8a + 4b \end{array}$
   2. $\begin{array}{r} c^2 - 6c \\ 3c^2 - 5c \\ \hline 4c^2 - 11c \end{array}$
   3. $\begin{array}{r} n^2 - 5ns + 6s^3 \\ n^2 \phantom{{}- 5ns} - 7s^3 \\ \hline 2n^2 - 5ns - s^3 \end{array}$

4. $\begin{array}{r} 8a + 7b - 3c \\ a - 3b \phantom{{}- 3c} \\ \hline 9a + 4b - 3c \end{array}$
   5. $\begin{array}{r} 1.3y^2 \phantom{{}+ 3xy} - 0.5x^2 \\ 0.7y^2 + 3xy + 4x^2 \\ \hline 2y^2 + 3xy + 3.5x^2 \end{array}$

6. $(z^2 + 5z - 7) + (3z^2 + 8 - 7z)$ $4z^2 - 2z + 1$

Add. Check by using $m = -1$, $n = 2$.

7. $\begin{array}{r} 6m - 3n - 5 \\ 4m + n - 9 \\ m + 5n + 8 \\ \hline 11m + 3n - 6 \end{array}$ $\begin{array}{rcl} 6(-1) - 3(2) - 5 &=& -17 \\ 4(-1) + 1(2) - 9 &=& -11 \\ 1(-1) + 5(2) + 8 &=& 17 \\ \hline 11(-1) + 3(2) - 6 &=& -11 \end{array}$

12-5 *Addition Properties*

TEACHING THESE PAGES

The assumption that addition of polynomials is both commutative and associative should be readily accepted. In fact, it might be difficult to persuade students *not* to make the assumption. The exercises, which involve demonstrations of the addition properties, also provide practice in combining terms and adding polynomials.

CHALKBOARD EXAMPLES

Simplify both members to show that they are equal.

1. $(3x - 5) + (x + 7) = (x + 7) + (3x - 5)$

$3x - 5 + x + 7$	$x + 7 + 3x - 5$
$3x + x - 5 + 7$	$x + 3x + 7 - 5$
$4x + 2$	$4x + 2$ $\checkmark$

2. $(7a + 4 + 5a) - 3 = 7a + (4 + 5a - 3)$

$(7a + 5a + 4) - 3$	$7a + (4 - 3 + 5a)$
$12a + 4 - 3$	$7a + 1 + 5a$
$12a + 1$	$7a + 5a + 1$
$12a + 1$	$12a + 1$ ✓

3. $(5n - 9) + (4n - 6) = 5n + [-9 + (4n - 6)]$

$5n - 9 + 4n - 6$	$5n + [-9 + 4n - 6]$
$5n + 4n - 9 - 6$	$5n + [4n - 15]$
$9n - 15$	$5n + 4n - 15$
$9n - 15$	$9n - 15$ ✓

Find both sums. Compare.

4.
$$\begin{array}{cc} 5a + 3 & a - 10 \\ a - 10 & 5a + 3 \\ \hline 6a - 7 & 6a - 7 \end{array}$$
The sums are the same.

5.
$$\begin{array}{cc} 6n + 8 & 6n + 8 \\ -7n & 3n - 11 \\ 3n - 11 & -7n \\ \hline 2n - 3 & 2n - 3 \end{array}$$
The sums are the same.

Find the value. Use $a = 2$, $b = 0$, $y = -1$, $x = 3$.

6. $-3y^2 + 4y - 3a$
$-3(-1)^2 + 4(-1) - 3(2) = -3 - 4 - 6 = -13$

7. $b^2 - y^2 + 2x^2$
$(0)^2 - (-1)^2 + 2(3)^2 = 0 - 1 + 18 = 17$

12-6 *Polynomials and Their Opposites*

TEACHING THESE PAGES

We can extend three other addition properties of numbers to addition of polynomials:

(1) Zero is the identity element of addition for polynomials.
(2) Every polynomial has an opposite, or additive inverse.
(3) The opposite of a polynomial is the sum of the opposites of its terms (from the property of the opposite of a sum).

This lesson provides practice in applying such properties and prepares the way for subtraction of polynomials, which follows in Section 12-7.

CHALKBOARD EXAMPLES

Give the opposite. Write your answer in standard form.

1. $3a^2 - 5a + 7$ $-3a^2 + 5a - 7$
2. $-4x + 5x^3 - 3x^2 - 5$ $-5x^3 + 3x^2 + 4x + 5$
3. $-(-7n^4 + 8n - 2n^2 - 3)$ $-7n^4 - 2n^2 + 8n - 3$

Write in standard form without using parentheses.

4. $-\left(\dfrac{s}{5} - \dfrac{s^3}{2} + 15\right)$ $\dfrac{s^3}{2} - \dfrac{s}{5} - 15$

5. $-\left(-\dfrac{h^4}{5} - \dfrac{h^2}{3} - \dfrac{h^3}{6} + 4\right)$ $\dfrac{h^4}{5} + \dfrac{h^3}{6} + \dfrac{h^2}{3} - 4$

Add. First arrange your work in vertical form.

6. $(m^3 + 3m^2)$ and $-(6m^2 - 6m)$

$$\begin{array}{l} m^3 + 3m^2 \\ - 6m^2 + 6m \\ \hline m^3 - 3m^2 + 6m \end{array}$$

7. $-(3x^2 - 5xy + y^2)$ and $-(4xy - 6y^2 - x^2)$

$$\begin{array}{l} -3x^2 + 5xy - y^2 \\ x^2 - 4xy + 6y^2 \\ \hline -2x^2 + xy + 5y^2 \end{array}$$

Write in standard form without grouping symbols.

8. $5 - (3c + 7) + (c^3 - 4c^2 - 3c)$
$5 - 3c - 7 + c^3 - 4c^2 - 3c = c^3 - 4c^2 - 6c - 2$

12-7 *Subtraction with Polynomials*

TEACHING THESE PAGES

Since we have already established subtraction as "adding the opposite," and students are now able both to add polynomials and name the opposite of a polynomial, it is only a short step further to subtraction of polynomials.

The simplest check for subtraction of polynomials is the same check we use for subtraction of numbers—to add the answer to the number subtracted and compare the result with the minuend:

Subtract:	Add:	Check:
$3x - 5y$	$3x - 5y$	$x - 4y$
$x - 4y$	$-x + 4y$	$2x - y$
	$2x - y$	$3x - 5y$ ✓

CHALKBOARD EXAMPLES

Subtract the second polynomial from the first. Check by addition.

1.
$$\begin{array}{l} 3a + 8b \\ a - 3b \\ \hline 2a + 11b \end{array}$$
Check:
$$\begin{array}{l} a - 3b \\ 2a + 11b \\ \hline 3a + 8b \end{array} ✓$$

2.
$$\begin{array}{l} 5x^2 + 3x \\ x^2 - 3x - 5 \\ \hline 4x^2 + 6x + 5 \end{array}$$
Check:
$$\begin{array}{l} x^2 - 3x - 5 \\ 4x^2 + 6x + 5 \\ \hline 5x^2 + 3x \end{array} ✓$$

3.
$$
\begin{array}{r}
4p + 7r - 9 \\
-\ p + 8r \\
\hline
5p - \ r - 9
\end{array}
$$

Check:
$$
\begin{array}{r}
-\ p + 8r \\
5p - \ r - 9 \\
\hline
4p + 7r - 9 \quad \checkmark
\end{array}
$$

Write the expression without parentheses. Do not combine similar terms.

4. $(y - 3) - (4y + 11)$ $y - 3 - 4y - 11$

5. $(-3x - 5n) - (x - 4n)$ $-3x - 5n - x + 4n$

Simplify. Write the answer in standard form.

6. $(3c^2 + 7) - (5 - 3c^2) = 3c^2 + 7 - 5 + 3c^2$
$$= 6c^2 + 2$$

12-8 Polynomials and Problem Solving

TEACHING THESE PAGES

In this lesson, further practice in the addition and subtraction of polynomials is provided, together with more experience in reading and interpreting word problems.

CHALKBOARD EXAMPLES

Write the answer as a polynomial in standard form.

1. Find the perimeter of a square if the length of each side is $11 - 3n$. $44 - 12n$

2. The length AC is $3m^3 + 5m^2 + 9$. If length BC is $5m^2 - 3$, find length AB. $3m^3 + 12$

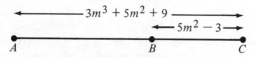

3. The area of square $ABCD$ is $4x^2 + 4x + 1$. The area of the shaded region is $3x^3 - 1$. Find the area of $AECD$. $-3x^3 + 4x^2 + 4x + 2$

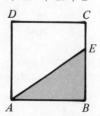

12-9 Polynomials and Solving Equations

TEACHING THESE PAGES

We have already developed skill in solving simple equations. We now use the simplification of a polyno-

mial expression as the first step in the approach toward solving slightly more difficult equations. Some of the equations in the exercises contain second degree terms, but, when simplified, these equations lead to linear equations.

CHALKBOARD EXAMPLES

Solve. Check your solution.

1.
$$
\begin{aligned}
(5x - 3) + (x - 7) &= 2 \\
5x - 3 + x - 7 &= 2 \\
5x + x - 3 - 7 &= 2 \\
6x - 10 &= 2 \\
x &= 2
\end{aligned}
$$

Check:
$$
\begin{aligned}
(5 \cdot 2 - 3) + (2 - 7) &\overset{?}{=} 2 \\
(10 - 3) + (-5) &\overset{?}{=} 2 \\
7 - 5 &\overset{?}{=} 2 \\
2 &= 2 \quad \checkmark
\end{aligned}
$$

2.
$$
\begin{aligned}
y + (y - 2) + (2y + 4) &= 6 \\
y + y - 2 + 2y + 4 &= 6 \\
y + y + 2y - 2 + 4 &= 6 \\
4y + 2 &= 6 \\
y &= 1
\end{aligned}
$$

Check:
$$
\begin{aligned}
1 + (1 - 2) + (2 \cdot 1 + 4) &\overset{?}{=} 6 \\
1 + (-1) + (2 + 4) &\overset{?}{=} 6 \\
1 - 1 + 6 &\overset{?}{=} 6 \\
6 &= 6 \quad \checkmark
\end{aligned}
$$

3.
$$
\begin{aligned}
11n - (7n + 3) &= 47 \\
11n - 7n - 3 &= 47 \\
4n - 3 &= 47 \\
4n &= 50 \\
n &= 12.5
\end{aligned}
$$

Check:
$$
\begin{aligned}
11(12.5) - (7 \cdot 12.5 + 3) &\overset{?}{=} 47 \\
137.5 - (87.5 + 3) &\overset{?}{=} 47 \\
137.5 - 90.5 &\overset{?}{=} 47 \\
47 &= 47 \quad \checkmark
\end{aligned}
$$

4.
$$
\begin{aligned}
43 &= (-z + 10) - (z - 5) \\
43 &= -z + 10 - z + 5 \\
43 &= -2z + 15 \\
28 &= -2z \\
-14 &= z
\end{aligned}
$$

Check:
$$
\begin{aligned}
43 &\overset{?}{=} [-(-14) + 10] - (-14 - 5) \\
43 &\overset{?}{=} (14 + 10) - (-19) \\
43 &\overset{?}{=} 24 + 19 \\
43 &= 43 \quad \checkmark
\end{aligned}
$$

13 Multiplication and Division of Polynomials

CHAPTER OVERVIEW

We begin by reviewing the use of a positive integral exponent to indicate the repetition of a factor in a product. Products of monomials lead to the product of a polynomial and a monomial, and then to the product of two polynomials. Two special products are introduced: (1) the square of a binomial, and (2) the product of the sum and the difference of the same two terms (the familiar "difference of two squares" pattern).

Division of polynomials is developed similarly, beginning with the quotient of two monomials, going on to division of a polynomial by a monomial, and then the division of a trinomial by a binomial. This last division corresponds to "long division" in arithmetic.

13-1 *Repeating Factors and Exponents*

TEACHING THESE PAGES

The lesson is a simple one, reviewing use of positive integral exponents as a short cut for writing repeating factors in a product. Point out that:

(1) xy^4 means $x \cdot y \cdot y \cdot y \cdot y$, but $(xy)^4$ means:

$$xy \cdot xy \cdot xy \cdot xy = x \cdot x \cdot x \cdot x \cdot y \cdot y \cdot y \cdot y$$

$$= x^4 y^4$$

(2) $m^3 \cdot m^2$ means $(m \cdot m \cdot m)(m \cdot m) = m^5$, or
$m^3 \cdot m^2 = m^{3+2} = m^5$.

In (2), note that the exponent of the product (5) is the sum of the exponents of the factors (3 and 2). Point out that when a multiplication involves two powers of the same factor, we can find the power of the product by adding the exponents of the factors. Remind the students that a means a^1 — the exponent 1 is not usually written. Thus $a \cdot a^3 = a^{1+3} = a^4$.

CHALKBOARD EXAMPLES

Give the value. Use $a = 2$ and $b = -3$.

1. $3ab^2$ 54   2. $(a \cdot b)^3$ -216 3. $b^2 \cdot a^3$ 72

Simplify.

4. $3 \cdot (-3)(n \cdot n \cdot n)$ $-9n^3$
5. $(-5)(6)(x^3 \cdot x)$ $-30x^4$
6. $7 \cdot 5 \cdot y^4 \cdot y^5$ $35y^9$
7. $-4 \cdot 5(p + 3)(p + 3)$ $-20(p + 3)^2$
8. $3^3 \cdot 3$ 3^4 or 81 9. $t \cdot t^3 \cdot t^4$ t^8

Find the correct replacement for x.

10. $5^7 = 5^x \cdot 5^5$ $x = 2$ 11. $n^x \cdot n^4 \cdot n^3 = n^{10}$ $x = 3$

13-2 *Products of Monomials*

TEACHING THESE PAGES

Here we use the commutative and associative properties of multiplication, together with the rule of exponents stated in the previous lesson, to simplify products of monomials.

CHALKBOARD EXAMPLES
Simplify.

1. $(3ab)(10ab^3)$ $30a^2b^4$
2. $(9n^3)(-7n^4)$ $-63n^7$
3. $(r)(-7r)(8s)$ $-56r^2s$
4. $(3xy)(-4yz)(-5y)$ $60xy^3z$
5. $(6cm)(-c)(-c^2m)$ $6c^4m^2$
6. $(3w)^2(tw)(5w)^2$ $225tw^5$
7. $\left(\frac{1}{4}n^3\right)\left(\frac{1}{3}ny\right)$ $\frac{1}{12}n^4y$
8. $\left(\frac{3}{4}mp\right)\left(\frac{1}{5}m^2\right)+\left(\frac{1}{2}p\right)\left(\frac{2}{5}m^3\right)$ $\frac{7}{20}m^3p$
9. $\left(\frac{2}{5}ad\right)\left(\frac{1}{3}a^3d\right)$ $\frac{2}{15}a^4d^2$
10. $(4x^3y)(-3xy^3)+(xy)(5x^3y^3)$ $-7x^4y^4$

13-3 *A Power of a Product*

TEACHING THESE PAGES
As we have seen, in simplifying a power of a product *each factor* of the product must be raised to the power. For example, $(5xy)^3$ means $(5xy)(5xy)(5xy)$, so:

$$(5xy)^3 = 5^3 \cdot x^3 \cdot y^3 = 125x^3y^3$$

When an expression that involves a power of one of its factors is raised to a power, such as $(5x^2y)^3$, the exponent of the factor (2) is multiplied by the power to which the product is raised (3). Thus:

$$(5x^2y)^3 = 5^3 \cdot x^{2\cdot3} \cdot y^3 = 125x^6y^3$$

CHALKBOARD EXAMPLES
Simplify.

1. $(5mn)^2$ $25m^2n^2$ 2. $(-2ab)^5$ $-32a^5b^5$
3. $(3abc)^3$ $27a^3b^3c^3$ 4. $(-y^3)^2$ y^6
5. $(-4y^4)^3$ $-64y^{12}$ 6. $(0\cdot10a)^3$ 0

Find the value. Use the given value of the variable.

7. $(3n^3)^2;\ n=-2$ 576
8. $(-3n^2)^3;\ n=-2$ -1728
9. $(0.5m^2)^3;\ m=10$ $125{,}000$
10. $(-4x^2)^3;\ x=\frac{1}{2}$ -1

Square the monomial.

11. $0.3x^2y^3$ $0.09x^4y^6$ 12. $-3x^3y^2$ $9x^6y^4$
13. $-10xyz$ $100x^2y^2z^2$

Simplify.

14. $(-a)(-3ab)^3+(9b)(2a^4b^2)=(-a)(-27a^3b^3)+$
$(18a^4b^3)=27a^4b^3+18a^4b^3=45a^4b^3$

13-4 *A Monomial Times a Polynomial*

TEACHING THESE PAGES
Multiplying a polynomial by a monomial is a direct application of the distributive property: $a(b+c)=ab+ac$. This lesson is a straightforward extension of the multiplication of monomials (Section 13-2), and should present no particular difficulty.

CHALKBOARD EXAMPLES
Multiply.

1. $3c(c^2-5c+3)$ $3c^3-15c^2+9c$
2. $(x^2+3xy-7y^2)(-x^3)$ $-x^5-3x^4y+7x^3y^2$

3.
$$\begin{array}{r} 4n^2-6n+5 \\ -0.5n \\ \hline -2n^3+3n^2-2.5n \end{array}$$

4.
$$\begin{array}{r} r^2-3s^2+5t^2 \\ 3r \\ \hline 3r^3-9rs^2+15rt^2 \end{array}$$

Simplify.

5. $b(3a-4b)+5(-ab)=3ab-4b^2-5ab=$
$-2ab-4b^2$

6. $-5m(m-2n)+n(2m-n)=-5m^2+10mn+$
$2mn-n^2=-5m^2+12mn-n^2$

7. $7xy(3x^2-xy)-5x^2(4xy+5y^2)=$
$21x^3y-7x^2y^2-20x^3y-25x^2y^2=$
$x^3y-32x^2y^2$

Multiply.

8. $8mn^2(5-mn^5-3m^4n^3-4n^8)=$
$40mn^2-8m^2n^7-24m^5n^5-32mn^{10}$

Solve.

9. $-4y+5(y-3)=-20$
$-4y+5y-15=-20$
$y=-5$

10. $5x+7(-x-1)=1$
$5x-7x-7=1$
$-2x=8$
$x=-4$

Express the area as a polynomial in simplest form.

11.

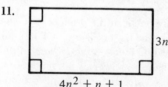

$A=lw=3n(4n^2+n+1)=12n^3+3n^2+3n$

13-5 *A Polynomial Times a Polynomial*

TEACHING THESE PAGES

Some students may be surprised to find in Example 1 on page 356 that they have been using the distributive property all along in multiplications in arithmetic.

Discuss the relative advantages of the horizontal and the vertical forms for multiplying polynomials. The vertical form most closely resembles multiplication in arithmetic, and is preferable for more complicated multiplications or when students are having difficulty combining like terms. The horizontal form is best suited for simpler multiplications and previews the form for the factorization of a trinomial. With practice, most students should be able to multiply two binomials mentally.

CHALKBOARD EXAMPLES

Multiply in two ways. Use both the vertical and the horizontal form.

1. $(x + 7)(x + 3) = x(x + 3) + 7(x + 3)$
   $= x^2 + 3x + 7x + 21$
   $= x^2 + 10x + 21$

$$\begin{array}{r} x + 7 \\ x + 3 \\ \hline 3x + 21 \\ x^2 + 7x \\ \hline x^2 + 10x + 21 \end{array}$$

Multiply. Use either form.

2. $(y + 5)(y - 5)$ $y^2 - 25$
3. $(a - 8)(a - 3)$ $a^2 - 11a + 24$
4. $(3z + 4)(5z - 7)$ $15z^2 - z - 28$
5. $(9 - m)(3 - 4m)$ $27 - 39m + 4m^2$

Multiply. Do it mentally if you can. Then substitute to check your answer, letting each variable equal 1.

6. $(x + 4)(4x + 3)$ $4x^2 + 19x + 12$
   Check: $(1 + 4)(4 \cdot 1 + 3) = 5(4 + 3) = 5(7) = 35$
   $4(1)^2 + 19(1) + 12 = 4(1) + 19 + 12 = 35$ ✓

7. $(3y - 5)(3y + 5)$ $9y^2 - 25$
   Check: $(3 \cdot 1 - 5)(3 \cdot 1 + 5) = (3 - 5)(3 + 5)$
   $= (-2)(8) = -16$
   $9(1)^2 - 25 = 9(1) - 25 = 9 - 25 = -16$ ✓

Multiply.

8. $(7 + 3m)(1 - 2m)$ $7 - 11m - 6m^2$
9. $(3k^2 + 5m)(k^2 - 2m)$ $3k^4 - k^2m - 10m^2$
10. $(a^2 - ab + b^2)(a + b)$ $a^3 + b^3$
11. $(3 + n)(5 - n + 4n^2)$ $15 + 2n + 11n^2 + 4n^3$

Express the area as a polynomial in simplest form.

12. A rectangle of length $3n - 2$ and width $n + 1$.
    $3n^2 + n - 2$

13-6 *Special Polynomial Products*

TEACHING THESE PAGES

Two special products students should learn to compute mentally, whenever possible, are (1) the square of a binomial, such as $(r + 4)^2$, and (2) the product of the sum and the difference of the same two terms, such as $(m + 2)(m - 2)$. Point out that the pattern $(a + b)^2 = a^2 + 2ab + b^2$ holds for *all directed* numbers, so a separate pattern for $(a - b)^2$ is not needed.

CHALKBOARD EXAMPLES

Multiply mentally. Then check your work on paper.

1. $(x + 7)^2$ $x^2 + 14x + 49$
2. $(m + 4n)^2$ $m^2 + 8mn + 16n^2$
3. $(z - 3)^2$ $z^2 - 6z + 9$
4. $(5a + 7)(5a - 7)$ $25a^2 - 49$
5. $(s - 4t)(s + 4t)$ $s^2 - 16t^2$

Expand.

6. $(7a + 3bc)^2$ $49a^2 + 42abc + 9b^2c^2$
7. $(x^2 - 5)^2$ $x^4 - 10x^2 + 25$
8. $\left(3n - \dfrac{1}{3}\right)^2$ $9n^2 - 2n + \dfrac{1}{9}$

9. Write a polynomial in simplest form to represent the area of a square when the length of one side is $5n + 3$. $A = s^2 = (5n + 3)^2 = 25n^2 + 30n + 9$

10. The square of a whole number is 45 less than the square of the next greater whole number. What are the numbers?
    Let n = greater number; $n - 1$ = lesser number.
    $(n - 1)^2 + 45 = n^2$; $n = 23$; $n - 1 = 22$.

13-7 *Multiplication Properties for Polynomials*

TEACHING THESE PAGES

Not too surprisingly, the order of two polynomials in a multiplication does not affect the result, nor does the way in which three or more polynomials are grouped. In other words, multiplication of polynomials is both commutative and associative, so we are free to select the most efficient order and grouping for a given multiplication.

The distributive property has already been used extensively in multiplying polynomials. The identity element for multiplication of polynomials is 1.

Every polynomial has a reciprocal, or multiplicative inverse. The product of a polynomial and its multiplicative inverse is 1.

CHALKBOARD EXAMPLES

Name the reciprocal. Assume that no numerator or denominator is zero.

1. $\dfrac{c-3d}{2cd}$ $\dfrac{2cd}{c-3d}$ 2. $\dfrac{-3x}{5(a^2+1)}$ $\dfrac{5(a^2+1)}{-3x}$

3. $\dfrac{a^2-b^2}{a+b}$ $\dfrac{a+b}{a^2-b^2}$ or $\dfrac{1}{a-b}$

Show that the sentence is correct.

4. $xy[(x+3)(x-3)] = (x^2y+3xy)(x-3)$

$xy(x^2-9)$	$(x^2y+3xy)x-(x^2y+3xy)(3)$
x^3y-9xy	$x^3y+3x^2y-3x^2y-9xy$
x^3y-9xy	x^3y-9xy $\checkmark$

13-8 *Dividing Monomials*

TEACHING THESE PAGES

To divide monomials, students should understand the three-part rule of exponents on pages 368–369. Informal statements like "Subtract the lesser exponent from the greater" can make this rule easier for students to remember.

CHALKBOARD EXAMPLES

Simplify. Assume no denominator is 0.

1. $\dfrac{a^7}{a^4}$ a^3 2. $\dfrac{a^4}{a^7}$ $\dfrac{1}{a^3}$ 3. $\dfrac{a^5}{a^5}$ 1 4. $\dfrac{6x^{10}}{2x^3}$ $3x^7$

5. $\dfrac{6a^3b^5}{15ab^6}$ $\dfrac{2a^2}{5b}$ 6. $\dfrac{-24z^8}{18z^5}$ $\dfrac{-4z^3}{3}$

7. $\dfrac{(10mn^3)^2}{-4m^5n}$ $\dfrac{-25n^5}{m^3}$ 8. $\dfrac{10(ab^4)^3}{15b^5}$ $\dfrac{2a^3b^7}{3}$

13-9 *Dividing a Polynomial by a Monomial*

TEACHING THESE PAGES

Although Example 1 on page 371 shows two ways of dividing a polynomial by a monomial, in general the second method proves to be more useful. Point out that $(6x^2+18x) \div 3$ can also be written $\frac{1}{3}(6x^2+18x)$. In other words, division as well as multiplication is distributive over addition.

CHALKBOARD EXAMPLES

Simplify. Assume no denominator is 0.

1. $\dfrac{1}{4}(24c-32)$ $6c-8$

2. $\dfrac{-3}{5}(10n+15)$ $-6n-9$

3. $(-6x^2+3x+21)\dfrac{1}{3}$ $-2x^2+x+7$

4. $(an^3-3n^4)\dfrac{1}{n^2}$ $an-3n^2$

Divide. Assume no denominator is 0.

5. $\dfrac{7z-42}{7}$ $z-6$ 6. $\dfrac{18m+45m^2}{9m}$ $2+5m$

7. $\dfrac{4x^6+8x^8}{x^5}$ $4x+8x^3$

8. $\dfrac{10a^3-15a^2+35a}{-5a}$ $-2a^2+3a-7$

13-10 *Dividing by a Binomial*

TEACHING THESE PAGES

The "long division" type of process used for division by a binomial demonstrates the value of writing a polynomial in standard form.

Finding the product of the divisor and the quotient is the time-honored check for division in arithmetic. We may use the same check for division of polynomials.

CHALKBOARD EXAMPLES

Divide and check.

1. $\dfrac{y^2+11y+18}{y+9}$ $y+2$

Check: $(y+9)(y+2) = y^2+11y+18$ $\checkmark$

2. $\dfrac{a^2-11a+24}{a-3}$ $a-8$

Check: $(a-3)(a-8) = a^2-11a+24$ $\checkmark$

3. $\dfrac{3n^2+16n+5}{3n+1}$ $n+5$

Check: $(3n+1)(n+5) = 3n^2+16n+5$ $\checkmark$

4. $\dfrac{4c^2-9}{2c-3}$ $2c+3$

Check: $(2c-3)(2c+3) = 4c^2-9$ $\checkmark$

Divide. First rewrite the polynomial in standard form.

5. $\dfrac{2b^2-25x^2+5bx}{b+5x}$ $2b-5x$

$$b+5x\overline{)2b^2+5bx-25x^2}$$
$$\underline{2b^2+10bx}$$
$$-5bx-25x^2$$
$$\underline{-5bx-25x^2}$$

14 Products and Factoring

CHAPTER OVERVIEW

Factoring can be thought of as the reverse of multiplication. In factoring, we look at the result of a multiplication and figure out all the prime expressions that were multiplied together to arrive at that result.

In this chapter, we begin by finding the prime factors of monomials, after which we factor selected polynomials (1) by using the GCF to remove common factors, where possible; (2) by recognizing products as the "difference of two squares;" (3) by identifying a trinomial as the square of a binomial; (4) by identifying trinomials as products of sums or differences of two binomials; and finally, (5) by recognizing trinomials as products of binomial sums and binomial differences.

14-1 *Factoring Whole Numbers*

TEACHING THESE PAGES

Point out, that by our definition, the number 1 is not a prime, because a prime is a number that has exactly two *different* factors. The factors we are dealing with here are all positive whole numbers. It is true that $(-2)(-3) = 6$ and $\frac{1}{2}(12) = 6$, but when we speak of the factors of 6, we mean the positive integral factors, 1, 2, 3, and 6. Prime factorization consists of expressing a whole number greater than 1 as a product of prime numbers. We use exponents to indicate repetitions of factors. Thus:

$$27 = 3 \cdot 3 \cdot 3 = 3^3$$
$$24 = 2 \cdot 2 \cdot 2 \cdot 3 = 2^3 \cdot 3$$

The prime factorization for any given number is unique. (This is called the Fundamental Theorem of Arithmetic.)

CHALKBOARD EXAMPLES

Name all the factors. Then express the number as a prime or as the product of primes.

1. 6 $1, 2, 3, 6$; $2 \cdot 3$
2. 63 $1, 3, 7, 9, 21, 63$; $3^2 \cdot 7$ **3.** 7 $1, 7$; prime

Complete the prime factorization.

4. $56 = 2 \cdot 2 \cdot 2 \cdot \underline{7} = \underline{?}$ $2^3 \cdot 7$
5. $26 = 2 \cdot \underline{?} = \underline{?}$ $2 \cdot 13$

Write the prime factorization.

6. 126 $2 \cdot 3^2 \cdot 7$ **7.** 333 $3^2 \cdot 37$ **8.** 90 $2 \cdot 3^2 \cdot 5$

14-2 *Factoring Monomials*

TEACHING THESE PAGES

We consider how to find the GCF (greatest common factor) of two monomials by comparing their complete factorizations. The GCF is the product of those factors which occur in *both* factorizations. For example:

$$36a^2 = \boxed{2 \cdot 2} \cdot 3 \cdot \boxed{3 \cdot a} \cdot a$$
$$48a = 2 \cdot \boxed{2 \cdot 2} \cdot 2 \cdot \boxed{3 \cdot a}$$

Thus, $36a^2 = 12a \cdot 3a$ and $48a = 12a \cdot 4$. The GCF of $36a^2$ and $48a$ is $12a$.

CHALKBOARD EXAMPLES

Complete.

1. $39n^2 = 3n(\underline{\ ?\ })$ $13n$
2. $-30a^3b = -2ab(\underline{\ ?\ })$ $15a^2$
3. $18x^3z^3 = 3x^3(\underline{\ ?\ })$ $6z^3$
4. $18x^3z^3 = -2z(\underline{\ ?\ })$ $-9x^3z^2$

Write two different factorizations.

5. $-15r^3s$ For example: $-15 \cdot r^3s$; $-5r \cdot 3r^2s$
6. $60\,mn^4$ For example: $5m \cdot 12n^4$; $-3n \cdot -20mn^3$

Write the complete factorization of each monomial. Then name the GCF.

7. $3x^3$ and $12x^2y^3$
$3x^3 = 3 \cdot x \cdot x \cdot x$
$12x^2y^3 = 2^2 \cdot 3 \cdot x \cdot x \cdot y \cdot y \cdot y$
GCF: $3x^2$

8. $24a$ and $6ab$
$24a = 2^3 \cdot 3 \cdot a$
$6ab = 2 \cdot 3 \cdot a$
GCF: $6a$

Name the GCF.

9. $32c^2d^4$ and $24c^3d$ $8c^2d$
10. $3x$, $4xy^3$, and $12xy$ x

SUGGESTED EXTENSION

Two or more expressions are *relatively prime* when their GCF is 1. Tell which of these pairs are relatively prime expressions. If they are not relatively prime, name the GCF.

1. $13a$ and $11a$ GCF: a
2. $5a^2$ and $3b^2$ relatively prime
3. $13a$ and $11b$ relatively prime
4. $6x^2y$ and $5xy^3$ GCF: xy
5. $14a$ and $12b$ GCF: 2
6. $8m^3$ and $7n^3$ relatively prime

14-3 *Factoring Polynomials*

TEACHING THESE PAGES

When a polynomial is to be factored, the first step is to identify the GCF of its terms. If the GCF is other than 1, write the polynomial as the product of the GCF and another polynomial. This is sometimes called "factoring out" the GCF. After the GCF has been factored out, it may or may not be possible to factor the polynomial further. The object of this lesson, however, is to identify the GCF and write the corresponding product of the GCF and the polynomial found by division.

CHALKBOARD EXAMPLES

Complete.

1. $15a^2 - 20ab = 5a(\underline{\ ?\ })$ $3a - 4b$
2. $28m^3n + 35mn^3 = 7mn(\underline{\ ?\ })$ $4m^2 + 5n^2$

Factor and check.

3. $6x^2 + 9xy$ $3x(2x + 3y)$
4. $5a^2 - 40b^2$ $5(a^2 - 8b^2)$
5. $8bcd - bc$ $bc(8d - 1)$
6. $4z^2 + 12az^2$ $4z^2(1 + 3a)$

Complete.

7. $ym^2 - 9xym + 16ym = ym(\underline{\ ?\ })$ $m - 9x + 16$
8. $18a^2 + 27a^2b - 54a^2b^2 = 9a^2(\underline{\ ?\ })$ $2 + 3b - 6b^2$

Factor.

9. $5n^3 - 15n^2 + n$ $n(5n^2 - 15n + 1)$
10. $3x^4 + 21x^3 - 51x^2$ $3x^2(x^2 + 7x - 17)$

14-4 *Factoring Polynomials by Grouping Terms*

TEACHING THESE PAGES

Sometimes there is no single common monomial factor for all terms of a polynomial, but the terms can be grouped so that each group of terms will have a common factor. For example:

$$2a^2 + 3a + 2ab + 3b = a(2a + 3) + b(2a + 3)$$
$$= (a + b)(2a + 3)$$

The same polynomial could also be factored as follows:

$$2a^2 + 3a + 2ab + 3b = 2a^2 + 2ab + 3a + 3b$$
$$= 2a(a + b) + 3(a + b)$$
$$= (2a + 3)(a + b)$$

Since multiplication of polynomials is commutative, the results of the two methods are equivalent.

Point out to the students that not every polynomial can be factored by grouping the terms, but it is one procedure to try.

CHALKBOARD EXAMPLES

Write the factored form.

1. $a(b + 5) + b(b + 5)$ $(a + b)(b + 5)$
2. $n(m - 4) - 3(m - 4)$ $(n - 3)(m - 4)$
3. $(3x - y)2a + (3x - y)b$ $(2a + b)(3x - y)$
4. $5w(a - 2) - 4(a - 2)$ $(5w - 4)(a - 2)$

Factor.

5. $(3rs + 5s) - (6rt + 10t) = s(3r + 5) - 2t(3r + 5)$
$$= (s - 2t)(3r + 5)$$

6. $(3a^2 - 9ab) + (a - 3b) = 3a(a - 3b) + 1(a - 3b)$
$$= (3a + 1)(a - 3b)$$

7. $m^2 + 4m - 5my - 20y = m(m + 4) - 5y(m + 4)$
$$= (m - 5y)(m + 4)$$

14-5 *Difference of Two Squares*

TEACHING THESE PAGES

One of two types of special products discussed in Section 13-6 was the product of the sum and the difference of the same two terms. In this lesson we reverse the process and recognize that a polynomial that is the difference of two squares must be the product of the sum and the difference of the same two terms. Such polynomials are easily factored:

$$a^2 - b^2 = (a + b)(a - b)$$

CHALKBOARD EXAMPLES

Complete.

1. $c^2 - 81 = (c + \underline{?})(c - \underline{?})$ 9
2. $25 - a^2 = (\underline{?} + a)(\underline{?} - a)$ 5, 5

Factor.

3. $n^2 - 36$ $(n + 6)(n - 6)$
4. $a^2b^2 - 4$ $(ab + 2)(ab - 2)$
5. $k^2 - 144$ $(k + 12)(k - 12)$
6. $4s^2 - 121$ $(2s + 11)(2s - 11)$
7. $100 - y^4$ $(10 + y^2)(10 - y^2)$
8. $81 - c^2d^4$ $(9 + cd^2)(9 - cd^2)$

Write as the difference of two squares. Then factor.

9. $-n^2 + a^4$ $a^4 - n^2 = (a^2 + n)(a^2 - n)$
10. $-4s^2 + 1$ $1 - 4s^2 = (1 + 2s)(1 - 2s)$
11. $-25 + b^2$ $b^2 - 25 = (b + 5)(b - 5)$

Factor completely.

12. $2a^2 - 8$ $2(a^2 - 4) = 2(a + 2)(a - 2)$

13. $5x^2 - 20$ $5(x^2 - 4) = 5(x + 2)(x - 2)$

14. $144 - 4n^2$
    $4(36 - n^4) = 4(6 + n^2)(6 - n^2)$ or
    $(12 - 2n)(12 + 2n) = 2 \cdot 2(6 - n)(6 + n) =$
    $4(6 - n)(6 + n)$

SUGGESTED EXTENSION

Note that when we factor $a^4 - b^4$ we get:

$$(a^2 + b^2)(a^2 - b^2)$$

As we already know:

$$(a^2 - b^2) = (a + b)(a - b)$$

So we can write $a^4 - b^4 = (a^2 + b^2)(a + b)(a - b)$. For this type of repeated factoring to be possible, exponents must be powers of 2, such as 4, 8, and 16, and the coefficients must be "squares of squares." Here are some suggested exercises:

Factor completely.

1. $x^4 - 16$ $(x^2 + 4)(x + 2)(x - 2)$
2. $16n^4 - 1$ $(4n^2 + 1)(2n + 1)(2n - 1)$
3. $a^4b^4 - 81$ $(a^2b^2 + 9)(ab + 3)(ab - 3)$
4. $n^8 - 1$ $(n^4 + 1)(n^2 + 1)(n + 1)(n - 1)$
5. $4c^4 - 400$ $4(c^2 + 10)(c^2 - 10)$
6. $8y^5 - 8y$ $8y(y^2 + 1)(y + 1)(y - 1)$

14-6 *Factoring Trinomial Squares*

TEACHING THESE PAGES

This lesson deals with squares of binomial *sums*, with squares of binomial differences reserved for the next lesson. Students should learn to recognize a trinomial as the square of a binomial sum by the following features:

(1) All three terms are positive.
(2) The first and the last term are squares.
(3) The middle term is twice the product of the square roots of the first and the last terms.

A quick mental check by squaring the binomial is always in order.

CHALKBOARD EXAMPLES

Complete.

1. $9a^2 = (\underline{?})^2$ $3a$ 2. $4n^2 = (\underline{?})^2$ $2n$
3. $121 = (\underline{?})^2$ 11

Name the missing term.

4. $x^2 + \underline{?} + 36 = (x + 6)^2$ $12x$
5. $49a^2 + \underline{?} + 1 = (7a + 1)^2$ $14a$

Expand.

6. $(c + 3d)^2$ $c^2 + 6cd + 9d^2$
7. $(5b + a)^2$ $25b^2 + 10ab + a^2$

Factor. Check by multiplying.

8. $a^2 + 2ab + b^2$ $(a + b)^2$
9. $n^2 + 14n + 49$ $(n + 7)^2$
10. $25x^2 + 30x + 9$ $(5x + 3)^2$
11. $9m^2 + 6m + 1$ $(3m + 1)^2$

14-7 *Square of a Binomial Difference*

TEACHING THESE PAGES

This lesson should present little difficulty, with Section 14-6 just completed. The square of a binomial difference always has a negative middle term, but otherwise, it resembles the square of a binomial sum:

(1) The first and the last terms are squares.
(2) The middle term is twice the product of the square roots of the first and third terms.

The same quick check by squaring the binomial mentally is recommended here, too.

CHALKBOARD EXAMPLES
Simplify.

1. $(-a)(-a)$ a^2 2. $(-3x)(-3x)$ $9x^2$
3. $(-5xy)(-5xy)$ $25x^2y^2$

Complete with positive factors and then with negative factors.

4. $x^2z^2 = \underset{xz \cdot xz}{\underline{?} \cdot \underline{?}} = \underset{-xz \cdot -xz}{\underline{?} \cdot \underline{?}}$
5. $16a^2 = \underset{4a \cdot 4a}{\underline{?} \cdot \underline{?}} = \underset{-4a \cdot -4a}{\underline{?} \cdot \underline{?}}$

Expand.

6. $(3n - a)^2$ $9n^2 - 6an + a^2$
7. $(5 - m)^2$ $25 - 10m + m^2$
8. $(2c - 3d)^2$ $4c^2 - 12cd + 9d^2$

Factor. Check by multiplication.

9. $s^2 - 6s + 9$ $(s - 3)^2$
10. $x^2 - 2x + 1$ $(x - 1)^2$
11. $4 - 12m + 9m^2$ $(2 - 3m)^2$
12. $a^2 - 14a + 49$ $(a - 7)^2$

SUGGESTED EXTENSION
Combine the ideas of Sections 14-5, 14-6, and 14-7 to factor expressions like:

$$a^2 + 6a + 9 - b^2 = (a + 3)^2 - b^2$$
$$= (a + 3 + b)(a + 3 - b), \text{ or}$$
$$(a + b + 3)(a - b + 3)$$

Have students factor polynomials like the following:

1. $n^2 - 10n + 25 - 4p^2$ $(n - 5 + 2p)(n - 5 - 2p)$
2. $25x^2 + 10x + 1 - 9y^2$
   $(5x + 1 + 3y)(5x + 1 - 3y)$
3. $9a^2 + 12ab + 4b^2 - 25$
   $(3a + 2b + 5)(3a + 2b - 5)$
4. $4c^2 - 20cd + 25d^2 - 1$
   $(2c - 5d + 1)(2c - 5d - 1)$
5. $16m^2 - 9a^2 - 12ab - 4b^2$
   $(4m + 3a + 2b)(4m - 3a - 2b)$

14-8 Product of Binomial Sums or Differences

TEACHING THESE PAGES
Point out that in a product of two binomial sums the sign of each term is positive while in a product of two binomial differences the first and last terms are positive, while the middle term is negative. This gives us a hint of what to look for in factoring trinomials with positive first and last terms. The sign of the middle term tells us whether the factors are sums or differences. As with all factoring, a final mental check by multiplication is strongly recommended.

CHALKBOARD EXAMPLES
Complete

1. $x^2 + 11x + 28 = (x + \underset{7}{\underline{?}})(x + \underset{4}{\underline{?}})$
2. $n^2 - 5n + 4 = (n - \underset{4}{\underline{?}})(n - \underset{1}{\underline{?}})$

Factor.

3. $t^2 - 11t + 24$ $(t - 8)(t - 3)$
4. $k^2 + 10k + 9$ $(k + 1)(k + 9)$
5. $18 + 11w + w^2$ $(9 + w)(2 + w)$
6. $x^2 - 9x + 14$ $(x - 7)(x - 2)$
7. $-8y + 7 + y^2$ $(y - 7)(y - 1)$
8. $12ab + a^2 + 11b^2$ $(11b + a)(b + a)$

14-9 More About Factoring Trinomials

TEACHING THESE PAGES
This final lesson of *Elementary Algebra, Part 1*, considers the factors of a trinomial with a negative third term. Obviously, if such a trinomial is the product of two binomials, one must be a sum and the other a difference. A mental check is very desirable here, to be sure that the factors selected do indeed have the given trinomial as a product.

CHALKBOARD EXAMPLES
Complete and check.

1. $x^2 + 5x - 24 = (x + \underset{8}{\underline{?}})(x - \underset{3}{\underline{?}})$
2. $n^2 + 10n - 24 = (n + \underset{12}{\underline{?}})(n - \underset{2}{\underline{?}})$
3. $s^2 - 2s - 24 = (s + \underset{4}{\underline{?}})(s - \underset{6}{\underline{?}})$

Factor.

4. $q^2 - q - 72$ $(q - 9)(q + 8)$
5. $b^2 + 3b - 18$ $(b + 6)(b - 3)$
6. $n^2 + 11n - 26$ $(n + 13)(n - 2)$
7. $s^2 - 2s - 15$ $(s - 5)(s + 3)$
8. $a^2 - 6ab - 27b^2$ $(a - 9b)(a + 3b)$
9. $x^2 + 11xy - 80y^2$ $(x + 16)(x - 5)$

Write in standard form and factor.

10. $4az - 45z^2 + a^2$
    $a^2 + 4az - 45z^2 = (a + 9z)(a - 5z)$

Richard A. Denholm

Robert G. Underhill

Mary P. Dolciani

EDITORIAL ADVISERS

Andrew M. Gleason

Albert E. Meder, Jr.

Elementary Algebra

New Edition

Part 1

HOUGHTON MIFFLIN COMPANY/BOSTON

Atlanta   Dallas   Geneva, Ill.   Lawrenceville, N.J.   Palo Alto   Toronto

ABOUT THE AUTHORS

Richard A. Denholm, Supervisor, Office of Teacher Education and Lecturer, Department of Mathematics, University of California at Irvine. Dr. Denholm has also served as Director of Curriculum for grades K-12, Orange County, California, and prior to that was Coordinator of Mathematics and Physical Science Instruction. In addition to his extensive authorship experience in mathematics texts and professional journals, he has made active contributions to the fields of mathematics curriculum planning, teacher preparation, and criteria reference testing.

Robert G. Underhill, Professor, Department of Curriculum and Instruction, College of Education, Kansas State University. Dr. Underhill has had extensive teaching experience in high school and university mathematics and teacher education. He has authored teacher-education books and audio-visual materials. In addition, he has contributed to professional journals and has been a speaker and presider at professional conferences.

Mary P. Dolciani, Professor of Mathematics, Hunter College, City University of New York. Dr. Dolciani has been a director and teacher in numerous National Science Foundation and New York State Education Department institutes for mathematics teachers. She was also Visiting Secondary School Lecturer for the Mathematical Association of America and a member of the United States Commission on Mathematical Instruction.

ABOUT THE EDITORIAL ADVISERS

Andrew M. Gleason, Hollis Professor of Mathematics and Natural Philosophy, Harvard University.

Albert E. Meder, Jr., Dean and Vice Provost and Professor of Mathematics, Emeritus, Rutgers, The State University of New Jersey.

Copyright © 1988, 1985, 1983, 1980, 1977 by Houghton Mifflin Company
All rights reserved. No part of this work may be reproduced or transmitted in any form or by any means, electronic or mechanical, including photocopying and recording, or by any information storage or retrieval system, except as may be expressly permitted by the 1976 Copyright Act or in writing by the Publisher. Requests for permission should be addressed in writing to: Permissions, Houghton Mifflin Company, One Beacon Street, Boston, Massachusetts 02108.

Printed in U.S.A.

ISBN: 0-395-43440-8

ABCDEFGHIJ-RM-8987

Contents

Subtraction; Functions

Features; Reviewing and Testing

10 *Multiplication and Division of Directed Numbers*   264

Multiplication

Division; Functions

Features; Reviewing and Testing

11 *Solving Equations and Inequalities*   292

Types of Equations

Multiplication of Polynomials

Division

Features; Reviewing and Testing

14 *Products and Factoring*   382

Factoring

Factoring Special Polynomials

Factoring Other Polynomials

Features; Reviewing and Testing

LIST OF SYMBOLS

		Page
=	equals (is equal to)	5
{ }	the set of	30
. . .	and so on	30
Ø	the empty set	30
$^-2$	negative two	34
$^+2$	positive two	34
π	pi	40
<	is less than	41
>	is greater than	41
(a,b)	ordered pair	47
$a:b$	ratio of a to b	68
%	percent	74
$\leftrightarrow$	is the same as	78

		Page
$\leq$	is less than or equal to	143
$\geq$	is greater than or equal to	143
$f(x)$	the value of the function f for x	175
$\neq$	does not equal	221
-3	the opposite of three	240
$\overset{?}{=}$	does it equal?	243
$|x|$	absolute value of x	261
P(e)	probability of event e	288
a^n	a to the nth power	346
$\sim$	is similar to	404
Δ	triangle	404
$\overline{PQ}$	line segment PQ	405
$\angle$	angle	405

LIST OF METRIC SYMBOLS

m	meter	g	gram
mm	millimeter	kg	kilogram
cm	centimeter	°C	degrees Celsius
km	kilometer	m^2	square meter

ABOUT THE CHAPTER OPENERS

There are two photographs at the beginning of each chapter. These pictures show some of the changes in a career or process which have taken place in the past century. Many of these changes are the result of developments in mathematics and technology.

Left: Pilots of 1913-model airplane.
This photo was taken just ten
years after the Wright Brothers'
flight.

Right: Pilot of modern jet.

1 Working with Integers

Numerical Expressions, Equations, Variables

1-1 *Numerical Expressions*

OBJECTIVE

Simplify expressions like $\dfrac{10 + 8}{9}$ and $5^2 - 10$.

A numerical expression may be very simple, like 15, or more complicated, like $\dfrac{10}{2 + 3}$. We may simplify an expression by doing the indicated operations.

EXAMPLE 1   Expression► $\underline{10 + 6} + 3$

Add (twice).► $\quad 16 \quad + 3 = \mathbf{19}$

EXAMPLE 2   Expression► $\dfrac{27 - 7}{2}$

Subtract. Then divide.► $\dfrac{20}{2} = \mathbf{10}$

EXAMPLE 3   Expression► $14 - 3^2$

Square. Then subtract.► $14 - 9 = \mathbf{5}$

EXAMPLE 4   Expression► $\dfrac{2 + (4 \times 2)}{5}$

Multiply.► $\dfrac{2 + 8}{5}$

Add.► $\dfrac{10}{5}$

Divide.► 2

Different expressions may name the same number. Each of these expressions names the number **3**.

EXAMPLE 5 $\dfrac{27}{5 + 4} = \dfrac{27}{9} = 3 \qquad \dfrac{3 \times 3 \times 3}{3^2} = \dfrac{27}{9} = 3$

$(10 \times 10) - 97 = 100 - 97 = 3$

Match to tell how to simplify.

1. $7 + 0$ c

2. $(5 \times 6) - 10$ A

3. $\dfrac{10}{2}$ E

4. $3^2 - 5$ B

5. $4 \times (3 + 6)$ D

A. Multiply 5 and 6. Subtract 10.
The result is 20.

B. Square 3. Subtract 5.
The result is 4.

C. Add 7 and 0.
The result is 7.

D. Add 3 and 6. Multiply by 4.
The result is 36.

E. Divide 10 by 2.
The result is 5.

Tell how to simplify.

Sample: $\dfrac{3 + 7}{5}$ *What you say:*   Add 3 and 7; the sum is 10.
Divide 10 by 5. The simplified form is 2.

6. $4 + 8 + 7$ 19   **7.** 4^2 16   **8.** $\dfrac{15}{5}$ 3

9. $46 - 9$ 37   **10.** 17×0 0   **11.** $\dfrac{32}{8}$ 4

12. $48 \div 6$ 8   **13.** 0×9 0   **14.** $\dfrac{23}{1}$ 23

For Extra Practice, see page 410.

Simplify.

Written
EXERCISES

1. $3 + (6 \times 7)$ 45   **2.** $4 \times (8 + 2)$ 40   **3.** $(4 \times 6) - 7$ 17

4. $4 \times 2 \times 9$ 72   **5.** 19×0 0   **6.** $25 - (2 \times 8)$ **A**

7. $100 \div 5$ 20   **8.** 1×37 37   **9.** 5^2 25 9

10. $50 + 50 + 50$ 150   **11.** $23 \div 1$ 23   **12.** $3^2 + 1$ 10

13. $\dfrac{45}{15}$ 3   **14.** $67 \div 67$ 1   **15.** $4 \times 4 \times 4$ 64

16. $10^2 + 6$ 106   **17.** $3^2 + 5^2$ 34   **18.** $426 \div 18$ $23\frac{2}{3}$ **B**

19. $\dfrac{14 + 11}{5}$ 5   **20.** $\dfrac{50}{6 + 19}$ 2   **21.** 1×7^2 49

22. $17 - 17$ 0   **23.** $4 - \dfrac{6}{2}$ 1   **24.** $\dfrac{48 + 12}{2^2 + 6}$ 6

Complete to name the given number.

$$\text{Sample:} \quad 5 \begin{cases} 5 \times \underline{\ ?\ } \\ 10 \div \underline{\ ?\ } \\ 17 - \underline{\ ?\ } \end{cases} \qquad \text{Solution:} \quad 5 \begin{cases} 5 \times 1 \\ 10 \div 2 \\ 17 - 12 \end{cases}$$

25. $40 \begin{cases} 1 \times 5 \times \underline{\ ?\ } \ 8 \\ 10 + \underline{\ ?\ } + 10 \ 20 \\ 2 \times 4 \times \underline{\ ?\ } \ 5 \end{cases}$

26. $64 \begin{cases} 7^2 + \underline{\ ?\ } \ 15 \\ 4 \times 4 \times \underline{\ ?\ } \ 4 \\ 50 + \underline{\ ?\ } + 3^2 \ 5 \end{cases}$

27. $18 \begin{cases} 3 \times 3 \times \underline{\ ?\ } \ 2 \\ 72 \div \underline{\ ?\ } \ 4 \\ 30 - \underline{\ ?\ } \ 12 \end{cases}$

28. $100 \begin{cases} 10 \times \underline{\ ?\ } \ 10 \\ \underline{\ ?\ } \div 1 \ 100 \\ 5 \times 5 \times \underline{\ ?\ } \ 4 \end{cases}$

29. $1 \begin{cases} 15 \div \underline{\ ?\ } \ 15 \\ 1 \times 1 \times \underline{\ ?\ } \ 1 \\ 0 + \underline{\ ?\ } \ 1 \end{cases}$

30. $75 \begin{cases} \underline{\ ?\ } + 75 \ 0 \\ 5^2 \times \underline{\ ?\ } \ 3 \\ 75 \times \underline{\ ?\ } \ 1 \end{cases}$

C

31. $9 \begin{cases} \dfrac{36}{?} \ 4 \\[2mm] \dfrac{50 + ?}{7} \ 13 \end{cases}$

32. $96 \begin{cases} 4^2 \times (5 + \underline{\ ?\ }) \ 1 \\ (\underline{\ ?\ } - 3^2) + 0 \ 105 \end{cases}$

33. $15 \begin{cases} \dfrac{?}{3} \ 45 \\[2mm] 3 + \underline{\ ?\ } + 3^2 \ 3 \end{cases}$

34. $38 \begin{cases} \underline{\ ?\ } + 1^2 \ 37 \\ 5^2 + (\underline{\ ?\ } + 3) \ 10 \end{cases}$

Christine Ladd-Franklin   1847–1930

Christine Ladd-Franklin received a degree from Vassar in 1869. Several of her articles on mathematics appeared in the *Educational Times* of Great Britain while she was teaching secondary school science. She attended Johns Hopkins University and later expanded her work in mathematics and logic to include the theory of color vision. At the age of 79 she wrote an article for the National Academy of Sciences on the "blue arcs" visual phenomenon, the origin of which is still not clearly understood.

1-2 *Expressions and Equations*

OBJECTIVES

Identify a statement as either true or false; $4 + 0 = 2 \times 2$ is **true**.

$18 - 6 = \dfrac{6}{2}$ is **false**.

Complete equations like $4 + 8 = 6 \times \underline{\ ?\ }$ to make true statements.

A number sentence that consists of two expressions joined by the "is equal to" symbol, $=$, is called an equation.

$$14 \times 3 = 6 \times 7$$
is equal to

$$15 + \underline{\ ?\ } = \dfrac{60}{3}$$
is equal to

An equation with no information missing makes a statement that is either true or false. The statement is true if both expressions name the same number. Otherwise it is false. If information is missing, the equation is neither true nor false.

True: $4 + 4 + 11 = 20 - 1$

$$19 = 19$$
same number

$$\dfrac{8 + 7}{3} = 2^2 + 1$$

$$5 = 5$$
same number

False: $15 + 7 = 12 \times 2$

$$22 = 24$$
not the same

$$\dfrac{15 - 3}{4} = 0 + 0$$

$$3 = 0$$
not the same

How can we make $18 - \underline{\ ?\ } = 4 + 9$ into a true statement? If we replace $\underline{\ ?\ }$ with 5, each expression names **13**. Then the statement is true.

$$18 - \underline{\ ?\ } = 4 + 9 \quad \blacktriangleright \quad 18 - 5 = 4 + 9$$
$$13 = 13$$

Tell why the statement is true or false.

Sample: $8 + 10 = 9 \times 2$ *What you say:*   True. Both $10 + 8$
and 9×2 name 18.

1. $2 + 2 = 4 \times 2$ False
2. $4 \times 4 \times 1 = 8 \times 3$ False
3. $10 - 7 = 3 + 0$ True
4. $8 \times 0 = 1 \times 8$ False
5. $10 \times 10 = 100$ True
6. $48 - 8 = 30 + 5$ False
7. $15 - 6 = 3 \times 3$ True
8. $\dfrac{15}{1} = \dfrac{15}{5}$ False
9. $\dfrac{6}{2} + 4 = 4 + 3$ True

Match expressions that name the same number.

10. $6 \times 7 \times 0$ D
11. $20 \div 5$ A
12. $\dfrac{12 + 6}{9}$ E
13. $6 + 7 + 0$ B
14. $10 + \dfrac{6}{3}$ C

A. 2^2
B. 13
C. $8 + 2^2$
D. $15 \times 0 \times 12$
E. $0 + 1 + 1$

Show whether each statement is true or false.

Sample: $4 \times 9 = 25 + 11$ *Solution:* $4 \times 9 = 36$ and
$25 + 11 = 36$; True

Sample: $18 + 7 = 14 + 10$ *Solution:* $18 + 7 = 25$ and
$14 + 10 = 24$; False

A

1. $12 + 6 = 6 \times 3$ True
2. $43 + 15 = 8 + 50$ True
3. $3 \times 4 \times 5 = 100 - 30$ False
4. $468 = 400 + 6 + 8$ False
5. $\dfrac{75}{3} = 10 + 10 + 10$ False
6. $1 \times 1 \times 0 = 1 + 1 + 0$ False
7. $9 \times 5 = 36 + 9$ True
8. $\dfrac{17}{17} = 7 - 5$ False
9. $\dfrac{16}{8} = \dfrac{18}{6}$ False
10. $4^2 = 1 + 3 + 5 + 7$ True

11. $10 + 3^2 = 4^2 + 8$ False

12. $2.5 \times 8 = 5 + 4$ False

13. $\dfrac{3 \times 6}{6 \times 3} = \dfrac{9 \times 2}{2 \times 9}$ True

14. $\dfrac{81}{9} + 11 = \dfrac{100}{5} + 0$ True

15. $\dfrac{80}{5} = 4 \times 4 \times 2$ False

16. $\dfrac{4 + 10}{7} = \dfrac{10 + 10}{4}$ False

Complete to make a true statement.

Sample: $35 + 4 = 6^2 + \underline{}$ *Solution:* $35 + 4 = 6^2 + 3$

17. $13 + 9 + 3 = 5 \times \underline{}$ 5

18. $3^2 = 1 + 3 + \underline{}$ 5

19. $\underline{} \times 18 \times 1 = 9 \times 4$ 2

20. $\dfrac{4 + 5 + 6}{3} = \underline{} + 0$ 5

21. $9^2 + \underline{} = 101$ 20

22. $6^2 = 11 + \underline{}$ 25

23. $145 + 35 = \underline{} \times 90$ 2

24. $\dfrac{48 + 204}{12} = \underline{}$ 21

25. $\dfrac{429}{3} + 7 = 5 \times 5 \times \underline{}$ 6

26. $\dfrac{1 + 4 + 5}{?} = 1$ 10

27. $73 - \underline{} = \dfrac{24 + 14}{2}$ 54

28. $5^2 - \underline{} = 5 + 3^2$ 11

29. $\underline{} + 52 = 10^2 \times 1$ 48

30. $(9 \times 2) - 8 = \dfrac{80}{?}$ 8

31. $16 - \underline{} = 3 \times 7 \times 0$ 16

32. $1^2 + 2^2 + 3^2 = \dfrac{?}{5}$ 70

Make a true statement. Use the same whole number to complete both expressions.

33. $4 + 6 + \underline{} = 6 \times \underline{}$ 2

34. $8 + 8 + \underline{} = 3 \times \underline{}$ 8

35. $2 \times \underline{} = 2 + \underline{}$ 2

36. $15 - \underline{} = \dfrac{50}{?}$ 10

37. $\underline{} \times 2^2 \times 2 = 21 + \underline{}$ 3

38. $10 + 5^2 + \underline{} = 6^2 \times \underline{}$ 1

39. $2 + 3 + \underline{} = \dfrac{50}{?}$ 5

40. $\dfrac{17 - ?}{5} = 5 - \underline{}$ 2

41. $\dfrac{? \times 6}{12} = \dfrac{?}{2}$ Any whole number

42. $3 + \underline{} = \dfrac{13 + ?}{2}$ 7

43. $\dfrac{9 - ?}{4} = 3 - \underline{}$ 1

44. $\underline{} + 3 = \dfrac{3 \times ?}{2}$ 6

1-3 *Expressions and Variables*

OBJECTIVE

Find the values of expressions that contain variables, using given replacements.

In algebra, letters are used to represent numbers that are not known. A letter used in this way is called a **variable**.

$$4 + n \qquad\qquad 6 + k + 12 \qquad\qquad m^2 + 7$$

variable $\qquad\qquad$ variable $\qquad\qquad$ variable

The value of an expression that contains one or more variables depends on the numbers used as replacements for the variables. Look at the examples:

Expression	Replacement	Value
$m - 6$	Let $m = 10$.	When $m = 10$, $\quad m - 6 = 10 - 6$ $\qquad\qquad = \mathbf{4}$
$2 \times b \times b$	Let $b = 5$.	When $b = 5$, $\quad 2 \times b \times b = 2 \times 5 \times 5$ $\qquad\qquad\qquad = \mathbf{50}$
$r + s + 6$	Let $r = 4$ and $s = 12$.	When $r = 4$ and $s = 12$, $\quad r + s + 6 = 4 + 12 + 6$ $\qquad\qquad\qquad = \mathbf{22}$
$36 \div a^2$	Let $a = 3$.	When $a = 3$, $\quad 36 \div a^2 = 36 \div 3^2$ $\qquad\qquad\quad = 36 \div 9$ $\qquad\qquad\quad = \mathbf{4}$

EXERCISES

Name the variables.

Sample: $\dfrac{t}{4} + b$ $\qquad$ *What you say:* The variables are t and b.

1. $m + 15\,m$
2. $w^2\ w$
3. $2 \times (t + n)$ t, n
4. $t \times 3 \times 5\ t$
5. $12 - y\ y$
6. $2 \times (c + a)$ c, a
7. $b + b + 0\ b$
8. $40 + d^2\ d$
9. $x + y + w$ x, y, w

Find the value.

Sample: $3 + q + r$, when $q = 6$ and $r = 20$

Solution: $3 + q + r = 3 + 6 + 20 = 29$

1. $6 + t + 18$, when $t = 12$ 36
2. e^2, when $e = 5$ 25
3. $3 \times 7 \times a$, when $a = 4$ 84
4. a^2, when $a = 8$ 64
5. $42 \div h$, when $h = 6$ 7
6. $20 + r^2$, when $r = 3$ 29
7. $13 - m$, when $m = 10$ 3
8. $3 \times (a + 5)$, when $a = 9$ 42
9. $x + 0$, when $x = 7$ 7
10. $\dfrac{6 + k}{3}$, when $k = 15$ 7
11. $\dfrac{15}{y}$, when $y = 5$ 3
12. $d + d + d + d$, when $d = 20$ 80
13. $\dfrac{26}{b}$, when $b = 2$ 13
14. $\dfrac{40}{s + 3}$, when $s = 7$ 4

Find the value. Use $a = 5$, $n = 7$, $t = 10$, and $m = 3$.

15. $\dfrac{a + n}{6}$ 2
16. $3 \times (n + m)$ 30
17. $a^2 + t$ 35
18. $a^2 \times m^2$ 225
19. $n^2 + m$ 52
20. $\dfrac{n + a}{m}$ 4
21. $t^2 + a^2$ 125
22. $t \times n \times a$ 350
23. $\dfrac{a^2 + n^2 + 1}{a^2}$ 3
24. $m \times (a + n)$ 36

25. $\dfrac{t^2}{n + m}$ 10
26. $(t + m) \times a^2$ 325
27. $a^2 \times a^2$ 625
28. $\dfrac{a + t}{m} + n$ 12

Answers may vary for Exs. 29–34. One possible answer is given.

Use at least two of the variables listed to write an expression having the value given. Use $m = 1$, $y = 2$, $w = 5$, and $z = 4$.

Sample: 10

Solution: $\dfrac{w \times z}{y}$, or $w + z + m$ (Other answers are possible.)

29. 8 $m + y + w$
30. 25 $(w \times z) + w$
31. 28 $w^2 + m + y$
32. 2 $m \times y$
33. 9 $(y \times w) - m$
34. 12 $(y \times w) + y$

1-4 *Special Equations: Formulas*

Formulas can help us to solve problems when we know the replacements for variables. Look at these examples:

Formula	Find the value of	Value
$A = l \times w$	A, when $l = 5$ $w = 7$	$A = l \times w$ $= 5 \times 7 = 35$
$r = \dfrac{d}{t}$	r, when $d = 48$ $t = 4$	$r = \dfrac{d}{t}$ $= \dfrac{48}{4} = 12$
$s = \dfrac{1}{2} \times a \times t^2$	s, when $a = 6$ $t = 5$	$s = \dfrac{1}{2} \times a \times t^2$ $= \dfrac{1}{2} \times 6 \times 5^2 = 75$

EXERCISES

Complete.

Sample: $V = 10 \times h$ *What you say:*   Replace h with 7.
      When $h = 7$, $V = \underline{\ ?\ }$ When $h = 7$,
                                        $V = 10 \times 7 = 70$.

1. $D = 4 \times s$ 2. $s = a \times 8$ 3. $T = \dfrac{s}{3}$

    When $s = 10$,   When $a = 32$,   When $s = 27$,
    $D = \underline{\ ?\ }$ 40 $s = \underline{\ ?\ }$ 256 $T = \underline{\ ?\ }$ 9

4. $V = b \times 30$ 5. $P = 6 \times l$ 6. $A = b - 14$
    When $b = 6$,   When $l = 9$,   When $b = 34$,
    $V = \underline{\ ?\ }$ 180 $P = \underline{\ ?\ }$ 54 $A = \underline{\ ?\ }$ 20

Complete.

Sample: $A = s \times s$
When $s = 15$,
$A = \underline{\ ?\ }$

Solution: $A = s \times s$
$= 15 \times 15 = 225$

A

1. $A = m + m$
When $m = 46$,
$A = \underline{\ ?\ }$ 92

2. $T = 6 \times a$
When $a = 135$,
$T = \underline{\ ?\ }$ 810

3. $A = \frac{1}{2} \times b \times h$
When $b = 16$
and $h = 4$,
$A = \underline{\ ?\ }$ 32

4. $V = e \times e \times e$
When $e = 14$,
$V = \underline{\ ?\ }$ 2744

5. $s = 16 \times t \times t$
When $t = 25$,
$s = \underline{\ ?\ }$ 10,000

6. $P = I \times E$
When $I = 10$
and $E = 1$,
$P = \underline{\ ?\ }$ 10

7. $d = r \times 6$
When $r = 45$,
$d = \underline{\ ?\ }$ 270

8. $A = l \times w$
When $l = 18$
and $w = 40$,
$A = \underline{\ ?\ }$ 720

9. $I = \frac{E}{R}$
When $E = 48$
and $R = 6$,
$I = \underline{\ ?\ }$ 8

10. $A = s \times s$
When $s = 24$,
$A = \underline{\ ?\ }$ 576

11. $V = B \times h$
When $B = 74$
and $h = 15$,
$V = \underline{\ ?\ }$ 1110

12. $A = s^2$
When $s = 13$,
$A = \underline{\ ?\ }$ 169

13. $S = \frac{1}{2} \times a \times t^2$
When $a = 32$
and $t = 5$,
$S = \underline{\ ?\ }$ 400

14. $V = l^2 \times l$
When $l = 12$,
$V = \underline{\ ?\ }$ 1728

15. $V = l^2 \times h$
When $l = 20$
and $h = 1$,
$V = \underline{\ ?\ }$ 400

B

16. $V = l \times w \times h$
When $l = 5$,
$w = 7$,
and $h = 12$,
$V = \underline{\ ?\ }$ 420

17. $A = (B + b) \times h$
When $B = 7$,
$b = 9$,
and $h = 16$,
$A = \underline{\ ?\ }$ 256

18. $B = 3L + W$
When $L = 9$
and $W = 10$,
$B = \underline{\ ?\ }$ 37

19. $C = s^2 \times l$
When $s = 3$
and $l = 17$,
$C = \underline{\ ?\ }$ 153

20. $A = \frac{22}{7} \times r^2$
When $r = 7$,
$A = \underline{\ ?\ }$ 154

21. $t = l^2 + e^2$
When $l = 5$
and $e = 10$,
$t = \underline{\ ?\ }$ 125

22. $a = \dfrac{v^2}{R + r}$

When $v = 16$,
$R = 8$,
and $r = 0$
$a = \underline{\ ?\ }$ 32

23. $k = \dfrac{m \times v^2}{2}$

When $m = 9$,
and $v = 4$,
$k = \underline{\ ?\ }$ 72

24. $G = \dfrac{3 + r^2}{m \times v}$

When $r = 9$,
$m = 3$,
and $v = 14$
$G = \underline{\ ?\ }$ 2

Problems

Complete. Use the given formula and dimensions.

1. $A = l \times w$

$A = \underline{\ ?\ }$ 204

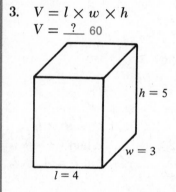

2. $A = \dfrac{1}{2} \times b \times h$

$A = \underline{\ ?\ }$ 100

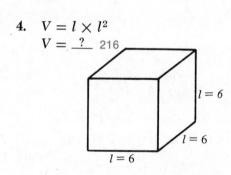

3. $V = l \times w \times h$
$V = \underline{\ ?\ }$ 60

4. $V = l \times l^2$
$V = \underline{\ ?\ }$ 216

5. $A = \dfrac{1}{2} \times (B + b) \times h$

$A = \underline{\ ?\ }$ 15

6. $A = \dfrac{22}{7} \times r^2$

$A = \underline{\ ?\ }$ 616

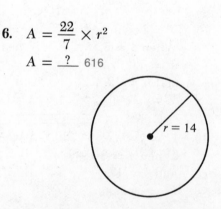

SELF-TEST 1

Be sure that you understand these terms and symbols.

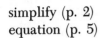

simplify (p. 2)   = (p. 5)
equation (p. 5)   variable (p. 8)

Simplify: **Section 1-1, p. 2**

1. $2 \times (3 + 4)$ 14   2. $\dfrac{24}{2^2 \times 3}$ 2

True or false?   **Section 1-2, p. 5**

3. $10 - 6 = 2 \times 3$ False   4. $14 - \dfrac{8}{2} = 2 \times 5$ True

Complete to make a true statement.   **Section 1-3, p. 8**

5. $5^2 = \underline{\ ?\ } + 17$ 8   6. $\dfrac{2 \times 8}{?} = 4$ 4

Find the value.   **Section 1-4, p. 10**

7. $A = l \times 5$ 8. $V = l^2 \times h$
    when $l = 4$,   When $l = 2$,
    $A = \underline{\ ?\ }$ 20   and $h = 3$, $V = \underline{\ ?\ }$ 12

Check your answers with those printed at the back of the book.

calculator corner

How can an arctic scientist go from one experiment site to another? You can find out by using your calculator. Begin with the
number of hours in one half-day. Multiply by the number of degrees
in two right angles. Divide by the number of months in a year.
Multiply by the number of centimeters in a meter. Add the number
of grams in 0.455 kilograms. Multiply by the number of huskies in
the scientist's kennel (25). To see the answer, turn your calculator
upside down. SLEIGH

Writing and Solving Equations

1-5 *Using Symbols for Words*

OBJECTIVE

State expressions and equations in words and in mathematical symbols.

Expressions and equations can be stated in words or in mathematical symbols.

EXAMPLE 1   Symbols: $20 - 4$

Words: The difference between twenty and four

EXAMPLE 2   Symbols: $5 + 8$

Words: The sum of five and eight

EXAMPLE 3   Symbols: $5 + 8$ $=$ 13

Words: The sum of five and eight   is equal to   thirteen.

When a variable is included, it can be referred to simply as "some number."

EXAMPLE 4   Symbols: $t - 2$

Words: The difference of some number and two

EXAMPLE 5   Symbols: $\dfrac{20}{n}$ $=$ 5

Words: The quotient of twenty and   is equal to   five.
some number (or twenty
divided by some number)

1. The difference of ten and seven   3. The difference between six and five
2. Twenty-four divided by six   4. The sum of six and twelve

EXERCISES

Give the meaning in words.

Sample: 8×6 *What you say:*   The product of six and eight

1. $10 - 7$ **2.** $\dfrac{24}{6}$ **3.** $6 - 5$

4. $6 + 12$ **5.** $18 \div 3$ **6.** 12×9

5. Eighteen divided by three
6. The product of twelve and nine

Write the meaning in words. See page A1 at the back of the book for Ex. 1–26.

Sample 1: $7 + 9$ *Solution:* The sum of seven and nine

Sample 2: $n \times 15$ *Solution:* The product of some number and fifteen

A

1. $6 + 9$ **2.** $s + 10$ **3.** $16 \div k$

4. $20 - 7$ **5.** $m - 6$ **6.** $4 \times a$

7. 5×8 **8.** $4 + t$ **9.** $r \times 10$

10. $\dfrac{18}{2}$ **11.** $25 - w$ **12.** $\dfrac{45}{n}$

Write the meaning in words.

Sample: $6 \times r = 42$ *Solution:* The product of six and some number is equal to forty-two.

13. $10 + x = 42$ **14.** $27 = m + 13$

15. $35 - t = 19$ **16.** $64 = w - 26$

17. $37 + a = 95$ **18.** $48 = 3 \times b$

19. $\dfrac{45}{9} = k$ **20.** $\dfrac{40}{b} = 5$

B

21. $t - 68 = 14$ **22.** $t = 111 \times 3$

23. $39 = n + 10$ **24.** $s \times 41 = 246$

25. $75 = 92 - y$ **26.** $300 = z \times 25$

Write an equation. Use a variable.

Sample: The sum of nineteen and some   *Solution:* $19 + b = 37$
number is equal to thirty-seven.

27. The sum of some number and sixteen is equal to seventeen. $n + 16 = 17$

28. The product of zero and some number is equal to zero. $0 \times n = 0$

29. Thirty-six is equal to the product of six and some number. $36 = 6 \times n$

30. The difference of some number and seven is equal to seventy-three. $n - 7 = 73$

31. Some number divided by fifteen is equal to ten. $\dfrac{n}{15} = 10$

32. Twelve is equal to ninety-six divided by some number. $12 = \dfrac{96}{n}$

33. The sum of zero and some number is equal to twelve. $0 + n = 12$

34. The product of six and some number is equal to the sum of nine and ten. $6 \times n = 9 + 10$

C

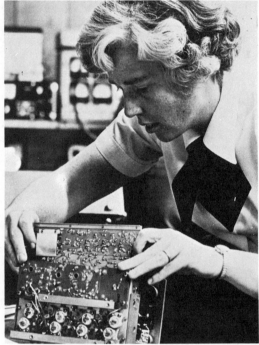

career capsule

Maintenance Electrician

Maintenance electricians keep electrical equipment in good working order. They detect and repair defects *before* equipment breaks down. They install wire and conduit (tubes which enclose electrical wire). Often they must make mathematical calculations. In emergencies, they advise management whether or not hazards require that equipment be shut down. Maintenance electricians who work in small plants and office buildings generally maintain all types of electrical equipment. Others may specialize.

Maintenance electricians need a background in algebra, basic science, electricity, and blueprint reading. They must have good color vision, good health, manual dexterity, and mechanical aptitude.

1-6 *Solving Equations: Addition and Subtraction*

OBJECTIVE

Solve equations like $x + 3 = 18$ and $15 = y - 3$.

The equation $4 + 8 = 6 + 6$ makes a true statement because the **left member,** $4 + 8$, names the same number as the **right member,** $6 + 6$. An equation may be thought of as a balanced scale. The left member "balances" (or names the same number as) the right member.

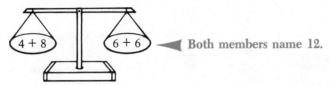

$(4 + 8)$ $(6 + 6)$ ◀ Both members name 12.

We solve an equation such as $x + 3 = 18$ by finding a replacement for the variable so that the left member and the right member name the same number.

| Equation: | $x + 3$ | $=$ | 18 |

Think: The sum of what number and 3   is equal to   18?

Solution: Since $15 + 3 = 18$, $x = 15$.

Both members name 18.

$(15 + 3)$ (18)

EXAMPLE 1   Equation: $46 + b$ $=$ 48

Think: 46 plus what number   is equal to   48?

Solution: Since $46 + 2 = 48$, $b = 2$.

EXAMPLE 2   Equation: 15 $=$ 20 $-$ t

Think: 15   equals   20   minus   what number?

Solution: Since $15 = 20 - 5$, $t = 5$.

EXAMPLE 3   Equation: $9 + 4$ $=$ K $+$ 3

Think: The sum of 9 and 4   equals   what number   plus   3?

Solution: Since $9 + 4 = 13$, and $10 + 3 = 13$, $K = 10$

Oral
EXERCISES

Match the equation with the question it suggests.

1. $14 + m = 20$ C
2. $20 - m = 10$ A
3. $20 = m - 10$ D
4. $m = 20 + 10$ B

A. What number subtracted from 20 equals 10?

B. What number is equal to the sum of 20 and 10?

C. 14 added to what number equals 20?

D. 20 equals the difference of what number and 10?

Use the true statement to solve each equation.

Sample: $15 + 10 = 25$ *What you say:*   In $15 + n = 25$, $n = 10$
$15 + n = 25$, $n = \underline{\ ?\ }$ In $s + 10 = 25$, $s = 15$
$s + 10 = 25$, $s = \underline{\ ?\ }$ In $15 + 10 = m$, $m = 25$
$15 + 10 = m$, $m = \underline{\ ?\ }$

5. $16 - 7 = 9$
$16 - b = 9$, $b = \underline{\ ?\ }$ 7
$16 - 7 = k$, $k = \underline{\ ?\ }$ 9
$t - 7 = 9$, $t = \underline{\ ?\ }$ 16

6. $13 = 18 - 5$
$b = 18 - 5$, $b = \underline{\ ?\ }$ 13
$13 = 18 - s$, $s = \underline{\ ?\ }$ 5
$13 = w - 5$, $w = \underline{\ ?\ }$ 18

7. $12 - 3 = 9$
$12 - b = 9$, $b = \underline{\ ?\ }$ 3
$12 - 3 = k$, $k = \underline{\ ?\ }$ 9
$t - 3 = 9$, $t = \underline{\ ?\ }$ 12

8. $3 = 8 - 5$
$b = 8 - 5$, $b = \underline{\ ?\ }$ 3
$3 = 8 - s$, $s = \underline{\ ?\ }$ 5
$3 = w - 5$, $w = \underline{\ ?\ }$ 8

9. $12 = 2 + 10$
$x = 2 + 10$, $x = \underline{\ ?\ }$ 12
$12 = 2 + z$, $z = \underline{\ ?\ }$ 10
$12 = h + 10$, $h = \underline{\ ?\ }$ 2

10. $18 = 74 - 56$
$18 = a - 56$, $a = \underline{\ ?\ }$ 74
$x = 74 - 56$, $x = \underline{\ ?\ }$ 18
$18 = 74 - d$, $d = \underline{\ ?\ }$ 56

For Extra Practice, see page 410.

Written
EXERCISES
A

Solve.

Sample: $15 + a = 22$ *Solution:*   Since $15 + 7 = 22$, $a = 7$.

1. $n + 7 = 19$ 12
2. $k - 5 = 13$ 18
3. $m - 6 = 30$ 36
4. $m + 10 = 45$ 35
5. $18 - x = 12$ 6
6. $20 = 25 - y$ 5
7. $12 + t = 20$ 8
8. $14 - r = 10$ 4
9. $t = 10 + 20 + 30$ 60
10. $6 + w = 20$ 14
11. $16 = 20 - x$ 4
12. $h + 2 = 36$ 34
13. $19 = 7 + b$ 12
14. $50 = s - 25$ 75
15. $50 - 10 = m$ 40
16. $34 = 24 + e$ 10
17. $35 = 36 - z$ 1
18. $3 + 8 = t$ 11

Sample: $h + 7 = 12 + 14$ *Solution:* $h + 7 = 12 + 14$
$$h + 7 = 26$$
Since $19 + 7 = 26,$
$$h = 19.$$

B

19. $n + 4 = 9 + 3$ 8

20. $m - 3 = 16 - 4$ 15

21. $10 - 7 = 1 + d$ 2

22. $t + 7 = 10 + 8$ 11

23. $c - 2 = 12 - 9$ 5

24. $18 - h = 5 + 10$ 3

25. $10 + b = 9 + 7$ 6

26. $12 + 3 = y + 2$ 13

27. $75 + 25 = 150 - q$ 50

28. $15 + k = 17 + 12$ 14

29. $n - 3 = 15 - 9$ 9

30. $23 - h = 7 + 10$ 6

Sample: $t - 10 = 47 + 4 + 9$ *Solution:* $t - 10 = 47 + 4 + 9$
$$t - 10 = 60$$
$$t = 70$$

C

31. $4 + s = 2 + 8 + 9$ 15

32. $15 = k + 4 + 0 + 8$ 3

33. $18 + 0 = t - 11$ 29

34. $m - 2 = 10 + 3 + 7$ 22

35. $n + 0 = 46 - 37$ 9

36. $6 + 8 - 7 = 15 - y$ 8

consumer notes *Nutrition*

Consumers must make important decisions when shopping for food. You must choose the foods which will give you the most food value for your money. Good nutrition is basic to good health.

Here's one way to tell if you are eating the right kind of food. You can use an encyclopedia, an almanac, or a cookbook to look up the recommended daily dietary allowances (RDA) established by nutritionists. The allowances vary according to age, height, weight and other factors. Look up *your* RDA for the following: protein, calcium, iron, vitamin A, thiamine, riboflavin, niacin, and ascorbic acid. The same information source should list the percentages of the RDA supplied by certain foods. Try to remember everything you ate yesterday. Use the charts to see if you met your nutritional needs. Now plan a menu for tomorrow.

1-7 Solving Equations: Multiplication and Division

OBJECTIVE
Solve equations like $4 \times m = 28$
and $\dfrac{30}{b} = 5$.

To solve the equation $4 \times m = 28$, we can use the idea of balancing a scale:

Think: The product of 4 and what number equals **28?**

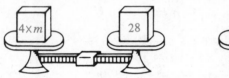 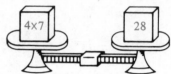

Solution: When $m = \mathbf{7}$, the scale balances.

EXAMPLE 1   Equation: $m \times 7 \qquad\qquad = \qquad 35$

Question: The product of what number and 7   equals   **35?**
Solution: Since $5 \times 7 = 35$, $m = \mathbf{5}$.

   ◀ **The scale is balanced.**

EXAMPLE 2   Equation: $\dfrac{18}{t} \qquad\qquad = \qquad 3$

Question: 18 divided by what number   equals   3?

Solution: Since $\dfrac{18}{6} = 3$, $t = 6$.

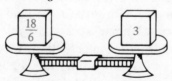

EXAMPLE 3   Equation: $m + 6 \qquad = \qquad 18 - 5$

Question: What number plus six   equals   18 minus 5?
Solution: Since $18 - 5 = 13$, and $7 + 6 = 13$, $m = 7$.

Match. For Extra Practice, see page 411.

1. $42 \div 6 = k$ B
2. $n \times 7 = 42$ C
3. $\dfrac{42}{m} = 7$ D
4. $42 \times 7 = t$ A
5. $\dfrac{r}{7} = 42$ E

A. The product of 42 and 7 equals what number?
B. 42 divided by 6 equals what number?
C. The product of what number and 7 equals 42?
D. 42 divided by what number equals 7?
E. What number divided by 7 equals 42?

6. The product of 4 and 5 equals 20.
Answer.
7. 40 is equal to the product of 5 and 8.

Sample: 24 divided by what number equals 8?
What you say: 24 divided by 3 equals 8.

6. The product of 4 and what number equals 20?
7. 40 is equal to the product of 5 and what number?
8. 10 is equal to 20 divided by what number?
9. What number divided by 7 equals 5?
10. The product of 7 and what number equals 49?

8. 10 is equal to 20 divided by 2.
9. 35 divided by 7 equals 5.
10. The product of 7 and 7 equals 49.

Write an equation.

Sample: The product of what number and 9 equals 36?
Solution: $n \times 9 = 36$

A

1. The product of 6 and what number equals 30? $6 \times n = 30$
2. What number divided by 2 equals 7? $n \div 2 = 7$
3. 8 equals 32 divided by what number? $8 = 32 \div n$
4. The product of what number and 50 equals 100? $n \times 50 = 100$
5. 42 equals the product of 6 and what number? $42 = 6 \times n$

Use the true statement to solve each equation.

Sample: $\dfrac{20}{2} = 10; \dfrac{c}{2} = 10, c = \underline{\ ?\ }$ *Solution:* $c = 20$

$\dfrac{20}{b} = 10, b = \underline{\ ?\ }$ $b = 2$

$\dfrac{20}{2} = a, a = \underline{\ ?\ }$ $a = 10$

6. $3 \times 7 = 21$
$3 \times t = 21, t = \underline{\ ?\ }$ 7
$s \times 7 = 21, s = \underline{\ ?\ }$ 3
$3 \times 7 = r, r = \underline{\ ?\ }$ 21

7. $5 \times 12 = 60$
$5 \times n = 60, n = \underline{\ ?\ }$ 12
$x \times 12 = 60, x = \underline{\ ?\ }$ 5
$5 \times 12 = z, z = \underline{\ ?\ }$ 60

8. $9 \times 8 = 72$
$m \times 8 = 72, m = \underline{\ ?\ }$ 9
$9 \times 8 = q, q = \underline{\ ?\ }$ 72
$9 \times n = 72, n = \underline{\ ?\ }$ 8

9. $42 = 7 \times 6$
$42 = c \times 6, c = \underline{\ ?\ }$ 7
$42 = 7 \times h, h = \underline{\ ?\ }$ 6
$d = 7 \times 6, d = \underline{\ ?\ }$ 42

10. $\dfrac{57}{3} = 19$

$\dfrac{57}{x} = 19, x = \underline{\ ?\ }$ 3

$\dfrac{w}{3} = 19, w = \underline{\ ?\ }$ 57

$\dfrac{57}{3} = z, z = \underline{\ ?\ }$ 19

11. $17 = \dfrac{153}{9}$

$a = \dfrac{153}{9}, a = \underline{\ ?\ }$ 17

$17 = \dfrac{n}{9}, n = \underline{\ ?\ }$ 153

$17 = \dfrac{153}{k}, k = \underline{\ ?\ }$ 9

Solve.

Sample: $\dfrac{m}{3} = 9$ *Solution:* Since $\dfrac{27}{3} = 9,\ m = 27.$

B

12. $t \times 5 = 40$ 8

13. $\dfrac{r}{2} = 25$ 50

14. $4 = \dfrac{28}{x}$ 7

15. $6 \times n = 30$ 5

16. $16 = k \times 1$ 16

17. $3 = \dfrac{z}{8}$ 24

18. $\dfrac{24}{m} = 2$ 12

19. $30 = 3 \times w$ 10

20. $y = \dfrac{45}{3}$ 15

Sample: $14 \times 2 = 7 \times k$　　*Solution:* $14 \times 2 = 7 \times k$
$$28 = 7 \times k$$
Since $28 = 7 \times 4,\ k = 4.$

C

21. $6 \times b = 4 \times 24$ 16

22. $3 \times 15 = 9 \times y$ 5

23. $\dfrac{20}{1} = \dfrac{80}{z}$ 4

24. $n \times 4 = \dfrac{40}{5}$ 2

25. $6 \times 10 = m \times 3$ 20

26. $b \times 3 = \dfrac{63}{7}$ 3

27. $\dfrac{x}{4} = 11$ 44

28. $3 + 9 = \dfrac{36}{c}$ 3

29. $4 \times k = \dfrac{60}{15}$ 1

1-8 Equations and Exponents

OBJECTIVE
Find a solution of an equation like $y^2 = 100$ or $n^3 = 125$.

We can use an exponent to shorten a multiplication expression when the same factor is used more than once.

10 used as a factor twice

$10 \times 10 = 10^2$ ◄ 10 squared
$10^2 = 100$

5 used as a factor three times

$5 \times 5 \times 5 = 5^3$ ◄ 5 cubed
$5^3 = 125$

We can find a solution of an equation when the variable has an exponent:

EXAMPLE 1 Equation: $\underbrace{s^2}_{\text{What number used as a factor twice}} \underbrace{=}_{\text{equals}} \underbrace{49}_{49?}$

$s \times s = 49$

Solution: Since $7 \times 7 = 49$, $s = 7$.

EXAMPLE 2 Equation: $\underbrace{m^3}_{\text{What number used as a factor three times}} \underbrace{=}_{\text{equals}} \underbrace{64}_{64?}$

$m \times m \times m = 64$

Solution: Since $4 \times 4 \times 4 = 64$, $m = 4$.

Complete.

Sample: When t is 10, the value of t^2 is __?__.

What you say: When t is 10, the value of t^2 is 10 times 10 or 100.

Sample: When m is 2, the value of m^3 is __?__.

What you say: When m is 2 the value of m^3 is 2 times 2 times 2 or 8.

Oral
EXERCISES

1. When n is 4, the value of n^2 is __?__. 16
2. When b is 6, the value of b^2 is __?__. 36
3. When k is 3, the value of k^3 is __?__. 27
4. When x is 1, the value of x^2 is __?__. 1
5. When s is 1, the value of s^3 is __?__. 1

Complete.

Sample 1: $3^2 = \underline{\ ?\ }$ *What you say:* Three squared equals nine.

Sample 2: $2^3 = \underline{\ ?\ }$ *What you say:* Two cubed equals eight.

6. $5^2 = \underline{\ ?\ }$ 25   **7.** $10^3 = \underline{\ ?\ }$ 1000   **8.** $1^3 = \underline{\ ?\ }$ 1

9. $8^2 = \underline{\ ?\ }$ 64   **10.** $6^2 = \underline{\ ?\ }$ 36   **11.** $7^2 = \underline{\ ?\ }$ 49

Written EXERCISES
A

Copy and complete the table of squares.

1.

n	n^2	
1	1	
2	4	
3	9	
4	?	16
5	?	25
6	?	36
7	?	49
8	?	64
9	?	81
10	?	100

2.

n	n^2	
11	121	
12	144	
13	?	169
14	?	196
15	?	225
16	?	256
17	?	289
18	?	324
19	?	361
20	?	400

3.

n	n^2	
21	441	
22	?	484
23	?	529
24	?	576
25	?	625
26	?	676
27	?	729
28	?	784
29	?	841
30	?	900

Copy and complete the table of cubes.

4.

n	n^3	
1	1	
2	8	
3	?	27
4	?	64
5	?	125

5.

n	n^3	
6	216	
7	?	343
8	?	512
9	?	729
10	?	1000

6.

n	n^3	
11	1331	
12	?	1728
13	?	2197
14	?	2744
15	?	3375

Complete. Use the tables of squares and cubes from Exercises 1–6.

Sample: $7^3 = \underline{\ ?\ } \times \underline{\ ?\ } \times \underline{\ ?\ }$ *Solution:* $7^3 = 7 \times 7 \times 7$
$\qquad\quad 7^3 = \underline{\ ?\ }$ $\qquad\qquad\qquad\qquad\quad 7^3 = 343$

7. $10^3 = \underset{10}{\underline{\ ?\ }} \times \underset{10}{\underline{\ ?\ }} \times \underset{10}{\underline{\ ?\ }}$
$\quad\ \ 10^3 = \underline{\ ?\ }$ 1000

8. $\underset{25}{\underline{\ ?\ }} \times \underset{25}{\underline{\ ?\ }} = 25^2$
$\quad 625\ \underline{\ ?\ } = 25^2$

9. $15^3 = \underline{\ ?\ }15 \times \underline{\ ?\ }15 \times \underline{\ ?\ }15$
$\quad\ \ 15^3 = \underline{\ ?\ }$ 3375

10. $6^3 = \underline{\ ?\ }6 \times \underline{\ ?\ }6 \times \underline{\ ?\ }6$
$\quad\ \ 6^3 = \underline{\ ?\ }$ 216

11. $\overset{28}{\underline{\ ?\ }} \times \overset{28}{\underline{\ ?\ }} = 28^2$

   $\underline{784}\ \underline{\ ?\ } = 28^2$

12. $12^2 = \overset{12}{\underline{\ ?\ }} \times \overset{12}{\underline{\ ?\ }}$

   $12^2 = \underline{\ ?\ }\ 144$

13. $19^2 = \underline{\ ?19\ } \times \underline{\ ?19\ }$

   $19^2 = \underline{\ ?\ }\ 361$

14. $1^3 = \underline{\ ?1\ } \times \underline{\ ?1\ } \times \underline{\ ?1\ }$

   $1^3 = \underline{\ ?1\ }$

15. $4^3 = \underline{\ ?4\ } \times \underline{\ ?4\ } \times \underline{\ ?4\ }$

   $4^3 = \underline{\ ?\ }\ 64$

Find a solution. Use the tables of squares and cubes.

Sample: $r^2 = 289$ *Solution:* Since $17 \times 17 = 289$, $r = 17$.

16. $y^2 = 81$ 9

17. $a^3 = 125$ 5

18. $m^2 = 324$ 18

B

19. $m^2 = 400$ 20

20. $r^3 = 1000$ 10

21. $k^2 = 576$ 24

22. $b^2 = 625$ 25

23. $1728 = z^3$ 12

24. $q^2 = 841$ 29

25. $h^2 = 64$ 8

26. $196 = x^2$ 14

27. $v^3 = 729$ 9

28. $900 = w^2$ 30

29. $p^3 = 2744$ 14

30. $c^2 = 529$ 23

Find a solution.

31. $r^2 = 20 + 5$ 5

32. $a^2 = 5^2 + 11$ 6

33. $m^3 = 0 + 8^{\overset{4}{2}}$

C

SELF-TEST 2

Be sure that you understand the term *exponent* (p. 23).

The product of three and some number is equal to twenty-one.

1. Write the meaning of $3 \times y = 21$ in words.

Section 1-5, p. 14

2. Write an equation for the sentence "The sum of some number and twelve is equal to forty-eight." $n + 12 = 48$

Section 1-6, p. 17

Solve.

Section 1-7, p. 20

3. $2 + n = 26$ 24

4. $20 - k = 3 + 6$ 11

5. $3 \times s = 33$ 11

6. $\dfrac{15}{t} = 5$ 3

Find a solution.

Section 1-8, p. 23

7. $x^2 = 16$ 4

8. $a^3 = 27$ 3

Check your answers with those printed at the back of the book.

chapter summary

1. An **expression** can be **simplified** by carrying out the indicated operations.

2. An **equation** consists of two expressions joined by the equality symbol ($=$). The expressions are called the **left member** and the **right member**. The statement made by an equation is **true** when the left member and the right member name the **same number.**

3. A letter used to represent an unknown number is called a **variable.**

4. To **solve** an equation that contains a variable, we find a number to replace the variable so that a **true** statement results.

chapter test

Simplify.

1. $\dfrac{3 \times 8}{6}$ ₄   2. $7^2 + 1$ 50   3. $4 \times 4 \times 4$ 64   4. $6 \times 19 \times 0$ 0

True or False?

5. $15 + 6 = 3 \times 7$ 6. $\dfrac{10}{10} = 5 - 3$ 7. $\dfrac{6}{3} + 0 = \dfrac{6}{3} \times 0$
      True   False   False

Name the value. Let $b = 5$, $t = 3$, and $m = 10$.

8. $b + 4$ 9   9. $b + t + 2$ 10   10. $m^2 + 35$ 135

11. $t \times b$ 15   12. $6 + t^2$ 15   13. $\dfrac{75}{t}$ 25

Write the equation in symbols.

14. The sum of twenty and sixteen equals thirty-six. $20 + 16 = 36$

15. The product of six and some number is equal to twenty-four. $6 \times n = 24$

Solve each equation.

16. $6 + t = 15$ 9   **17.** $14 - k = 8$ 6   **18.** $\dfrac{20}{s} = 5$ 4

19. $w \times 8 = 80$ 10   **20.** $40 = n + 30$ 10   **21.** $12 = m - 3$ 15

22. $6 = \dfrac{36}{b}$ 6   **23.** $34 = c \times 2$ 17

Find the value.

24. 4^2 16   **25.** 3^3 27   **26.** 5^2 25   **27.** 10^3 1000

challenge topics

Cutting Up

A. Duplicate this figure three times on squared paper.
   1. Cut one shape into **two** congruent pieces.
   2. Cut one shape into **three** congruent pieces.
   3. Cut one shape into **four** congruent pieces. Then try to put the four pieces back together to form the original shape.

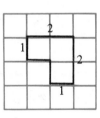

B. Duplicate figures I and II on squared paper.

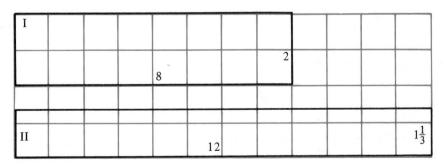

Cut shape II into two congruent pieces that will exactly cover shape I.

Review of Skills

Complete. Use $<$, $>$, or $=$.

1. 909 _?_ 990 $<$
2. $2\frac{1}{2}$ _?_ $\frac{5}{2}$ $=$
3. 0 _?_ 0.1 $<$
4. $1\frac{1}{2}$ _?_ 4 $<$
5. $5\frac{1}{2}$ _?_ $5\frac{1}{4}$ $>$
6. 4.50 _?_ 4.8 $<$

Express in words.

7. Forty-seven hundredths
8. Five and eight hundredths
9. Three and one hundred thirty-five thousandths

Sample: 2.35   *Solution:* Two and thirty-five hundredths.

7. 0.47
8. 5.08
9. 3.125
10. 10.3   ten and three tenths
11. 0.01   one hundredth
12. 136.1   one hundred thirty-six and one tenth

Express as a decimal.

Sample: one and twenty-three hundredths   *Solution:* 1.23

13. ten and sixty-five hundredths 10.65
14. five tenths 0.5
15. five hundredths 0.05
16. seven and sixteen thousandths 7.016

Add or subtract.

17. 425
    816
    102
    ――
    1343

18. 6.04
    2.15
    17.33
    ―――
    25.52

19. $5\frac{7}{10}$
    $6\frac{1}{10}$
    ――
    $11\frac{4}{5}$

20. $106.25
    35.00
    19.37
    ―――
    $160.62

21. 396
    $-$127
    ――
    269

22. 861.23
    $-$104.08
    ―――
    757.15

23. $9\frac{7}{8}$
    $-2\frac{1}{8}$
    ――
    $7\frac{3}{4}$

24. $643.85
    $-$62.79
    ―――
    $581.06

Multiply or divide.

25. 146×35 5110
26. 9.2×8 73.6
27. $\frac{3}{5} \times \frac{1}{2}$ $\frac{3}{10}$
28. 275×1000 275,000
29. 2.5×7.1 17.75
30. $2\frac{7}{8} \times \frac{1}{3}$ $\frac{23}{24}$
31. $27\overline{)216}$ 8
32. $23 \div 4.6$ 5
33. $\frac{2}{3} \div 2\frac{1}{3}$
34. $\frac{184}{46}$ 4
35. $230 \div 4.6$ 50
36. $\frac{5}{8} \div \frac{7}{8}$ $\frac{5}{7}$

28

Left: Linotype operator, 1919.

Right: Operator using video display terminal, part of computer typesetting system.

2 Positive and Negative Numbers

Whole Numbers and Integers

2-1 *The Whole Numbers*

OBJECTIVES

Complete patterns of consecutive whole numbers.

Name the coordinate of a point on the number line.

The whole numbers are the numbers 0, 1, 2, 3, 4, 5, and so on. The set of whole numbers can be written:

$$\{0, 1, 2, 3, 4, 5, 6, 7, 8, 9, 10, 11, \ldots\}$$

{ } means "the set of"   The three dots show that the set goes on and on.

Whole numbers are pictured here in consecutive order on the number line.

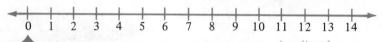

0 is the first whole number. The point marked 0 is called the origin.

Note that the marks are equally spaced.

The number line can help us understand ideas about whole numbers.

EXAMPLE 1 The set of whole numbers greater than 5 is $\{6, 7, 8, 9, 10, \ldots\}$.

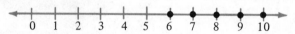

EXAMPLE 2 The set of whole numbers between 2 and 6 is $\{3, 4, 5\}$.

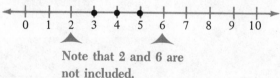

Note that 2 and 6 are not included.

EXAMPLE 3 The set of whole numbers between 6 and 7 has no members. Such a set is called the empty set. The symbol for the empty set is **∅**.

The number matched with a point on the line is called the **coordinate** of that point. Sometimes letters as well as numbers are used to name the points.

EXAMPLE 4

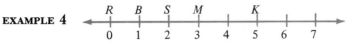

The coordinate of point S is 2.

EXAMPLE 5

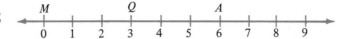

The point half the distance from point M to point A is point Q. Its coordinate is 3.

Name the set of whole numbers described. Look at the number line if you need help.

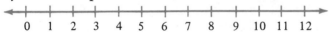

Sample: Between 3 and 9   *What you say:* The set 4, 5, 6, 7, 8

1. Between 1 and 6 {2, 3, 4, 5}   2. Less than 1 {0}

3. Between 2 and 9 {3, 4, 5, 6, 7, 8} 4. Greater than 6

5. Between 0 and 5 {1, 2, 3, 4}   6. Greater than 5 {6, 7, 8, 9}

7. Less than 3 {0, 1, 2}   8. Less than 0 ∅

{7, 8, 9, 10, 11, . . .}

Complete the whole number pattern.

Sample: 33, 34, _?_, _?_, _?_, 38   *What you say:* 33, 34, 35, 36, 37, 38.

9. 67, 68, _?_, _?_, _?_, _?_, 73 67, 68, 69, 70, 71, 72, 73

10. 495, 496, _?_, _?_, _?_, _?_, 501 495, 496, 497, 498, 499, 500, 501

11. 96, 97, 98, _?_, _?_, _?_, _?_ 96, 97, 98, 99, 100, 101, 102

12. 997, 998, 999, _?_, _?_, _?_, _?_ 997, 998, 999, 1000, 1001, 1002, 1003

13. _?_, _?_, _?_, 302, _?_, 304, 305 299, 300, 301, 302, 303, 304, 305

14. _?_, _?_, _?_, _?_, 5002, 5003, 5004
   4998, 4999, 5000, 5001, 5002, 5003, 5004

Written
EXERCISES

Name the coordinate.

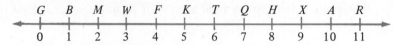

$$\begin{array}{ccccccccccccc} G & B & M & W & F & K & T & Q & H & X & A & R \\ 0 & 1 & 2 & 3 & 4 & 5 & 6 & 7 & 8 & 9 & 10 & 11 \end{array}$$

Sample: W   *Solution:* 3

A
1. *H* 8   2. *G* 0   3. *Q* 7
4. *X* 9   5. *F* 4   6. *A* 10
7. *B* 1   8. *K* 5   9. *M* 2
10. The point between *W* and *K*. F
11. The point between *H* and *T*. Q
12. The point between *K* and *Q*. T
13. The point between *X* and *Q*. H
14. The point half the distance from *W* to *K*. F
15. The point half the distance from *G* to *F*. M
16. The point half the distance from *M* to *H*. K

B
17. The point one-third the distance from *B* to *F*. M
18. The point one-third the distance from *H* to *R*. X
19. The point one-fourth the distance from *G* to *F*. B
20. The point one-fourth the distance from *B* to *X*. W

Name the set of whole numbers described.

Sample: Between 7 and 11   *Solution:* {8, 9, 10}

21. Between 0 and 2 {1}   22. Less than 0 ∅
23. Between 5 and 8 {6, 7}   24. Between 999 and 1001 {1000}
25. Less than 6 {0, 1, 2, 3, 4, 5}   26. Between 9998 and 10,001
27. Less than 9 {0, 1, 2, 3, 4, 5, 6, 7, 8}   28. Between 0 and 1 {9999, 10,000} ∅

Describe the numbers listed.

Sample: 9, 10, 11, 12 and 13   *Solution:* Whole numbers be-
Ex. 29–36 are the set of whole numbers: tween 8 and 14

C
29. 3, 4, 5, and 6 between 2 and 7   30. 0, 1, 2, 3, and 4 less than 5
31. 10, 11, and 12 between 9 and 13   32. 0, 1, 2, 3, 4, 5, 6, 7, and 8 less than 9
33. 7, 8, 9, 10, 11, . . . greater than 6   34. 0, 1, 2, 3, 4, . . . the whole numbers
35. 21, 22, 23, 24, . . . greater than 20   36. 1000, 1001, 1002, 1003, . . . greater than 999

Find the average.

Sample: 5, 6 and 7   *Solution:* $\dfrac{5 + 6 + 7}{3} = 6$

37. 2, 3, and 4 $\dfrac{2 + 3 + 4}{3} = 3$ **38.** 53, 54, and 55

39. 7, 8, and 9 $\dfrac{7 + 8 + 9}{3} = 8$ **40.** 99, 100, and 101

41. 12, 13, and 14   **42.** 405, 406, and 407

43. What pattern do you see in Exercises 37–42? Explain.

44. Name two consecutive whole numbers whose sum is 15. 7, 8

45. Name two consecutive whole numbers whose sum is 27. 13, 14

46. Name two consecutive whole numbers whose product is 12.   3, 4

47. Name three consecutive whole numbers whose sum is 30.   9, 10, 11

38. $\dfrac{53 + 54 + 55}{3} = 54$ **40.** $\dfrac{99 + 100 + 101}{3} = 100$ **41.** $\dfrac{12 + 13 + 14}{3} = 13$

42. $\dfrac{405 + 406 + 407}{3} = 406$ **43.** The average of three consecutive whole numbers is the middle consecutive whole number.

consumer notes *Electricity*

Do you turn off the radio when you're not listening to it? If you do, you are saving electricity. Electricity usage is measured in kilowatt-hours (KWH) by an electric meter. To read a meter, note the position of the pointer on each of the four dials. If the pointer is between numbers, read the smaller number. This meter reads 4726 KWH.

THOUSANDS	HUNDREDS	TENS	ONES

KILOWATT HOURS

Find the electric meter in your house or apartment. Make readings at 9 A.M., 1 P.M., 6 P.M., and 10 P.M. Calculate the amount of electricity used during each period. Over which of these periods was the most electricity used? Have your family think of and use ways of saving electricity. After a week make another set of readings. Do you notice a difference?

2-2 *The Integers*

OBJECTIVES

Read symbols for positive and negative integers.

Name the opposite of an integer.

Complete patterns of consecutive integers.

When the number line shows points on both sides of zero, positive numbers name points to the **right** of 0. Negative numbers name points to the **left** of 0.

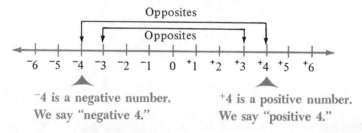

$^-4$ is a negative number. We say "negative 4."

$^+4$ is a positive number. We say "positive 4."

Each positive number can be matched with a negative number that is its opposite. **Zero** is its own opposite.

$$^+5 \text{ and } ^-5$$
$$\text{Opposites} \blacktriangleright \quad ^-17 \text{ and } ^+17$$
$$0 \text{ and } 0$$

The whole numbers (including zero) and their opposites are called the integers.

$$\{\text{the integers}\} = \{\ldots\ ^-4,\ ^-3,\ ^-2,\ ^-1,\ 0,\ ^+1,\ ^+2,\ ^+3,\ ^+4, \ldots\}$$

The number line can be used as a model for ideas about integers.

EXAMPLE 1

The set of integers between $^-3$ and $^+2$ is $\{^-2,\ ^-1,\ 0,\ ^+1\}$.

EXAMPLE 2

The coordinate of point B is $^-1$.

EXAMPLE 3

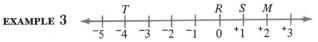

The point half the distance from point R to point M is point S. Its coordinate is $^+1$.

Read the symbol.

Sample: $^-6$ *What you say:*   Negative six

Sample: $^+3$ *What you say:*   Positive three

1. $^-8$ negative eight   **2.** $^+15$ positive fifteen   **3.** $^-19$ negative nineteen

4. $^-10$ negative ten   **5.** $^+3$ positive three   **6.** $^+12$ positive twelve

7. $^+7$ positive seven   **8.** $^-75$ negative seventy-five   **9.** $^-12$ negative twelve

Name the opposite.

Sample: $^+20$ *What you say:*   The opposite of $^+20$ is $^-20$.

10. $^+5$ opposite of $^+5$ is $^-5$ **11.** $^-21$ opposite of $^-21$ is $^+21$ **12.** $^+11$ opposite of $^+11$ is $^-11$

13. $^+1$ opposite of $^+1$ is $^-1$ **14.** 0 opposite of 0 is 0   **15.** $^+100$ opposite of $^+100$ is $^-100$

Name the coordinate.

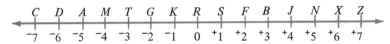

Sample: G *Solution:* $^-2$

A

1. J $^+4$ **2.** B $^+3$ **3.** Z $^+7$

4. D $^-6$ **5.** R 0 **6.** M $^-4$

7. C $^-7$ **8.** K $^-1$ **9.** S $^+1$

10. The point between D and M. A

11. The point between N and Z. X

12. The point between A and T. M

13. The point between K and S. R

14. The point half the distance between S and B. F

15. The point half the distance between R and M. G

Complete the pattern.

Sample: ⁻8, ⁻7, ⁻6, _?_, _?_, _?_

Solution: ⁻8, ⁻7, ⁻6, ⁻5, ⁻4, ⁻3

16. ⁻15, ⁻14, ⁻13, _?_, _?_, _?_ ⁻15, ⁻14, ⁻13, ⁻12, ⁻11, ⁻10

17. _?_, _?_, ⁻1, 0, ⁺1, _?_, _?_ ⁻3, ⁻2, ⁻1, 0, ⁺1, ⁺2, ⁺3

18. ⁻4, ⁻3, _?_, _?_, _?_, ⁺1 ⁻4, ⁻3, ⁻2, ⁻1, 0, ⁺1

19. ⁻2, _?_, _?_, _?_, _?_, ⁺3, ⁺4 ⁻2, ⁻1, 0, ⁺1, ⁺2, ⁺3, ⁺4

20. ⁻1, 0, ⁺1, _?_, _?_, _?_ ⁻1, 0, ⁺1, ⁺2, ⁺3, ⁺4

21. ⁻5, ⁻4, ⁻3, _?_, _?_, _?_ ⁻5, ⁻4, ⁻3, ⁻2, ⁻1, 0

22. ⁻3, ⁻2, _?_, _?_, _?_, ⁺2 ⁻3, ⁻2, ⁻1, 0, ⁺1, ⁺2

23. ⁻2, ⁻1, _?_, _?_, _?_, ⁺3 ⁻2, ⁻1, 0, ⁺1, ⁺2, ⁺3

25. {⁻5, ⁻4, ⁻3, ⁻2, ⁻1, 0}
27. {⁺1, ⁺2, ⁺3, ⁺4, ⁺5, ...}

Name the set of integers described.

Sample: Between ⁻6 and ⁻2   *Solution:* {⁻5, ⁻4, ⁻3}

B

24. Between ⁻7 and ⁻4 {⁻6, ⁻5}   25. Between ⁺1 and ⁻6

26. Between ⁻2 and ⁺2 {⁻1, 0, ⁺1}   27. The positive integers

28. Between ⁻1 and ⁺5   29. The negative integers

30. Between ⁻1 and ⁺10   31. Between ⁻1 and ⁺1 {0}

32. Between ⁻5 and ⁺5   33. The integer that is neither

28. {0, ⁺1, ⁺2, ⁺3, ⁺4}   positive nor negative {0}

29. {... ⁻5, ⁻4, ⁻3, ⁻2, ⁻1}   30. {0, ⁺1, ⁺2, ⁺3, ⁺4, ⁺5, ⁺6, ⁺7, ⁺8, ⁺9}

32. {⁻4, ⁻3, ⁻2, ⁻1, 0, ⁺1, ⁺2, ⁺3, ⁺4}

Describe the integers listed.

Sample: ⁻5, ⁻4, and ⁻3   *Solution:* The integers between ⁻6

The integers between ⁺4 and ⁺9   and ⁻2

34. ⁺5, ⁺6, ⁺7, and ⁺8   The integers between ⁻3 and ⁺1

                             35. ⁻2, ⁻1, and 0

36. ⁺1, ⁺2, and ⁺3   37. ⁻1, 0, and ⁺1

The integers between 0 and ⁺4   The integers between ⁻2 and ⁺2

The number-line graph of a number is described. Tell whether the number is positive or negative.

C

38. Four units to the left of the graph of ⁺1 negative

39. Two units to the right of the graph of ⁻5 negative

40. Five units to the left of the graph of ⁺4 negative

41. Three units to the right of the graph of ⁻2 positive

42. Two units to the left of the graph of ⁻2 negative

2-3 Graphing Integers on the Number Line

OBJECTIVES

Graph a set of integers on the number line.

Name a set of integers by stating a rule or listing the members.

We can graph a set of integers on the number line. We put a dot at each point that corresponds to an integer to be graphed.

EXAMPLE 1 $\{-1, 0, {}^{+}1, {}^{+}2\}$ ◀ This set is named by listing the members.

EXAMPLE 2 {the integers between ${}^{+}2$ and ${}^{-}3$} ◀ This set is named by stating a rule, or description.

Note that ${}^{+}2$ and ${}^{-}3$ are *not* included in the graph.

EXAMPLE 3 {the integers between ${}^{+}1$ and ${}^{-}3$, *inclusive*}

Note that ${}^{+}1$ and ${}^{-}3$ *are* included in the graph.

Match.

EXERCISES

1. {the integers between ${}^{-}1$ and ${}^{+}3$} B

    A.

2. {the positive integers less than ${}^{+}4$} A

    B.

3. $\{{}^{-}1, 0, {}^{+}1\}$ E

    C.

4. {the integers between ${}^{-}3$ and 0, inclusive} D

    D.

5. $\{0, {}^{+}2\}$ C

    E.

Written EXERCISES

Name the set graphed. List the members.

Sample:

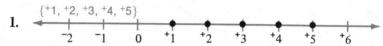

$$\begin{array}{cccccccccc} & \overline{6} & \overline{5} & \overline{4} & \overline{3} & \overline{2} & \overline{1} & 0 & ^+1 & ^+2 \end{array}$$

Solution: $\{^-5, ^-4, ^-3\}$

A

1. $\{^+1, ^+2, ^+3, ^+4, ^+5\}$

   $$\begin{array}{cccccccccc} & \overline{2} & \overline{1} & 0 & ^+1 & ^+2 & ^+3 & ^+4 & ^+5 & ^+6 \end{array}$$

2. $\{^+1, ^+3, ^+5\}$

   $$\begin{array}{cccccccccc} & \overline{2} & \overline{1} & 0 & ^+1 & ^+2 & ^+3 & ^+4 & ^+5 & ^+6 & ^+7 \end{array}$$

3. $\{^-3, ^-2, ^+2, ^+3\}$

   $$\begin{array}{cccccccccc} & \overline{4} & \overline{3} & \overline{2} & \overline{1} & 0 & ^+1 & ^+2 & ^+3 & ^+4 & ^+5 \end{array}$$

4. $\{^-2, ^-1, 0\}$

   $$\begin{array}{cccccccccc} & \overline{4} & \overline{3} & \overline{2} & \overline{1} & 0 & ^+1 & ^+2 & ^+3 & ^+4 \end{array}$$

5. $\{^-6, ^-4, ^-2\}$

   $$\begin{array}{cccccccccc} & \overline{7} & \overline{6} & \overline{5} & \overline{4} & \overline{3} & \overline{2} & \overline{1} & 0 & ^+1 & ^+2 \end{array}$$

Graph the set of numbers on the number line.

Sample: {the integers between ⁻6 and 0}

Solution:

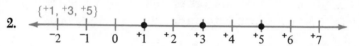

$$\begin{array}{cccccccccc} & \overline{6} & \overline{5} & \overline{4} & \overline{3} & \overline{2} & \overline{1} & 0 & ^+1 & ^+2 \end{array}$$

Check students' graphs.

6. $\{^-9, ^-7, ^-5, ^-3\}$

7. {the integers between ⁻5 and ⁺5}

8. $\{^-1, 0, ^+1, ^+2, ^+3\}$

9. {the integers between ⁻4 and ⁺2, inclusive}

10. {the integers between ⁻3 and 0}

11. {the integers between ⁻1 and ⁺4, inclusive}

B

12. {the negative integers greater than ⁻3}

13. {the positive integers between ⁻1 and ⁺8}

14. {the negative integers greater than ⁻7}

15. {the integers between ⁻1 and ⁺1}

16. {the positive integers less than ⁺5}

17. {⁻5, ⁻2, and their opposites}

Name each set with a rule. See page A1 at the back of the book
for Ex. 18–31.

Sample: {⁻9, ⁻8, ⁻7, ⁻6}   *Solution:* {the integers between ⁻10 and ⁻5}

18. {0, ⁺1, ⁺2, ⁺3}   **19.** {⁺100, ⁺101, ⁺102, . . .}

20. {⁺1, ⁺2, 0, ⁻1, ⁻2}   **21.** {⁺10, ⁺11, ⁺12, ⁺13}

22. {⁻5, ⁻6, ⁻7, ⁻8}   **23.** {. . . , ⁻3, ⁻2, ⁻1, 0}

24. {0}   **25.** {⁻4, ⁻3, ⁻2, ⁻1}

26. ∅   **27.** {⁻9, ⁻8, ⁺8, ⁺9}

28. {⁺1, ⁺2, ⁻1, ⁻2}   **29.** {0, ⁺2, ⁺4, ⁺6}

30. {⁻1, ⁻2, ⁻3, ⁻4, . . .}   **31.** {⁺5, ⁺6, ⁺7, ⁺8, ⁺9, . . .}

SELF-TEST 1

Be sure that you understand these terms and symbols.

whole numbers (p. 30)   { } (p. 30)   . . . (p. 30)
origin (p. 30)   ∅ (p. 30)   coordinate (p. 31)
⁻ (p. 34)   ⁺ (p. 34)   integers (p. 34)

1. Name the whole numbers between 793 and 798. 794, 795, 796, 797   Section 2-1 p. 30

2. Name the coordinates of *V, S, Q, P,* and *W.* 0; 5; 1; 9; 7

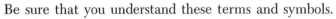

3. Name coordinates of *A, G, J, Y,* and *C,* and their   Section 2-2 p. 34
    opposites.

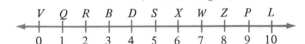   ⁻1; ⁺1; ⁻4; ⁺4;
⁺2; ⁻2; ⁺4; ⁻4; ⁻2; ⁺2

4. Complete the pattern. ⁻11, ⁻10, ⁻9, __?__ , __?__ , __?__ . ⁻9, ⁻8, ⁻7, ⁻6

5. Graph the set on the number line and list the members.   Section 2-3 p. 37
    {the integers between ⁻2 and ⁺3} {⁻1, 0, ⁺1, ⁺2}

6. Name {⁺3, ⁺4, ⁺5, ⁺6} with a rule. The integers between ⁺2 and ⁺7

Check your answers with those printed at the back of the book.

Graphs of Numbers and of Pairs of Integers

2-4 *Numbers on the Number Line*

OBJECTIVES

Assign numbers written as fractions and decimals to points on the number line.

Use the symbols > and < to compare numbers.

Arrange positive and negative numbers in order.

Every point on the number line can be matched with a number. Here are some examples.

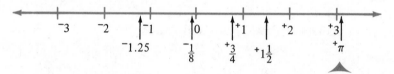

The value of $^+\pi$ is about $3\frac{1}{7}$.

Every number is the coordinate of a point on the number line. This is why we can speak of "a number on the number line." For example, we say that every number on the number line has an opposite.

$$\blacktriangleright\ \frac{^+1}{2}\text{ and }\frac{^-1}{2}$$

Opposites $\blacktriangleright\ ^-0.25\text{ and }^+0.25$

$$\blacktriangleright\ \frac{^+\pi}{2}\text{ and }\frac{^-\pi}{2}$$

The number line can help us compare numbers. Let's agree to speak of the position of a number on the number line when we mean the position of its graph.

The graph of $^-1$ is to the left of the graph of $^+1$. $\blacktriangleright$ $^-1$ is to the left of $^+1$.

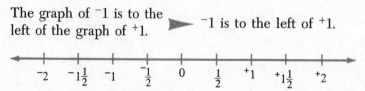

EXAMPLE 1 Compare $^-1$ and $^+\frac{1}{2}$.

$^-1$ **is to the left of** $^+\frac{1}{2}$ ▶ $^-1$ **is less than** $^+\frac{1}{2}$

We write $^-1 < {}^+\frac{1}{2}$.

▲
is less than

EXAMPLE 2 Compare 0 and $^-2$.
0 **is to the right of** $^-2$ ▶ 0 **is greater than** $^-2$
We write $0 > {}^-2$.

▲
is greater than

Give the meaning.

Sample: $\dfrac{^-2}{3}$ *What you say:* Negative two-thirds

Sample: $^+0.35$ *What you say:* Positive thirty-five hundredths

1. $\dfrac{^-1}{4}$ negative one fourth

2. $\dfrac{^+2}{5}$ positive two fifths

3. $^+3.8$ positive three and eight tenths

4. $^+1\frac{3}{4}$

5. $\dfrac{^-7}{8}$ negative seven-eighths

6. $\dfrac{^+9}{10}$ positive nine-tenths

7. $^-4.7$ negative four and seven-tenths

8. $^-2\frac{1}{2}$

positive one and three-fourths

negative two and one-half

Name the opposite.

Sample: $\dfrac{^+5}{8}$ *What you say:* The opposite of $\dfrac{^+5}{8}$ is $\dfrac{^-5}{8}$.

9. $\dfrac{^+3}{4}\ \dfrac{^-3}{4}$

10. $\dfrac{^-1}{3}\ \dfrac{^+1}{3}$

11. $^+2\frac{3}{5}\ {}^-2\frac{3}{5}$

12. $^-6.25$

$^+6.25$

13. $\dfrac{^+1}{8}\ \dfrac{^-1}{8}$

14. $\dfrac{^-7}{10}\ \dfrac{^+7}{10}$

15. $^-1\frac{1}{10}\ {}^+1\frac{1}{10}$

16. $^+1.5\ {}^-1.5$

Read the sentence. Tell whether it is true or false.

17. $^-2 > {}^+1$
False

18. $\dfrac{^+2}{3} < {}^+1$
True

19. $^+1\frac{1}{2} < {}^+2$
True

20. $\dfrac{^-1}{3} > {}^-3$
True

ORal
EXERCISES

Written EXERCISES

Name the coordinate of each point.

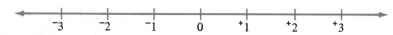

A T R   S K   M N X P   Z
$^-3$ $^-2$ $^-1$ 0 $^+1$ $^+2$ $^+3$

Sample: *M*   *Solution:* $\dfrac{^+1}{4}$

A

1. R $^-1\frac{1}{2}$
2. X $^+1\frac{1}{4}$
3. Z $^+2\frac{3}{4}$
4. K $\dfrac{^-1}{2}$
5. T $^-1\frac{3}{4}$
6. A $^-2\frac{1}{2}$
7. S $\dfrac{^-3}{4}$
8. P $^+1\frac{3}{4}$
9. N $\dfrac{^+3}{4}$

Complete. Use **left** or **right** and $<$ or $>$.

$^-3$ $^-2$ $^-1$ 0 $^+1$ $^+2$ $^+3$

Sample: $^-2$ is to the __?__ of $^+1$, so $^-2$ __?__ $^+1$.

Sample: $^-2$ is to the **left** of $^+1$, so $^-2 < {^+1}$.

10. $^+1$ is to the __?__ of $^-1$, so $^+1$ __?__ $^-1$. right; $>$
11. $^-3$ is to the __?__ of $^+3$, so $^-3$ __?__ $^+3$. left; $<$
12. $\dfrac{^+2}{3}$ is to the __?__ of 0, so $\dfrac{^+2}{3}$ __?__ 0. right; $>$
13. $\dfrac{^-2}{3}$ is to the __?__ of 0, so $\dfrac{^-2}{3}$ __?__ 0. left; $<$
14. $^-2$ is to the __?__ of $\dfrac{^+1}{3}$, so $^-2$ __?__ $\dfrac{^+1}{3}$. left; $<$
15. $^-1\frac{2}{3}$ is to the __?__ of $\dfrac{^-2}{3}$, so $^-1\frac{2}{3}$ __?__ $\dfrac{^-2}{3}$. left; $<$
16. $^-3$ is to the __?__ of $\dfrac{^-1}{3}$, so $^-3$ __?__ $\dfrac{^-1}{3}$. left; $<$

Arrange in order, least to greatest. Use the number line if needed.

Sample: $^-6, {^+7}, {^-3\frac{1}{2}}, 0, \dfrac{^+3}{4}$ *Solution:* $^-6, {^-3\frac{1}{2}}, 0, \dfrac{^+3}{4}, {^+7}$

17. $0, {^-5}, {^-7}, {^+2}, {^+8}$ $^-7, {^-5}, 0, {^+2}, {^+8}$
18. $0, \dfrac{^-1}{2}, \dfrac{^-9}{10}, \dfrac{^+1}{2}, \dfrac{^+9}{10}$

19. $^-2, {^-6}, {^+1}, 0, {^-3}$ $^-6, {^-3}, {^-2}, 0, {^+1}$
20. $^+5, {^-3}, \dfrac{^-1}{10}, {^-10}, 0$

18. $\dfrac{^-9}{10}, \dfrac{^-1}{2}, 0, \dfrac{^+1}{2}, \dfrac{^+9}{10}$
20. $^-10, {^-3}, \dfrac{^-1}{10}, 0, {^+5}$

21. $-1\frac{1}{2}$, $^{+}1$, $^{+}5$, $^{-}6$, 0
$^{-}6$, $-1\frac{1}{2}$, 0, $^{+}1$, $^{+}5$

22. $^{-}4$, $-3\frac{3}{4}$, $-3\frac{1}{2}$, $-3\frac{1}{4}$, $^{-}3$
$-3\frac{3}{4}$, $^{-}3$, $-3\frac{1}{2}$, $-3\frac{1}{4}$, $^{-}4$

23. $^{-}3$, $^{-}5$, $^{-}7$, $^{-}1$, $^{-}4$
$^{-}7$, $^{-}5$, $^{-}4$, $^{-}3$, $^{-}1$

24. $^{-}1$, $^{+}1$, $^{+}2\frac{1}{2}$, $-2\frac{1}{2}$, 0 $-2\frac{1}{2}$, $^{-}1$, 0, $^{+}1$, $^{+}2\frac{1}{2}$

25. $\frac{-2}{3}$, $\frac{^{+}1}{2}$, $-1\frac{1}{2}$, $\frac{^{+}4}{5}$
$-1\frac{1}{2}$, $\frac{-2}{3}$, $\frac{^{+}1}{2}$, $\frac{^{+}4}{5}$

26. $^{+}1.5$, $^{+}2.0$, 0, $^{-}3.7$, $^{-}0.5$
$^{-}3.7$, $^{-}0.5$, 0, $^{+}1.5$, $^{+}2.0$

27. $^{+}5$, $\frac{-3}{4}$, $^{-}75$, 0, $^{+}1$
$^{-}75$, $\frac{-3}{4}$, 0, $^{+}1$, $^{+}5$

28. $^{-}3.4$, $^{+}3.4$, $^{-}3.3$, $^{+}3.3$, $^{+}1$, $^{-}1$
$^{-}3.4$, $^{-}3.3$, $^{-}1$, $^{+}1$, $^{+}3.3$, $^{+}3.4$

B

Draw a number line. Locate each point as accurately as you can.

Check students' drawings.

Sample: $\frac{-4}{5}$, $^{+}1\frac{1}{3}$, $-1\frac{1}{2}$, $^{+}\pi$

Solution:

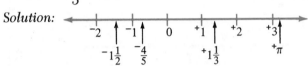

29. $\frac{^{+}2}{5}$, $^{+}1\frac{3}{5}$, $-1\frac{2}{5}$, $-4\frac{1}{2}$

30. $-2\frac{1}{3}$, $\frac{^{+}5}{6}$, $^{+}1\frac{2}{3}$, $\frac{-1}{6}$

31. $\frac{-1}{2}$, $\frac{^{+}1}{2}$, $-1\frac{3}{4}$, $^{+}1\frac{3}{4}$

32. $-1\frac{5}{8}$, $\frac{-3}{8}$, $\frac{^{+}7}{8}$, $^{+}2\frac{1}{8}$

33. $\frac{^{+}6}{10}$, $^{+}1\frac{6}{10}$, $\frac{-6}{10}$, $-1\frac{6}{10}$

34. $\frac{^{+}7}{10}$, $\frac{^{+}3}{10}$, $\frac{^{+}9}{10}$, $\frac{-5}{10}$, $\frac{-1}{10}$

35. $^{-}1.5$, $^{+}1.9$, $^{+}0.3$, $^{-}0.7$

36. $^{-}0.5$, $^{-}5.0$, $^{-}0.05$, $^{+}5.0$, $^{+}0.5$

37. 0, $^{-}2.5$, $^{+}3.5$, $^{+}2.5$

38. $^{-}\pi$, $^{-}3$, $^{+}\pi$, $^{+}3$, $^{-}1$, $^{+}1$

C

Complete. Use $<$ or $>$.

39. $\frac{^{+}3}{4}$ _?_ $\frac{^{+}1}{2}$ $>$

40. $^{-}4.5$ _?_ $^{+}4.8$ $<$

41. $-1\frac{1}{2}$ _?_ $^{+}1\frac{1}{4}$ $<$

42. $^{+}3.7$ _?_ $^{-}3.9$ $>$

Time out

A horse trader sold a horse that had four shoes and six nails in each shoe. The price was set in this way: the buyer was to pay 1¢ for the first nail, 2¢ for the second, 4¢ for the third, and so on, doubling the amount for each nail until all were paid for. What was the price of the horse? $83,886.08

2-5 *More about Graphing Numbers*

OBJECTIVE

Graph sets of numbers on the number line.

A set such as {the numbers greater than ⁺2}, contains infinitely many members. Notice how the graph of such a set is drawn.

EXAMPLE 1 {the numbers greater than ⁺2}

The arrow shows that the set continues on and on.

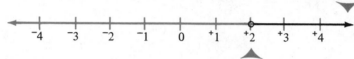

The hollow dot shows that ⁺2 is not included.

EXAMPLE 2 {the numbers between ⁺1 and ⁻3}

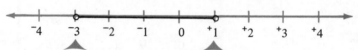

The hollow dots show that ⁻3 and ⁺1 are not included.

The graph of a set such as {the numbers between ⁻2 and ⁺1, inclusive} is just a little different.

EXAMPLE 3 {the numbers between ⁻2 and ⁺1, inclusive.}

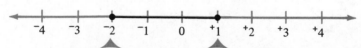

The solid dots show that ⁻2 and ⁺1 are included.

EXAMPLE 4 {⁺1 and the numbers less than ⁺1}

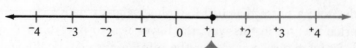

The solid dot shows that ⁺1 is included.

Match each set with its graph below.

1. {the numbers less than ⁻1} C
2. {the numbers between ⁻1 and ⁺2} D
3. {⁺3 and the numbers greater than ⁺3} A
4. {the numbers between ⁻1 and ⁺2, inclusive} E
5. {the numbers greater than 0} B

A.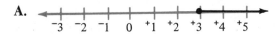

B.

C.

D.

E.

Graph. Check students' graphs.

Sample 1: {the numbers between ⁺2 and 0}
Solution:

Sample 2: {the numbers between ⁺1 and ⁺4, inclusive}
Solution:

A

1. {the numbers between ⁻2 and ⁻1}
2. {the numbers between ⁻1 and ⁺2, inclusive}
3. {the numbers between 0 and ⁺3}
4. {the numbers between ⁻3 and 0, inclusive}
5. {the numbers between ⁺2 and ⁻2}
6. {the numbers between ⁻2 and ⁺2, inclusive}
7. {the numbers between ⁻1 and ⁺4}
8. {the numbers between ⁻4 and ⁻2, inclusive}

Graph. Check students' graphs.

Sample 1:   {the numbers greater than $^+1$}

Solution: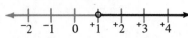

Sample 2:   {the numbers less than $^-2$}

Solution:

9.   {the numbers greater than $^+4$}
10. {the numbers less than $^+5$}
11. {the numbers greater than $^-2$}
12. {$^-4$ and the numbers less than $^-4$}
13. {the numbers greater than 0}
14. {the numbers less than $^-1$}
15. {$^-3$ and the numbers greater than $^-3$}
16. {0 and the numbers less than 0}

Graph. Check students' graphs.

Sample: {the numbers between $^-3$ and $^+1$, including $^+1$}

Solution:

17. {the numbers between $^-2$ and $^+2$, including $^+2$}
18. {$^-3$ and the numbers between $^-3$ and 0}
19. {the numbers between $^-3$ and $^+2$, including $^-3$}
20. {$^-1$ and the numbers between $^-1$ and $^+1$}

B 21. {the numbers between $^-2$ and $^+1\frac{1}{4}$}
22. {the numbers between $^-1$ and $^+2\frac{1}{2}$}
23. {the numbers between $^-1\frac{1}{2}$ and $^+3\frac{1}{2}$, inclusive}
24. {the numbers between $^-4\frac{1}{3}$ and $^+3$, inclusive}
25. {$^+2\frac{3}{4}$ and the numbers between 0 and $^+2\frac{3}{4}$}
26. {$^-1\frac{1}{2}$ and the numbers between $^-1\frac{1}{2}$ and $^+4$}
27. {the numbers between $^-3.5$ and $^+3.5$}
28. {the numbers between $^-\pi$ and $^+\pi$}
29. {$\dfrac{^-2}{3}$ and the numbers between $\dfrac{^-2}{3}$ and 0}

2-6 *Graphing Pairs of Integers*

OBJECTIVES

Use coordinate axes to graph ordered pairs of integers.

Name number pairs as coordinates of points in the plane.

Two perpendicular number lines called coordinate axes may be used to graph an ordered pair of numbers, such as ($^+$2, $^-$3). The numbers in the pair are called coordinates.

($^+$2, $^-$3)

first coordinate second coordinate

To graph ($^+$2, $^-$3), begin at the point (0, 0) where the number lines intersect. This point is called the origin. Count **two** units to the **right**, then **three** units **down**.

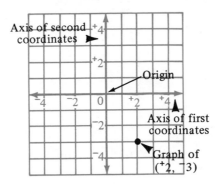

EXAMPLE 1 Graph the ordered pairs:
($^+$4, $^+$1)   ($^-$2, $^+$3)
($^+$3, 0)   ($^-$2, $^-$4)

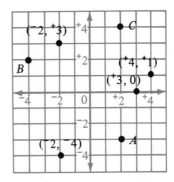

EXAMPLE 2 Name the ordered pair for each lettered point.
A, ($^+$2, $^-$3)
B, ($^-$4, $^+$2)
C, ($^+$2, $^+$4)

Oral
EXERCISES

Tell how to locate the graph.

Sample: ($^+$3, $^-$4): _?_ units to the right, _?_ units down.
What you say: **Three** units to the **right, four** units **down.**

1. ($^+$1, $^+$5): _?_ unit to the right, _?_ units up. one, five
2. ($^+$2, $^+$4): _?_ units to the right, _?_ units up. two, four
3. ($^-$3, $^+$1): _?_ units to the left, _?_ units up. three, one
4. ($^-$5, $^+$2): _?_ units to the left, _?_ units up. five, two
5. ($^-$2, $^-$7): _?_ units to the left, _?_ units down. two, seven
6. (0, $^-$5): _?_ units to left or right, _?_ units down. zero, five

Written
EXERCISES

A

Match the ordered pair with the letter that names its graph.

1. ($^+$4, $^+$2) B
2. ($^-$1, $^-$2) C
3. ($^-$1, $^+$2) D
4. ($^+$2, $^-$4) E
5. ($^-$4, $^+$2) A

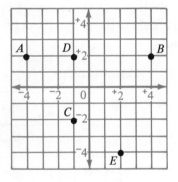

Name the coordinates of each point.

Sample: R   *Solution:* ($^+$5, $^-$2)

6. P ($^-$1$\frac{1}{2}$, $^+$5)
7. N ($^+$1, $^+$5)
8. M ($^+$3, $^+$4)
9. T ($^-$4, 0)
10. Q ($^-$2, $^+$1)
11. W ($^+$5, 0)
12. L ($^-$4, $^-$3)
13. K ($^-$2, $^-$5)
14. B (0, $^-$3)
15. H ($^+$2, $^-$2)

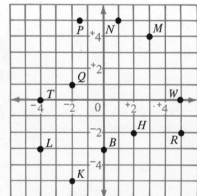

Draw axes and graph. Check students' graphs.

Sample: (⁻2, ⁺3)   *Solution:*

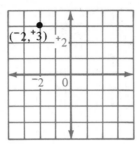

16. (⁺1, ⁺3)   **17.** (0, 0)
18. (⁺1, ⁻5)   **19.** (0, ⁻3)
20. (⁻2, ⁺3)   **21.** (0, ⁺6)
22. (⁻4, ⁻4)   **23.** (⁺4, 0)
24. (⁻8, ⁺2)   **25.** (⁻7, 0)

Name the coordinates.

Sample: 8 units right, 2 units down   *Solution:* (⁺8, ⁻2)

B

26. 1 unit right, 5 units down (⁺1, ⁻5)
27. 2 units right, 0 units up or down (⁺2, 0)
28. 5 units left, 1 unit down (⁻5, ⁻1)
29. 8 units left, 5 units up (⁻8, ⁺5)
30. 3 units right, 2 units down (⁺3, ⁻2)
31. 0 units left or right, 6 units up (0, ⁺6)
32. 4 units right, 1 unit up (⁺4, ⁺1)
33. 0 units left or right, 6 units down (0, ⁻6)

Name the points. Describe the pattern.

Sample:

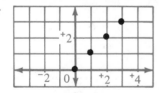

Solution: (0, 0), (⁺1, ⁺1),
       · (⁺2, ⁺2) and (⁺3, ⁺3)

The first and second coordinates are the same positive number.

34.

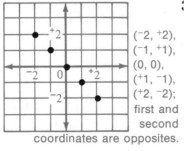

(⁻2, ⁺2),
(⁻1, ⁺1),
(0, 0),
(⁺1, ⁻1),
(⁺2, ⁻2);
first and
second
coordinates are opposites.

35.

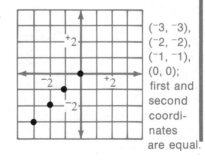

(⁻3, ⁻3),
(⁻2, ⁻2),
(⁻1, ⁻1),
(0, 0);
first and
second
coordinates
are equal.

Name the points. Describe the pattern.

36.

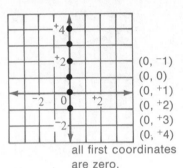

(0, ⁻1)
(0, 0)
(0, ⁺1)
(0, ⁺2)
(0, ⁺3)
(0, ⁺4)

all first coordinates
are zero.

37.

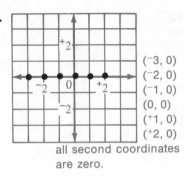

(⁻3, 0)
(⁻2, 0)
(⁻1, 0)
(0, 0)
(⁺1, 0)
(⁺2, 0)

all second coordinates
are zero.

SELF-TEST 2

Be sure that you understand these terms and symbols.

$<$ (p. 41) $>$ (p. 41)   hollow dot (p. 44)
coordinate axes (p. 47)   ordered pair (p. 47)   origin (p. 47)

Section 2-4 p. 40   **1.** Name the coordinates of A, V, J, L and T.

$\frac{-3}{4}$; ⁺3$\frac{1}{4}$; ⁺2$\frac{1}{4}$; ⁺1$\frac{1}{2}$; ⁻2

(number line with points: T, A, L, J, V marked above; ⁻2, ⁻1, 0, ⁺1, ⁺2, ⁺3 below)

2. Complete using $<$ or $>$. $\frac{-3}{4}$ __?__ $\frac{-1}{4}$ $<$

3. Arrange ⁻2$\frac{3}{4}$, ⁻5, $\frac{+9}{10}$, ⁺1, $\frac{+1}{2}$ in order from least to the greatest.

⁻5, ⁻2$\frac{3}{4}$, $\frac{+1}{2}$, $\frac{+9}{10}$, ⁺1

Section 2-5 p. 44   **4.** Graph {the numbers between ⁻2 and ⁺3}.

Section 2-6 p. 47   **5.** Name the coordinates of P, Q, Z, and T.

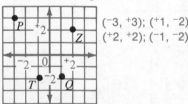

(⁻3, ⁺3); (⁺1, ⁻2);
(⁺2, ⁺2); (⁻1, ⁻2)

6. Draw axes and graph (⁻3, ⁺2).

Check your answers with those printed at the back of the book.

chapter summary

1. The set of **whole numbers** is written: $\{0, 1, 2, 3, 4, 5, \ldots\}$.

2. The set of **integers** includes both positive and negative numbers. Positive numbers are indicated by small raised plus signs $(^+)$. Negative numbers are indicated by small raised minus signs $(^-)$.

3. The set of integers includes 0, the positive whole numbers, and their **opposites.** The set of integers is written:

$$\{\ldots\ ^-3,\ ^-2,\ ^-1,\ 0,\ ^+1,\ ^+2,\ ^+3,\ \ldots\}.$$

Zero is neither positive nor negative.

4. Integers may be matched with equally spaced points on the number line. An integer is the coordinate of the point with which it is matched. The point is marked with a dot to show the **graph** of the number.

5. Every point on the number line can be matched with a positive number, a negative number, or zero. Every number has an opposite. Zero is its own opposite.

6. The symbol $>$ means **is greater than** and the symbol $<$ means **is less than.** These symbols are used to compare numbers.

7. Positive and negative numbers can be graphed on the number line.

8. An ordered number pair can be graphed on a plane, using coordinate axes consisting of two perpendicular number lines.

calculator corner

The decimal form of $6\frac{2}{3}$ or $\frac{20}{3}$ is a repeating decimal, $6.\overline{6}$, larger than any calculator display can show. You can use this decimal to tell if your calculator rounds when it performs division. Divide 20 by 3. If the display shows 6.666 . . . . 7, your calculator rounds. If the display shows 6.666 . . . , your calculator "truncates" or merely cuts off any digits not displayed.

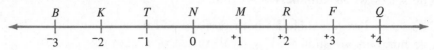

chapter test

Name the coordinate.

	B	K	T	N	M	R	F	Q
	$^-3$	$^-2$	$^-1$	0	$^+1$	$^+2$	$^+3$	$^+4$

1. B $^-3$ 2. R $^+2$ 3. Q $^+4$

4. M $^+1$ 5. K $^-2$ 6. The point half the distance from T to M N

7. T $^-1$ 8. F $^+3$ 9. The point half the distance from K to N T

Name the opposite.

10. $^+6$ $_-6$ 11. $^-8$ $_+8$ 12. 0 $_0$ 13. $\dfrac{^+3}{4}$ $_{\frac{-3}{4}}$

14. $^-18$ $_{+18}$ 15. $^+10$ $_{-10}$ 16. $\dfrac{^-1}{2}$ $_{\frac{+1}{2}}$ 17. $^+1.5$ $_{-1.5}$

Complete the pattern.
18. $^-4, ^-3, ^-2, ^-1, 0, ^+1, ^+2$ **19.** $^-5, ^-4, ^-3, ^-2, ^-1, 0$
20. $^-1, 0, ^+1, ^+2, ^+3, ^+4, ^+5$ **21.** $^-26, ^-25, ^-24, ^-23, ^-22, ^-21$

18. $^-4, ^-3, ^-2, \underline{\ ?\ }, \underline{\ ?\ }, \underline{\ ?\ }, ^+2$ 19. $\underline{\ ?\ }, \underline{\ ?\ }, ^-3, ^-2, ^-1, \underline{\ ?\ }$

20. $^-1, 0, \underline{\ ?\ }, \underline{\ ?\ }, \underline{\ ?\ }, \underline{\ ?\ }, ^+5$ 21. $^-26, ^-25, \underline{\ ?\ }, \underline{\ ?\ }, \underline{\ ?\ }, ^-21$

Graph. Use the number line. Check students' graphs.

22. The integers between $^+1$ and $^-3$.

23. The integers between 0 and $^+4$, inclusive.

24. The numbers greater than $^-1$.

25. The numbers between $^-2$ and $^+2$.

Complete. Use $<$ or $>$.

26. $^+5 \underline{\ ?\ } ^+2$ $>$ 27. $^+3 \underline{\ ?\ } ^-1$ $>$ 28. $0 \underline{\ ?\ } ^-10$ $>$

29. $0 \underline{\ ?\ } ^+4$ $<$ 30. $^-5 \underline{\ ?\ } ^-2$ $<$ 31. $^-4 \underline{\ ?\ } ^-7\frac{1}{2}$ $>$

Draw axes and graph. Check students' graphs.

32. $(^-3, ^+3)$, $(^-2, ^+4)$, $(^+5, ^+4)$, $(^+4, ^-2)$, and $(^+1, ^+3)$.

33. $(^-1, ^-6)$, $(0, ^+3)$, $(0, ^-3)$, $(^+5, 0)$, and $(^-5, 0)$.

challenge topics

Sugar Cubes

Each one of eight sugar cubes is 1 centimeter on an edge. The eight cubes can be arranged in several different ways to form three-dimensional shapes. One way is shown here.

1 cm
1 cm
8 cm

How many faces of cubes could be shown in this shape? Imagine covering the shape with graph paper that is divided into centimeter squares. Make a diagram of the graph paper needed to cover the shape. Count the number of squares in your diagram. The number of square centimeters is the surface area of the shape. Draw sketches of a cube and other shapes you can make with the eight sugar cubes. Make a diagram of the graph paper needed to cover each shape. Copy the chart below and use it to record your findings. List the shapes in order from least to greatest, according to the surface area.

Sketch Of Shape	Surface Area	Faces Showing

Do you notice any similarities in your completed chart? Explain them. Which shape is listed first? Which is listed second? Do you notice any physical similarities between the two shapes? Which shape is listed last? Is this shape different from those listed first? How? Make a statement about relationships between shapes and surface areas.

The surface area is the same as the total number of faces showing. The figure listed first should be a cube. The figure listed second should be the one most nearly shaped like a cube. The figure listed last should be the one shown in the diagram on page 53. The figure made with 8 one-centimeter cubes having the least surface area is a cube. For a given volume, the three-dimensional rectangular figure with least surface area is a cube.

Review of Skills

Complete.

1. $2 \times \underline{\ ?\ } = 18$ 9

2. $\underline{\ ?\ } \times 5 = 35$ 7

3. $7 \times \underline{\ ?\ } = 28$ 4

4. $4 \times \underline{\ ?\ } = 32$ 8

5. $\dfrac{8}{1} = \underline{\ ?\ }$ 8

6. $9 \times \underline{\ ?\ } = 45$ 5

7. $\underline{\ ?\ } \times 7 = 28$ 4

8. $\dfrac{?}{2} = 4$ 8

9. $\underline{\ ?\ } = \dfrac{4}{1}$ 4

Name the place value of the underlined digit.

Sample: 34<u>8</u>.25   *Solution:*   8 is in the ones place.

10.   2.<u>9</u>3 tenths

11.   <u>6</u>39.1 hundreds

12.   31.89<u>5</u> thousandths

13.   45.0<u>7</u> hundredths

14.   0.004<u>2</u> ten thousandths

15.   5.<u>0</u> tenths

16.   5.<u>7</u>2 tenths

17.   1.23<u>3</u> thousandths

18.   0.000<u>1</u>
ten thousandths

Simplify.

Sample: $(2 \times 9) + 1$ *Solution:* $(2 \times 9) + 1 = 18 + 1 = 19$

19. $(2 \times 5) + 3$ $_{10 + 3 = 13}$

20. $(5 \times 4) + 1$ $_{20 + 1 = 21}$

21. $3 + (4 \times 4)$ $_{3 + 16 = 19}$

22. $(4 \times 6) + 1$ $_{24 + 1 = 25}$

23. $(2 \times 7) + 1$ $_{14 + 1 = 15}$

24. $(9 \times 2) + 1$ $_{18 + 1 = 19}$

25. $(2 \times 2) + 2$ $_{4 + 2 = 6}$

26. $3 \times (1 + 1)$ $_{3 \times 2 = 6}$

27. $1 \times (1 + 1)$ $_{1 \times 2 = 2}$

Divide. Write the quotient as a decimal.

28. $5\overline{)2}$ 0.4

29. $2\overline{)5}$ 2.5

30. $5\overline{)4}$ 0.8

31. $5\overline{)3}$ 0.6

32. $4\overline{)3}$ 0.75

33. $8\overline{)4}$ 0.5

34. $8\overline{)7}$ 0.875

35. $4\overline{)1}$ 0.25

36. $8\overline{)5}$ 0.625

Left: Actors producing sound effects for 1925 radio program "Rip Van Winkle."

Right: Television sound engineer.

3 Factors and Multiples; Decimals and the Metric System

Factors and Multiples

3-1 *Multiples and Common Multiples*

OBJECTIVES

Name the multiples of a number.

Name the common multiples and the least common multiple of two numbers.

From now on, we will not use $^+$ signs for positive numbers. For example, $^+1$ will be written 1.

The counting numbers are the numbers 1, 2, 3, 4, 5, 6, 7, and so on. The multiples of a whole number are found by multiplying the number by the counting numbers. That is, to find the multiples of 3, we replace n in $3 \times n$ with counting numbers. (Remember 0 is *not* a counting number.)

Replace n with counting numbers.

EXAMPLE 1

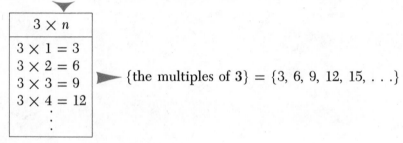

$$3 \times n$$

$3 \times 1 = 3$
$3 \times 2 = 6$
$3 \times 3 = 9$
$3 \times 4 = 12$
$\vdots$

{the multiples of 3} = {3, 6, 9, 12, 15, . . .}

To find the multiples of 5 we replace n in $5 \times n$ with counting numbers.

EXAMPLE 2

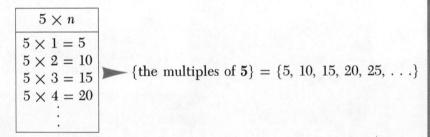

$$5 \times n$$

$5 \times 1 = 5$
$5 \times 2 = 10$
$5 \times 3 = 15$
$5 \times 4 = 20$
$\vdots$

{the multiples of 5} = {5, 10, 15, 20, 25, . . .}

The set of common multiples of 3 and 5 is {15, 30, 45, 60, . . .}. The least common multiple (LCM) of 3 and 5 is **15**.

Complete.

EXERCISES

Sample: {the multiples of 10} = {10, 20, 30, _?_, _?_, _?_, ...}
What you say: The set 10, 20, 30, 40, 50, 60, and so on.

1. {the multiples of 2} = {2, 4, 6, _?8_, _?10_, _?12_, _?14_, ...}
2. {the multiples of 6} = {6, 12, 18, _?24_, _?30_, _?36_, _?42_, ...}
3. {the multiples of 4} = {4, 8, 12, _?16_, _?20_, _?24_, _?28_, ...}
4. {the multiples of 100} = {100, 200, 300, $\frac{?}{400}$, $\frac{?}{500}$, $\frac{?}{600}$, $\frac{?}{700}$, ...}

Complete. Use 2, 3, 4, or 5.

Sample: 12 is divisible by _?_, _?_, and _?_.
12 is a multiple of _?_, _?_, and _?_.
What you say: 12 is divisible by 2, 3, and 4.
12 is a multiple of 2, 3, and 4.

5. 10 is divisible by _?_ and _?_. 2; 5
10 is a multiple of _?_ and _?_. 2; 5
6. 15 is divisible by _?_ and _?_. 3; 5
15 is a multiple of _?_ and _?_. 3; 5
7. 30 is divisible by _?_, _?_ and _?_. 2; 3; 5
30 is a multiple of _?_, _?_ and _?_. 2; 3; 5

Solve each equation. Then write the set of multiples.

EXERCISES

Sample: $1 \times 7 = s$ *Solution:* $s = 7$
$2 \times 7 = b$ $b = 14$
$3 \times 7 = n$ $n = 21$
$4 \times 7 = y$ $y = 28$
{the multiples of 7} = _?_   {7, 14, 21, 28, ...}

A

1. $1 \times 9 = a$ 9
$2 \times 9 = h$ 18
$3 \times 9 = m$ 27
$4 \times 9 = w$ 36
{the multiples of 9} = _?_

2. $1 \times 12 = c$ 12
$2 \times 12 = k$ 24
$3 \times 12 = x$ 36
$4 \times 12 = s$ 48
{the multiples of 12} = _?_

3. $1 \times 4 = t$ 4
$2 \times 4 = r$ 8
$3 \times 4 = p$ 12
$4 \times 4 = n$ 16
{the multiples of 4} = _?_

4. $1 \times 8 = f$ 8
$2 \times 8 = z$ 16
$3 \times 8 = y$ 24
$4 \times 8 = n$ 32
{the multiples of 8} = _?_

Complete the table.

	Numbers	Sets of Multiples	Common Multiples	LCM
5.	3 2	{3, 6, 9, 12, 15, 18, ...} {2, 4, 6, 8, 10, 12, ...}	{6, 12, 18, ...}	? 6
6.	4 5	{4, 8, 12, 16, ?20, 24. . .} {5, 10, 15, 20,? 25, 30, . . .}	? {20, 40, 60, . . .}	? 20
7.	3 4	{3, 6, 9, 12, ?15, 18, . . .} {4, 8, 12, 16,? 20, 24, . . .}	? {12, 24, 36. . .}	? 12
8.	6 10	{6, 12, 18, 24,? 30, 36, . . .} {10, 20, 30, ?40, 50, 60. . .}	? {30, 60, 90, . . .}	? 30

Name the least common multiple (LCM).

Sample 1: 4 and 8   *Solution:* 8
Sample 2: 4 and 9   *Solution:* 36

9. 3 and 5 15   10. 2 and 5 10   11. 3 and 4 12
12. 3 and 6 6   13. 12 and 4 12   14. 4 and 6 12
15. 5 and 10 10   16. 2 and 10 10   17. 8 and 16 16

B

18. 5 and 8 40   19. 6 and 9 18   20. 10 and 15 30
21. 12 and 15 60   22. 7 and 9 63   23. 15 and 20 60
24. 7 and 8 56   25. 5 and 11 55   26. 9 and 10 90

Complete. Name the greatest multiple of 10 that makes the statement true.

Sample: _?_ < 48   *Solution:* 40 < 48

27. _?_ < 14 10   28. _?_ < 21 20   29. _?_ < 47 40
30. _?_ < 86 80   31. _?_ < 29 20   32. _?_ < 31 30
33. _?_ < 43 40   34. _?_ < 125 120   35. _?_ < 279 270

Complete. Name the greatest multiple of 8 that makes the statement true.

Sample: _?_ < 63   *Solution:* 56 < 63

C

36. _?_ < 75 72   37. _?_ < 100 96   38. _?_ < 130 128
39. _?_ < 86 80   40. _?_ < 203 200   41. _?_ < 148 144

3-2 *Common Factors*

OBJECTIVES

Name the whole number factors of a number.

Name the common factors and the greatest common factor of two numbers.

We can use either multiplication or division to find the whole number factors of a number.

$$8 = 8 \times 1 \qquad\qquad 8 \div 1 = 8$$
$$8 = 2 \times 4 \qquad\qquad 8 \div 2 = 4$$

factors of 8 $\qquad$ factors of 8

The set of factors of 8 is $\{1, 2, 4, 8\}$.

$$20 = 1 \times 20 \qquad 20 \div 1 = 20$$
$$20 = 2 \times 10 \qquad 20 \div 2 = 10$$
$$20 = 4 \times 5 \qquad 20 \div 4 = 5$$

The set of factors of 20 is $\{1, 2, 4, 5, 10, 20\}$. We can use a diagram to show the **common factors** of 8 and 20.

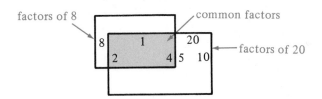

The set of common factors of 8 and 20 is $\{1, 2, 4\}$. The **greatest common factor** (GCF) of 8 and 20 is 4.

EXAMPLE 1 Factors of 6: $\{1, 2, 3, 6\}$ $\qquad$ Factors of 18: $\{1, 2, 3, 6, 9, 18\}$
Common factors of 6 and 18: $\{1, 2, 3, 6\}$
The GCF of 6 and 18 is 6.

EXAMPLE 2 Factors of 7: $\{1, 7\}$ $\qquad$ Factors of 9: $\{1, 3, 9\}$
Common factor of 7 and 9: $\{1\}$
The GCF of 7 and 9 is 1.

Oral EXERCISES

Name the common factors and the GCF.

Sample:

Numbers	Factors
6	{**1**, 2, **3**, 6}
15	{**1**, **3**, 5, 15}

What you say:
The common factors are
1 and 3. The GCF is 3.

1.

Numbers	Factors
3	{1, 3}
5	{1, 5}

{1}; 1

2.

Numbers	Sets of Factors
12	{1, 2, 3, 4, 6, 12}
20	{1, 2, 4, 5, 10, 20}

{1, 2, 4}; 4

3.

Numbers	Factors
7	{1, 7}
8	{1, 2, 4, 8}

{1}; 1

4.

Numbers	Factors
9	{1, 3, 9}
6	{1, 2, 3, 6}

{1, 3}; 3

Written EXERCISES

A

Match.

1. {the factors of 18} D
2. {the factors of 21} E
3. {the factors of 11} A
4. {the factors of 22} C
5. {the factors of 9} B

A. {1, 11}
B. {1, 3, 9}
C. {1, 2, 11, 22}
D. {1, 2, 3, 6, 9, 18}
E. {1, 3, 7, 21}

Replace the variable in each equation. Then list the factors.

Sample 1: $1 \times r = 8$
$2 \times s = 8$
{the factors of 8} = ___?___

Solution: $1 \times 8 = 8$
$2 \times 4 = 8$
{the factors of 8} = {1, 2, 4, 8}

Sample 2: $\dfrac{10}{1} = W$

$\dfrac{10}{2} = Z$

{the factors of 10} = ___?___

Solution: $\dfrac{10}{1} = 10$

$\dfrac{10}{2} = 5$

{the factors of 10} = {1, 2, 5, 10}

6. $1 \times a = 16$ 16
  $2 \times b = 16$ 8
  $4 \times c = 16$ 4
  {the factors of 16} = _?_ {1, 2, 4, 8, 16}

7. $1 \times V = 35$ 35
  $5 \times W = 35$ 7
  {the factors of 35} = _?_
  {1, 5, 7, 35}

8. $\dfrac{6}{1} = m$ 6

$\dfrac{6}{2} = n$ 3

{the factors of 6} = _?_ {1, 2, 3, 6}

9. $\dfrac{17}{1} = k$ 17

{the factors of k} = _?_
{1, 17}

List the numbers in each set.

Sample: {the factors of 27}   *Solution:* 1, 3, 9, 27

10. {the factors of 4} 1, 2, 4
11. {the factors of 26} 1, 2, 13, 26
12. {the factors of 24}
13. {the factors of 14} 1, 2, 7, 14
14. {the factors of 30}
15. {the factors of 13} 1, 13
16. {the factors of 15} 1, 3, 5, 15
17. {the factors of 50} 1, 2, 5, 10, 25, 50

12. 1, 2, 3, 4, 6, 8, 12, 24   **14.** 1, 2, 3, 5, 6, 10, 15, 30

Name the common factors and the greatest common factor (GCF).

20. {1, 3}; 3

B

18. 6 and 4 {1, 2}; 2
19. 8 and 12 {1, 2, 4}; 4
20. 21 and 27
21. 16 and 20 {1, 2, 4}; 4
22. 5 and 15 {1, 5}; 5
23. 9 and 18
24. 20 and 25 {1, 5}; 5
25. 5 and 7 {1}; 1
26. 2 and 11

{1, 3, 9}; 9
{1}; 1

Name the GCF.

27. 9 and 12 3
28. 4 and 10 2
29. 14 and 18 2
30. 27 and 45 9
31. 15 and 16 1
32. 24 and 30 6

True or false?

33. 1 is a factor of every counting number. True

34. Every counting number is a factor of itself. True

35. The GCF of two consecutive counting numbers is always 1. True

36. The GCF of two odd numbers is always 1. False

C

3-3 *Special Sets of Whole Numbers*

OBJECTIVES

Express an even number in the form 2 × *n*.

Express an odd number in the form (2 × *n*) + 1.

Identify prime numbers.

Beginning with zero, we count by 2's to name the even numbers. Each even number can be expressed in the form $2 \times n$, where n is a whole number.

$2 \times n$
$2 \times 0 = 0$
$2 \times 1 = 2$
$2 \times 2 = 4$
$2 \times 3 = 6$
$\vdots$

▶ {the even numbers} = $\{0, 2, 4, 6, 8, 10, 12, \ldots\}$

A whole number that is not even is an odd number. Each odd number can be expressed in the form $(2 \times n) + 1$, where n is a whole number.

$(2 \times n) + 1$
$(2 \times 0) + 1 = 1$
$(2 \times 1) + 1 = 3$
$(2 \times 2) + 1 = 5$
$(2 \times 3) + 1 = 7$
$\vdots$

▶ {the odd numbers} = $\{1, 3, 5, 7, 9, 11, 13, \ldots\}$

A whole number that has exactly two different factors, itself and 1, is a prime number.

EXAMPLES

Number	Set of Factors	Prime/Not Prime
2	$\{1, 2\}$	Prime
5	$\{1, 5\}$	Prime
1	$\{1\}$	Not Prime
15	$\{1, 3, 5, 15\}$	Not Prime

Simplify. Tell whether the result is an odd or an even number.

Sample 1: 2×7 *What you say:* $2 \times 7 = 14$; 14 is an even number.

Sample 2: $(2 \times 4) + 1$ *What you say:* $(2 \times 4) + 1 = 9$; 9 is an odd number.

1. 2×5 even
2. 2×8 even
3. $(2 \times 8) + 1$ | odd

4. $(2 \times 3) + 1$ odd
5. $(2 \times 25) + 1$ odd
6. 2×25 even

7. 2×100 even
8. $(2 \times 100) + 1$ odd
9. $(2 \times 30) + 1$ | odd

Write in the form $2 \times n$ if possible, where n is a whole number.

Sample 1: 60   *Solution:* $60 = 2 \times 30$

Sample 2: 19   *Solution:* Not possible.

1. 14 2×7
2. 20 2×10
3. 17 not possible

4. 30 2×15
5. 7 not possible
6. 41 not possible

7. 72 2×36
8. 0 2×0
9. 11 not possible

Write in the form $(2 \times n) + 1$ if possible, where n is a whole number.

Sample 1: 15   *Solution:* $15 = 14 + 1 = (2 \times 7) + 1$

10. 9 $(2 \times 4) + 1$
11. 28 not possible
12. 5 $(2 \times 2) + 1$

13. 23 $(2 \times 11) + 1$
14. 27 $(2 \times 13) + 1$
15. 41 $(2 \times 20) + 1$

16. 50 not possible
17. 55 $(2 \times 27) + 1$
18. 17 $(2 \times 8) + 1$

Add. Label each number *odd* or *even*.

Sample: $40 + 12$ *Solution:* $40 + 12 = 52$
even + even = even

e = even
o = odd

19. $6 + 8$ e + e = e
20. $12 + 10$ e + e = e
21. $4 + 0$ e + e = e

22. $2 + 16$ e + e = e
23. $7 + 9$ o + o = e
24. $11 + 11$ | o + o = e

25. $5 + 1$ o + o = e
26. $3 + 15$ o + o = e
27. $8 + 9$ e + o = o

Complete.

28. The last digit of an even number is 0, 2, _?4_, _?6_, or _?8_.

29. The last digit of an odd number is 1, 3, _?5_, _?7_, or _?9_.

Write the set of factors. Tell which numbers are prime.

Sample: 17: {_?_, _?_}   *Solution:* 17: {1, 17}; 17 is prime.

B

30. 5: {_?_, _?_} **31.** 14: {_?_, _?_, _?_, _?_} **32.** 7: {_?_, _?_}
{1, 5}; 5   {1, 2, 7, 14}; 2, 7   {1, 7}; 7

33. 11 {1, 11}; 11 **34.** 2 {1, 2}; 2   **35.** 4 {1, 2, 4}; 2

36. 23 {1, 2, 3}; 23 **37.** 57 {1, 3, 19, 57}; 3, 19   **38.** 61 {1, 61}; 61

Complete. Use odd or even.

39. even + even = _?_ even   **40.** odd + odd = _?_ even

41. odd + even = _?_ odd   **42.** even + even + even = _?_
                                                                            even

43. even × 1 = _?_ even   **44.** even × 3 = _?_ even

Answers may vary for Exs. 46–50. One possible solution is given.

Write each even number as the sum of two prime numbers.

C

45. 8 3 + 5   **46.** 78 71 + 7   **47.** 50 31 + 19

48. 12 7 + 5   **49.** 100 97 + 3   **50.** 84 79 + 5

SELF-TEST 1

Be sure that you understand these terms.

counting numbers (p. 56)   multiple (p. 56)
common multiple (p. 56)   least common multiple (p. 56)
factor (p. 59)   greatest common factor (p. 59)
even number (p. 62)   odd number (p. 62)
prime number (p. 62)

Section 3-1, p. 56 Solve for each variable. Then write the set of multiples.

1. 1 × 7 = r 7   **2.** 1 × 3 = d 3
   2 × 7 = s 14   2 × 3 = e 6
   3 × 7 = t 21   3 × 3 = f 9

3. Name the least common multiple of 7 and 11. 77

Section 3-2, p. 59 **4.** List the numbers in {the factors of 10}. {1, 2, 5, 10}

5. Name the common factors and the GCF of 6 and 3. {1, 3}; 3

Section 3-3, p. 62 **6.** Write 18 in the form 2 × n, where n is a whole number. 2 × 9

7. Write 15 in the form (2 × n) + 1, where n is a whole number.
                                                          (2 × 7) + 1

Check your answers with those printed at the back of the book.

Fractions, Decimals, and Percents

3-4 *Fractions*

OBJECTIVES

Write equivalent forms of a fraction: $\dfrac{2}{3} = \dfrac{4}{6} = \dfrac{6}{9}, \ldots$

Write a fraction in lowest terms.

Fractions that name the same number are called equivalent fractions. To write a fraction in the form of an equivalent fraction, we either multiply or divide both the numerator and the denominator by the same number (except 0).

EXAMPLE 1 $\dfrac{5}{8} = \dfrac{5 \times 3}{8 \times 3} = \dfrac{15}{24}$ ▶ $\dfrac{5}{8}$ and $\dfrac{15}{24}$ are equivalent fractions.

EXAMPLE 2 $\dfrac{24}{30} = \dfrac{24 \div 6}{30 \div 6} = \dfrac{4}{5}$ ▶ $\dfrac{24}{30}$ and $\dfrac{4}{5}$ are equivalent fractions.

A fraction is in lowest terms when the greatest common factor (GCF) of the numerator and denominator is 1.

EXAMPLE 3 $\dfrac{2}{3}, \dfrac{5}{8}, \dfrac{3}{4},$ and $\dfrac{10}{7}$ are fractions in lowest terms.

$\dfrac{8}{24}, \dfrac{12}{16},$ and $\dfrac{20}{15}$ are fractions *not* in lowest terms.

We can divide the numerator and denominator by common factors to write a fraction in lowest terms.

EXAMPLE 4 $\dfrac{8}{24} = \dfrac{8 \div 2}{24 \div 2} = \dfrac{4}{12} = \dfrac{4 \div 2}{12 \div 2} = \dfrac{2}{6} = \dfrac{2 \div 2}{6 \div 2} = \dfrac{1}{3}$

$\dfrac{8}{24}$ in lowest terms is $\dfrac{1}{3}$.

The quickest way is to divide the numerator and denominator by their GCF.

EXAMPLE 5 $\dfrac{8}{24} = \dfrac{8 \div 8}{24 \div 8} = \dfrac{1}{3}$ ◀ The GCF of 8 and 24 is 8.

EXERCISES

Tell how to complete the pattern of equivalent fractions.

Sample: $\dfrac{1}{3}, \dfrac{2}{6}, \dfrac{3}{9}, \dfrac{?}{\rule{1em}{0.4pt}}, \dfrac{?}{\rule{1em}{0.4pt}}, \dfrac{?}{\rule{1em}{0.4pt}}$ *What you say:* $\dfrac{4}{12}, \dfrac{5}{15}, \dfrac{6}{18}$

1. $\dfrac{1}{5}, \dfrac{2}{10}, \dfrac{3}{15}, \dfrac{?4}{?20}, \dfrac{?5}{?25}, \dfrac{?6}{?30}$ 2. $\dfrac{2}{3}, \dfrac{4}{6}, \dfrac{6}{9}, \dfrac{?8}{?12}, \dfrac{?10}{?15}, \dfrac{?12}{?18}$

3. $\dfrac{3}{2}, \dfrac{6}{4}, \dfrac{9}{6}, \dfrac{?12}{?8}, \dfrac{?15}{?10}, \dfrac{?18}{?12}$ 4. $\dfrac{6}{12}, \dfrac{7}{14}, \dfrac{8}{16}, \dfrac{?9}{?18}, \dfrac{?10}{?20}, \dfrac{?11}{?22}$

5. $\dfrac{7}{8}, \dfrac{14}{16}, \dfrac{21}{24}, \dfrac{?28}{?32}, \dfrac{?35}{?40}, \dfrac{?42}{?48}$ 6. $\dfrac{2}{5}, \dfrac{?4}{?10}, \dfrac{?6}{?15}, \dfrac{?8}{?20}, \dfrac{10}{25}, \dfrac{12}{30}$

Tell whether or not the fraction is in lowest terms. Explain.

Sample 1: $\dfrac{7}{8}$ *What you say:* $\dfrac{7}{8}$ is in lowest terms because the GCF of 7 and 8 is 1.

Sample 2: $\dfrac{9}{12}$ *What you say:* $\dfrac{9}{12}$ is not in lowest terms because the GCF of 9 and 12 is 3.

7. $\dfrac{3}{4}$ GCF is 1.   8. $\dfrac{2}{10}$ GCF is 2.   9. $\dfrac{14}{15}$ GCF is 1.   10. $\dfrac{6}{8}$ GCF is 2.

11. $\dfrac{10}{16}$ GCF is 2.   12. $\dfrac{15}{36}$ GCF is 3.   13. $\dfrac{7}{8}$ GCF is 1.   14. $\dfrac{5}{4}$ GCF is 1.

For Extra Practice, see page 411.

Written EXERCISES

Solve.

Sample 1: $\dfrac{3 \times 2}{10 \times 2} = t$ *Solution:* $t = \dfrac{6}{20}$

Sample 2: $\dfrac{15 \div 5}{20 \div 5} = y$ *Solution:* $y = \dfrac{3}{4}$

A

1. $\dfrac{1 \times 5}{2 \times 5} = s \quad \dfrac{5}{10}$ 2. $\dfrac{1 \times 3}{4 \times 3} = b \quad \dfrac{3}{12}$ 3. $\dfrac{4 \times 2}{5 \times 2} = k \quad \dfrac{8}{10}$

4. $\dfrac{2 \times 5}{3 \times 5} = y \quad \dfrac{10}{15}$ 5. $\dfrac{2 \div 2}{16 \div 2} = t \quad \dfrac{1}{8}$ 6. $\dfrac{7 \div 7}{7 \div 7} = n \quad 1$

7. $\dfrac{9 \div 3}{24 \div 3} = m \quad \dfrac{3}{8}$ 8. $\dfrac{10 \div 10}{20 \div 10} = r \quad \dfrac{1}{2}$ 9. $\dfrac{5 \div 5}{5 \div 5} = a \quad \dfrac{1}{2}$

10. $\dfrac{8 \div 8}{8 \div 8} = f \quad 1$ 11. $\dfrac{3 \times 1}{4 \times 1} = h \quad \dfrac{3}{4}$ 12. $\dfrac{2 \times 1}{7 \times 1} = d \quad \dfrac{2}{7}$

Write in lowest terms.

13. $\dfrac{14}{20}$ $\dfrac{7}{10}$ **14.** $\dfrac{33}{66}$ $\dfrac{1}{2}$ **15.** $\dfrac{50}{100}$ $\dfrac{1}{2}$ **16.** $\dfrac{20}{100}$ $\dfrac{1}{5}$

17. $\dfrac{6}{10}$ $\dfrac{3}{5}$ **18.** $\dfrac{18}{16}$ $\dfrac{9}{8}$ **19.** $\dfrac{75}{100}$ $\dfrac{3}{4}$ **20.** $\dfrac{100}{1000}$ $\dfrac{1}{10}$

21. $\dfrac{4+10}{20}$ $\dfrac{7}{10}$ **22.** $\dfrac{5+5}{4}$ $\dfrac{5}{2}$ **23.** $\dfrac{4 \times 5}{100}$ $\dfrac{1}{5}$ **24.** $\dfrac{10 \times 9}{100}$ $\dfrac{9}{10}$

Replace the variable to name the equivalent fraction.

Sample: $\dfrac{4}{5} = \dfrac{n}{20}$ *Solution:* $\dfrac{4}{5} = \dfrac{16}{20}$

25. $\dfrac{3}{10} = \dfrac{x}{70}$ 21 **26.** $\dfrac{2}{9} = \dfrac{s}{45}$ 10 **27.** $\dfrac{9}{10} = \dfrac{a}{100}$ 90

28. $\dfrac{72}{48} = \dfrac{y}{2}$ 3 **29.** $\dfrac{25}{100} = \dfrac{m}{20}$ 5 **30.** $\dfrac{5}{9} = \dfrac{w}{63}$ 35

31. $\dfrac{5}{12} = \dfrac{r}{48} = \dfrac{t}{36}$ **32.** $\dfrac{10}{15} = \dfrac{m}{30} = \dfrac{n}{3}$ **33.** $\dfrac{20}{32} = \dfrac{h}{8} = \dfrac{a}{80}$ **B**

34. $\dfrac{s}{5} = \dfrac{40}{100} = \dfrac{c}{35}$ **35.** $\dfrac{x}{10} = \dfrac{5}{2} = \dfrac{y}{16}$ **36.** $\dfrac{m}{21} = \dfrac{z}{35} = \dfrac{4}{7}$

31. $r = 20; t = 15$ **32.** $m = 20; n = 2$ **33.** $h = 5; a = 50$
34. $s = 2; c = 14$ **35.** $x = 25; y = 40$
36. $m = 12; z = 20$

Time out

You can use lined 6 × 9 cards to make a "Multiple Sorter."
Beginning at the left-hand edge of the long side of each card, mark
22 one–cm intervals. Using a hole puncher, punch holes which are
just about centered in each interval. The bottom edge of the hole
should just touch the top line on the card.

Now number the cards from 1 through 22. On the first card,
number the holes from 1 to 22. On each card cut out the top part
of each hole whose number corresponds to a factor of the number
named on the card. For example, on the card labeled 16, cut out
the tops of the first, second, fourth, eighth, and sixteenth holes.

To find the multiples of 3 which are less than 22, straighten out
a paper clip and insert it in the third hole, all the way through the
stack. Lift. The cards naming the multiples of 3 will drop off. To
find the common multiples of 2 and 3 which are less than 22, insert
paper clips in both holes and lift.

3-5 Ratios

Two quantities can be compared by stating the **ratio** of one to the other.

EXAMPLE 1

4 out of 10 of the shapes are triangles.

Ratio of triangles to all the shapes ▶ 4 to 10, or 4:10

Fraction form: $\frac{4}{10}$ or, in lowest terms, $\frac{2}{5}$

Ratio of triangles to squares ▶ 4 to 6, or 4:6

Fraction form: $\frac{4}{6}$ or $\frac{2}{3}$

EXAMPLE 2 There are 15 boys and 12 girls in an algebra class.
Ratio of boys to girls ▶ 15 to 12, or 15:12

Fraction form: $\frac{15}{12} = \frac{5}{4}$

Ratio of girls to boys ▶ 12 to 15, or 12:15

Fraction form: $\frac{12}{15} = \frac{4}{5}$

Oral EXERCISES

Express each ratio as a fraction.

Sample: 8 out of 10 *What you say:* $\frac{8}{10}$

1. 13 out of 40 $\frac{13}{40}$
2. 20 out of 50 $\frac{20}{50}$
3. 95 out of 100 $\frac{95}{100}$
4. 9 out of 12 $\frac{9}{12}$
5. 36 out of 72 $\frac{36}{72}$
6. 5 out of 9 $\frac{5}{9}$
7. 15 out of 60 $\frac{15}{60}$
8. 17 out of 20 $\frac{17}{20}$
9. 10 out of 15 $\frac{10}{15}$
10. 65 out of 80 $\frac{65}{80}$
11. 13 out of 30 $\frac{13}{30}$
12. 17 out of 20 $\frac{17}{20}$

Write each ratio as a fraction in lowest terms.

Sample: 8 out of 12   *Solution:* $\dfrac{8}{12} = \dfrac{2}{3}$

1. 10 out of 15 $\frac{2}{3}$ 2. 28 out of 100 $\frac{7}{25}$ 3. 19 out of 100 $\frac{19}{100}$

4. 20 out of 40 $\frac{1}{2}$ 5. 10 out of 12 $\frac{5}{6}$ 6. 15 out of 25 $\frac{3}{5}$

7. 9 out of 5 $\frac{9}{5}$ 8. 6 out of 20 $\frac{3}{10}$ 9. 25 out of 100 $\frac{1}{4}$

10. 10 out of 16 $\frac{5}{8}$ 11. 8 out of 40 $\frac{1}{5}$ 12. 7 out of 9 $\frac{7}{9}$

Complete.

Sample: □   _?_ out of _?_ shapes are squares.
              □△□

Solution: 3 out of 4 shapes are squares.

13. ◆ ◆ ◇ ◇ ◇   _2?_ out of _5?_ shapes are shaded.

14. ◯ ◯ ▲ ◯   _1?_ out of _4?_ shapes are triangles.

15. ■△△△■   _3?_ out of _5?_ of the shapes are triangles.

16. ⬡ ⬡ ⬢ ⬡   _2?_ out of _4?_ of the shapes are shaded.

17. A, B, C, d, e, f, g   _3?_ out of _7?_ letters are capitals.

Solve. Write all ratios as fractions in lowest terms.

1. A pitcher of lemonade is made from lemon concentrate and water. The ratio of concentrate to water is 1 to 4. What fraction of the pitcher of lemonade is concentrate? What fraction is water? $\frac{1}{5}, \frac{4}{5}$

2. If the vote on an issue is 3725 for and 1250 against, what is the ratio of votes "for" to the total vote cast? $\frac{149}{199}$

3. For the rectangle shown here, what is the ratio of the length to the width? of the sum of the length and width to the length? $\frac{7}{3}, \frac{10}{7}$

   14 cm
   [rectangle] 6 cm

4. The monthly income of the Jones family is $650. If they pay $130 a month for rent, what is the ratio of their rent to their total income? $\frac{1}{5}$

5. The pitch or slope of a drain pipe is shown in the picture. State the ratio $\frac{rise}{run}$ that describes the pitch of the drain pipe.

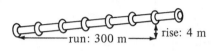

run: 300 m   rise: 4 m

Hint: $\frac{rise}{run} = \frac{4}{300} = \underline{\ ?\ }$ $\frac{1}{75}$

6. A construction company is building a new highway. The road bed is to have a 2 meter rise for every 90 meters of run. The picture shows that the road bed rises 12 meters over a distance of 540 meters. Tell whether or not the grading has been done correctly for this part of the road bed.

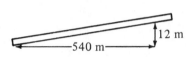

540 m   12 m

(*Hint:* are the fractions $\frac{2}{90}$ and $\frac{12}{540}$ equivalent fractions?)

$\frac{2}{90} = \frac{12}{540}$; the grading has been done correctly.

7. The front sprocket of a bicycle has 42 teeth and the rear sprocket has 14 teeth. Use $\frac{\text{teeth in front sprocket}}{\text{teeth in rear sprocket}}$ to find the ratio of the sprockets. $\frac{3}{1}$

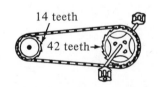

14 teeth   42 teeth

8. A stretch of railroad track going up-grade through the mountains rises 80 meters over a distance of 3200 meters. Over another stretch 10,000 meters in length the rise is 250 meters. How do the grades of these two stretches of track compare?

$\frac{80}{3200} = \frac{250}{10,000}$; the grades are the same

3-6 *Fractions and Decimals*

OBJECTIVES

Name the decimal equivalent of a fraction.

Read decimal numerals.

Name the fraction equivalent of a decimal.

Recall that a fraction indicates division.

EXAMPLE 1 $\dfrac{3}{4}$ means $4\overline{)3.00}$ (0.75) so $\dfrac{3}{4} = 0.75$.

EXAMPLE 2 $\dfrac{9}{10}$ means $10\overline{)9.0}$ (0.9) so $\dfrac{9}{10} = 0.9$.

EXAMPLE 3 $\dfrac{3}{2}$ means $2\overline{)3.0}$ (1.5) so $\dfrac{3}{2} = 1\dfrac{1}{2} = 1.5$.

In Examples 1–3, we have found the decimal equivalent of each fraction.

EXAMPLE 4   The decimal equivalent of $\dfrac{8}{10}$ is 0.8. ◀ Read "eight tenths."

EXAMPLE 5 $4\dfrac{3}{10} = 4.3$

▲   ▲

four and three tenths

Compare these fractions and decimals.

$$\dfrac{3}{100} = \dfrac{30}{1000} = \dfrac{300}{10,000} = \ldots \quad \blacktriangleright \quad 0.03 = 0.030 = 0.0300 = \ldots$$

Once the decimal point is placed in a numeral, extra zeros do not change the value.

EXAMPLE 6 $53 = 53.0 = 53.00 = 53.000 = \ldots$

▲

The decimal point is understood to be to the right of the ones' place.

EXERCISES

Give the division meaning.

Sample: $\dfrac{3}{7}$ *What you say:*   Three divided by seven.

1. $\dfrac{9}{10}$ nine divided by ten
2. $\dfrac{4}{5}$ four divided by five
3. $\dfrac{7}{8}$ seven divided by eight

4. $\dfrac{10}{2}$ ten divided by two
5. $\dfrac{7}{4}$ seven divided by four
6. $\dfrac{9}{1}$ nine divided by one

Name the fraction or decimal that is not equivalent to the others.

Sample: 0.60, 0.6, 0.06
What you say: 0.06 is not equivalent to either 0.60 or 0.6.

7. $\dfrac{3}{10}, \dfrac{30}{10}, \dfrac{30}{100}$ $\dfrac{30}{10}$

8. 0.5, 0.05, 0.050 0.5

9. 1.5, 0.15, 1.50 0.15

10. 40.0, 4.000, 4.0 40.0

Complete.

11. $0.6 = \dfrac{?}{10}$ 6
12. $0.9 = \dfrac{?}{10}$ 9
13. $0.50 = \dfrac{?}{100}$ 50

14. $0.38 = \dfrac{?}{100}$ 38
15. $5.7 = 5\dfrac{7}{?}$ 10
16. $0.2 = \dfrac{2}{?}$ 10

For Extra Practice, see page 412.

Written
EXERCISES

Write the meaning in words.

4. six and sixty-five hundredths
5. eight and forty hundredths
7. twenty-five thousandths

Sample 1: $\dfrac{7}{1000}$ *Solution:*   seven thousandths

Sample 2: 0.15   *Solution:*   fifteen hundredths

A

1. 0.5 five tenths
2. 0.03 three hundredths
3. $\dfrac{23}{100}$ twenty-three hundredths
4. 6.65

5. 8.40
6. $\dfrac{40}{100}$ forty hundredths
7. 0.025
8. 0.25

9. $\dfrac{8}{10}$ eight tenths
10. 10.8 ten and eight tenths
11. 20.00 twenty and zero hundredths
12. $\dfrac{147}{1000}$

8. twenty-five hundredths
12. one hundred forty-seven thousandths

Write as a decimal. Use division.

Sample: $\dfrac{3}{10}$ *Solution:* $10\overline{)3.0}$ $\dfrac{0.3}{}$; $\dfrac{3}{10} = 0.3$

13. $\dfrac{1}{5}$ 0.2 **14.** $\dfrac{1}{4}$ 0.25 **15.** $\dfrac{2}{5}$ 0.4 **16.** $\dfrac{6}{10}$ | 0.6

17. $\dfrac{4}{5}$ 0.8 **18.** $\dfrac{3}{4}$ 0.75 **19.** $\dfrac{7}{10}$ 0.7 **20.** $\dfrac{5}{5}$ 1.0

21. $\dfrac{18}{100}$ 0.18 **22.** $\dfrac{35}{100}$ 0.35 **23.** $\dfrac{1}{8}$ 0.125 **24.** $\dfrac{3}{8}$ | 0.375

Complete.

Sample 1: $\dfrac{3}{10} = \dfrac{?}{100} = \dfrac{?}{1000}$ *Solution:* $\dfrac{3}{10} = \dfrac{30}{100} = \dfrac{300}{1000}$

Sample 2: $0.3 = 0.30 = \underline{\ ?\ } = \underline{\ ?\ }$
Solution: $0.3 = 0.30 = 0.300 = 0.3000$

25. $\dfrac{2}{10} = \dfrac{?}{100} = \dfrac{?}{1000}$ 20; 200 **26.** $\dfrac{?}{10} = \dfrac{?}{100} = \dfrac{700}{1000}$ 7; 70

27. $\dfrac{13}{100} = \dfrac{?}{1000} = \dfrac{?}{10,000}$ 130; 1300 **28.** $\dfrac{3}{1} = \dfrac{?}{10} = \dfrac{?}{100}$ 30; 300

29. $0.7 = 0.70 = \underline{\ ?\ } = \underline{\ ?\ }$ 0.700; 0.7000

30. $4.3 = 4.30 = \underline{\ ?\ } = \underline{\ ?\ }$ 4.300; 4.3000

31. $0.12 = 0.120 = \underline{\ ?\ } = \underline{\ ?\ }$ 0.1200; 0.12000

32. $\underline{\ ?\ } = \underline{\ ?\ } = 2.700 = 2.7000$ 2.7; 2.70

Write as a fraction in lowest terms.

Sample: 0.6   *Solution:* $0.6 = \dfrac{6}{10} = \dfrac{3}{5}$

33. 0.4 $\frac{2}{5}$ **34.** 0.25 $\frac{1}{4}$ **35.** 0.75 $\frac{3}{4}$

36. 0.039 $\frac{39}{1000}$ **37.** 0.875 $\frac{7}{8}$ **38.** 3.5 $3\frac{1}{2}$ B

39. 0.025 $\frac{1}{40}$ **40.** 0.625 $\frac{5}{8}$ **41.** 1.05 $1\frac{1}{20}$

Find the decimal equivalent.

Sample: $\dfrac{1}{3}$ *Solution:* $3{\overline{\smash{)}1.000\ldots}}^{\,0.333\ldots}$ $\dfrac{1}{3} = 0.3333\ldots = 0.\overline{3}$

42. $\dfrac{2}{6}$ $0.\overline{3}$ **43.** $\dfrac{1}{6}$ $0.1\overline{6}$ **44.** $\dfrac{2}{3}$ $0.\overline{6}$ **45.** $\dfrac{5}{9}$ | $0.\overline{5}$ C

46. $\dfrac{5}{11}$ $0.\overline{45}$ **47.** $\dfrac{5}{6}$ $0.8\overline{3}$ **48.** $1\frac{4}{5}$ 1.8   **49.** $2\frac{3}{4}$ | 2.75

3-7 *Fractions, Decimals, and Percents*

> **OBJECTIVES**
>
> **Express fractions as percents.**
>
> **Express decimals as percents.**

You are familiar with percent expressions and the percent symbol, %.

$$50\% \blacktriangleright 50 \text{ percent} \qquad 100\% \blacktriangleright 100 \text{ percent}$$

Percent means *hundredths*. A number given in hundredths can be expressed directly as a percent.

EXAMPLE 1 $\dfrac{4}{100}$ ►four hundredths ► 4%

0.04 ► four hundredths ► 4%

EXAMPLE 2 $\dfrac{15}{100} = 15\%$ $\qquad 0.15 = 15\%$

EXAMPLE 3 $\dfrac{7}{10} = \dfrac{70}{100} = 70\%$ $\qquad 0.7 = 0.70 = 70\%$

A fraction such as $\dfrac{1}{2}$ or $\dfrac{17}{25}$ can first be written as hundredths, then as a percent.

EXAMPLE 4 $\dfrac{1}{2}$ ► $2\overline{)\begin{array}{c} 0.50 \\ 1.00 \end{array}}$ ► $0.50 = 50\%$

EXAMPLE 5 $\dfrac{17}{25}$ ► $25\overline{)\begin{array}{c} 0.68 \\ 17.00 \end{array}}$ ► $0.68 = 68\%$

Important Pattern ► To write a decimal as a percent, "move" the decimal point two places to the *right* and add the %. To write a percent as a decimal, "move" the decimal point two places to the *left* and drop the %.

Decimal ► Percent		Percent ► Decimal	
0.62	= 62%	85.%	= 0.85
0.08	= 8%	250.%	= 2.50

Express as a percent.

Sample 1: six hundredths   *What you say:* 6%

Sample 2: 0.45   *What you say:* 45%

1. eight hundredths 8%
2. 0.27 27%
3. 0.55 | 55%
4. ninety hundredths 90%
5. 0.93 93%
6. 0.01 | 1%
7. sixty-one hundredths 61%
8. 0.04 4%
9. $\dfrac{17}{100}$ | 17%

Complete.

10. 0.75 = _?_ % 75
11. 0.9 = _?_ % 90
12. 1.00 = _?_ % 100
13. 1.85 = _?_ % 185
14. 2.00 = _?_ % 200
15. 0.04 = _?_ % 4

For Extra Practice, see page 412.

Write as a decimal and as a percent.

Sample 1: $\dfrac{24}{100}$ *Solution:* $\dfrac{24}{100} = 0.24 = 24\%$

Sample 2: $\dfrac{2}{5}$ *Solution:* $\dfrac{2}{5} = 0.40 = 40\%$

1. $\dfrac{37}{100}$ 37%
2. $\dfrac{49}{100}$ 49%
3. $\dfrac{7}{10}$ 70%
4. $\dfrac{4}{10}$ | 40% A
5. $\dfrac{1}{4}$ 25%
6. $\dfrac{3}{4}$ 75%
7. $\dfrac{5}{8}$ 62.5%
8. $\dfrac{1}{8}$ | 12.5%
9. $\dfrac{4}{4}$ 100%
10. $\dfrac{300}{1000}$ 30%
11. $\dfrac{450}{1000}$ 45%
12. $\dfrac{3}{5}$ | 60%

Write as a fraction in lowest terms.

Sample: 20%   *Solution:* $20\% = \dfrac{20}{100} = \dfrac{1}{5}$

13. 15% $\frac{3}{20}$
14. 40% $\frac{2}{5}$
15. 10% $\frac{1}{10}$
16. 50% $\frac{1}{2}$
17. 25% $\frac{1}{4}$
18. 30% $\frac{3}{10}$
19. 42% $\frac{21}{50}$
20. 75% $\frac{3}{4}$
21. 100% 1

Sample: 23.6%   *Solution:* $23.6\% = 0.236 = \dfrac{236}{1000} = \dfrac{59}{250}$

B

22. 37.5% $\frac{3}{8}$ **23.** 12.5% $\frac{1}{8}$ **24.** 32.4% $\frac{81}{250}$

25. 87.5% $\frac{7}{8}$ **26.** 2% $\frac{1}{50}$ **27.** 5% $\frac{1}{20}$

28. 1% $\frac{1}{100}$ **29.** 325% $3\frac{1}{4}$ **30.** 140% $1\frac{2}{5}$

Write as a decimal.

Sample: $4\frac{1}{2}\%$ *Solution:* $4\frac{1}{2}\% = 4.5\% = 0.045$

C

31. $3\frac{1}{2}\%$ 0.035   **32.** $5\frac{1}{4}\%$ 0.0525   **33.** $2\frac{3}{4}\%$ 0.0275

34. $\dfrac{8}{10}\%$ 0.008   **35.** $\dfrac{3}{4}\%$ 0.0075   **36.** $\dfrac{1}{3}\%$ $0.00\overline{3}$

Problems

Complete. Use the information that is given in the accompanying circle graph. Give fractions in lowest terms.

1. The graph shows that 60% of the students at Lincoln High School are girls and 40% are boys. What fractional part of the student body are the boys? What fractional part of the student body are the girls? boys, $\dfrac{2}{5}$; girls, $\dfrac{3}{5}$

LINCOLN HIGH STUDENT BODY

2. This graph shows the percent of students enrolled at each grade level at Harbor High School. What fraction tells what part of Harbor High students are in grade 12? $\dfrac{1}{5}$
What fraction tells what part of the students are in grade 9? $\dfrac{3}{10}$
What percent tells what part of the students are in grades 9 through 12? 100%

HARBOR HIGH ENROLLMENT

3. What fractional part of the books are non-fiction? $\dfrac{1}{2}$
What fractional part of the books are fiction? $\dfrac{1}{4}$
What percent of the books are reference? $\dfrac{1}{8}$
Which two types of books together make up three-fourths of all the books in the library? fiction and non-fiction

BOOKS IN LIBRARY

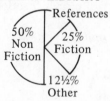

4. Copy and complete the table.

Vehicles	Decimal	Percent
Vans	0.125	?12½%
Compacts	?0.125	? 30%
Full Size	?0.30	?37½%
Pickups	?0.20	? 20%
TOTALS	1.000	100%

VEHICLES IN SCHOOL PARKING LOT

Compact $\frac{3}{10}$ / $\frac{1}{5}$ Pickups

Vans $\frac{1}{8}$ / $\frac{3}{8}$ Full Size

SELF-TEST 2

Be sure that you understand these terms and symbols.

equivalent fractions (p. 65)   lowest terms (p. 65)
ratio (p. 68)   decimal equivalent (p. 71)
% (p. 74)

1. Replace the variable to name the equivalent fraction: Section 3-4, p. 65
   $\frac{5}{7} = \frac{x}{42}$. 30

2. Write $\frac{6}{69}$ in lowest terms. $\frac{2}{23}$

3. Write this ratio as a fraction in lowest terms: Section 3-5, p. 68
   16 out of 28 $\frac{4}{7}$

4. Write the meaning of $\frac{7}{8}$ in words. seven eighths; 0.875 Then write $\frac{7}{8}$ as a decimal.   Section 3-6, p. 71

5. Write the decimal equivalents of $2\frac{6}{100} = 2\frac{60}{1000} = 2\frac{600}{10,000}$. 2.06 = 2.060 = 2.0600

Write as a decimal and as a percent.   Section 3-7, p. 74

6. $\frac{2}{5}$ 0.40; 40%   7. $\frac{4}{20}$ 0.20; 20%

Check your answers with those printed at the back of the book.

3-8 *Metric Measurement*

> **OBJECTIVES**
>
> **Express linear measures in meters, centimeters, and millimeters.**
>
> **Use a metric ruler to draw and to measure line segments.**

The **meter** (m) is the basic unit of length in the metric system. The meter is divided into 100 equal parts called **centimeters** (cm). The centimeter is divided into 10 equal parts called **millimeters** (mm).

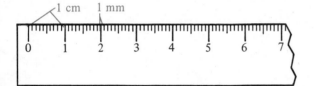 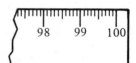

The arrow (↔) means "is the same as."

Equivalent Measures	
1 m ↔ 100 cm ↔ 1,000 mm	1 mm ↔ 0.1 cm ↔ 0.001 m
2 m ↔ 200 cm ↔ 2,000 mm	10 mm ↔ 1 cm ↔ 0.01 m
2.5 m ↔ 250 cm ↔ 2,500 mm	25 mm ↔ 2.5 cm ↔ 0.025 m

Study these examples.

EXAMPLE 1 256 cm ↔ 200 cm and 56 cm ↔ 2 m 56 cm

EXAMPLE 2 3.85 m ↔ 3 m and 0.85 m ↔ 3 m 85 cm

EXAMPLE 3 150 cm ↔ 1 m 50 cm ↔ 1.50 m

EXAMPLE 4 65 mm ↔ 60 mm and 5 mm ↔ 6 cm 5 mm

EXAMPLE 5 278 mm ↔ 270 mm and 8 mm ↔ 27 cm 8 mm
                                    ↔ 27.8 cm

Tell how to complete each statement.

Oral
EXERCISES

1. 3 m ↔ _?_ cm 300
2. 1.5 m ↔ _?_ cm 150
3. 2.2 m ↔ _?_ cm 220
4. 10 cm ↔ _?_ m 0.1
5. 25 cm ↔ _?_ m 0.25
6. 200 cm ↔ _?_ m 2
7. 15 cm ↔ _?_ mm 150
8. 10 cm ↔ _?_ mm 100
9. 100 cm ↔ _?_ mm 1000
10. 60 mm ↔ _?_ cm 6
11. 10 m ↔ _?_ cm 1000
12. 140 mm ↔ _?_ cm 14

Find the length. Use a metric ruler.

Written
EXERCISES

Sample: ━━━━━━━━━━

Solution: ━━━━━━━━━━ 3.5 cm

A

1. ━━━━━━━━━━━━━━━━━━━━━━ 9.0 cm
2. ━━━━━━━━━━━━━━━━━━━━━━━━ 10.5 cm
3. ━━━━━━━━━━ 4.5 cm
4. ━━━━━━━ 3.2 cm
5. ━━━━━━━━━━━━━━ 6.7 cm
6. ━━━━━━━━━━━━━━━━━ 8.0 cm
7. ━━━━━━━━━━━━━━━━━━━━ 9.5 cm
8. ━━━━━━━━━━━━ 6.0 cm

Draw line segments. Use a metric ruler. Check students' drawings.

9. 10 cm
10. 12 cm
11. 1 cm
12. 4.5 cm
13. 100 mm
14. 75 mm
15. 8 cm 3 mm
16. 5 cm
17. 6.5 cm

Complete.

Sample 1: 145 cm ↔ _?_ m _?_ cm
Solution: 145 cm ↔ 1 m 45 cm

Sample 2: 64 mm ↔ _?_ cm _?_ mm
Solution: 64 mm ↔ 6 cm 4 mm

Complete.

B

18. 134 cm ↔ ? m ? cm   **19.** 95 cm ↔ ? m 0.95

20. 240 cm ↔ ? m ? cm   **21.** 60 cm ↔ ? m 0.6

22. 395 cm ↔ ? m ? cm   **23.** 409 cm ↔ ? m ? cm
                                                             4 m 9 cm

24. 27 mm ↔ ? cm ? mm   **25.** 50 mm ↔ ? cm 5

26. 86 mm ↔ ? cm ? mm   **27.** 100 mm ↔ ? cm 10

28. 31 mm ↔ ? cm ? mm   **29.** 125 mm ↔ ? cm ? mm
                                                             12 cm 5 mm

C Complete.

30. 59 mm ↔ ? cm 5.9   **31.** 26 mm ↔ ? cm 2.6

32. 152 mm ↔ ? cm 15.2   **33.** 9.4 cm ↔ ? mm 94

34. 21.8 cm ↔ ? mm 218   **35.** 0.6 cm ↔ ? mm 6

36. 75 cm ↔ ? m 0.75   **37.** 15 cm ↔ ? m 0.15

38. 8 cm ↔ ? m 0.08   **39.** 0.75 m ↔ ? cm 75

40. 0.52 m ↔ ? cm 52   **41.** 0.6 m ↔ ? cm 60

18. 1 m 34 cm   **20.** 2 m 40 cm   **22.** 3 m 95 cm
24. 2 cm 7 mm   **26.** 8 cm 6 mm   **28.** 3 cm 1 mm

SELF-TEST 3

Be sure that you understand these terms.

meter (p. 78)   centimeter (p. 78)   millimeter (p. 78)

Complete.

1. 590 cm ↔ ? m ? cm 5 m 90 cm

2. 87 cm ↔ ? m 0.87

Complete.

3. 39 mm ↔ ? cm 3.9

4. 720 mm ↔ ? cm ? mm 72 cm 0 mm

5. Draw a line segment 7 cm 3 mm long. Use a metric ruler.

Check your answers with those printed at the back of the book.

chapter summary

1. The product of a number and any counting number is a **multiple** of the first number.

2. The least number that is a multiple of two given numbers is called their **least common multiple** (LCM).

3. The **divisors** of a number are the factors of that number.

4. The greatest divisor of two numbers is called their **greatest common factor** (GCF).

5. A number that can be expressed in the form $2 \times n$, where n is a whole number, is an **even number.**

6. A number that can be expressed in the form $(2 \times n) + 1$, where n is a whole number, is an **odd number.**

7. A whole number that has exactly two factors, itself and 1, is a **prime** number.

8. Fractions that name the same number are called **equivalent fractions.**

9. When the GCF of the numerator and the denominator of a fraction is 1, the fraction is in lowest terms.

10. A ratio may be expressed as a fraction.

11. The **decimal equivalent** of a fraction can be found by dividing the numerator by the denominator.

calculator corner

You can use your calculator to divide, even if the result is a quotient with remainder. (This will show up as a decimal.) For example, divide 416 by 59. Think of the steps you would use to figure out the division problem without a calculator. Now divide 416 by 59. Multiply the whole number in the quotient by 59. Subtract this new result from 416. The number which is displayed is the remainder. 416 divided by 59 is 7 with a remainder of 3. You can use this method to solve other division problems. Try $728 \div 15$ or $963 \div 38$.

chapter test

Name the sets of multiples and the LCM.

1. 3 and 4 {3, 6, 9, 12, . . .}; {4, 8, 12, . . .}; 12

2. 2 and 8 {2, 4, 6, 8, . . .}; {8, 16, 24, . . .}; 8

3. 6 and 8 {6, 12, 18, 24, . . .}; {8, 16, 24, . . .}; 24

4. 3 and 5 {3, 6, 9, 12, 15, . . .}; {5, 10, 15, . . .}; 15

Name the sets of factors and the GCF.

5. 30 and 24 {1, 2, 3, 5, 6, 15, 30}; {1, 2, 3, 4, 6, 8, 12, 24}; 6

6. 12 and 15 {1, 2, 3, 4, 6, 12}; {1, 3, 5, 15}; 3

7. 4 and 9 {1, 2, 4}; (1, 3, 9); 1

8. 5 and 10 {1, 5}; {1, 2, 5, 10}; 5

Write in the form $2 \times n$ or $(2 \times n) + 1$.

9. 15 $(2 \times 7) + 1$ **10.** 12 2×6 **11.** 34 2×17 **12.** 51 $(2 \times 25) + 1$

Complete.

13. $\frac{1}{2} = \frac{?}{10}$ 5

14. $\frac{12}{15} = \frac{?}{5}$ 4

15. $\frac{7}{8} = \frac{?}{16}$ 14

16. $\frac{8}{12} = \frac{4}{?}$ 6

Write as a fraction in lowest terms.

17. 9 out of 12 $\frac{3}{4}$

18. 6 out of 30 $\frac{1}{5}$

19. 3 out of 5 $\frac{3}{5}$

20. 10 out of 100 $\frac{1}{10}$

Write the decimal equivalent.

21. $\frac{3}{4}$ 0.75 **22.** $1\frac{1}{2}$ 1.5 **23.** $\frac{9}{10}$ 0.9 **24.** $\frac{25}{100}$ 0.25

Write as a percent.

25. $\frac{15}{100}$ 15% **26.** $\frac{95}{100}$ 95% **27.** 0.04 4% **28.** 0.34 34%

Complete.

29. 1 m ↔ _?_ cm 100

30. 50 cm ↔ _?_ m 0.5

31. 1 cm ↔ _?_ mm 10

32. 10 cm ↔ _?_ mm 100

33. 1 m ↔ _?_ mm 1000

34. 450 mm ↔ _?_ m 0.45

challenge topics *Symmetry*

Study the figure of a moth shown below. Do you agree that the broken line is a **line of symmetry?** That is, if we fold the shape along this line, the two parts will fit over each other exactly. We can test this conclusion by folding the figure along this line to see if the parts fit over each other exactly. Another way to test for a line of symmetry is to place a mirror along the broken line, as shown. If the

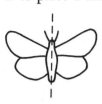  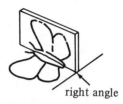

right angle

mirror image, the part of the moth that you see in the mirror, is exactly like the part that is hidden behind the mirror, the figure is symmetric about the line determined by the edge of the mirror.

Which of the following figures have at least one line of symmetry?

1.  yes   **2.**  yes   **3.** 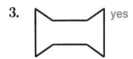 yes

How many lines of symmetry can you find in each figure? Make a copy of each to illustrate your answer.

4.  2   **5.**  2   **6.** TOOT 1

7. NOON 0   **8.** 6   **9.** ◯ infinitely many

Copy each figure and sketch its mirror image, using the broken line as a line of symmetry for the completed figure. Check students' drawings.

10.   **11.**   **12.**   **13.**

Review of Skills

Write as a mixed numeral or whole number.

1. $\dfrac{7}{2}$ $3\frac{1}{2}$

2. $\dfrac{15}{6}$ $2\frac{1}{2}$

3. $\dfrac{21}{5}$ $4\frac{1}{5}$

4. $\dfrac{8}{2}$ 4

Write as a fraction.

5. $2\frac{1}{3}$ $\dfrac{7}{3}$

6. $1\frac{1}{8}$ $\dfrac{9}{8}$

7. $1\frac{1}{7}$ $\dfrac{8}{7}$

8. $2\frac{3}{5}$ $\dfrac{13}{5}$

Add or subtract.

9. $\begin{array}{r} 5.067 \\ +1.300 \\ \hline 6.367 \end{array}$

10. $\begin{array}{r} 8.46 \\ -2.7 \\ \hline 5.76 \end{array}$

11. $\begin{array}{r} 46.3 \\ -11.8 \\ \hline 34.5 \end{array}$

12. $\begin{array}{r} 3.05 \\ +0.7 \\ \hline 3.75 \end{array}$

Multiply or divide.

13. $\dfrac{2}{3} \times \dfrac{3}{2}$ 1

14. $10 \times \dfrac{1}{10}$ 1

15. $\dfrac{1}{8} \times \dfrac{8}{7}$ $\dfrac{1}{7}$

16. $\begin{array}{r} 4.3 \\ \times 6.8 \\ \hline 29.24 \end{array}$

17. $\begin{array}{r} 16.4 \\ \times 3.2 \\ \hline 52.48 \end{array}$

18. $\begin{array}{r} 1.25 \\ \times 64 \\ \hline 80 \end{array}$

19. $5\overline{)86.15}$ 17.23

20. $5\overline{)8.615}$ 1.723

21. $34\overline{)57.8}$ 1.7

22. $34\overline{)578}$ 17

23. $0.8\overline{)184}$ 230

24. $8\overline{)1840}$ 230

Find the area. Use the given formula.

25.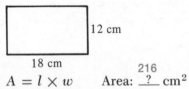

12 cm

18 cm

$A = l \times w$ Area: $\underset{216}{\underline{\ ?\ }}$ cm^2

26.

6 m

9 m

$A = \dfrac{1}{2} \times b \times h$ Area: $\underset{27}{\underline{\ ?\ }}$ m^2

Name the number pair graphed by the point.

27. A (3, 2)

28. B (1, 5)

29. C (5, 4)

30. D (4, 5)

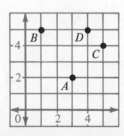

84

Left: General store, early 1900's.

Right: Clerk using optical scanner. It scans a printed code on the package and records the price automatically.

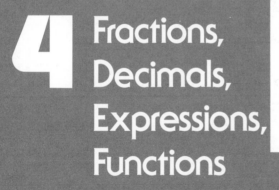

4 Fractions, Decimals, Expressions, Functions

Working with Fractions and Decimals

4-1 *Fractions: Addition and Subtraction*

OBJECTIVE

Add and subtract with fractions.

Adding or subtracting with like fractions is easy. **Like fractions** have a common denominator.

EXAMPLE 1 $\dfrac{1}{10} + \dfrac{8}{10}$ ▶—Add the numerators.
▶—Use the common denominator. ▶ $\dfrac{1+8}{10} = \dfrac{9}{10}$

EXAMPLE 2 $\dfrac{7}{8} - \dfrac{3}{8}$ ▶—Subtract the numerators.
▶—Use the common denominator. ▶ $\dfrac{7-3}{8} = \dfrac{4}{8} = \dfrac{1}{2}$

Recall that $4\frac{1}{5}$ means $4 + \dfrac{1}{5}$.

EXAMPLE 3 $4\frac{1}{5} + 2\frac{3}{5} = 4 + 2 + \dfrac{1}{5} + \dfrac{3}{5}$ ◀ First add the whole numbers.
Then add the fractional numbers.

$$= \quad 6 \quad + \quad \dfrac{4}{5} \quad = 6\tfrac{4}{5}$$

EXAMPLE 4 $5\frac{5}{8} - 4\frac{2}{8} = 5 - 4 + \dfrac{5}{8} - \dfrac{2}{8}$ ◀ First subtract the whole numbers.
Then subtract the fractional numbers.

$$= \quad 1 \quad + \quad \dfrac{3}{8} \quad = 1\tfrac{3}{8}$$

To add or subtract with unlike fractions, rename the fractions so they have a common denominator.

EXAMPLE 5 $\dfrac{1}{4} + \dfrac{1}{6} = \dfrac{6}{24} + \dfrac{4}{24} = \dfrac{6+4}{24} = \dfrac{10}{24} = \dfrac{5}{12}$

We can use the LCM of 4 and 6 to rename the fractions. Then our work is simpler.

$$\dfrac{1}{4} + \dfrac{1}{6} = \dfrac{3}{12} + \dfrac{2}{12} = \dfrac{3+2}{12} = \dfrac{5}{12}$$

Name the solution.

Sample: $b = \dfrac{4 + 2}{7}$ *What you say:* $b = \dfrac{6}{7}$

1. $c = \dfrac{2 + 1}{8}$ $\tfrac{3}{8}$

2. $m = \dfrac{3 + 6}{9}$ 1

3. $h = \dfrac{3 + 4}{10}$ $\tfrac{7}{10}$

4. $n = \dfrac{4 - 3}{5}$ $\tfrac{1}{5}$

5. $k = \dfrac{1 + 1}{3}$ $\tfrac{2}{3}$

6. $s = \dfrac{7 - 6}{2}$ $\tfrac{1}{2}$

Complete.

Sample: $\dfrac{1}{2} = \dfrac{?}{6}$ *What you say:* $\dfrac{1}{2} = \dfrac{3}{6}$

7. $\dfrac{1}{5} = \dfrac{?}{10}$ 2

8. $\dfrac{1}{2} = \dfrac{?}{4}$ 2

9. $\dfrac{2}{3} = \dfrac{?}{9}$ 6

10. $\dfrac{2}{5} = \dfrac{?}{15}$ 6

11. $\dfrac{1}{3} = \dfrac{?}{6}$ 2

12. $\dfrac{5}{10} = \dfrac{?}{2}$ 1

For Extra Practice, see page 413.

Add or subtract.

Written
EXERCISES

Sample: $\dfrac{7}{8} - \dfrac{5}{8}$ *Solution:* $\dfrac{7 - 5}{8} = \dfrac{2}{8} = \dfrac{1}{4}$

A

1. $\dfrac{2}{9} + \dfrac{5}{9}$ $\tfrac{7}{9}$

2. $\dfrac{1}{4} + \dfrac{2}{4}$ $\tfrac{3}{4}$

3. $\dfrac{1}{4} + \dfrac{1}{4} + \dfrac{1}{4}$ $\tfrac{3}{4}$

4. $\dfrac{3}{5} + \dfrac{1}{5}$ $\tfrac{4}{5}$

5. $\dfrac{9}{10} - \dfrac{3}{10}$ $\tfrac{3}{5}$

6. $\dfrac{3}{5} - \dfrac{3}{5}$ 0

7. $\dfrac{5}{12} + \dfrac{1}{12}$ $\tfrac{1}{2}$

8. $\dfrac{3}{4} - \dfrac{1}{4}$ $\tfrac{1}{2}$

9. $1\tfrac{1}{2} - \dfrac{1}{2}$ 1

Sample: $2\tfrac{1}{5} + 5\tfrac{2}{5}$ *Solution:* $2\tfrac{1}{5} + 5\tfrac{2}{5} = 2 + 5 + \dfrac{1}{5} + \dfrac{2}{5} = 7\tfrac{3}{5}$

10. $4\tfrac{1}{3} + 2\tfrac{1}{3}$ $6\tfrac{2}{3}$

11. $5\tfrac{1}{2} - 1\tfrac{1}{2}$ 4

12. $\dfrac{1}{10} + 2\tfrac{7}{10}$ $2\tfrac{4}{5}$

13. $6\frac{3}{10} + 1\frac{4}{10}$ $7\frac{7}{10}$ **14.** $7\frac{1}{3} + 1\frac{2}{3}$ 9   **15.** $3\frac{1}{4} - 3$ $\frac{1}{4}$

16. $8\frac{7}{8} - 6\frac{2}{8}$ $2\frac{5}{8}$ **17.** $4\frac{3}{4} - 3\frac{1}{4}$ $1\frac{1}{2}$ **18.** $7\frac{2}{5} + 2$ $9\frac{2}{5}$

Complete.

Sample: $b = \dfrac{5}{8} + \dfrac{1}{4}$ *Solution:* $b = \dfrac{5}{8} + \dfrac{1}{4}$

$\qquad\qquad b = \dfrac{5}{8} + \dfrac{?}{8} \qquad\qquad\qquad b = \dfrac{5}{8} + \dfrac{2}{8}$

$\qquad\qquad b = \underline{\ ?\ } \qquad\qquad\qquad\qquad b = \dfrac{7}{8}$

19. $t = \dfrac{1}{8} + \dfrac{1}{2}$ **20.** $x = \dfrac{1}{2} + \dfrac{1}{3}$ **21.** $m = \dfrac{1}{5} + \dfrac{3}{10}$

$\quad t = \dfrac{1}{8} + \dfrac{?}{8}$ 4 $x = \dfrac{?}{6} + \dfrac{?}{6}$ 3; 2 $m = \dfrac{?}{10} + \dfrac{3}{10}$ 2

$\quad t = \underline{\ ?\ }$ $\frac{5}{8}$ $x = \underline{\ ?\ }$ $\frac{5}{6}$ $m = \underline{\ ?\ }$ $\frac{1}{2}$

Solve.

B

22. $s = \dfrac{3}{4} + \dfrac{1}{8}$ $\frac{7}{8}$ **23.** $\dfrac{7}{8} - \dfrac{3}{5} = k$ $\frac{11}{40}$ **24.** $h = \dfrac{1}{2} + \dfrac{1}{4} + \dfrac{1}{4}$ 1

25. $n = \dfrac{3}{4} - \dfrac{1}{2}$ $\frac{1}{4}$ **26.** $5\frac{3}{4} - 2\frac{1}{2} = t$ $3\frac{1}{4}$ **27.** $q = 4\frac{1}{2} - 2\frac{1}{3}$ $2\frac{1}{6}$

28. $z = \dfrac{4}{5} - \dfrac{1}{3}$ $\frac{7}{15}$ **29.** $6\frac{1}{8} + 5\frac{1}{4} = c$ $11\frac{3}{8}$ **30.** $r = 5\frac{3}{10} - 2\frac{6}{10}$ $2\frac{7}{10}$

Find the value when $x = 2$, $y = 3$, and $z = 5$.

Sample: $\dfrac{2}{y} + \dfrac{1}{z}$ *Solution:* $\dfrac{2}{y} + \dfrac{1}{z} = \dfrac{2}{3} + \dfrac{1}{5}$

$\qquad\qquad\qquad\qquad\qquad\qquad = \dfrac{10}{15} + \dfrac{3}{15} = \dfrac{13}{15}$

31. $\dfrac{1}{y} + \dfrac{2}{y}$ 1   **32.** $\dfrac{1}{x} - \dfrac{1}{z}$ $\frac{3}{10}$ **33.** $\dfrac{x}{y} + \dfrac{z}{y}$ $2\frac{1}{3}$

34. $\dfrac{1}{x} + \dfrac{3}{z}$ $1\frac{1}{10}$ **35.** $\dfrac{x}{10} + \dfrac{y}{10}$ $\frac{1}{2}$ **36.** $\dfrac{x + z}{y}$ $2\frac{1}{3}$

37. $\dfrac{1}{y} + \dfrac{1}{y} + \dfrac{1}{y}$ 1   **38.** $\dfrac{y}{2} + \dfrac{y}{4}$ $2\frac{1}{4}$ **39.** $\dfrac{y - x}{z}$ $\frac{1}{5}$

career capsule *Market Researcher*

Market researchers collect, study, and relate facts about company products and services, advertising, sales policies, and consumer opinions. They collect information from company records, personal opinion surveys, and statistics on changes in population.

A market researcher must have an ability to work with figures and an interest and flair for research. A college degree in economics or business administration is generally required. Courses in mathematics and social studies are useful.

4-2 *Fractions: Multiplication and Division*

EXAMPLE 1 $\dfrac{3}{8} \times \dfrac{1}{2}$ ► Multiply the numerators. $\dfrac{3 \times 1}{8 \times 2} = \dfrac{3}{16}$
► Multiply the denominators.

EXAMPLE 2 $2\frac{1}{2} \times 2\frac{1}{4} = \dfrac{5}{2} \times \dfrac{9}{4} = \dfrac{5 \times 9}{2 \times 4} = \dfrac{45}{8} = 5\frac{5}{8}$

To divide with fractions, we use reciprocals. Two numbers are **reciprocals** of each other if their product is 1.

EXAMPLE 3 $\dfrac{2}{3}$ and $\dfrac{3}{2}$ $\qquad \dfrac{2}{3} \times \dfrac{3}{2} = \dfrac{6}{6} = 1$

reciprocals $\qquad\qquad$ The product is 1.

EXAMPLE 4 5 and $\dfrac{1}{5}$ $\qquad \dfrac{5}{1} \times \dfrac{1}{5} = \dfrac{5}{5} = 1$

reciprocals $\qquad\qquad$ The product is 1.

EXAMPLE 5 $1\frac{1}{4}$ and $\dfrac{4}{5}$ $\qquad 1\frac{1}{4} \times \dfrac{4}{5} = \dfrac{5}{4} \times \dfrac{4}{5} = 1$

Dividing by a number is the same as multiplying by its reciprocal.

EXAMPLE 6 $\dfrac{1}{2} \div \dfrac{3}{4} = \dfrac{1}{2} \times \dfrac{4}{3}$ ◄ $\dfrac{4}{3}$ is the reciprocal of $\dfrac{3}{4}$.

$\qquad\qquad = \dfrac{1 \times 4}{2 \times 3} = \dfrac{4}{6} = \dfrac{2}{3}$

EXAMPLE 7 $\dfrac{2}{3} \div 1\frac{3}{4} = \dfrac{2}{3} \div \dfrac{7}{4} = \dfrac{2}{3} \times \dfrac{4}{7}$ ◄ $\dfrac{4}{7}$ is the reciprocal of $1\frac{3}{4}$.

$\qquad\qquad = \dfrac{2 \times 4}{3 \times 7} = \dfrac{8}{21}$

Name the reciprocal.

Sample: $\dfrac{4}{3}$ *What you say:* $\dfrac{3}{4}$

1. $\dfrac{3}{2}$ $\tfrac{2}{3}$ 2. $\dfrac{1}{3}$ 3 3. $\dfrac{7}{10}$ $\tfrac{10}{7}$

4. $\dfrac{4}{5}$ $\tfrac{5}{4}$ 5. $\dfrac{5}{3}$ $\tfrac{3}{5}$ 6. $\dfrac{6}{1}$ $\tfrac{1}{6}$

Express as a fraction. Then name the reciprocal.

7. $1\tfrac{1}{2}$ $\tfrac{3}{2};\tfrac{2}{3}$ 8. $4\tfrac{1}{8}$ $\tfrac{33}{8};\tfrac{8}{33}$ 9. $2\tfrac{1}{2}$ $\tfrac{5}{2};\tfrac{2}{5}$

10. $1\tfrac{2}{3}$ $\tfrac{5}{3};\tfrac{3}{5}$ 11. $2\tfrac{1}{3}$ $\tfrac{7}{3};\tfrac{3}{7}$ 12. $1\tfrac{1}{8}$ $\tfrac{9}{8};\tfrac{8}{9}$

For Extra Practice, see page 414.

Multiply.

Sample: $\dfrac{2}{5} \times \dfrac{1}{3}$ *Solution:* $\dfrac{2}{5} \times \dfrac{1}{3} = \dfrac{2 \times 1}{5 \times 3} = \dfrac{2}{15}$

1. $\dfrac{1}{9} \times \dfrac{2}{3}$ $\tfrac{2}{27}$ 2. $\dfrac{5}{2} \times \dfrac{1}{4} \times \dfrac{1}{4}$ $\tfrac{5}{32}$ 3. $2\tfrac{1}{5} \times 1\tfrac{1}{2}$ $3\tfrac{3}{10}$

4. $\dfrac{3}{5} \times \dfrac{1}{2}$ $\tfrac{3}{10}$ 5. $\dfrac{2}{1} \times \dfrac{1}{2} \times \dfrac{1}{2}$ $\tfrac{1}{2}$ 6. $1\tfrac{1}{3} \times 3$ 4

7. $\dfrac{1}{2} \times \dfrac{3}{2} \times \dfrac{1}{5}$ $\tfrac{3}{20}$ 8. $\dfrac{1}{9} \times \dfrac{1}{9}$ $\tfrac{1}{81}$ 9. $\dfrac{2}{3} \times 1\tfrac{1}{4}$ $\tfrac{5}{6}$

Copy and complete.

10. $\dfrac{3}{10} \div \dfrac{2}{3}$

 $\dfrac{3}{10} \times \dfrac{3}{2}$

 $\underline{\quad?\quad}$ $\tfrac{9}{20}$

11. $\dfrac{1}{4} \div \dfrac{2}{3}$

 $\dfrac{1}{4} \times \dfrac{3}{2}$

 $\underline{\quad?\quad}$ $\tfrac{3}{8}$

12. $\dfrac{4}{5} \div \dfrac{3}{1}$

 $\dfrac{4}{5} \times \underline{\quad?\quad}$ $\tfrac{1}{3}$

 $\underline{\quad?\quad}$ $\tfrac{4}{15}$

13. $\dfrac{1}{2} \div \dfrac{3}{5}$

 $\dfrac{1}{2} \times \dfrac{5}{3}$

 $\underline{\quad?\quad}$ $\tfrac{5}{6}$

14. $\dfrac{7}{8} \div \dfrac{2}{1}$

 $\dfrac{7}{8} \times \underline{\quad?\quad}$ $\tfrac{1}{2}$

 $\underline{\quad?\quad}$ $\tfrac{7}{16}$

15. $\dfrac{6}{1} \div \dfrac{2}{1}$

 $\dfrac{6}{1} \times \underline{\quad?\quad}$ $\tfrac{1}{2}$

 $\underline{\quad?\quad}$ 3

Name the reciprocal.

Sample: $\dfrac{1}{5} : \underline{\ ?\ }$ *Solution:* $\dfrac{1}{5} : \dfrac{5}{1}$ or 5

B

16. $\dfrac{1}{4} : \underline{\ ?\ }$ 4

17. $\dfrac{8}{1} : \underline{\ ?\ }$ $\dfrac{1}{8}$

18. $6 : \underline{\ ?\ }$ $\dfrac{1}{6}$

19. $\dfrac{5}{2} : \underline{\ ?\ }$ $\dfrac{2}{5}$

20. $3 : \underline{\ ?\ }$ $\dfrac{1}{3}$

21. $\dfrac{1}{9} : \underline{\ ?\ }$ 9

22. $\dfrac{10}{1} : \underline{\ ?\ }$ $\dfrac{1}{10}$

23. $7 : \underline{\ ?\ }$ $\dfrac{1}{7}$

24. $15 : \underline{\ ?\ }$ $\dfrac{1}{15}$

25. 1 1

26. $\dfrac{x}{3}$ (x is not zero) $\dfrac{3}{x}$

27. $\dfrac{m}{n}$ (m and n not 0) $\dfrac{n}{m}$

28. $\dfrac{5}{5}$ $\dfrac{5}{5}$ or 1

29. $\dfrac{a}{5}$ (a is not 0) $\dfrac{5}{a}$

30. $\dfrac{r}{s}$ (r and s not 0) $\dfrac{s}{r}$

Solve.

Sample 1: $x = \dfrac{3}{4} \div \dfrac{1}{2}$ *Solution:* $x = \dfrac{3}{4} \div \dfrac{1}{2} = \dfrac{3}{4} \times \dfrac{2}{1}$

$$x = \dfrac{3}{2} \ (\text{or } 1\tfrac{1}{2})$$

31. $k = \dfrac{4}{5} \div \dfrac{2}{3}$ $1\tfrac{1}{5}$

32. $t = 2 \div 1\tfrac{1}{2}$ $1\tfrac{1}{3}$

33. $y = 1\tfrac{1}{2} \times 1\tfrac{1}{2}$ $2\tfrac{1}{4}$

34. $b = \dfrac{2}{3} \div \dfrac{1}{2}$ $1\tfrac{1}{3}$

35. $r = \dfrac{1}{2} \times 2$ 1

36. $z = 7 \div 2\tfrac{1}{3}$ 3

37. $w = \dfrac{3}{4} \div \dfrac{7}{8}$ $\tfrac{6}{7}$

38. $x = \dfrac{3}{4} \times 1\tfrac{1}{2}$ $1\tfrac{1}{8}$

39. $c = 6 \div \dfrac{3}{1}$ 2

40. $\dfrac{1}{2} \div \dfrac{7}{9} = n$ $\tfrac{9}{14}$

41. $2\tfrac{1}{8} \times \dfrac{8}{10} = a$ $1\tfrac{7}{10}$

42. $10 \times \dfrac{1}{10} = m$ 1

Find the value when $a = 1$, $b = 2$, $c = 3$, and $d = 4$.

Sample: $\dfrac{a}{c} \times \dfrac{b}{c}$ *Solution:* $\dfrac{1}{3} \times \dfrac{2}{3} = \dfrac{2}{9}$

43. $\dfrac{a}{b} \times \dfrac{a}{b}$ $\tfrac{1}{4}$

44. $\dfrac{c}{d} \div \dfrac{c}{d}$ 1

45. $\dfrac{a}{b} \times \dfrac{a}{c} \times \dfrac{a}{d}$ $\tfrac{1}{24}$

46. $\dfrac{b}{c} \times \dfrac{c}{b}$ 1

47. $\dfrac{d}{a} \div a$ 4

48. $\dfrac{b}{c} \times \dfrac{c}{d} \times b$ 1

4-3 *Decimals: Addition and Subtraction*

OBJECTIVES

Add and subtract numbers expressed as decimals, including money.

Estimate sums and differences of numbers expressed as decimals by rounding.

When you work with decimals, it is very important to have decimal points positioned correctly.

EXAMPLE 1 $2.075 + 3.4 = \underline{\ ?\ }$

$$
\begin{array}{r}
2.075 \\
+\,3.400 \\
\hline
5.475
\end{array}
$$

◀ Include extra zeros. Be sure the decimal points are in line.

$2.075 + 3.4 = \mathbf{5.475}$

EXAMPLE 2 $9.67 - 1.5 = \underline{\ ?\ }$

$$
\begin{array}{r}
9.67 \\
-\,1.50 \\
\hline
8.17
\end{array}
$$

◀ Include extra zeros. Be sure the decimal points are in line.

$9.67 - 1.5 = \mathbf{8.17}$

It's a good idea to check an operation with decimals by rounding to whole numbers and estimating the answer. The exact answer and the estimate should be about the same.

EXAMPLE 3 $24.2 - 5.73 = \underline{\ ?\ }$

$$
\begin{array}{r}
24.20 \\
-\ 5.73 \\
\hline
\text{exact answer} \blacktriangleright \quad 18.47
\end{array}
\qquad
\begin{array}{r}
\text{rounds to} \quad 24 \\
\text{rounds to} \quad -\ 6 \\
\hline
\text{estimate} \blacktriangleright \quad 18
\end{array}
$$

EXAMPLE 4 $14.79 + 8.26 = \underline{\ ?\ }$

$$
\begin{array}{r}
14.79 \\
+\ 8.26 \\
\hline
\text{exact answer} \blacktriangleright \quad 23.05
\end{array}
\qquad
\begin{array}{r}
\text{rounds to} \quad 15 \\
\text{rounds to} \quad +\ 8 \\
\hline
\text{estimate} \blacktriangleright \quad 23
\end{array}
$$

EXAMPLE 5 $30.40 - 26.19 = \underline{\ ?\ }$

$$
\begin{array}{r}
30.40 \\
-\ 26.19 \\
\hline
\text{exact answer} \blacktriangleright \quad 4.21
\end{array}
\qquad
\begin{array}{r}
\text{rounds to} \quad 30 \\
\text{rounds to} \quad -\ 26 \\
\hline
\text{estimate} \blacktriangleright \quad 4
\end{array}
$$

Round to the nearest whole number.

Sample 1:   68.35   *What you say:*   68

Sample 2:   1.821   *What you say:*   2

1. 9.3 9   **2.** 7.99 8   **3.** 0.2 0

4. 10.7 11   **5.** 3.09 3   **6.** 0.539 1

7. 8.5 9   **8.** 0.8 1   **9.** 127.099 127

Round to the nearest dollar.

Sample: $6.89   *What you say:* $7.00

10. $ 5.17 $5.00   **11.** $7.50 $8.00   **12.** $10.09 $10.00

13. $49.95 $50.00   **14.** $ .89 $1.00   **15.** $19.89 $20.00

For Extra Practice, see page 415.

Written
EXERCISES

A

Add or subtract.

Sample: $6.72 + 5.09 = ?$ *Solution:* $6.72 + 5.09 = 11.81$

1. $12.38 + 6.8 = \underline{\ ?\ }$ 19.18   **2.** $10.75 - 3.44 = \underline{\ ?\ }$ 7.31

3. $100 - 42.3 = \underline{\ ?\ }$ 57.7   **4.** $1.84 + 0.95 = \underline{\ ?\ }$ 2.79

5. $0.008 - 0.003 = \underline{\ ?\ }$ 0.005   **6.** $4.06 - 0.03 = \underline{\ ?\ }$ 4.03

7. $6.5 + 2.1 + 3.007 = \underline{\ ?\ }$ 11.607 **8.** $9.07 - 2.005 = \underline{\ ?\ }$ 7.065

9. $6.703 - 2.5 = \underline{\ ?\ }$ 4.203   **10.** $7.0 + 0.3 + 0.024 = \underline{\ ?\ }$ 7.324

11. $9 + 34 + 5.6 = \underline{\ ?\ }$ 48.6   **12.** $0.5 + 0.5 = \underline{\ ?\ }$ 1.0

Add or subtract. Estimate to check your work.

Sample: $94.27 - 13.88 = \underline{\ ?\ }$ *Solution:* $94.27 - 13.88 = 80.39$
                                                     Check: $94 - 14 = 80$

13. $2.04 + 8.93 = \underline{\ ?\ }$ 10.97   **14.** $3.25 + 6.9 = \underline{\ ?\ }$ 10.15

15. $10.00 - 3.75 = \underline{\ ?\ }$ 6.25   **16.** $38.1 + 1.5 = \underline{\ ?\ }$ 39.6

17. $1.007 + 5.901 = \underline{\ ?\ }$ 6.908   **18.** $8.0039 + 7.001 = \underline{\ ?\ }$ 15.0049

Solve.

Sample: $n = 4.07 + 3.5 + 0.007$

Solution: $n = 4.07 + 3.5 + 0.007 = 7.577$

B

19. $t = 0.08 + 436.2$ 436.28   **20.** $60.9 - 18 = K$ 42.9

21. $m = 7.1 + 0.004 + 0.3$ 7.404   **22.** $0.3 + 0.3 + 0.3 = n$ 0.9

23. $y = 17.4 - 6.004$ 11.396   **24.** $8.204 - 3.204 = a$ 5

25. $149.75 + 99.02 = b$ 248.77   **26.** $s = 1.0 + 1.11 + 0.010$ 2.120

27. $4.8 = x + 2.3$ 2.5   **28.** $4.25 + 1.07 + r = 8.5$ 3.18   C

29. $9.04 - w = 6.5$ 2.54   **30.** $y - 3.01 = 6.275$ 9.285

31. $38.004 + h = 50$ 11.996   **32.** $c + 246.30 = 422.09$ 175.79

Estimate the answer by rounding each amount to the nearest dollar. Then find the exact answer.

Problems

1. A carpenter bought these tools. How much did they cost altogether? $13.00; $13.16

   hammer: $3.19   screw driver: $.87
   saw: $6.95   wrench: $2.15

2. Walter's car expenses last week were as follows: gasoline, $19.70; oil, $1.19; new tire, $24.30. What were his total expenses? $45.00; $45.19

3. Arlene took some friends to lunch. The total bill was $13.86. She paid the waiter with a $20 bill. How much change did she receive? $6.00; $6.14

Estimate the answer by rounding each amount to the nearest ten dollars. Then find the exact answer.

4. Harmon bought a new suit for $129.95, a pair of shoes for $32, and a raincoat for $37.50. What was the total cost? $200.00; $199.45

5. Jade bought four new tires for her car. If they cost $39.95 each, how much did she pay for all four tires? $160.00; $159.80

Time out

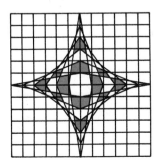

You can make curved designs using straight lines. Take a piece of graph paper. Label a horizontal line from 0 to 6. Starting at 0 label the vertical line through 0 from 1 to 6. With straight lines, connect 1 on the horizontal with 6 on the vertical, 2 on the horizontal with 5 on the vertical, etc. Color in your design. Experiment by combining designs and changing numbers.

4-4 Decimals: Multiplication and Division

OBJECTIVES

Multiply and divide numbers expressed as decimals.

Find a percent of a number.

When you multiply or divide with decimals, estimating the answer will help you locate the decimal point.

EXAMPLE 1 $3.2 \times 4.9 = \underline{\ ?\ }$ ▶ Estimate: $3 \times 5 = 15$
Exact answer: $3.2 \times 4.9 = 15.68$

EXAMPLE 2 $16.96 \div 5.3 = \underline{\ ?\ }$ ▶ Estimate: $17 \div 5$ is about 3.

product factor factor
▼   ▼   ▼
Exact answer: $16.96 \div 5.3 = \mathbf{3.2}$

Important Pattern ▶ The number of digits after the decimal point in the product is the sum of the numbers of digits after the decimal points in the factors.

$1.6 \times 3.4 = 5.44$ $5.46 \div 0.3 = 18.2$
▲   ▲   ▲   ▲   ▲   ▲
$1 + \quad 1 = \quad 2$ $2 = \quad 1 + \quad 1$

To find a percent of a number, first write the percent as a decimal. Then multiply.

EXAMPLE 3   15% of 80 is $\underline{\ ?\ }$ ▶ $0.15 \times 80 = 12.00 = 12$
15% of 80 is **12.**

EXAMPLE 4   125% of 3.6 is $\underline{\ ?\ }$ ▶ $1.25 \times 3.6 = 4.500 = 4.5$
125% of 3.6 is **4.5.**

EXERCISES

Match. Use estimates.

1. 4.8×31 B   **A.** 0.1488
2. 4.8×3.1 D   **B.** 148.8
3. 48×31 C   **C.** 1488
4. 0.48×0.31 A   **D.** 14.88

Tell where the decimal point should be placed in the answer.

Sample 1: $42.9 \div 3 = 143$ *What you say:* $43 \div 3$ is about 14, so $42.9 \div 3 = \mathbf{14.3}$

Sample 2: $9.0 \div 3.6 = 250$ *What you say:* $9 \div 4$ is about 2, so $9.0 \div 3.6 = \mathbf{2.50}$

5. $11.0 \div 8.8 = 125$ 1.25   **6.** $25.96 \div 5.5 = 472$ 4.72

7. $118.75 \div 0.95 = 125$ 125.   **8.** $81.6508 \div 3.08 = 2651$ 26.51

9. $3847.2 \div 1.2 = 32060$ 3206.0   **10.** $46.452 \div 9.8 = 474$ 4.74

For Extra Practice, see page 415.

Solve. Check by estimating the solution.

Sample: $t = 9.75 \times 6.3$ *Solution:* $t = 9.75 \times 6.3 = 61.425$
Estimate: $10 \times 6 = 60$

Written
EXERCISES

A

1. $n = 4.87 \times 0.95$ 4.6265   **2.** $40.1 \times 6.85 = y$ 274.685

3. $b = 24.68 \times 2.35$ 57.998   **4.** $8.001 \times 10.1 = w$ 80.8101

5. $m = 1.04 \times 0.008$ 0.00832   **6.** $1.8 \times 1.8 = s$ 3.24

7. $x = 3.75 \times 3.0$ 11.25   **8.** $8.0 \times 0.8 = t$ 6.4

Solve. Use an estimate to check your work.

Sample: $19.5 \div 5 = r$ *Solution:* $r = 19.5 \div 5 = 3.9$
Estimate: $20 \div 5 = 4$

9. $87.64 \div 2 = y$ 43.82   **10.** $42.03 \div 9 = b$ 4.67

11. $37.2 \div 12 = n$ 3.1   **12.** $46.05 \div 15 = z$ 3.07

13. $101.01 \div 3 = t$ 33.67   **14.** $43.01 \div 5 = s$ 8.602

15. $\dfrac{5.73}{3} = a$ 1.91   **16.** $22.12 \div 7 = w$ 3.16

Complete.

Sample: $15\% \times 54 = t$ *Solution:* $15\% \times 54 = t$
$0.15 \times 54 = t$
$8.1 = t$

17. $24\% \times 75 = m$ **18.** $12.5\% \times 104 = a$
$0.24 \times 75 = m$ $0.125 \times 104 = a$
$\underline{\quad?\quad} = m$ 18 $\underline{\quad?\quad} = a$ 13

19. $200\% \times 56 = s$
$2.00 \times 56 = s$
$\underline{\ ?\ } = s$ 112

20. $4\% \times 32 = n$
$0.04 \times 32 = n$
$\underline{\ ?\ } = n$ 1.28

Solve

B

21. $12.2 \times 0.95 = s$ 11.59

22. $a = 9.24 \div 0.3$ 30.8

23. $0.75 \times 3.4 = n$ 2.55

24. $k = 16.02 \div 0.09$ 178

25. $16 \times \$19.50 = y$ $312

26. $w = \dfrac{41.04}{0.8}$ 51.3

27. $24 \times \$40.35 = x$ $968.40

28. $t = \dfrac{\$64.92}{12}$ $5.41

29. $16\% \times 85 = r$ 13.6

30. $w = 0.4 \times 0.4 \times 0.4$ 0.064

31. $75\% \times 1024 = b$ 768

32. $z = 1.0 \times 0.1 \times 0.01$ 0.001

Complete.

33. $5.0275 \times 1 = \underline{\ ?\ }$ 5.0275
$5.0275 \times 10 = \underline{\ ?\ }$ 50.275
$5.0275 \times 100 = \underline{\ ?\ }$ 502.75
$5.0275 \times 1000 = \underline{\ ?\ }$ 5027.5

34. $1674.2 \times 1 = \underline{\ ?\ }$ 1674.2
$1674.2 \times 0.1 = \underline{\ ?\ }$ 167.42
$1674.2 \times 0.01 = \underline{\ ?\ }$ 16.742
$1674.2 \times 0.001 = \underline{\ ?\ }$ 1.6742

35. $34.28 \times 1 = \underline{\ ?\ }$ 34.28
$34.28 \times 10 = \underline{\ ?\ }$ 342.8
$34.28 \times 100 = \underline{\ ?\ }$ 3428
$34.28 \times 1000 = \underline{\ ?\ }$ 34,280

36. $43.79 \times 1 = \underline{\ ?\ }$ 43.79
$43.79 \times 0.1 = \underline{\ ?\ }$ 4.379
$43.79 \times 0.01 = \underline{\ ?\ }$ 0.4379
$43.79 \times 0.001 = \underline{\ ?\ }$ 0.04379

Problems

1. Sarah earns $15,000 each year. She pays 18% of her earnings in income taxes. How much income tax does she pay in a year? $2700

2. An industrial storage bin holds 5000 kilograms of salt. Salt is about 40% sodium. How many kilograms of sodium are in the bin? About 2000 kg of sodium

3. The total surface area of a cube is 4302.06 square centimeters. What is the area of each of the six faces? 717.01 cm²

4. Tom works part time. He earns $2.87 per hour. Last week he worked 25 hours. How much did he earn? $71.75

5. A three-kilogram roast cost $14.19. What was the cost per kilogram? $4.73/kg

Be sure that you understand these terms.

like fractions (p. 86)   unlike fractions (p. 86)
reciprocal (p. 90)

Add or subtract.   Section 4-1, p. 86

1. $\dfrac{5}{6} - \dfrac{2}{6}$ $\dfrac{1}{2}$ 2. $\dfrac{1}{2} + \dfrac{2}{5}$ $\dfrac{9}{10}$

Multiply or divide.   Section 4-2, p. 90

3. $\dfrac{2}{11} \times \dfrac{3}{5}$ $\dfrac{6}{55}$ 4. $\dfrac{3}{7} \div \dfrac{1}{4}$ $\dfrac{12}{7}$ or $1\frac{5}{7}$

Add or subtract. Estimate to check your work.   Section 4-3, p. 93

5. $11.072 - 0.23 = \underline{\ ?\ }$ 10.842   6. $\$5.24 + \$2.57 = \underline{\ ?\ }$ \$7.81

Solve.   Section 4-4, p. 96

7. $2.03 \times 4.7 = x$ 9.541   8. $69.04 \div 2 = y$ 34.52

9. $10\% \times 60 = z$ 6

Check your answers with those printed at the back of the book.

consumer notes

Discount and Sales Tax

Maria and Phil went shopping at a discount record shop. They each bought an album which was marked down 25 percent of the original list price of $6.00. Phil went to a cashier who charged a sales tax of 3 percent on the discount price. Maria went to another cashier, who charged a sales tax of 3 percent on the list price and then deducted 25 percent. Maria argued that she was charged sales tax on money that she didn't spend. Phil and Maria then compared costs. How much did Phil spend? How much did Maria spend? What happened? Phil, $4.64; Maria, $4.63

4-5 *Order of Operations*

OBJECTIVES

Simplify expressions that contain parentheses, brackets, and fraction bars.

Use grouping symbols in expressions to show the order of operations.

Grouping symbols such as parentheses, brackets, and fraction bars are used to show the order in which operations are to be done.

EXAMPLE 1 $4 \times (5 + 7)$ ► First add 5 and 7. ► $4 \times (5 + 7) = 4 \times 12$
Then multiply 4×12. ► $= 48$

EXAMPLE 2 $\dfrac{10 + 8}{3}$ ► First add 10 and 8. ► $\dfrac{10 + 8}{3} = \dfrac{18}{3}$
Then divide 18 by 3. ► $= 6$

When no operation sign, such as $+$, $-$, $\times$, or $\div$, is used between a numeral and an expression within parentheses, the operation intended is multiplication.

EXAMPLE 3 $7(10 + 3)$ ► First add 10 and 3. ► $7(10 + 3) = 7(13)$
Then multiply 7 and 13. ► $= 91$

EXAMPLE 4 $(12 - 5)8$ ► First subtract 5 from 12. ► $(12 - 5)(8) = (7)8$
Then multiply 7 and 8. ► $= 56$

Sometimes two different grouping symbols are used in one expression. We simplify the innermost expression first. Then we simplify the expression in the outer symbol.

EXAMPLE 5 $5[3 + (12 - 8)]$

First subtract 8 from 12. ► $5[3 + (12 - 8)] = 5[3 + 4]$
Then add 3 and 4. ► $= 5[7]$
Then multiply 5 and 7. ► $= 35$

When no grouping symbols are used, first do all multiplications and divisions in order from left to right. Then do all additions and subtractions in order from left to right.

EXAMPLE 6 $3 \times 4 + 7 \times 5$

First multiply 3×4 and 7×5. ► $3 \times 4 + 7 \times 5 = 12 + 35$
Then add 12 and 35. ► $12 + 35 = 47$

Tell how to simplify the expression.

Oral
EXERCISES

Sample: $3 + (10 \times 5)$ *What you say:* Multiply 10 and 5, then add 3. The result is 53.

1. $18 \div (2 \times 3)$ 3   **2.** $(15 + 8) - 10$ 13
3. $(4 + 7)5$ 55   **4.** $[3(4)] + 6$ 18
5. $(10 - 2)(3)$ 24   **6.** $\dfrac{10 + 14}{3}$ 8

Tell what number is named.

Sample 1: $(10 + 20)3$ *What you say:* $(10 + 20)3 = (30)3 = 90$

Sample 2: $\dfrac{6 + 3}{18}$ *What you say:* $\dfrac{6 + 3}{18} = \dfrac{9}{18} = \dfrac{1}{2}$

7. $10 - (4 + 2)$ 4   **8.** $(6 + 4)(3 + 1)$ 40
9. $3(5 + 6)$ 33   **10.** $(2 \times 3 \times 5) + 9$ 39
11. $(2 + 7)5$ 45   **12.** $6 + [3 + (5 + 2)]$ 16
13. $7(10)$ 70   **14.** $15 - [2(4 + 1)]$ 5

For Extra Practice, see page 416.

Simplify.

Written
EXERCISES

A

1. $(4 + 5)(2 + 6)$ 72   **2.** $(2 \times 5 \times 4) + 9$ 49
3. $(1 \times 2 \times 3) + 25$ 31   **4.** $(17 + 8)6$ 150
5. $4(5 + 3)$ 32   **6.** $(24 - 6)4$ 72
7. $18 + (34 - 6)$ 46   **8.** $(9)(4 + 8)$ 108
9. $6\left(\dfrac{1}{2} + \dfrac{1}{4}\right)$ $4\frac{1}{2}$ **10.** $(6 + 7)(10 - 3)$ 91

11. $\dfrac{4 + 9}{3}$ $4\frac{1}{3}$

12. $\dfrac{2 + 6}{(3)(3)}$ $\frac{8}{9}$

13. $\dfrac{5 + 7 + 8}{4}$ 5

14. $40 \div (6 + 4)$ 4

15. $(6 + 71) - 10$ 67

16. $[(5)(9)] \div 3$ 15

Solve.

Sample: $(18 - 2) - 6 = n$ *Solution:* $(18 - 2) - 6 = n$
$$16 - 6 = n$$
$$10 = n$$

B

17. $(6 + 8)15 = w$ 210

18. $h = (8 + 6)(5 - 1)$ 56

19. $8 + (9 \times 7) = m$ 71

20. $k = 37 - \dfrac{8 + 4}{9 + 6}$ $36\frac{1}{5}$

21. $(5)(7)(4 + 1) = t$ 175

22. $\dfrac{(45 \div 9) + 3}{6} = b$ $1\frac{1}{3}$

23. $5 + \dfrac{3(8 - 7)}{4} = y$ $5\frac{3}{4}$

24. $(15)\left(\dfrac{6}{2}\right)\left(\dfrac{3}{3}\right) = s$ 45

25. $z = \dfrac{6 + 4}{5} + \dfrac{10 + 4}{10 + 6}$ $2\frac{7}{8}$

26. $3^2 + (36 \div 9) = r$ 13

27. $a = \dfrac{10(3 + 4 + 2)}{9}$ 10

28. $(10^2 + 8^2) - 14 = n$ 150

29. $4[2 + (8 - 1)] = x$ 36

30. $(51 - 27) \div [2(9 + 3)] = b$ 1

31. $5^2 + [(3 + 6) - 4] = m$ 30

32. $\dfrac{[(25 \times 4) + (15 \div 5)]}{3} = t$ $34\frac{1}{3}$

C

33. $4\left[\dfrac{72 - 24}{6^2}\right] = k$ $5\frac{1}{3}$

34. $\left[\dfrac{9 \times 6}{(3)(3)(2)}\right]5 = a$ 15

Use grouping symbols to show the correct order of operations.

Sample 1: $5 \times 3 + 4 = 19$ *Solution:* $(5 \times 3) + 4 = 19$

Sample 2: $3 + 12 \div 2 = 9$ *Solution:* $3 + (12 \div 2) = 9$

37. $17 = (3 \times 5) + 2$

35. $6 \times 4 - 4 = 20$
$(6 \times 4) - 4 = 20$

36. $4 \times \dfrac{1}{2} \div \dfrac{1}{2} - 1 = 3$

37. $17 = 3 \times 5 + 2$

38. $32 \div 8 + 2 = 6$
$(32 \div 8) + 2 = 6$

39. $4 + 2 \times 3 \times 1 = 10$

40. $2 \times 3 + 4 \times 1 = 10$

41. $12 \div 4 - 3 = 0$

42. $12 - 2 \times 6 = 0$
$12 - (2 \times 6) = 0$

43. $8 - 2 \times 3 = 28 - (2 \times 3) = 2$

44. $8 \div 2 - 3 = 1$
$(8 \div 2) - 3 = 1$

36. $4 \times \left(\dfrac{1}{2} \div \dfrac{1}{2}\right) - 1 = 3$ or $\left(4 \times \dfrac{1}{2}\right) \div \dfrac{1}{2} - 1 = 3$

39. $4 + (2 \times 3 + 1) = 10$ **40.** $(2 \times 3) + (4 \times 1) = 10$

41. $(12 \div 4) - 3 = 0$

calculator corner

You can use a calculator and your knowledge of algebra to "guess" a friend's birthday. Give your friend the calculator with instructions to hold it so that you can't see the display. Then have him or her do the following:

1. Punch in the number of the month of his or her birth.
2. Multiply by 5.
3. Add 6.
4. Multiply by 4.
5. Add 9.
6. Multiply by 5.
7. Add the day of birth.
8. Subtract 165.

The display will show your friend's birthday. 1201, for example, means December 1; 207 means February 7.

Maggie Lena Walker   1867–1934

Maggie Lena Walker first entered her career as an insurance and banking executive as an agent for the Woman's Union, an insurance company in Richmond, Virginia. After taking business courses in accounting and sales, she became the executive secretary-treasurer of the Independent Order of St. Luke, a Black fraternal society and cooperative insurance venture. Under her leadership the organization grew in 25 years from 3,400 to 50,000 members. In 1902, Walker began publishing the *St. Luke Herald* which reported the order's affairs. In 1903 the St. Luke Penny Savings Bank was established at her initiative. It grew to become the Consolidated Bank and Trust Company in 1929–1930, with Maggie Lena Walker as president.

4-6 *Evaluating Expressions*

OBJECTIVE

Evaluate an expression by replacing each of the variables by the members of a replacement set.

The **replacement set** for a variable in an expression is the set of numbers that the variable may represent. We **find the value** of an expression by substituting members of the replacement set for each variable.

EXAMPLE 1 Find the value of $2m + 1$. The replacement set for m is $\{0, 2, 4\}$.

$$\text{If } m = 0 \blacktriangleright 2(0) + 1 = 0 + 1 = \mathbf{1}$$
$$\text{If } m = 2 \blacktriangleright 2(2) + 1 = 4 + 1 = \mathbf{5}$$
$$\text{If } m = 4 \blacktriangleright 2(4) + 1 = 8 + 1 = \mathbf{9}$$

EXAMPLE 2 Find the value of $2t + y$. Let $t = 3$ and $y = 7$. (We *could* say "The replacement set for t is $\{3\}$ and for y is $\{7\}$.")

$$2t + y = 2(3) + 7 = 6 + 7 = \mathbf{13}$$

It's important to remember that you must give a variable the same value each time it appears in the expression.

EXAMPLE 3 Find the value of $\dfrac{2x + 2}{x}$. The replacement set for x is $\{1, 2\}$.

$$\text{If } x = 1 \blacktriangleright \frac{2x + 2}{x} = \frac{2(1) + 2}{1} = \frac{2 + 2}{1} = \mathbf{4}$$

$$\text{If } x = 2 \blacktriangleright \frac{2x + 2}{x} = \frac{2(2) + 2}{2} = \frac{4 + 2}{2} = \mathbf{3}$$

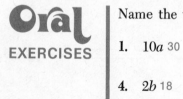

Oral EXERCISES

Name the value of the expression. Let $a = 3$ and $b = 9$.

1. $10a$ 30

2. $\dfrac{a}{b}$ $\frac{1}{3}$

3. $4(b)$ 36

4. $2b$ 18

5. $\dfrac{b}{a}$ 3

6. $\dfrac{1}{3}b$ 3

7. $3a$ 9

8. $2(a)(b)$ 54

9. a^2 9

Find all values of the expression. The replacement set for m is $\{1, 3, 5, 7\}$.

Sample: $4 + m$ Solution: $4 + 1 = 5$
$$4 + 3 = 7$$
$$4 + 5 = 9$$
$$4 + 7 = 11$$

A

1. $m + 10$ 11; 13; 15; 17
2. $5m + 0$ 5; 15; 25; 35
3. $(m + 10) - 8$ 3; 5; 7; 9

4. $16 + m$ 17; 19; 21; 23
5. $\dfrac{m}{2}$ $\tfrac{1}{2}$; $1\tfrac{1}{2}$; $2\tfrac{1}{2}$; $3\tfrac{1}{2}$
6. $m(m + 3)$ 4; 18; 40; 70

7. $3m$ 3; 9; 15; 21
8. $\dfrac{m}{5}$ $\tfrac{1}{5}$; $\tfrac{3}{5}$; 1; $1\tfrac{2}{5}$
9. m^2 1; 9; 25; 49

10. $8m$ 8; 24; 40; 56
11. $2(m + 4)$ 10; 14; 18; 22
12. m^3 1; 27; 125; 343

Find the value of the expression. Let $r = 2$, $s = 3$, and $t = 5$.

Sample: $rs + t$ Solution: $(2)(3) + 5 = 6 + 5 = 11$

13. $rs + rt$ 16
14. $(r \times s) + t^2$ 31

15. rst 30
16. $s^2 + rt$ 19

17. $\dfrac{1}{2}rt$ 5
18. $(r + s) - t$ 0

B

19. $r^2 + s^2 + t^2$ 38
20. $t^2 + \dfrac{r^2}{s^2}$ $25\tfrac{4}{9}$

21. $2r\left(s + \dfrac{t}{4}\right)$ 17
22. $(6 + r)(t)(s)$ 120

23. $(s - r)(s + r)$ 5
24. $\dfrac{s + s + s}{(rs)(rs)}$ $\tfrac{1}{4}$

25. $rs + \dfrac{t + s}{r}$ 10
26. $\dfrac{3rt}{5rs} + \left(\dfrac{s}{t}\right)\left(\dfrac{r}{t}\right)$ $1\tfrac{6}{25}$

Find the value of the expression. Let $x = 2$ and $y = 7$.

C

27. $x + [y(3 + x)]$ 37
28. $\left[\left(\dfrac{y}{x}\right) + xy\right] - 12$ $5\tfrac{1}{2}$

29. $y + [2x + (y - x)] + x$ 18
30. $\left[\dfrac{xy}{x^2} + y\right] - \dfrac{5}{x}$ 8

31. $y - \left[2\left(\dfrac{x + y}{3}\right)\right] + 4$ 5
32. $[(x^2 + y^2)(xy)] \div 10$ $74\tfrac{1}{5}$

4-7 Factors, Coefficients, and Terms

OBJECTIVES

Factor expressions like $\dfrac{3n}{2}$.

Name the terms of an expression.

Simplify expressions involving indicated products, such as
$(3)(4)(t) = 12t$.

There are many ways to indicate multiplication. In each case, the numbers to be multiplied are called factors. The result is called the product.

$$4 \times 7 = 28 \qquad 4 \cdot 7 = 28 \qquad (4)(7) = 28$$

factors   product   factors   product   factors   product

We factor a number or an expression by writing it as an indicated multiplication, called a factorization. A number or expression may have many factorizations.

EXAMPLE 1   Some factorizations of 10 ► $1 \cdot 10 \qquad 2 \cdot 5 \qquad \dfrac{1}{2} \cdot 20$

Any factor of an expression can be called the coefficient of the remaining factors. Often the numerical part of a multiplication expression containing variables is called *the* coefficient.

EXAMPLE 2 $6ab$ ► $6a(b)$ ► $6a$ is the coefficient of b.
$\qquad\qquad\qquad\quad 3(2ab)$ ► 3 is the coefficient of $2ab$.
$\qquad\qquad\qquad\quad 6(ab)$ ► 6 is the coefficient of ab.

EXAMPLE 3 $5x$ ► 5 is the coefficient of x.

EXAMPLE 4 m ► 1 is understood to be the coefficient of m.

The parts of an expression like $3m + 2t$ that are separated by a plus or minus sign are called terms.

EXAMPLE 5 $3m + 2t \qquad 4a + \dfrac{b}{2} - c \qquad 3ab$

    two terms   three terms   one term

Name the coefficient of the variables.

Sample: $\frac{1}{5}y$ *What you say:* $\frac{1}{5}$ is the coefficient of y.

1. $5m$ 5

2. $2.6n$ 2.6

3. $\frac{z}{2}$ $\frac{1}{2}$

4. $7t$ 7

5. $3ab$ 3

6. $0.75r$ 0.75

7. $\frac{1}{3}b$ $\frac{1}{3}$

8. $10mn$ 10

9. $6w^2$ 6

Name the terms.

Sample: $2x + 5$ *What you say:* $2x$ is a term and 5 is a term.

10. $b + c$ b; c

11. $\frac{a}{b} + \frac{1}{2}$ $\frac{a}{b}$; $\frac{1}{2}$

12. $2t + n$ 2t; n

13. $\frac{ab}{c}$ $\frac{ab}{c}$

14. $7k - 5$ 7k; 5

15. $w \div y$ w ÷ y

Complete each factorization for the given product.

Sample: Product: $8ab$; $8(\underline{})$, $4(\underline{})$, $(\underline{})4a$
Solution: $8(ab)$, $4(2ab)$, $(2b)4a$

1. Product: $10s$; $10(\underline{})$, $5(\underline{})$, $2(\underline{})$ (s); (2s); (5s)

2. Product: $9b$; $3(\underline{})$, $9(\underline{})$, $1(\underline{})$ (3b); (b); (9b)

3. Product: $\frac{1}{8}t$; $(\underline{})t$, $\frac{1}{2}(\underline{})$, $\frac{1}{4}(\underline{})$ $\left(\frac{1}{8}\right)$; $\left(\frac{1}{4}t\right)$; $\left(\frac{1}{2}t\right)$

4. Product: $5ab$; $\underline{}(ab)$, $5a(\underline{})$, $(\underline{})a$ 5; (b); (5b)

5. Product: xyz; $x(\underline{})$, $y(\underline{})$, $(\underline{})z$ (yz); (xz); (xy)

6. Product: $\frac{rs}{2}$; $\frac{1}{2}(\underline{})$, $\frac{r}{2}(\underline{})$, $\frac{s}{2}(\underline{})$ (rs); (s); (r)

7. Product: $\frac{2m}{3}$; $\frac{1}{3}(\underline{})$, $\frac{2}{3}(\underline{})$, $(\underline{})\frac{m}{3}$ (2m); (m); (2)

8. Product: $2abc$; $2a(\underline{})$, $(\underline{})c$, $(\underline{})b$ (bc); (2ab); (2ac)

Show that each factorization has the same value as the given product. Use $k = 4$ and $t = 6$.

Sample: Product: $\frac{3}{4}t$ Factorizations: $\frac{1}{4}(3t)$; $3\left(\frac{t}{4}\right)$

Solution: $\frac{3}{4}t = \frac{3}{4} \cdot 6 = 4\frac{1}{2}$

$\frac{1}{4}(3t) = \frac{1}{4} \cdot 18 = 4\frac{1}{2}$

$3\left(\frac{t}{4}\right) = 3 \cdot \frac{6}{4} = 4\frac{1}{2}$

9. Product: $8k$ Factorizations: $4(2k)$; $2(4k)$ 32
10. Product: $3tk$ Factorizations: $3t(k)$; $3k(t)$ 72
11. Product: $\frac{kt}{3}$ Factorizations: $\frac{1}{3}(kt)$; $\frac{t}{3}(k)$ 8
12. Product: $\frac{9k}{t}$ Factorizations: $9 \cdot \frac{k}{t}$; $\frac{1}{t} \cdot 9k$ 6
13. Product: $\frac{1}{2}kt$ Factorizations: $\frac{k}{2} \cdot t$; $k \cdot \frac{t}{2}$ 12
14. Product: $\frac{4k}{5t}$ Factorizations: $\frac{4}{5} \cdot \frac{k}{t}$; $4k \cdot \frac{1}{5t}$ $\frac{8}{15}$

Name the product.

Sample 1: $(7)(5)(m)$ *Solution:* $35m$

Sample 2: $c = 2 \times \pi \times r$ *Solution:* $2\pi r$

15. $(3)(2)(x)$ 6x
16. $3(t)(n)$ 3tn
17. $(9)(a)(b)$ 9ab
18. $(0.5)(5.0)(c)(d)$ 2.5 cd

B

19. $p = l + l + w + w$ 2l + 2w
20. $A = \pi \times r \times r$ πr^2
21. $A = \frac{1}{2} \times b \times h$ $\frac{1}{2}bh$
22. $v = \frac{4}{3} \times \pi \times r \times r \times r$ $\frac{4}{3}\pi r^3$
23. $(6)(m)(m)(9)$ 54m²
24. $5(t)(r)\left(\frac{s}{9}\right)$ $\frac{5}{9}trs$
25. $(3)(b)\left(\frac{c}{5}\right)$ $\frac{3}{5}bc$
26. $\left(\frac{a}{2}\right)\left(\frac{b}{4}\right)\left(\frac{3}{c}\right)$ $\frac{3ab}{8c}$
27. $\left(\frac{1}{2}\right)(p)(5q)$ $\frac{5}{2}pq$
28. $a\left(\frac{a}{2}\right)\left(\frac{a}{3}\right)$ $\frac{a^3}{6}$

In Exs. 29–40 answers may vary.

Write three different factorizations.

Sample: $12pq$ **Solution:** $12p(q)$; $(3p)(4q)$; $(6q)(2p)$ (Other answers are possible.)

29. $8rs$
$8(rs)$; $(8r)s$; $8r(s)$

30. $2ab$
$2(ab)$; $(2a)b$; $2a(b)$

31. $\dfrac{3rs}{t}$
$3\left(\dfrac{rs}{t}\right)$; $3\left(\dfrac{r}{t}\right)s$; $\left(\dfrac{3}{t}\right)rs$

32. abc
$(a)(b)(c)$; $a(bc)$; $(ab)c$

33. $2d$
$2(d)$; $\dfrac{1}{2}(4d)$; $6\left(\dfrac{d}{3}\right)$

34. $\dfrac{ab}{cd}$
$ab\left(\dfrac{1}{cd}\right)$; $a\left(\dfrac{b}{cd}\right)$; $\left(\dfrac{a}{cd}\right)b$

35. $\dfrac{3t}{10}$
$3t\left(\dfrac{1}{10}\right)$; $\dfrac{3}{10}(t)$; $3\left(\dfrac{t}{10}\right)$

36. r^2
$(r)(r)$; $\dfrac{1}{2}(2r^2)$; $(2r)\left(\dfrac{1}{2}r\right)$

37. $\dfrac{a}{cd}$
$\left(\dfrac{a}{c}\right)\left(\dfrac{1}{d}\right)$; $a\left(\dfrac{1}{cd}\right)$; $a\left(\dfrac{1}{c}\right)\left(\dfrac{1}{d}\right)$

38. $\dfrac{xy}{3}$
$\left(\dfrac{x}{3}\right)y$; $x\left(\dfrac{y}{3}\right)$; $\dfrac{1}{3}(xy)$

39. $\dfrac{4cd}{5}$
$\dfrac{4}{5}(cd)$; $4c\left(\dfrac{d}{5}\right)$; $4d\left(\dfrac{c}{5}\right)$

40. $\dfrac{1}{rst}$
$\left(\dfrac{1}{r}\right)\left(\dfrac{1}{s}\right)\left(\dfrac{1}{t}\right)$; $\left(\dfrac{1}{rs}\right)\left(\dfrac{1}{t}\right)$; $\left(\dfrac{1}{r}\right)\left(\dfrac{1}{st}\right)$

C

Solve.

Problems

Sample: The area of a rectangle is $5ab$. If the length is $5a$, what is the width?

Solution Use $A = lw$, where $A = 5ab$ and $l = 5a$.
$5ab = (5a)(b)$, so the width is b.

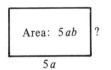
Area: $5ab$?
$5a$

1. The area of a rectangle is $18cd$. If the length is $9c$, what is the width? Use $A = lw$. 2d

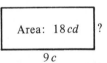
Area: $18cd$?
$9c$

2. The area of a triangle is $8rs$. If the length of the base is $4r$, what is the height? Use $A = \dfrac{1}{2}bh$. 4s

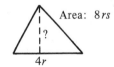
Area: $8rs$
?
$4r$

3. The area of a triangle is $24abc$. If the height is $8c$, what is the base? Use $A = \dfrac{1}{2}bh$. 6ab

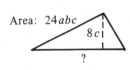
Area: $24abc$
$8c$
?

4. The dimensions of two rectangles are shown. Compare their areas. Use $A = lw$.
Area of I $= \dfrac{9ab}{16}$
Area of II $= \dfrac{9ab}{16}$
The areas are equal.

I
$3a$ Area: ?
$\dfrac{3b}{16}$

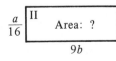
II
$\dfrac{a}{16}$ Area: ?
$9b$

4-8 Functions

OBJECTIVES

Use an algebraic expression as a rule for finding a set of ordered pairs that is a function.

Graph a function.

In Chapter 2 you learned about ordered pairs of numbers and their graphs. Now we use those ideas in working with mathematical functions.

Suppose you have a "function machine." When a card bearing a number is fed in, the machine names another number according to a stated rule. Let's call the number on the card an **input** number and the machine's response an **output** number. We can write a set of ordered pairs of the form (**input, output**).

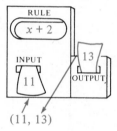

(11, 13)

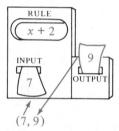

(7, 9)

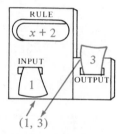

(1, 3)

EXAMPLE 1   Let's use a function machine with the rule $x + 2$. The table shows the ordered pairs that result when the counting numbers are fed in.

Input Number: x	Output Number: $x + 2$	Ordered Pair: $(x, x + 2)$
1	3	(1, 3)
2	4	(2, 4)
3	5	(3, 5)
4	6	(4, 6)
.	.	.
.	.	.
.	.	.

▲

Notice the three dots. They show that the set continues on indefinitely.

We get the set of ordered pairs {(1, 3), (2, 4), (3, 5), (4, 6), ...}. This set is called a **function**.

EXAMPLE 2 Let us use the "function machine" with the rule $\dfrac{n}{3}$ to generate a different set of ordered pairs.

Input Number: n	Output Number: $\dfrac{n}{3}$	Ordered Pair: $\left(n, \dfrac{n}{3}\right)$
1	$\dfrac{1}{3}$	$\left(1, \dfrac{1}{3}\right)$
2	$\dfrac{2}{3}$	$\left(2, \dfrac{2}{3}\right)$
3	1	$(3, 1)$
4	$1\frac{1}{3}$	$(4, 1\frac{1}{3})$
5	$1\frac{2}{3}$	$(5, 1\frac{2}{3})$

The function that results is $\left\{\left(1, \dfrac{1}{3}\right), \left(2, \dfrac{2}{3}\right), (3, 1), (4, 1\frac{1}{3}), (5, 1\frac{2}{3})\right\}$.

Many rules can be used to form sets of ordered pairs that are functions. Notice that for a set of ordered pairs to be a function, no two different pairs can have the same first number.

Functions	*Not Functions*
$\{(0, 7), (1, 8), (2, 9), (3, 10)\}$	$\{(2, 0), (2, 5), (3, 6), (4, 9)\}$
	same first member
$\{(1, 2), (2, 3), (3, 5), (5, 7), \ldots\}$	$\{(2, 2), (4, 4), (2, 6), (4, 6), \ldots\}$

We can graph a function by locating on a set of axes the points that represent the ordered pairs in the set.

EXAMPLE 3

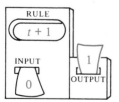

Input Number t	Output Number $t + 1$	Ordered Pair $(t, t + 1)$
0	1	$(0, 1)$
1	2	$(1, 2)$
2	3	$(2, 3)$
3	4	$(3, 4)$

Graph:

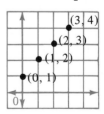

Function: $\{(0, 1), (1, 2), (2, 3), (3, 4)\}$

Tell how to complete the table.

1. Rule: $s - 3$

Input Number	Output Number	Number Pair	
3	0	(3, 0)	
5	2	(?, ?)	(5, 2)
7	4	(?, ?)	(7, 4)
9	?	(9, 6)	6
11	?	(11, 8)	8

2. Rule: $\dfrac{w}{2}$

Input Number	Output Number	Number Pair	
2	1	(2, 1)	
3	$1\frac{1}{2}$	(?, ?)	$(3, 1\frac{1}{2})$
5	?	$(5, 2\frac{1}{2})$	$2\frac{1}{2}$
7	?	$(7, 3\frac{1}{2})$	$3\frac{1}{2}$
9	?	(?, ?)	$4\frac{1}{2}$

Tell whether or not the set of ordered pairs is a function.

3. $\{(0, 4), (1, 8), (2, 16), (3, 32)\}$ function

4. $\{(1, 2), (1, 3), (1, 4), (2, 6), (2, 8), (2, 10)\}$ not a function

5. $\{(1, 5), (2, 5), (3, 5), (4, 6), (5, 6), (6, 6), \ldots\}$ function

6. $\left\{\left(\frac{1}{2}, \frac{1}{3}\right), \left(\frac{1}{3}, \frac{1}{4}\right), \left(\frac{1}{4}, \frac{1}{5}\right), \left(\frac{1}{5}, \frac{1}{6}\right)\right\}$ function

Complete the set of ordered pairs by using the given rule.

Sample: $r + 3$: $\{(5, 8), (7, 10), (9, \underline{\ ?\ }), (11, \underline{\ ?\ }), (13, \underline{\ ?\ })\}$
Solution: $\{(5, 8), (7, 10), (9, 12), (11, 14), (13, 16)\}$

1. $b - 5$: $\{(5, 0), (6, 1), (7, \underline{\ ?2\ }), (8, \underline{\ ?3\ }), (9, \underline{\ ?4\ }), (10, 5)\}$

2. $4 + s$: $\{(1, 5), (2, 6), (3, \underline{\ ?7\ }), (4, \underline{\ ?8\ }), (5, \underline{\ ?9\ }), (6, \underline{10?\ }), (7, \underline{11?\ })\}$

3. $\dfrac{t}{5}$: $\left\{(0, 0), \left(4, \frac{4}{5}\right), (8, \underline{\ ?\ }^{1\frac{3}{5}}), (12, \underline{\ ?\ }^{2\frac{2}{5}}), (16, \underline{\ ?\ }^{3\frac{1}{5}}), (20, \underline{\ ?\ }^{4}), (24, \underline{\ ?\ }^{4\frac{4}{5}})\right\}$

4. $k + 0$: $\{(0, 0), (1, 1), (2, 2), (3, \underline{\ ?3\ }), (4, \underline{\ ?4\ }), (5, \underline{\ ?5\ }), (6, \underline{\ ?6\ })\}$

5. w^2: $\{(1, 1), (2, \underline{\ ?4\ }), (3, \underline{\ ?9\ }), (4, 16), (5, 25), (6, \underline{36?\ }), (7, \underline{49?\ })\}$

6. $2(x + 1)$: $\{(10, 22), (20, 42), (30, 62), (40, \underline{\ ?\ }^{82}), (50, \underline{\ ?\ }^{102}), (60, \underline{\ ?\ }^{122})\}$

Graph the function. Check students' graphs.

Sample: {(0, 4), (1, 3), (2, 2), (3, 1)}   *Solution:*

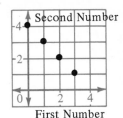

7. {(0, 0), (1, 1), (2, 2), (3, 3), (4, 4), (5, 5)}
8. {(1, 0), (2, 1), (3, 2), (4, 3), (5, 4), (6, 5)}
9. $\left\{\left(0, \frac{1}{2}\right), (1, 1\frac{1}{2}), (1\frac{1}{2}, 2), (2, 2\frac{1}{2}), (2\frac{1}{2}, 3), (3, 3\frac{1}{2}), (3\frac{1}{2}, 4)\right\}$

Match each set of ordered pairs with the correct rule.

10. {(0, 2), (1, 4), (2, 6), (3, 8), (4, 10)} B   **A.** $n + 0.3$

11. $\left\{\left(\frac{1}{7}, 1\right), \left(\frac{2}{7}, 2\right), \left(\frac{3}{7}, 3\right), \left(\frac{4}{7}, 4\right), \left(\frac{5}{7}, 5\right)\right\}$ C   **B.** $2(x + 1)$

12. {(1, 1.3), (2, 2.3), (3, 3.3), (4, 4.3)} A   **C.** $7w$

13. {(0, 0), (2, 1), (4, 2), (6, 3), (8, 4)} E   **D.** $\dfrac{s}{3}$

14. {(6, 2), (7\frac{1}{2}, 2\frac{1}{2}), (9, 3), (10\frac{1}{2}, 3\frac{1}{2})} D   **E.** $y \div 2$

Complete.

	Input Number m	Output Number 0.5 + m	Ordered Pair (m, 0.5 + m)
15.	0	0.5	(0, 0.5) ?
16.	1.0	1.5	(1.0, ? 1.5)
17.	1.5	? 2.0	(1.5, ? 2.0)
18.	2.0	? 2.5	(2.0, ? 2.5)
19.	2.5	? 3.0	(2.5, ? 3.0)

RULE
0.5 + m
INPUT 0 0.5 OUTPUT

B

	n	$\dfrac{n + 2}{3}$	$\left(n, \dfrac{n + 2}{3}\right)$
20.	2	$1\frac{1}{3}$	(2, 1\frac{1}{3}) ?
21.	4	? 2	(4, 2) ?
22.	6	? $2\frac{2}{3}$	(6, 2\frac{2}{3}) ?
23.	8	? $3\frac{1}{3}$	(8, 3\frac{1}{3}) ?
24.	10	? 4	(10, 4) ?
25.	12	? $4\frac{2}{3}$	(12, 4\frac{2}{3}) ?

Write an expression that is a function rule for the set of ordered pairs.

Sample: $\{(2, 4\frac{1}{2}), (3, 5\frac{1}{2}), (4, 6\frac{1}{2}), (6, 8\frac{1}{2})\}$ *Solution:* $x + 2\frac{1}{2}$

C

26. $\{(2, 10), (3, 15), (4, 20), (5, 25), (6, 30)\}$ $5x$

27. $\left\{(0, 0), \left(1, \dfrac{1}{2}\right), (2, 1), (3, 1\frac{1}{2}), (4, 2), (5, 2\frac{1}{2})\right\}$ $\frac{1}{2}x$

28. $\{(0, 0), (1, 1), (2, 2), (3, 3), (8, 8), (9, 9), (10, 10)\}$ x

29. $\{(1, 1), (2, 4), (3, 9), (4, 16), (5, 25), (6, 36)\}$ x^2

30. $\{(1, 0.75), (2, 1.5), (3, 2.25), (4, 3), (5, 3.75)\}$ $0.75x$

31. $\left\{\left(1, \dfrac{1}{10}\right), \left(2, \dfrac{1}{5}\right), \left(3, \dfrac{3}{10}\right), \left(4, \dfrac{2}{5}\right), \left(5, \dfrac{1}{2}\right), \left(6, \dfrac{3}{5}\right)\right\}$ $\frac{1}{10}x$

SELF-TEST 2

Be sure that you understand these terms.

replacement set (p. 104)   coefficient (p. 106)
terms (p. 106)   function (p. 110)

Section 4-5, p. 100   1.   Simplify $15 \div (3 - 2)$. 15

Section 4-6, p. 104   2.   Show that both expressions have the same value when $x = 2$
and $y = 7$: $3(xy) = 3x(y)$ $3(xy) = 3(2)(7) = 42; 3x(y) = 3(2)(7) = 42$

3.   Find the value of $(r + s^2)t$ if $r = 5$, $s = 3$, and $t = 2$. 28

Section 4-7, p. 106   4.   Complete the following factorizations of $\dfrac{2s}{7}$:

$\dfrac{1}{7}(\underset{?}{\overset{2s}{\_\_}}), \dfrac{2}{7}(\underset{?}{\overset{s}{\_\_}}), (\underset{?}{\overset{2}{\_\_}})\dfrac{s}{7}$

5.   Simplify $(3)(x)(y)(2)$. $6xy$

Section 4-8, p. 110   6.   Complete the set of number pairs by using the rule $t + 7$:
$\{(0, 7), (1, \underset{8}{\_?\_}), (2, \underset{9}{\_?\_}), (3, \underset{10}{\_?\_}), (4, \underset{11}{\_?\_}), (5, \underset{12}{\_?\_}), (6, \underset{13}{\_?\_})\}$.

7.   Graph $\{(0, 0), (1, 1), (2, 4), (3, 9), (4, 16)\}$. Check students' graphs.

Check your answers with those printed at the back of the book.

chapter summary

1. To add (or subtract) with fractions that have a common denominator, add (or subtract) the numerators and use the common denominator.

2. To add (or subtract) with fractions that do not have a common denominator, first rename one or both fractions so they have a common denominator. Then add or subtract.

3. To multiply with fractions, multiply the numerators and the denominators.

4. Two numbers are **reciprocals** of each other if their product is 1.

5. Dividing by a number is the same as multiplying by its reciprocal.

6. When we multiply or divide with decimals, the number of digits after the decimal point in the product is the sum of the numbers of digits after the decimal points in the factors.

7. To find a percent of a number, multiply the number by the percent written in decimal form.

8. Values of an expression containing a variable are found when the variable is replaced with the members of the specified replacement set.

9. Any factor of an expression can be called the **coefficient** of the remaining factors.

chapter test

Add or subtract.

1. $\dfrac{5}{8} + \dfrac{1}{8}$ $\dfrac{3}{4}$

2. $\dfrac{9}{10} - \dfrac{8}{10}$ $\dfrac{1}{10}$

3. $\dfrac{1}{2} + \dfrac{3}{8}$ $\dfrac{7}{8}$

4. $2\frac{2}{5} + 4\frac{1}{5}$ $6\frac{3}{5}$

5. $7\frac{3}{4} - 4\frac{1}{4}$ $3\frac{1}{2}$

6. $\dfrac{4}{5} - \dfrac{2}{3}$ $\dfrac{2}{15}$

Multiply or divide.

7. $\dfrac{5}{8} \times \dfrac{1}{3}$ $\dfrac{5}{24}$

8. $1\frac{1}{2} \div \dfrac{1}{4}$ 6

9. $2\frac{1}{3} \div 4$ $\dfrac{7}{12}$

10. $\dfrac{2}{3} \div \dfrac{3}{5}$ 1$\frac{1}{9}$

11. $\dfrac{1}{2} \times 4\frac{1}{3}$ 2$\frac{1}{6}$

12. $6 \times 1\frac{1}{8}$ $\frac{27}{4}$

Find the value when $a = 1$, $b = 2$, and $c = 3$.

13. $\dfrac{a}{6} + \dfrac{a}{c}$ $\frac{1}{2}$

14. $\dfrac{c}{b} \times \dfrac{a}{b}$ $\frac{3}{4}$

15. $\dfrac{a}{b} \times \dfrac{a}{c} \times \dfrac{b}{c}$ $\frac{1}{9}$

Solve.

16. $t = 27.1 + 3.8$ 30.9

17. $x = 58.09 - 12.25$ 45.84

18. $s = 13.04 \times 5.7$ 74.328

19. $t = 119.7 \div 0.95$ 126

Find the percent.

20. 32% $\times$ 65 20.8

21. 8% $\times$ $125 $10

Simplify.

22. $(10 - 6)(3 + 7)$ 40

23. $[(3)(4)]$ 12

24. $\dfrac{8 + 7}{3}$ 5

25. $5a(b)(c)$ 5abc

Find the values of the expression. The replacement set for t is $\{2, 4, 6\}$.

26. $3t + 15$ 21; 27; 33

27. $5(t + 3)$ 25; 35; 45

Complete each factorization of the given product.

28. Product: $6bc$; $6(\underline{\ ?\ })$, $3(\underline{\ ?\ })$, $2b(\underline{\ ?\ })$ (bc); $(2bc)$; $(3c)$

29. Product: $\dfrac{3t}{5}$; $\dfrac{3}{5}(\underline{\ ?\ })$, $3(\underline{\ ?\ })$, $\dfrac{1}{5}(\underline{\ ?\ })$ (t); $\left(\dfrac{t}{5}\right)$; $(3t)$

Complete the table and graph the function.

30.

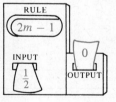

Input Number	Output Number	Ordered Pair
$\dfrac{1}{2}$	0	$\left(\dfrac{1}{2}, 0\right)$
1	1	(1, 1) ?
$1\frac{1}{2}$	2	$(1\frac{1}{2}, ? \ 2)$
2	? 3	(2, 3) ?
$2\frac{1}{2}$	? 4	$(2\frac{1}{2}, ? \ 4)$

challenge topics

Decimal Slide Ruler

Make a pair of scales from heavy paper, as shown below. Each scale is marked off in centimeters and millimeters which will be used to add and subtract with decimals.

These scales are arranged to show that $3.5 + 4.3 = 7.8$

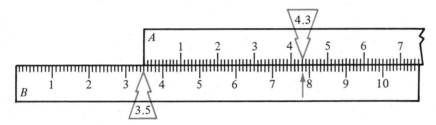

Read the answer 7.8 from the *B* scale.

To complete the subtraction $8.9 - 2.4 = \underline{\ ?\ }$ the scales are arranged like this:

Read the answer 6.5 from the *A* scale.

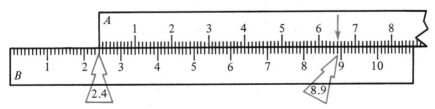

Add or subtract. Use the decimal slide ruler.

1. $2.8 + 6.9$ 9.7
2. $4.0 + 5.5$ 9.5
3. $0.6 + 3.7$ 4.3
4. $6.2 - 3.1$ 3.1
5. $10.0 - 4.7$ 5.3
6. $7.3 - 2.4$ 4.9

Write using exponents.

Sample: $8 \cdot 8 \cdot 8$ *Solution:* $8 \cdot 8 \cdot 8 = 8^3$

1. $10 \cdot 10$ 10^2

2. $2 \cdot 2 \cdot 2 \cdot 2 \cdot 2 \cdot 2$ 2^6

3. $5 \cdot 5 \cdot 5 \cdot 5 \cdot 5$ 5^5

4. $3 \cdot 3 \cdot 3 \cdot 3$ 3^4

5. $n \cdot n \cdot n$ n^3

6. $7 \cdot 7 \cdot 7$ 7^3

Multiply.

7. $9 \cdot 5$ 45

8. $9 \cdot 9$ 81

9. $7 \cdot 5 \cdot 14$ 490

10. $6 \cdot 7$ 42

11. $7 \cdot 8 \cdot 2 \cdot 0$ 0

12. $5 \cdot 12 \cdot 9$ 540

13. $4 \cdot 4 \cdot 4$ 64

14. $10 \cdot 10 \cdot 15$ 1500

15. $7 \cdot 8 \cdot 18$ 1008

Tell whether the statement is true or false.

16. $8 + 3 < 11$ False

17. $(2 \cdot 5) + 1 > 10$ True

18. $3 \times 5 < 10$ False

19. $\dfrac{7}{3} + 2 > 3$ True

Divide.

20. $36 \div 9$ 4

21. $56 \div 8$ 7

22. $64 \div 8$ 8

23. $63 \div 7$ 9

24. $72 \div 9$ 8

25. $40 \div 5$ 8

Add.

26.
$$\begin{array}{r} 477 \\ 209 \\ \underline{83} \\ 769 \end{array}$$

27.
$$\begin{array}{r} \$\ 47.10 \\ 65.28 \\ \underline{300.00} \\ \$412.38 \end{array}$$

28. $\$9.24 + \$18.37 + \$15.80$ $43.41

Write the fraction in lowest terms.

29. $\dfrac{15}{20}$ $\dfrac{3}{4}$

30. $\dfrac{3}{9}$ $\dfrac{1}{3}$

31. $\dfrac{12}{48}$ $\dfrac{1}{4}$

32. $\dfrac{50}{100}$ $\dfrac{1}{2}$

33. $\dfrac{18}{30}$ $\dfrac{3}{5}$

34. $\dfrac{19}{57}$ $\dfrac{1}{3}$

35. $\dfrac{36}{48}$ $\dfrac{3}{4}$

36. $\dfrac{75}{100}$ $\dfrac{3}{4}$

Left: Telephone switchboard, 1908.

Right: Modern electronic switching system.

5 Equations and Inequalities

Solution Sets and Strategies

5-1 *Equations and Inequalities*

OBJECTIVES

Classify number sentences as equations or inequalities.

Use the symbols =, ≠, <, and > to write number sentences.

You recall that a number sentence which contains the symbol $=$ is called an **equation**. A number sentence which contains one of the symbols $<$, $>$, or $\neq$ is called an **inequality**.

EXAMPLE 1   The sum of 10 and 9   is not equal to   20.

$$10 + 9 \qquad\qquad \neq \qquad\quad 20$$

This statement is **true**.

EXAMPLE 2   The product of 5 and 9   is less than   40.

$$5 \cdot 9 \qquad\qquad\qquad < \qquad\quad 40$$

This statement is **false**.

An equation or inequality that contains a variable is neither true nor false until the variable is replaced by a member of a specified replacement set.

EXAMPLE 3   The sum of 6 and some number   is less than   15.

$$6 + n \qquad\qquad\qquad < \qquad\quad 15$$

This sentence is **neither** true nor false.

EXAMPLE 4   The quotient of 20 and 5   is greater than   some number.

$$20 \div 5 \qquad\qquad\qquad > \qquad\qquad x$$

This sentence is neither true nor false.

EXAMPLE 5   The product of 2 and some number   is less than   2.

$$2m \qquad\qquad\qquad\quad < \qquad\quad 2$$

This sentence is neither true nor false.

Tell whether the sentence is an equation or an inequality.

Oral
EXERCISES

1. $\frac{5}{t} + 3 = 8$ equation

2. $2w + 3 = 20$ equation

3. $7 + 4 < 18$ inequality

4. $7 + 4x \neq 15$ inequality

5. $9 \neq 3 \cdot 3 \cdot 3$ inequality

6. $4 + 0 > 4 \cdot 0$ inequality

Tell whether the sentence is *true*, *false*, or *neither*.

Sample 1: $9 - 2 > 4$ *What you say:* True

Sample 2: $2 \cdot k = 14$ *What you say:* Neither

7. $4 \cdot 7 > 7 \cdot 4$ false

8. $x - 1 = 6$ neither

9. $3 + 2 < 8$ true

10. $2(1 + 5) = 12$ true

11. $16 \div 4 \neq 4$ false

12. $1.01 < 1.001$ false

Match each number sentence in Column 1 with the correct item in Column 2.

Written
EXERCISES

A

COLUMN 1

1. $x^2 + 2 = 3$ C

2. $14 > 2x$ G

3. $26 + x < 4x$ F

4. $10 < 5 \cdot x^3$ A

5. $6 + x^2 < 20$ H

6. $15^2 - x = 5$ D

7. $4x \neq 6^2$ B

8. $2.1 - x = 16.2$ E

COLUMN 2

A. 10 is less than 5 times x^3.

B. 4 times x is not the same as 6^2.

C. 2 plus the square of some number is three.

D. 5 is x less than the square of 15.

E. 16.2 is x less than 2.1.

F. x added to 26 is less than 4 times x.

G. 2 times x is less than 14.

H. 6 plus the square of x is less than 20.

Write a number sentence.

Sample: Some number is greater than the product of 6 and 7.
Solution: $k > 6 \cdot 7$.

9. 8 more than some number is 23. $n + 8 = 23$

10. 8 less than some number is 23. $n - 8 = 23$

11. 3 times the cube of some number is greater than 7. $3x^3 > 7$

12. Some number times 9 is not equal to 13. $9n \neq 13$

13. The product of 3.4 and some number is greater than 100. $3.4n > 100$

14. Three-fourths of some number is not equal to 3. $\frac{3}{4}n \neq 3$

B **15.** Twice the square of some number is greater than 5. $2n^2 > 5$

16. The sum of 8 and 7 is greater than the product of some number and 3. $8 + 7 > n \cdot 3$

17. One more than four times some number is less than 20. $4n + 1 < 20$

18. Four less than the cube of some number is not equal to 62. $x^3 - 4 \neq 62$

Write a word statement. Answers for Exs. 19–26 may vary. One possible answer is given.

Sample: $m - 5 > 7$ *Solution:* m minus 5 is greater than 7. (Other answers are possible.)

19. $7 < t - 2$ **20.** $16 \neq 1.8b + 1$

21. $5^2 + s \neq 8$ **22.** $22 < m - 6 \cdot 7$

23. $3 + 2r < 11$ **24.** $r^2 - 1 \neq 0$

25. $a^2 - 3 > 10$ **26.** $5 + c^2 < 24$

19. 7 is less than some number minus 2.

20. 16 is not equal to 1 more than the product of 1.8 and b.

21. The sum of 5 squared and s is not equal to 8.

22. 22 is less than m minus the product of 6 and 7.

23. The product of 2 and r added to 3 is less than 11.

consumer notes *Packaging*

Packages are designed to attract consumers visually. Many times bigger looking packages or bottles do not contain larger amounts of items.

Go into a food store and compare soft drink bottles. Often two bottles *appear* to be the same size. Choose two brands. Which bottle seems to contain more? Read the labels and compare the volumes of the bottles. Which bottle actually contains more beverage? You can compare potato chips in the same way. Choose a bag of potato chips which contains two or more separate parts. Choose a single bag of chips of the same size. Read both labels. Which bag contains more? Which has the lower unit price? Think of other items you can compare. A wise consumer reads labels, rather than trusts appearances.

24. r squared minus 1 is not equal to 0.

25. 3 subtracted from the square of a is greater than 10.

26. 5 plus c squared is less than 24.

career capsule

Instrument Maker

Instrument makers redesign or build mechanical or electronic instruments. Often they work in cooperation with engineers and scientists in developing laboratory equipment and experimental models. They work from blueprints, rough sketches, or verbal instructions. Instrument makers must have an interest in mechanics and the ability to work with their hands. They must study each individual part and understand its relationship to the whole machine.

A four or five year apprenticeship is usually required. High school courses in algebra, geometry, trigonometry, science and machine shop are very useful.

5-2 Equations and Solution Sets

OBJECTIVES

Use tables to find solutions of equations.

Write and graph solution sets of equations.

We have seen that an equation such as $4 + t = 19$ is neither true nor false until a value is given to t. If a replacement set for t is specified, we can substitute members to see which, if any, make the equation true. Sometimes a table is helpful.

EXAMPLE 1 $4 + t = 19$; replacement set: $\{5, 10, 15, 20\}$

Replacement	$4 + t = 19$	True/False
5	$4 + 5 = 19$	False
10	$4 + 10 = 19$	False
15	$\mathbf{4 + 15 = 19}$	**True**
20	$4 + 20 = 19$	False

15 is a solution. It is the only replacement that makes the equation true. The solution set is $\{15\}$.

EXAMPLE 2 $2m = 7 + 1$; replacement set: $\{0, 2, 5\}$

Replacement	$2m = 7 + 1$	True/False	Solution?
0	$0 = 7 + 1$	False	No
2	$4 = 7 + 1$	False	No
5	$10 = 7 + 1$	False	No

There are no solutions. Solution set: $\emptyset$ ◄ the empty set

Once an equation is solved, we can graph its solution set.

EXAMPLE 3 $x \cdot 0 = 0$; replacement set: $\{1, 1\frac{1}{2}, 2, 3\}$

Replacement	$x \cdot 0 = 0$	True/False	Solution?
1	$1 \cdot 0 = 0$	True	YES
$1\frac{1}{2}$	$1\frac{1}{2} \cdot 0 = 0$	True	YES
2	$2 \cdot 0 = 0$	True	YES
3	$3 \cdot 0 = 0$	True	YES

All the replacements are solutions. Solution set: $\{1, 1\frac{1}{2}, 2, 3\}$

Graph: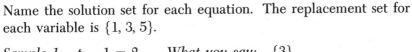

Name the solution set for each equation. The replacement set for each variable is $\{1, 3, 5\}$.

Sample 1: $t - 1 = 2$ *What you say:* $\{3\}$

Sample 2: $6 + k = 4$ *What you say:* $\emptyset$

1. $4 + n = 9$ $\{5\}$

2. $2t = 2$ $\{1\}$

3. $m - 5 = 2$ $\emptyset$

4. $s \div 3 = 1$ $\{3\}$

5. $5 + 4 = 3r$ $\{3\}$

6. $10 \div b = 2$ $\{5\}$

7. $16 + d = 0$ $\emptyset$

8. $n^2 \cdot 0 = 0$ $\{1, 3, 5\}$

For Extra Practice, see page 416.

Make a table to find the solution set. Use the given replacement set.

Sample: $m^2 - 1 = 8$; $\{3, 4, 5\}$

Solution:

Replacement	$m^2 - 1 = 8$	True/False
3	$3^2 - 1 = 8$	True
4	$4^2 - 1 = 8$	False
5	$5^2 - 1 = 8$	False

The solution set is $\{3\}$.

1. $m - 4 = 6 + 2$; $\{10, 12, 14\}$ False, True, False; $\{12\}$

2. $n^2 + 1 = 26$; $\{3, 4, 5\}$ False, False, True; $\{5\}$

3. $2m + 5 = 17$; $\{6, 7, 8\}$ True, False, False; $\{6\}$

A

4. $10 - b = 2$; $\{6, 8, 10\}$ False, True, False; $\{8\}$

5. $4 - 2b = 0$; $\{0, 2, 4\}$ False, True, False; $\{2\}$

6. $7.2 + m = 10.2$; $\{1, 3, 5, 7\}$ False, True, False, False; $\{3\}$

7. $3(1 + s) = 12$; $\{0, 1, 2, 3\}$ False, False, False, True; $\{3\}$

8. $t \cdot 0 + 1 = t$; $\{0, 1, 2, 3\}$ False, True, False, False; $\{1\}$

9. $d + 3d = 8$; $\{0, 1, 2, 3\}$ False, False, True, False; $\{2\}$

10. $2t + 4 = 2(t + 2)$; $\{2, 4, 6, 8\}$ True, True, True, True; $\{2, 4, 6, 8\}$

Find the solution set and graph it. Use the specified replacement set.

Sample: $7k - 14 = 7(k - 2)$; $\{3, 4\}$

Solution:

Replacement	$7k - 14 = 7(k - 2)$	True/False
3	$21 - 14 = 7(1)$	True
4	$28 - 14 = 7(2)$	True

Solution set: $\{3, 4\}$

Graph:

B

11. $10 - a = 20 - 3a$; $\{0, 5, 10\}$ False, True, False; $\{5\}$

12. $2m + 5 = 6$; $\{1, 2, 3\}$ False, False, False; $\emptyset$

13. $2s = 3s - 7$; $\{3, 5, 7\}$ False, False, True; $\{7\}$

14. $m(3 + 5) = 7m + m$; $\{0, 1, 2\}$ True, True, True; $\{0, 1, 2\}$

15. $5k + 2 = k + 10$; $\{0, 2, 4, 6\}$ False, True, False, False; $\{2\}$

16. $b + b + 2b = 4b$; $\{1, 2, 3\}$ True, True, True; $\{1, 2, 3\}$

17. $n(n - 1) = 6$; $\{0, 1, 2\}$ False, False, False; $\emptyset$

18. $2t + 0 \cdot t = t + t$; $\{0, 1, 2, 3\}$ True, True, True, True; $\{0, 1, 2, 3\}$

19. $r + r = r^2$; $\{0, 1, 2\}$ True, False, True; $\{0, 2\}$

20. $t + t = t^2$; $\{0, 3, 5\}$ True; False; False; $\{0\}$

21. $4n - 5 = 3$; $\{2, 3, 4, 5\}$ True; False; False; False; $\{2\}$

22. $2m - 1 = 5$; $\{1, 2, 3, 4\}$ False; False; True; False; $\{3\}$

23. $b + 2b = 4b$; $\{0, 2, 4\}$ True; False; False; $\{0\}$

24. $4x - 2 = 0$; $\left\{0, \dfrac{1}{2}, 1, 1\frac{1}{2}\right\}$ False; True; False; False; $\left\{\dfrac{1}{2}\right\}$

25. $\dfrac{2}{3}t + 4 = 6$; $\{2, 4, 6, 8\}$ False, False, False, False; $\emptyset$

26. $m^2 - 1 = 0$; $\{0, 1, 2, 3\}$ False, True, False, False; $\{1\}$

5-3 *Equation Solving Strategies*

OBJECTIVES

Solve an equation by adding the same number to both members.

Solve an equation by subtracting the same number from both members.

The fact that an equation is like a balanced scale can be used in solving equations.

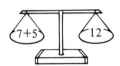
Add 3 to both sides.
The scale is still balanced.
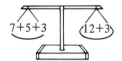

You will agree that $7 + 5 = 12$ is a true statement. Note that when the same number is added to both members of the equation, another true statement results. A true statement also results if we subtract the same number from both members.

$$7 + 5 = 12 \quad \blacktriangleleft \text{True}$$
$$7 + 5 + 3 = 12 + 3 \blacktriangleleft \text{Also true}$$
$$15 = 15$$

$$7 + 5 = 12 \quad \blacktriangleleft \text{True}$$
$$7 + 5 - 2 = 12 - 2 \blacktriangleleft \text{Also true}$$
$$10 = 10$$

These ideas can help us solve an equation that contains a variable. The basic strategy is to get the variable to stand alone as one member of the equation. In working with the following equations, let us agree to use as the replacement set {0 and all numbers to the right of zero on the number line}. We shall call this set {the numbers of arithmetic}.

EXAMPLE 1
$$x + 9 = 14$$
$$x + 9 - 9 = 14 - 9 \blacktriangleleft \text{Subtract 9 from both members.}$$
$$x = 5$$

Solution: 5

EXAMPLE 2
$$8 = m - 1$$
$$8 + 1 = m - 1 + 1 \blacktriangleleft \text{Add 1 to both members.}$$
$$9 = m$$

Solution: 9

True or false?

True   False

1. $5 - 5 = 0$ True   **2.** $3\frac{1}{2} - 3\frac{1}{2} = 0$ **3.** $9 + 3 - 3 = 3$

4. $10 + 5 - 5 = 10$ **5.** $2m - m = m$ **6.** $4.7 + 2 - 2 = 4.7$
           True   True   True

Simplify.

Sample: $23 + 2 - 2$ *What you say:* 23

7. $15 + 7 - 7$ 15   **8.** $17 - 1 + 1$ 17

9. $15 - c + c$ 15   **10.** $32 + 2 - 2$ 32

11. $10 + b - b$ 10   **12.** $m + 3 - 3$ m

For Extra Practice, see page 417.

Written EXERCISES

Add the given number to both members. Simplify the result.

Sample: $11 + 5 = 16$; add 3   *Solution:* $11 + 5 + 3 = 16 + 3$
                                          $19 = 19$

A

1. $2 + 8 = 10$; add 3 13   **2.** $4 = 1 + 3$; add 10 14

3. $15 = 6 + 9$; add 1 16   **4.** $4 + 1 = 5$; add $\frac{1}{2}$ $5\frac{1}{2}$

5. $4.5 = 1 + 3.5$; add 1.5 6   **6.** $6 - 2 = 3 + 1$; add 0 4

Subtract the given number from both members. Simplify the result.

7. $4 + 8 = 12$; subtract 5 7

8. $3 + 5 = 8$; subtract 2 6

9. $35 = 10 + 25$; subtract 4 31

10. $5.0 + 2.5 = 7.5$; subtract 2.5 5.0

11. $3 + 9 = 3 + 4 + 5$; subtract 0 12

12. $18 - 1 = 10 + 7$; subtract 7 10

Add or subtract. Simplify the result.

Sample: $m + 2$; subtract 2   *Solution:* $m + 2 - 2 = m$

13. $x + 10$; subtract 10 x **14.** $w + 9$; subtract 9 w

15. $n - 17$; add 17 n **16.** $y - \frac{1}{2}$; add $\frac{1}{2}$ y

17. $10 + z$; subtract 10 z **18.** $s - 2.7$; add 2.7 s

19. $3.9 + m$; subtract 3.9 m **20.** $a - b$; add b a

Complete the solution. The replacement set is {the numbers of arithmetic}.

21. $w + 1 = 8$
$w + 1 - 1 = 8 - 1$
$w = \underline{\ ?\ }$ 7

22. $k + 10 = 40$
$k + 10 - 10 = 40 - 10$
$k = \underline{\ ?\ }$ 30

23. $a - 5 = 25$
$a - 5 + 5 = 25 + 5$
$a = \underline{\ ?\ }$ 30

24. $t - 6 = 18$
$t - 6 + 6 = 18 + 6$
$t = \underline{\ ?\ }$ 24

25. $14 = n + 6\frac{1}{2}$
$14 - 6\frac{1}{2} = n + 6\frac{1}{2} - 6\frac{1}{2}$
$\underline{\ ?\ } = n$ 7$\frac{1}{2}$

26. $18.7 = y - 4.1$
$18.7 + 4.1 = y - 4.1 + 4.1$
$\underline{\ ?\ } = y$ 22.8

B

27. $2x = 5 + x$
$2x - x = 5 + x - x$
$x = \underline{\ ?\ }$ 5

28. $s + 2 = 2$
$s + 2 - 2 = 2 - 2$
$s = \underline{\ ?\ }$ 0

29. $r + \dfrac{2}{3} = 4$

$r + \dfrac{2}{3} - \dfrac{2}{3} = 4 - \dfrac{2}{3}$

$r = \underline{\ ?\ }$ 3$\frac{1}{3}$

30. $7 + m = 2m$
$7 + m - m = 2m - m$
$\underline{\ ?\ } = m$ 7

31. $21 = m^2 - 4$
$21 + 4 = m^2 - 4 + 4$
$\underline{\ ?\ } = m^2$ 25
$\underline{\ ?\ } = m$ 5

32. $x^2 + 3 = 19$
$x^2 + 3 - 3 = 19 - 3$
$x^2 = \underline{\ ?\ }$ 16
$x = \underline{\ ?\ }$ 4

C

33. $t^2 - 7 = 29$
$t^2 - 7 + 7 = 29 + 7$
$t^2 = \underline{\ ?\ }$ 36
$t = \underline{\ ?\ }$ 6

34. $2n^2 = n^2 + 100$
$2n^2 - n^2 = n^2 - n^2 + 100$
$n^2 = \underline{\ ?\ }$ 100
$n = \underline{\ ?\ }$ 10

Solve. Add or subtract as indicated.

35. $a + 12 = 33$; subtract 12 from both members 21

36. $27 = y + 14$; subtract 14 from both members 13

37. $b - 29 = 50$; add 29 to both members 79

38. $m - 8 = 19 + 10$; add 8 to both members 37

39. $r - \dfrac{1}{2} = 6$; add $\dfrac{1}{2}$ to both members 6$\frac{1}{2}$

40. $8 = t + \dfrac{5}{8}$; subtract $\dfrac{5}{8}$ from both members 7$\frac{3}{8}$

5-4 More Equation Solving Strategies

OBJECTIVES

Solve an equation by multiplying both members by the same number.

Solve an equation by dividing both members by the same number.

Let's use the true statement $2 + 6 = 8$ to see what happens when both members are multiplied by the same number or divided by the same number.

Multiply both members by 3:

$$2 + 6 = 8 \qquad \blacktriangleleft \text{ True } \blacktriangleright$$

$$3(2 + 6) = 3 \times 8 \qquad \blacktriangleleft \text{ Also true } \blacktriangleright$$

$$24 = 24$$

Divide both members by 4:

$$2 + 6 = 8$$

$$\frac{2 + 6}{4} = \frac{8}{4}$$

$$2 = 2$$

When both members of a true equation are multiplied or divided by the same number, the result is another true equation. We can use this idea to solve equations. Again our strategy is based on getting the variable to stand alone as one member.

EXAMPLE 1

$$\frac{1}{3} \cdot x = 4$$

$$3 \cdot \frac{1}{3} \cdot x = 3 \cdot 4 \quad \blacktriangleleft \text{Multiply both members by 3.}$$

$$x = 12$$

EXAMPLE 2

$$\frac{m}{2} = 8 \quad \blacktriangleleft \text{ Recall that } \frac{m}{2} \text{ means } \frac{1}{2} \cdot m.$$

$$2 \cdot \frac{m}{2} = 2 \cdot 8 \quad \blacktriangleleft \text{Multiply both members by 2.}$$

$$m = 16$$

EXAMPLE 3

$$5t = 35$$

$$\frac{5t}{5} = \frac{35}{5} \quad \blacktriangleleft \text{Divide both members by 5.}$$

$$t = 7$$

True or false?

1. $3 \cdot \dfrac{1}{3} = 1$ True

2. $\dfrac{4}{4} = 0$ False

3. $2 \cdot \dfrac{1}{2} = 1$ True

4. $\dfrac{3n}{3} = n$ True

5. $\dfrac{x}{10} = 10x$ False

6. $3m = m \cdot 3$ True

7. $\dfrac{1}{2} \cdot k = \dfrac{k}{2}$ True

8. $\dfrac{2t}{2} = 1$ False

9. $5 \cdot \dfrac{s}{5} = 5s$ False

For Extra Practice, see page 417.

Multiply or divide both members. Simplify the result.

Sample: $3 + 12 = 15$; divide by 3

Solution: $3 + 12 = 15$
$$\dfrac{3 + 12}{3} = \dfrac{15}{3}$$
$$5 = 5$$

A

For Ex. 1–12, only the last step is given.

1. $10 + 8 = 18$; divide by 9 $2 = 2$
2. $15 = 9 + 6$; multiply by 3 $45 = 45$
3. $6 = 2 \cdot 3$; multiply by 5 $30 = 30$
4. $\dfrac{9}{2} = 4.5$; multiply by 2 $9 = 9$
5. $24 = 3 \cdot 8$; divide by 6 $4 = 4$
6. $3 + 9 = 6 + 6$; divide by 4 $3 = 3$
7. $\dfrac{16}{2} = \dfrac{24}{3}$; multiply by 3 $24 = 24$
8. $6 \cdot 5 = 3 \cdot 10$; divide by 5 $6 = 6$
9. $6^2 = 36$; divide by 6 $6 = 6$
10. $1 = \dfrac{7}{7}$; multiply by 0 $0 = 0$
11. $2 \cdot 8 = 4 \cdot 4$; multiply by $\dfrac{1}{2}$ $8 = 8$
12. $21 = 14 + 7$; multiply by $\dfrac{1}{3}$ $7 = 7$

Complete.

Sample: $\dfrac{3y}{3} = \underline{\ ?\ }$ *Solution:* $\dfrac{3y}{3} = y$

13. $\dfrac{8n}{8} = \underline{\ ?\ }$ n

14. $\dfrac{5m}{5} = \underline{\ ?\ }$ m

15. $\dfrac{4x}{4} = \underline{\ ?\ }$ x

16. $3 \cdot \dfrac{d}{3} = \underline{\ ?\ }$ d

17. $7 \cdot \dfrac{h}{7} = \underline{\ ?\ }$ h

18. $\dfrac{6s}{?} = s$

Copy and complete.

Sample: $\qquad \dfrac{w}{7} = 1 + 5$

$\qquad\qquad 7 \cdot \dfrac{w}{7} = 7(1 + 5)$

$\qquad\qquad\qquad w = \underline{\ ?\ }$

Solution: $\qquad \dfrac{w}{7} = 1 + 5$

$\qquad\qquad 7 \cdot \dfrac{w}{7} = 7(1 + 5)$

$\qquad\qquad\qquad w = 42$

19. $\qquad \dfrac{t}{3} = 7$

$\qquad 3 \cdot \dfrac{t}{3} = 3 \cdot 7$

$\qquad\qquad t = \underline{\ ?\ }$ 21

20. $\qquad \dfrac{b}{10} = 8$

$\qquad 10 \cdot \dfrac{b}{10} = 10 \cdot 8$

$\qquad\qquad b = \underline{\ ?\ }$ 80

21. $\qquad 4m = 72$

$\qquad \dfrac{4m}{4} = \dfrac{72}{4}$

$\qquad\qquad m = \underline{\ ?\ }$ 18

22. $\qquad 45 = 9k$

$\qquad \dfrac{45}{9} = \dfrac{9k}{9}$

$\qquad\qquad \underline{\ ?\ } = k$ 5

23. $\qquad \dfrac{n}{7} = 40 + 2$

$\qquad 7 \cdot \dfrac{n}{7} = 7(40 + 2)$

$\qquad\qquad n = \underline{\ ?\ }$ 294

24. $\qquad 15 = \dfrac{x}{3}$

$\qquad 3 \cdot 15 = 3 \cdot \dfrac{x}{3}$

$\qquad\qquad \underline{\ ?\ } = x$ 45

To solve, multiply or divide as indicated.

Sample: $10t = 65$; divide by 10

Solution: $10t = 65$

$\qquad \dfrac{10t}{10} = \dfrac{65}{10}$

$\qquad\qquad t = 6.5$

B

25. $8w = 104$; divide by 8 13

26. $1.5t = 7.5$; divide by 1.5 5

27. $80 = \dfrac{y}{5}$; multiply by 5 400

28. $17 = \dfrac{a}{3}$; multiply by 3 51

29. $\dfrac{1}{5} \cdot x = 6.5$; multiply by 5 32.5

30. $n \cdot 4 = 280$; divide by 4 70

31. $3.2z = 16$; divide by 3.2 5   **32.** $0.5w = 9$; divide by 0.5 18

33. $\dfrac{2m}{3} = 15$; multiply by 3, then divide by 2 $22\frac{1}{2}$

34. $\dfrac{4t}{5} = 20$; multiply by 5, then divide by 4 25

35. $\dfrac{9r}{2} = 27$; multiply by 2, then divide by 9 6

C

SELF-TEST 1

Be sure that you understand these terms and symbols.

equation (p. 120)   inequality (p. 120)
solution (p. 124)   solution set (p. 124)
Ø (p. 124)

Write an open sentence for each of the following.   **Section 5-1, p. 120**

1. k increased by 2 is less than 14. $k + 2 < 14$

2. The product of r and 7 is greater than 5. $r \cdot 7 > 5$

Write a word statement for each of the following. Answers may vary.

3. $3m > 1$ The product of 3 and m is greater than 1.   **4.** $4(n - 3) < 10$ 4 multiplied by n minus 3 is less than 10.

Make a table to find the solution set for the given replacement set. T = True; F = False   **Section 5-2, p. 124**

5. $a + 1 = 9$; $\{6, 7, 8\}$ **6.** $b^2 - 1 = 24$; $\{4, 5, 6\}$ F; T; F; $\{5\}$
      F; F; T; $\{8\}$

Solve. Add or subtract as indicated.   **Section 5-3, p. 127**

7. $x + 6 = 10.5$; subtract 6 4.5   **8.** $y - 4 = 6$; add 4 10

Solve. Multiply or divide as indicated.   **Section 5-4, p. 130**

9. $\dfrac{r}{3} = 4$; multiply by 3 12   **10.** $7s = 56$; divide by 7 8

Check your answers with those printed at the back of the book.

5-5 *Writing Expressions Using Variables*

> **OBJECTIVE**
> Write expressions containing variables to represent given information.

The following steps are helpful in preparing to solve an algebraic problem:

1. Read the problem carefully, two or three times if necessary.
2. Identify all the information given.
3. Make a sketch of the problem if you can.
4. Use symbols of algebra to express the given information.

EXAMPLE 1 The large container holds 8 liters of water. Each of the other containers holds t liters of water. How can you express the total amount of water?

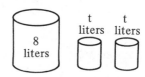

Solution: $8 + 2t$

EXAMPLE 2 Two cars are parked bumper to bumper in a parking space 14 meters long. Each car is x meters long. How much of the space is not being used?

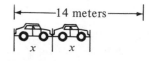

Solution: $14 - 2x$

EXERCISES

State an algebraic expression.

1. Preston has one bank for saving nickels and one for saving dimes. He has n nickels and d dimes.
   a. What is the value in cents of the coins in the nickel bank? $5n$
   b. What is the value in cents of the coins in the dime bank? $10d$

2. Maria had k hockey pucks and lost 2 pucks. How many pucks were left? $k - 2$

3. Frank could jump b centimeters last year. This year he can jump one centimeter less than twice as far as last year. How far can he jump this year? $2b - 1$

4. The Stevenson High baseball team spends d hours each week at practice. Batting practice takes up one-half that time. How much time is left for other activities? $\frac{1}{2}d$ or $\frac{d}{2}$

5. Larry has d dimes and Karen has n nickels. How much money do they have altogether? $10d + 5n$

6. The width of a playing field is y meters. The length is 80 meters. What is the area? $80ym^2$

Write an algebraic expression to answer the question.

Sample: Given the number z, what is 1 less than half that number?

Solution: $\frac{1}{2}z - 1$ or $\frac{z}{2} - 1$

Written EXERCISES

A

1. Given the number a, what is two-thirds that number? $\frac{2}{3}a$

2. Given the number b, what is 7 more than that number? $b + 7$

3. Given the number c, what is one less than the square of that number? $c^2 - 1$

4. Given the number d, what is one more than five times that number? $5d + 1$

5. Given the number m, what is two more than three-fourths that number? $\frac{3}{4}m + 2$

6. Given the number $9n$, what is one less than half that number?
$\frac{1}{2} \cdot 9n - 1$ or $\frac{9n}{2} - 1$

Write an algebraic expression.

7. 2 less than x $x - 2$

8. z less than t $t - z$

9. 5 more than y $y + 5$

10. $5k$ more than $4j$ $4j + 5k$

11. 1 less than one-third z $\frac{1}{3}z - 1$ or $\frac{z}{3} - 1$

12. 2 less than 4 times f $4f - 2$

13. 5 more than $10a$ $10a + 5$

14. three-fourths the product of 3 and r $\frac{3}{4} \cdot 3r$

15. b more than t $\quad t + b$

16. one-half the sum of x and y $\quad \frac{1}{2}(x + y)$ or $\frac{x + y}{2}$

17. $5r$ less than 100 $\quad 100 - 5r$

18. $2m$ more than j $\quad j + 2m$

19. $2k$ less than 5 times n $\quad 5n - 2k$

20. one-third the sum of $2m$ and n $\quad \frac{1}{3}(2m + n)$ or $\frac{2m + n}{3}$

Write the algebraic expression indicated. Use a variable to stand for any missing values.

B **21.** The area of the smaller rectangle pictured is one-fourth the area of the other rectangle. Name the area of the smaller rectangle. $\frac{1}{4}mn$ or $\frac{mn}{4}$

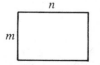

22. If the width of each volume of an encyclopedia is k centimeters, how much shelf space is needed for all 24 volumes? $24k$

23. Let n represent an odd number. Name the odd number which precedes n. $n - 2$

24. Let m represent an even number. Name the next two even numbers greater than m. $m + 2$; $m + 4$

25. The members of the swimming team scored a, b, and c points. What was the team average? $\frac{a + b + c}{3}$

C **26.** A table cloth x centimeters long and w centimeters wide is hemmed on all four sides. How long is the hem? $2x + 2w$ or $2(x + w)$

27. What is the total value of d dimes, n nickels, and p pennies? $10d + 5n + p$

Evariste Galois   1811–1832

Evariste Galois began his short mathematical career at an early age. When he was 16 he began to develop theories about solving algebraic equations. Galois met with many misfortunes during his lifetime. He failed his school entrance examinations twice. He wrote two papers involving significant mathematical discoveries which were lost before being read. He died at the age of 21 in a duel. At the time of his death he had begun work on a theory of functions which was completed years later by another famous mathematician, Bernhard Riemann.

5-6 *Using Formulas to Solve Problems*

OBJECTIVE

Use perimeter and area formulas to solve problems.

Formulas, which are special kinds of equations, can often be used in solving problems.

EXAMPLE The mainsail on a boat has the shape of a triangle, with a base 6 meters in length. The height of the sail is 8.5 meters. What is the area of the sail in square meters (m^2)?

The area formula for triangles is $A = \dfrac{bh}{2}$.

Replacements for variables ►— A is unknown.
$$b = 6$$
$$h = 8.5$$

Equation ► $A = \dfrac{bh}{2} = \dfrac{6(8.5)}{2}$

$$= \dfrac{51}{2} = 25.5$$

The area of the sail is $25.5\,m^2$.

Match each description in Column 1 with the correct formula in Column 2.

O͡ral
EXERCISES

COLUMN 1

1. Perimeter of a square C
2. Area of a square F
3. Perimeter of a rectangle H
4. Area of a rectangle A
5. Perimeter of a triangle B
6. Area of a triangle D
7. Circumference of a circle E
8. Area of a circle G

COLUMN 2

A. $A = lw$

B. $P = a + b + c$

C. $P = 4s$

D. $A = \dfrac{1}{2}bh$

E. $C = \pi d$

F. $A = s^2$

G. $A = \pi r^2$

H. $P = 2l + 2w$

For Extra Practice, see page 418.

Written
EXERCISES

Tell which formula is required. Then complete the solution. Where π is used, let $\pi = \frac{22}{7}$.

Sample: Find the perimeter of a triangle with sides of length 6 centimeters, 8 centimeters, and 12 centimeters.

Solution: $P = a + b + c$
$P = 6 + 8 + 12$
$P = 26$ centimeters

A

1. Find the area of a rectangle with width 10 centimeters and length 11.5 centimeters. 115 cm²

2. Find the perimeter of a square if the length of a side is 22 millimeters. 88 mm

3. Find the area of a triangle with base of length 13 centimeters and height 10 centimeters. 65 cm²

4. The radius of a circle is 49 millimeters. Find the area. 7546 mm²

5. The lengths of the sides of a triangle are 1.5 meters, 0.9 meters and 2 meters. Find the perimeter. 4.4 m

6. The length of the side of a square is 15 meters. Find the area. 225 m²

7. The perimeter of a square is 144 centimeters. Find the length of each side. 36 cm

8. A circle has radius 35 centimeters. Find the circumference. 220 cm

9. The width of a rectangle is 8.2 centimeters and the length is 41 centimeters. Find the area. 336.2 cm²

B

10. Find the perimeter of a square if the area is 64 square centimeters. 32 cm

11. The circumference of a circle is 132 centimeters. Find the radius. 21 cm

C

12. The circumference of a circle is 88 centimeters. Find the area. 616 cm²

13. Find the distance around the figure.

a.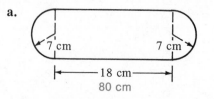

7 cm   7 cm

|←——— 18 cm ———→|
80 cm

b.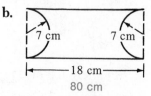

7 cm   7 cm

|←——— 18 cm ———→|
80 cm

14. Find the area of the shaded part of the figure.

a.

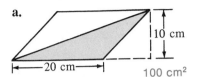

10 cm

20 cm

100 cm²

b.

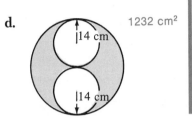

14 cm

30 cm
266 cm²

c.

1848 cm²

28 cm

d.

1232 cm²

|14 cm

|14 cm

SELF-TEST 2

Write an algebraic expression.

1. 1.3 less than 4 times k $4k - 1.3$ Section 5-5, p. 134

2. Let m represent a multiple of 7. What are the multiples of 7 which precede and follow m? $m - 7, m + 7$

Tell what formula is needed to solve the problem. Then complete the solution.   Section 5-6, p. 137

3. The width of a rectangle is 10 centimeters and the length is 13 centimeters. What is the perimeter?   46 cm

4. The length of the side of a square is 7.5 meters. What is the area?   56.25 m

Check your answers with those printed at the back of the book.

5-7 *Solving Inequalities: The $<$ and $>$ Relationships*

> **OBJECTIVES**
>
> Solve inequalities that involve the $<$ and the $>$ relationships.
>
> Write and graph the solution set of an inequality.

For solving and graphing inequalities, we can use tables and the number line as we did with equations.

EXAMPLE 1 $n + 3 < 8$; replacement set: $\{1, 3, 5, 7\}$

Replacement	$n + 3 < 8$	True/False	Solution?
1	$1 + 3 < 8$	True	YES
3	$3 + 3 < 8$	True	YES
5	$5 + 3 < 8$	False	No
7	$7 + 3 < 8$	False	No

The solutions are 1 and 3. Solution set: $\{1, 3\}$

EXAMPLE 2 $3x + 1 > 7$; replacement set: $\{$the counting numbers$\}$

Replacement	$3x + 1 > 7$	True/False	Solution?
1	$3(1) + 1 > 7$	False	No
2	$3(2) + 1 > 7$	False	No
3	$3(3) + 1 > 7$	True	YES
4	$3(4) + 1 > 7$	True	YES
5	$3(5) + 1 > 7$	True	YES
.	.	.	.
.	.	.	.
.	.	.	.

3, 4, and 5 are solutions. You can see that, in fact, any counting number greater than 3 is a solution. The solution set is $\{3, 4, 5, \ldots\}$.

It is often easiest to show the solution set of an inequality with a number line graph.

> Recall that this set includes 0 and all the numbers to the right of 0 on the number line.

▼

EXAMPLE 3 $3y > 9$; replacement set: {the numbers of arithmetic}

$3y > 9$ is a true statement when y is replaced by any number greater than 3.

Solution set: {the numbers of arithmetic greater than 3}

Graph:

```
  ←——+——+——+——○——+——+——+——→
      0   1   2   3   4   5   6
```

▲

The hollow dot shows that 3 is *not* included.

1. Which numbers of arithmetic are less than 9?
2. Which numbers of arithmetic, when divided by 2, are greater than 10?
3. Which numbers of arithmetic, when multiplied by 5, are less than 10?

State the question that is suggested by the inequality. The replacement set is {the numbers of arithmetic}.

Oral EXERCISES

Sample 1: $m > 10$ *What you say:* Which numbers of arithmetic are greater than 10?

Sample 2: $\dfrac{k}{2} < 8$ *What you say:* Which numbers of arithmetic, when divided by 2, are less than 8?

4. Which numbers of arithmetic, when decreased by 5, are less than 2?

1. $r < 9$ 2. $\dfrac{1}{2}a > 10$

3. $5s < 10$ 4. $b - 5 < 2$

5. $t + 1 > 14$ 6. $c + 1 > 5$

5. Which numbers of arithmetic, when increased by 1, are greater than 14?
6. Which numbers of arithmetic, when 1 is added, are greater than 5?

For Extra Practice, see page 419.

Tell whether the resulting statement is true or false when these replacements are used for the variables: $a = 1$, $b = 3$, $c = 5$.

Written EXERCISES

A

Sample: $2a - 1 < 10$ *What you say:* $2 - 1 < 10$
$1 < 10$; True

1. $5a < 9$ true   2. $a + b < 10$ true

3. $2b > 6$ false   4. $6a + 1 > 2$ true

5. $4c < 16$ false   6. $7b - 8 > 15$ false

7. $2a > 4$ false   8. $b + b > b^2$ false

For each member of the replacement set, show whether a true or a false statement results when the variable in the sentence is replaced by the number. T = True; F = False

Sample: $k - 7 < 10$; $\{16, 16.5, 17\}$
Solution: $16 - 7 < 10$ ▶ $9 < 10$ True
  $16.5 - 7 < 10$ ▶ $9.5 < 10$ True
  $17 - 7 < 10$ ▶ $10 < 10$ False

9. $5 < a + 1$; $\{0, 5, 10\}$ F, T, T   10. $10 > b - 8$; $\{10, 15, 20\}$ T, T, F

11. $3c + 5 < 14$; $\{0, 1, 2, 3\}$ T, T, T, F   12. $21 > m \cdot m$; $\{3, 4, 5, 6\}$ T, T, F, F

13. $5n < 14$; $\{0, 1, 2, 3\}$ T, T, T, F   14. $3p > 10$; $\{1, 3, 5, 7\}$ F, F, T, T

B 15. $m^2 > m + m$; $\{0, 1, 2, 3\}$ F, F, F, T   16. $k \div 3 < 7$; $\{0, 5, 10, 15\}$ T, T, T, T

17. $4.2 + h > 4$; $\{0, 1, 2\}$ T, T, T   18. $4.2 - h < 4$; $\{0, 1, 2, 3\}$ F, T, T, T

19. $6k < 2$; $\left\{0, \dfrac{1}{2}, \dfrac{1}{3}, \dfrac{1}{4}\right\}$ T, F, F, T 20. $m^3 < 10$; $\{0, 1, 2\}$ T, T, T

Find the solution set if the replacement set is {the numbers of arithmetic}. Graph the solution set on the number line.

Sample 1: $2a > 8$
Solution: {the numbers of arithmetic greater than 4}

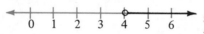

See page A1 at the back of the book for Ex. 21–38.

Sample 2: $16 + b < 20$
Solution: {the numbers of arithmetic less than 4}

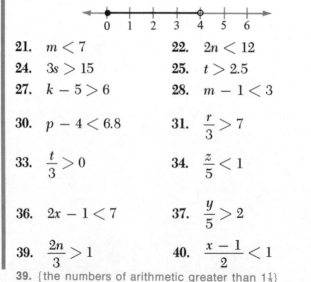

21. $m < 7$ 22. $2n < 12$ 23. $r > 4$

24. $3s > 15$ 25. $t > 2.5$ 26. $6 > 2a$

27. $k - 5 > 6$ 28. $m - 1 < 3$ 29. $n + 7 > 5.2$

C 30. $p - 4 < 6.8$ 31. $\dfrac{r}{3} > 7$ 32. $\dfrac{s}{4} < 10.1$

33. $\dfrac{t}{3} > 0$ 34. $\dfrac{z}{5} < 1$ 35. $x^2 + 3 > 50$

36. $2x - 1 < 7$ 37. $\dfrac{y}{5} > 2$ 38. $x^2 > 1$

39. $\dfrac{2n}{3} > 1$ 40. $\dfrac{x - 1}{2} < 1$ 41. $m^2 > 4$

39. {the numbers of arithmetic greater than $1\frac{1}{2}$}
40. {the numbers of arithmetic less than 3}

41. {the numbers of arithmetic greater than 2}.

5-8 *Solving Inequalities: The ≤ and ≥ Relationships*

OBJECTIVES

Solve inequalities that involve the ≤ and ≥ relationships.

Write and graph the solution set of an inequality.

When the symbols ≤ and ≥ are used in sentences

> ≤ means is less than or equal to
> ≥ means is greater than or equal to.

EXAMPLE 1 $m + 3 \geq 7$; replacement set: {the numbers of arithmetic}

This statement is true when m is replaced by 4 or any number greater than 4.

Solution set: {the numbers of arithmetic greater than or equal to 4} Graph:

The solid dot shows that 4 is included.

EXAMPLE 2 $t + 1\frac{1}{2} \leq 4$; replacement set: {the numbers of arithmetic}

This statement is true when t is replaced by $2\frac{1}{2}$ or any number less than $2\frac{1}{2}$.

Solution set: {the numbers of arithmetic less than or equal to $2\frac{1}{2}$} Graph:

Solid dots at 0 and $2\frac{1}{2}$.

EXAMPLE 3 $h \leq \frac{1}{2}$; replacement set: {all the numbers on the number line}

This statement is true when n is replaced by $\frac{1}{2}$ or any number less than $\frac{1}{2}$.

Solution set: {all the numbers on the number line less than or equal to $\frac{1}{2}$} Graph:

State the question that is suggested by the inequality. The replacement set is {the numbers of arithmetic}.

Sample: $\frac{1}{3}k \leq 6$ *What you say:* Which numbers of arithmetic, when divided by 3, are less than or equal to 6?

1. $a \leq 5$ 2. $b \geq 10$ 3. $c \leq 1.1$
4. $3k \leq 20$ 5. $4m \geq 18$ 6. $n + 1 \geq 9$

1. Which numbers of arithmetic are less than or equal to 5?

Tell whether the resulting statement is true or false when these replacements are used for the variables: $r = 2$, $s = 3$, $t = 6$.

7. $r \leq 10$ true   8. $s \geq 5$ false   9. $t \leq 6$ true
10. $s + 1 \geq 3$ true   11. $t - 1 \geq 6$ false   12. $2r + 1 \leq 5$ true

2. Which numbers of arithmetic are greater than or equal to 10?
3. Which numbers of arithmetic are less than or equal to 1.1?
4. Which numbers of arithmetic, when multiplied by 3, are less than or equal to 20?
5. Which numbers of arithmetic, when multiplied by 4, are greater than or equal to 18?
6. Which numbers of arithmetic, when increased by 1, are greater than or equal to 9?

For Extra Practice, see page 419.

Written EXERCISES

For each member of the replacement set, show whether a true or a false statement results when the variable is replaced by the number. T = True; F = False

Sample: $m + 4 \geq 10$; $\{4, 6, 8\}$
$4 + 4 \geq 10 \blacktriangleright 8 \geq 10$ False
$6 + 4 \geq 10 \blacktriangleright 10 \geq 10$ True
$8 + 4 \geq 10 \blacktriangleright 12 \geq 10$ True

A

1. $p \leq 4$; $\{2, 4, 6\}$ T; T; F   2. $q \geq 5$; $\{3, 5, 7\}$ F; T; T
3. $r \geq 1.5$; $\{0, 1, 2\}$ F; F; T   4. $s \leq 7.8$; $\{7, 8, 9\}$ T; F; F
5. $x + 1 \leq 8$; $\{5, 6, 7\}$ T; T; T   6. $y + 7 \geq 10$; $\{1, 2, 3\}$ F; F; T
7. $z - 10 \leq 6$; $\{16, 17, 18\}$ T; F; F   8. $2a \leq 11$; $\{2, 4, 6\}$ T; T; F
9. $3b \geq 14$; $\{3, 5, 7\}$ F; T; T   10. $c + 6 \neq 8$; $\{1, 2, 3\}$ T; F; T
11. $2d + 1 \leq 13$; $\{6, 7, 8\}$ T; F; F   12. $5n + 6 \neq 18$; $\{2, 3, 4\}$ T; T; T

B

13. $3b + 6 \leq 80$; $\{10, 20, 30\}$ T; T; F 14. $4j - 9 \geq 37$; $\{10, 20, 30\}$ F; T; T
15. $t \leq {}^-1$; $\{2, 3, {}^-2, {}^-3\}$ F; F; T; T   16. ${}^-2 \geq x$; $\{{}^-3, {}^-2, {}^-1, 0\}$ T; T; F; F

Match each inequality in Column 1 with the graph of its solution in Column 2. The replacement set is {the numbers of arithmetic}.

17. $3a \leq 4$ B

A. (number line graph 0–5)

18. $\frac{1}{2}b \geq 1$ E

B. (number line graph 0–5)

19. $3c \leq 6$ A

C. (number line graph 0–5)

20. $4d \neq 8$ C

D. (number line graph 0–5)

21. $k \leq 1$ D

E. (number line graph 0–5)

See page A1 at the back of the book for Ex. 22–27.
Describe the set of numbers represented by the number line graph.

Sample: (number line graph 0–5)

Solution: {the numbers of arithmetic less than or equal to $3\frac{1}{2}$}

22. (number line graph 0–5)   **23.** (number line graph 0–5)

24. (number line graph 0–5)   **25.** (number line graph 0–5)

26. (number line graph 0–5)   **27.** (number line graph 0–5)

Describe the solution set. The replacement set is {all the numbers on the number line}. **28.** {the numbers on the number line greater than or equal to $2\frac{1}{2}$}

Sample: $k \div 3 \leq 10$
Solution: {the numbers on the number line less than or equal to 30}

28. $g \geq 2\frac{1}{2}$ **29.** $h \leq 0$ C

30. $k \geq {}^-1$ **31.** $m \leq {}^-3$

32. $3s - 4 \geq 10.1$ **33.** $9x + 1 \neq 28$
29. {the numbers on the number line less than or equal to 0}

Graph the solution on the number line. The replacement set is {all the numbers on the number line}.

Sample: $r \leq 1$ Check students' graphs.
Solution:

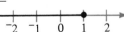

30. {the numbers on the number line greater than or equal to ${}^-1$}
34. $s \leq 2\frac{1}{2}$ **35.** $t \geq 0$

36. $x \geq {}^-2$ **37.** $y \neq {}^-1$

38. $4z \leq 6$ **39.** $3a \geq 2$
31. {the numbers on the number line less than or equal to ${}^-3$}
32. {the numbers on the number line greater than or equal to 4.7}
33. {the numbers on the number line   **EQUATIONS AND INEQUALITIES / 145**
except 3}

5-9 *Using Inequalities to Solve Problems*

OBJECTIVE

Solve problems by writing inequalities and finding their solution sets.

When a problem involves the idea of *less than* or *greater than,* an inequality can be used to solve it.

EXAMPLE Ted has two pieces of string. Their combined length is less than 25 centimeters. The longer piece is 16 centimeters long. What can we say about the length of the shorter piece?

Let s = the length of the shorter piece.
  16 = the length of the longer piece.

Then $s + 16$ = the combined length.

Equation: $\underbrace{\text{The combined length}}$ $\underbrace{\text{is less than}}$ $\underbrace{\text{25 centimeters.}}$
              $s + 16$ $<$ 25
                        $s < 9$

This statement is true when
s is replaced by any number
of arithmetic less than 9.

Solution set: {the numbers of arithmetic < 9}
The length of the shorter piece is less than 9 cm.

EXERCISES

Name the solution set for the open sentence suggested. The replacement set is {the numbers of arithmetic}.

Sample: If a number is added to 6, the sum is greater than or equal to 13. What numbers meet this condition? (*Hint:* $6 + m \geq 13$)

Solution: Since $6 + 7 = 13$, we see that if m is any number greater than 7, $6 + m > 13$. Thus the solution set is

{the numbers of arithmetic ≥ 7}.

1. The difference between 9 and a number is at least 6. What numbers can meet this condition? (*Hint:* $k - 9 \geq 6$) ≥ 15

2. The sum of the ages of a brother and sister is no more than 21 years. If one is 16, what can we say about the age of the other? (*Hint:* $m + 16 \leq 21$) ≤ 5

3. An auditorium has a capacity of 600 people. If 289 people have been seated, at most how many more may be seated? (*Hint:* $289 + n \leq 600$) ≤ 311

4. The perimeter of a square must be less than 44 meters. What is the range of values for the length of one side? (*Hint:* $4s < 44$) < 11

5. The perimeter of a given rectangle cannot exceed 240 meters. If the width is 20 meters, what is the range of values for the length? (*Hint:* $(2 \cdot 20) + (2 \cdot l) \leq 240$) ≤ 100

6. A carrying case for records is designed to hold 48 records. If 29 records are in the case, what is the possible number of records which can be added? (*Hint:* $29 + z \leq 48$) ≤ 19

7. The perimeter of a square table must be at least 320 centimeters but it cannot exceed 400 centimeters. What are the possible lengths of one side? (*Hint:* $4s \geq 320$ and $4s \leq 400$)

≥ 80 and ≤ 100

Problems

Write an inequality. Then write the solution set if the replacement set is {the numbers of arithmetic}.

Sample: A square has an area which is at most 100 square meters. What numbers might describe the length of one side?

Solution: Let s stand for the length of one side in meters. Then

$$s \cdot s \leq 100$$
$$s \leq 10$$

The solution set is {the numbers of arithmetic ≤ 10}.

A

1. The perimeter of a square is less than 50 meters. What numbers might describe the length of one side? $< 12\frac{1}{2}$

2. 41 minus some number is less than 13. What are the possible values of the other number? > 28

3. Aurelia lost her wallet. She knew that she had no more than $18. She also remembered that she had a $10 bill. What numbers represent the amount which may have been lost? between 10 and 18

4. The sum of 12 and some number is greater than 40. What are the possible values of the other number? > 28

5. When $2\frac{1}{3}$ is increased by some number, the resulting number is less than 14. What numbers meet these conditions? $< 11\frac{2}{3}$

6. Joe is having a party. There will be 15 people there. He plans on each person eating at most three pieces of pizza. What numbers represent the number of pieces of pizza Joe may buy? ≤ 45

B 7. The perimeter of a rectangle is at most 100 centimeters. If the width is 10 centimeters, what numbers represent the possible values of the length? ≤ 40

8. If the area of a rectangle is less than 100 square centimeters and the length is 15 centimeters, what numbers may represent the width in centimeters? $< 6.\overline{6}$

9. The circumference of a circle is at least 44 centimeters. What are the possible values of the diameter? Radius? $d \geq 14$ $r \geq 7$

10. A rectangle has a perimeter which is at least 120 centimeters. If the length is 40 centimeters, what are the possible values of the width? ≥ 20

SELF-TEST 3

Be sure that you understand these symbols.

A hollow dot on a number line graph (p. 141)
A solid dot on a number line graph (p. 143)
$\leq$ (p. 143) $\geq$ (p. 143)

Section 5-7, p. 140 1. Indicate whether each replacement makes the statement true or false: $3k - 1 < 14$; $\{1, 3, 5\}$ T; T; F

2. Solve $3r > 12$ and graph the solution set. The replacement set is {the numbers of arithmetic}. {the numbers of arithmetic greater than 4}

Section 5-8, p. 143 3. Indicate whether each replacement makes the statement true or false: $\frac{1}{2}t + 1 \geq 10$; $\{0, 9, 18\}$ F; F; T

4. Graph the solution set of this inequality: $2s - 1 \leq 7$.

Write an inequality for each problem. Then write the solution set Section 5-9, p. 146 if the replacement set is {the numbers of arithmetic}.

5. The area of a rectangle with width 8 meters is at most 120 $8l = 120$; square meters. What are the possible values of the length? ≤ 15

6. When 7 is increased by 2 times a number, the total is less than 19. What are the possible values of the number? $7 + 2n < 19$; < 6

Check your answers with those printed at the back of the book.

---

chapter summary

1. Number sentences that contain the symbol $=$ are called **equations**. Those that contain the symbols $\neq$, $<$, $>$, $\leq$, or $\geq$ are called **inequalities**.

2. A number sentence that contains a variable is neither true nor false until the variable is replaced by a member of a specified **replacement set**.

3. To solve an equation or inequality the variable in the equation or inequality is replaced by members of the replacement set. The result is either a true statement or a false statement. Each replacement that results in a true statement is a **solution** and all such replacements make up the **solution set**.

4. A solution set that contains no members is called the **empty set** and is indicated by the symbol $\emptyset$.

5. When the same number is added to both members of an equation, or subtracted from both members of an equation, the meaning of the equation remains unchanged.

6. When both members of an equation are multiplied by the same number, or both members are divided by the same number, the meaning of the equation remains unchanged.

7. **Variables** are used to represent missing values in equations.

8. Equations used to solve problems that occur frequently, are called **formulas**.

chapter test

Write a number sentence. Do not determine the solution set.

1. If some number is increased by 10, the result is 13. $n + 10 = 13$
2. If 18 is decreased by some number, the difference is 13. $18 - n = 13$
3. If some number is multiplied by 4, the product is less than 15. $n \cdot 4 < 15$

Find the solution set. The replacement set is given.

4. $m + 8 = 2m + 4$; $\{0, 2, 4, 6\}$ $\{4\}$ 5. $n^2 = 2n$; $\{0, 1, 2, 3\}$ $\{0, 2\}$
6. $2k < 10$; $\{1, 3, 5, 7\}$ $\{1, 3\}$ 7. $2h + 4 \geq 12$; $\{2, 4, 6, 8\}$ $\{4, 6, 8\}$

To solve, perform the operation as indicated.

8. $3k = 12$; divide by 3 4 9. $m + 4 = 8$; subtract 4 4

10. $n - 8 = 1$; add 8 9 11. $\dfrac{r}{3} = 7$; multiply by 3 21

12. $s + 7 = 11$; subtract 7 4 13. $t - \dfrac{1}{3} = 4$; add $\dfrac{1}{3}$ $4\frac{1}{3}$

Graph the solution set on the number line. The replacement is {the numbers of arithmetic}.

14. $2a > 6$ 15. $b - 4 < 10$

Write an algebraic expression.

16. The sum of a number and 2, multiplied by 3 $3(n + 2)$
17. 4 times a given number, increased by 3 $4n + 3$
18. The perimeter of a rectangle with width n centimeters and length 12 centimeters $2 \cdot 12 + 2n$

Write a number sentence. Then solve it to answer the question.

19. If the length of a rectangle is 10 meters and the width is 8 meters, what is the perimeter? $P = 2(10) + 2(8) = 20 + 16 = 36m$

20. Teresa's mother is more than twice as old as she. If Teresa is 17, what numbers represent how old her mother might be? $n > 2(17)$; $n > 34$

challenge topics

Triangular Numbers

   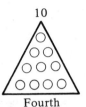

	1	3	6	10
	First	Second	Third	Fourth

Study the triangular arrays of dots shown above. The number of dots in each array is called a **triangular** number. Can you find a pattern that will help you predict the next triangular number?

1. What are the first fifteen numbers in the set of triangular numbers? Use this pattern: 1, 3, 6, 10, 15, 21, 28, 36, 45, 55, 66, 78, 91, 105, 120

1st	2nd	3rd	4th	. . .	15th
1	3	6	10	. . .	?
1	(1 + 2)	(3 + 3)	(6 + 4)		

2. The set of counting numbers is {1, 2, 3, 4, 5, . . .}.
    From your completed pattern in Question 1, notice that:
    The **second** triangular number is the sum of the first **two** counting numbers.
    The **third** triangular number is the sum of the first _three_ counting numbers.
    The **fourth** triangular number is the sum of the first _four_ counting numbers.
    The **fifth** triangular number is the sum of the first _five_ counting numbers.

3. In general, to find the nth triangular number, you find the sum of the first _n_ counting numbers.
    A short cut for finding this sum is to find the value of $\dfrac{n \cdot (n + 1)}{2}$, where n stands for the number of counting numbers. Thus the sum of the first 5 counting numbers is $\dfrac{5 \cdot (5 + 1)}{2} = \dfrac{5 \cdot 6}{2} = 15$. Then the **fifth** triangular number is **15**.

    What is the fifteenth triangular number? 120
    What is the twentieth triangular number? 210
    What is the fifty-first triangular number? 1326

Review of Skills

Simplify.

1. $6 \cdot (10 + 2)$ 72

2. $8 \cdot (9 - 2)$ 56

3. $6(7 \cdot 5)$ 210

4. $2\frac{1}{2} + 3\frac{1}{4}$ $5\frac{3}{4}$

5. $\frac{2}{5} + \left(\frac{1}{4} + \frac{1}{2}\right)$ $\frac{23}{20}$ or $1\frac{3}{20}$

6. $8.6 + 9.3$ 17.9

7. $\frac{30 - 18}{2}$ 6

8. $\frac{1.8}{6} + 4.2$ 4.5

9. $\frac{1}{2}(15 + 7)$ 11

10. $2 \cdot 5\frac{3}{8} \cdot 1$ $\frac{43}{4}$ or $10\frac{3}{4}$

11. $25 + \frac{300}{3}$ 125

12. $0 + \frac{5}{8}$ $\frac{5}{8}$

Name the value of the expression. Let $m = 3$ and $n = 4$.

13. $4 + 7m$ 25

14. $m^2 + 4$ 13

15. $3n - 4m$ 0

16. $n(6 + 2)$ 32

17. $(m + 8)m$ 33

18. $n^2 \cdot n$ 64

True or false?

19. $4 + 6 = 10$, so $6 + 4 = 10$ true

20. $5.7 + 2.1 = 7.8$, so $2.1 + 5.7 = 7.8$ true

21. $\frac{18}{6} = 3$, so $\frac{6}{18} = 3$ false

22. $\frac{42}{6} = \frac{12}{6} + \frac{30}{6}$ false

23. $647 = 600 + 40 + 7$ true

24. $10 - 3 = 7$, so $3 - 10 = 7$ false

25. $4 \cdot (5 \cdot 1) = (4 \cdot 5) \cdot 1$ true

26. $18 - (6 - 2) = (18 - 6) - 2$ false

27. $20 + (8 - 3) = (20 + 8) - 3$ true

28. $8^2 \cdot 1 = 1 \cdot 8^2$ true

Complete.

29. $8 + 0 = \underline{\ ?\ }$ 8

30. $0 + 15 = \underline{\ ?\ }$ 15

31. $0 \cdot 43 = \underline{\ ?\ }$ 0

32. $1 \cdot 19 = \underline{\ ?\ }$ 19

33. $0 + 2.5 = \underline{\ ?\ }$ 2.5

34. $0 \cdot 6.7 = \underline{\ ?\ }$ 0

35. $\frac{0}{8} = \underline{\ ?\ }$ 0

36. $1 \cdot 4\frac{7}{8} = \underline{\ ?\ }$ $4\frac{7}{8}$

37. $\frac{12}{1} = \underline{\ ?\ }$ 12

Name the solution set. Use the given replacement set.

38. $m \cdot 0 = 0$; $\{1, 2, 3, 4\}$ {1, 2, 3, 4}

39. $0 \cdot 8 = t$; $\{0, 2, 4, 6, 8\}$ {0}

40. $x = \frac{15}{3}$; $\{1, 3, 5, 7\}$ {5}

41. $\frac{s}{4} = 1$; $\{2, 4, 6, 8\}$ {4}

Left: Farmer with tractor, early 1900's.

Right: Farmer using modern combine to harvest wheat.

6 Axioms and Properties

Some Basic Properties

6-1 *Basic Axioms of Equality*

OBJECTIVES

Identify and apply these properties of equality:

1. The reflexive property
2. The symmetric property
3. The transitive property

In mathematics there are certain ideas that we must assume to be true. They provide the basis for rules which we develop about numbers and operations with numbers. For example, any number is equal to itself. This property can be stated in terms of a variable.

The Reflexive Property of Equality ▶ For any number r, $r = r$.

EXAMPLES
$15 = 15$
$42 + 3 = 42 + 3$
$a + 7 = a + 7$ for every number a.

The left and right members of any equation may be interchanged.

The Symmetric Property ▶ For any numbers r and s, if $r = s$,
of Equality   then $s = r$.

EXAMPLES
$5 + 7 = 12$, so $12 = 5 + 7$.
If $m + 2 = 8$, then $8 = m + 2$.
If $t^2 - 6 = m + 7$, then $m + 7 = t^2 - 6$.

If one number is equal to a second number, and the second number is equal to a third, the first and third numbers are equal.

The Transitive Property ▶ For any numbers r, s, and t,
of Equality   if $r = s$ and $s = t$, then $r = t$.

EXAMPLES
$4 + 5 = 3 + 6$ and $3 + 6 = 7 + 2$, so $4 + 5 = 7 + 2$.
If $a - 3 = b + 1$ and $b + 1 = c$, then $a - 3 = c$.

Tell which property of equality is illustrated.

Sample 1: $12 \div 4 = 3$, so $3 = 12 \div 4$.
What you say: The symmetric property of equality

Sample 2: $3 + 8 = 3 + 8$
What you say: The reflexive property of equality

1. $3 + 4 = 3 + 4$ reflexive
2. $6 + 2 = 8$ and $8 = 4 + 4$, so $6 + 2 = 4 + 4$. transitive
3. $3 + 3 + 3 = 9$, so $9 = 3 + 3 + 3$. symmetric
4. $4 \cdot 4 \cdot 4 = 64$, so $64 = 4 \cdot 4 \cdot 4$. symmetric
5. If $a = 9$, then $9 = a$. symmetric
6. If $a + 2 = b$ and $b = 7$, then $a + 2 = 7$. transitive
7. $a \div k = a \div k$ reflexive
8. If $2 + a = b$, then $b = 2 + a$. symmetric
9. If $14m = 10n$, then $10n = 14m$. symmetric
10. $r + 7 = r + 7$ reflexive

Use the expression to illustrate the reflexive property of equality.

Written
EXERCISES

A

Sample: $4 + k$ *Solution:* $4 + k = 4 + k$

1. m $m = m$
2. $9 + n$ $9 + n = 9 + n$
3. $a + b^3$
   $a + b^3 = a + b^3$
4. $x^2 + 5x + 8$ $x^2 + 5x + 8 = x^2 + 5x + 8$

Assume the equation is true. Use the symmetric property of equality to write another equation.

Sample: $b + 2 = 11$ *Solution:* $11 = b + 2$

5. $m + 4 = 10$ $10 = m + 4$
6. $a + k = 15$ $15 = a + k$
7. $b + 2c = p$ $p = b + 2c$
8. $4(q - 8) = 4q - 32$ $4q - 32 = 4(q - 8)$
9. $k^3 + k^2 + k = k(k^2 + k + 1)$
   $k(k^2 + k + 1) = k^3 + k^2 + k$
10. $12 - m + n = b + c$
    $b + c = 12 - m + n$

Use the transitive property of equality to complete the statement.

Sample: $3 = 2 + 1$ and $2 + 1 = 3 + 0$, so _?_.
Solution: $3 = 2 + 1$ and $2 + 1 = 3 + 0$, so $3 = 3 + 0$.

11. $7 = 6 + 1$ and $6 + 1 = 5 + 2$, so _?_. $7 = 5 + 2$
12. $13 + 4 = 17$ and $17 = 10 + 7$, so _?_. $13 + 4 = 10 + 7$

13. $48 \div 12 = 4$ and $4 = 2^2$, so __?__. $48 \div 12 = 2^2$

B

14. $3 + 4 = 7$ and __?__, so $3 + 4 = 6 + 1$. $7 = 6 + 1$

15. If $k - 8 = q$ and __?__, then $k - 8 = 13$. $q = 13$

16. If __?__ and $r = s - 1$, then $10 = s - 1$. $r = 0$

17. If $x^2 + 1 = 10$ and $10 = p^3$, then __?__. $x^2 + 1 = p^3$

18. If $m + n = 100$ and $100 = rs$, then __?__. $m + n = rs$

19. If __?__ and $r^2 + t = 8$, then $m = 8$. $m = r^2 + t$

20. If $v^2 + w^2 = 1$ and __?__, then $v^2 + w^2 = rst$. $1 = rst$

Tell which property of equality is used to reach each conclusion.

C

21. $6 \div 2 = 3$ and $9 \div 3 = 3$, so
   (1) $3 = 9 \div 3$ symmetric
   (2) and $6 \div 2 = 9 \div 3$. transitive

22. If $rk = 10$ and $m + n = 10$, then
   (1) $10 = m + n$ symmetric
   (2) and $rk = m + n$ transitive

23. $2(6 + 2) = 2(8)$, and $2(8) = 16$, and $16 = 10 + 6$, so
   (1) $2(6 + 2) = 16$ transitive
   (2) and $2(6 + 2) = 10 + 6$ transitive

24. $12 \times \dfrac{1}{3} = 12 \div 3$, and $12 \div 3 = 0 + 4$, and $0 + 4 = 2^2$, so
   (1) $12 \div 3 = 2^2$ transitive
   (2) and $12 \times \dfrac{1}{3} = 2^2$ transitive

25. If $mn = 4$ and $m + n = 4$, then
   (1) $4 = m + n$ symmetric
   (2) and $mn = m + n$ transitive

26. If $2xy = 21 - x$ and $21 - x = 9$, then
   (1) $2xy = 9$ transitive
   (2) and $9 = 2xy$ symmetric

27. $4(3 + 2) = 4(5)$, and $4(5) = 20$, and $20 = \frac{100}{5}$, so
   (1) $4(3 + 2) = 20$ transitive
   (2) and $4(3 + 2) = \frac{100}{5}$ transitive

28. If $rs = 100$ and $100 = 2r + 8s$, then
   (1) $rs = 2r + 8s$ transitive
   (2) and $2r + 8s = rs$ symmetric

6-2 *The Closure Property and the Substitution Principle*

OBJECTIVES

Determine whether sets of numbers are closed under specified operations.

Use the substitution principle.

If you add any two whole numbers, the result is always a whole number. We say the set of whole numbers is **closed under addition.** If you multiply any two whole numbers, the product is always a whole number. We say the set of whole numbers is **closed under multiplication.**

The Closure Property ▶— A set of numbers is closed under an operation if performing the operation on any two members of the set results in a member of the set.

EXAMPLE 1   The sum of any two even numbers is an even number. The set of even numbers is closed under addition.

$$2 + 2 = 4 \qquad 220 + 80 = 300$$

EXAMPLE 2   The sum of any two odd numbers is an even number. The set of odd numbers is **not** closed under addition.

$$13 + 9 = 22 \qquad 71 + 23 = 94$$

You know that any given number may have many names. We use this idea to simplify an expression.

$$\underline{3 + 6} + 10$$
$$\underline{9 + 10} \quad \blacktriangleleft \text{ Substitute 9 for } 3 + 6.$$
$$19 \quad \blacktriangleleft \text{ Substitute 19 for } 9 + 10.$$

The Substitution Principle ▶— A numeral may be substituted for any other numeral that names the same number.

EXAMPLE 3   Simplify $\dfrac{2}{5} + \dfrac{1}{5} + 4.$ ▶ $\dfrac{2}{5} + \dfrac{1}{5} + 4$

$$\underline{\dfrac{3}{5} + 4}$$
$$4\dfrac{3}{5}$$

Tell whether or not the specified set of numbers is closed under the indicated operation. If it is *not closed,* give at least one example to support your answer.

Sample {0, 1}; addition
What you say: Not closed; $1 + 1 = 2$, and 2 is not a member of {0, 1}.

1. {2, 4, 6, 8, 10, 12}; addition not closed; $2 + 12 = 14$
2. {1, 2, 3, 4, 5, 6, 7, 8}; multiplication not closed; $2 \cdot 8 = 16$
3. {1, 2, 3, 4, ...}; division not closed; $1 \div 2 = \frac{1}{2}$
4. {0, 1, 2}; addition not closed; $1 + 2 = 3$
5. {4, 8, 12, 16, ...}; addition closed

Describe the substitutions made in simplifying the expression.

Sample: $5 + 4 + 1 + 2$ *What you say:*

$9 + 1 + 2$ 9 is substituted for $5 + 4$

$10 + 2$ 10 is substituted for $9 + 1$

12 12 is substituted for $10 + 2$

6. $8 + 1 + 2 + 5$ 7. $2 \cdot 3 \cdot 4 \cdot 5$

$9 + 2 + 5$ 9 for $8 + 1$ $6 \cdot 4 \cdot 5$ 6 for $2 \cdot 3$

$11 + 5$ 11 for $9 + 2$ $24 \cdot 5$ 24 for $6 \cdot 4$

16 16 for $11 + 5$ 120 120 for $24 \cdot 5$

8. $3 \cdot 3 \cdot 3 \cdot 3$ 9. $5 + 5 + 5 + 5$

$9 \cdot 3 \cdot 3$ 9 for $3 \cdot 3$ $10 + 5 + 5$ 10 for $5 + 5$

$27 \cdot 3$ 27 for $9 \cdot 3$ $15 + 5$ 15 for $10 + 5$

81 81 for $27 \cdot 3$ 20 20 for $15 + 5$

C = closed; NC = not closed; A = Add; S = sub; M = mult.; D = div.
Tell whether or not the set is closed under addition, subtraction, multiplication, and division.

1. {2, 4, 6, 8, 10, ...} C: A, M; NC: S, D   2. {multiples of 3} C: A, M; NC: S, D
3. {1, 2, 3, 4, ...} C: A, M; NC: S, D   4. {prime numbers} NC: A, S, M, D
5. {1, 3, 5, 7, 9, ...} C: M; NC: A, S, D   6. {multiples of 10} C: A, M; NC: S, D
7. {1} C: M, D; NC: A, S   8. {2, 4, 6, 8} NC: A, S, M, D
9. {1, 2} NC: A, S, M, D   10. {1, 3, 5} NC: A, S, M, D

Simplify. Use the substitution principle as indicated.

Sample: $\dfrac{1}{2}\cdot\dfrac{1}{3}\cdot\dfrac{1}{4}\cdot\dfrac{1}{5}$

$\underline{\quad?\quad}\cdot\dfrac{1}{4}\cdot\dfrac{1}{5}$

$\underline{\quad?\quad}\cdot\dfrac{1}{5}$

$\underline{\quad?\quad}$

Solution: $\dfrac{1}{2}\cdot\dfrac{1}{3}\cdot\dfrac{1}{4}\cdot\dfrac{1}{5}$

$\dfrac{1}{6}\cdot\dfrac{1}{4}\cdot\dfrac{1}{5}$

$\dfrac{1}{24}\cdot\dfrac{1}{5}$

$\dfrac{1}{120}$

11. $\dfrac{1}{5}\cdot 5\cdot\dfrac{1}{2}\cdot 2$

$\underline{\quad?\quad}\cdot\dfrac{1}{2}\cdot 2 \;\; 1$

$\underline{\quad?\quad}\cdot 2 \;\; \dfrac{1}{2}$

$\underline{\quad?\quad} \;\; 1$

12. $\dfrac{2}{5}+\dfrac{4}{5}+1+\dfrac{4}{5}$

$\dfrac{2}{5}+\underline{\quad?\quad}+\dfrac{4}{5} \;\; 1\frac{4}{5}$

$\dfrac{2}{5}+\underline{\quad?\quad} \;\; 2\frac{3}{5}$

$\underline{\quad?\quad} \;\; 3$

13. $(2\cdot 9)+(2\cdot 7)$

$\underline{\quad?\quad}+\underline{\quad?\quad} \;\; 18;\ 14$

$\underline{\quad?\quad} \;\; 32$

14. $(12-7)+(12-5)$

$\underline{\quad?\quad}+\underline{\quad?\quad} \;\; 5;\ 7$

$\underline{\quad?\quad} \;\; 12$

15. $\dfrac{1}{4}+\dfrac{2}{4}+\dfrac{3}{4}+1$

$\underline{\quad?\quad}+\dfrac{3}{4}+1 \;\; \dfrac{3}{4}$

$\underline{\quad?\quad}+1 \;\; 1\frac{1}{2}$

$\underline{\quad?\quad} \;\; 2\frac{1}{2}$

16. $3\cdot 6\cdot 10\cdot\dfrac{1}{2}$

$\underline{\quad?\quad}\cdot 10\cdot\dfrac{1}{2} \;\; 18$

$\underline{\quad?\quad}\cdot\dfrac{1}{2} \;\; 180$

$\underline{\quad?\quad} \;\; 90$

17. $1.3+4.8+7.7+5.6$

$\underline{\quad?\quad}+7.7+5.6 \;\; 6.1$

$\underline{\quad?\quad}+5.6 \;\; 13.8$

$\underline{\quad?\quad} \;\; 19.4$

18. $(5\cdot 8)+(5\cdot 4)$

$\underline{\quad?\quad}+\underline{\quad?\quad} \;\; 40;\ 20$

$\underline{\quad?\quad} \;\; 60$

19. $\left(\dfrac{1}{3}\cdot\dfrac{2}{3}\right)+\left(\dfrac{1}{9}\cdot 4\right)$

$\underline{\quad?\quad}+\underline{\quad?\quad} \;\; \dfrac{2}{9};\ \dfrac{4}{9}$

$\underline{\quad?\quad} \;\; 2\frac{1}{2}$

20. $(18-2)+8-10$

$\underline{\quad?\quad}+8-10 \;\; 16$

$\underline{\quad?\quad}-10 \;\; 24$

$\underline{\quad?\quad} \;\; 14$

B

21.
$$\underbrace{\frac{1}{4} \cdot 8}_{\ ?\ } + \underbrace{\frac{1}{4} \cdot 2}_{\ ?\ } \ 2; \frac{1}{2}$$
$$\underbrace{}_{\ ?\ } \ 2\frac{1}{2}$$

22.
$$\underbrace{\frac{3}{4} \cdot \frac{1}{2}}_{\ ?\ } + \underbrace{\frac{1}{4} \cdot \frac{3}{2}}_{\ ?\ } \ \frac{3}{8}; \frac{3}{8}$$
$$\underbrace{}_{\ ?\ } \ \frac{3}{4}$$

SELF-TEST 1

Be sure that you understand these terms.

Reflexive property of equality (p. 154)
Symmetric property of equality (p. 154)
Transitive property of equality (p. 154)
Substitution principle (p. 157)
Closure (p. 157)

Section 6-1, p. 154 Use the property indicated to complete the statement.

Reflexive: **1.** $6 + x = \underline{\ ?\ }\ 6 + x$ **2.** $m(n + 3) = \underline{\ ?\ }$ $\overset{m(n + 3)}{}$

Symmetric: **3.** $4 + a = 9$, so $\underline{\ ?\ }\ 9 = 4 + a$ **4.** $x^2 \cdot 6 = z$, so $\underline{\ ?\ }$ $\underset{z = x^2 \cdot 6}{}$

Transitive: **5.** $3 - k = 1$ and $1 = m \cdot 5$, so $\underline{\ ?\ }\ 3 - k = m \cdot 5$

6. $\underline{\ ?\ }$ and $m = 2n - 1$, so $k + 3 = 2n - 1$. $k + 3 = m$

Section 6-2, p. 157 Tell whether the set is closed under addition, subtraction, multiplication and division.

7. $\{5, 10, 15, \ldots\}$ C: A, M; NC: S, D   **8.** $\{1\}$ C: M, D; NC: A, S

Simplify these expressions using the substitution principle.

9. $\underbrace{6 + 7}_{\ ?\ } + 8 + 9$ 13
$$\underbrace{ + 8}_{\ ?\ } + 9 \ 21$$
$$\underbrace{}_{\ ?\ } \ 30$$

10. $\underbrace{\frac{2}{3} \cdot \frac{3}{4}}_{\ ?\ } \cdot \frac{4}{5} \cdot \frac{1}{6}$
$$\underbrace{ \cdot \frac{4}{5}}_{\ ?\ } \cdot \frac{1}{6} \ \frac{1}{2}$$
$$\underbrace{ \cdot \frac{1}{6}}_{\ ?\ } \ \frac{2}{5}$$
$$\underbrace{}_{\ ?\ } \ \frac{1}{15}$$

Check your answers with those printed at the back of the book.

career capsule *Landscape Architect*

Landscape architects work on parks, gardens, housing developments, school campuses, airports, roads, recreational areas and industrial parks, planning the best design for the land and the objects on it. They study the site, map the slope of the land, and determine the existing soil type. Architects also check building codes and develop blueprints of plans and materials to be used. Many landscape architects also supervise construction.

Landscape architects must have mathematical and artistic ability, an interest in art and nature, and a desire to work outdoors. A college degree in landscape architecture or a related field is required.

More Properties

6-3 *The Commutative and Associative Properties*

OBJECTIVES

Identify and apply the commutative and associative properties of addition.

Identify and apply the commutative and associative properties of multiplication.

The order in which two numbers are added or multiplied does not affect the result. We say that addition and multiplication are commutative operations.

$$12 + 6 = 18 \qquad 9 \times 4 = 36$$
$$6 + 12 = 18 \qquad 4 \times 9 = 36$$

The Commutative Property ▶ For every number r and every number s,
of Addition $\quad r + s = s + r.$

The Commutative Property ▶ For every number r and every number s,
of Multiplication $\quad r \cdot s = s \cdot r.$

EXAMPLE 1 $\quad n + 4 = 4 + n$ for every number n
$\quad n \cdot 10 = 10 \cdot n$ for every number n

Subtraction and division are *not* commutative operations.

EXAMPLE 2 $\quad 15 - 8 \neq 8 - 15$
$\quad 12 \div 6 \neq 6 \div 12$

When three or more numbers are added or multiplied, the way in which we group the numbers does not affect the result. We say that addition and multiplication are associative operations.

$$2 + 8 + 7 \blacktriangleright (2 + 8) + 7 = 17 \qquad 5 \times 6 \times 2 \blacktriangleright (5 \times 6) \times 2 = 60$$
$$2 + (8 + 7) = 17 \qquad\qquad 5 \times (6 \times 2) = 60$$

The Associative Property ▶ For every number r, every number s, and every
of Addition $\quad$ number t, $r + (s + t) = (r + s) + t.$

The Associative Property ▶── For every number r, every number s, and every
  of Multiplication   number t, $r(st) = (rs)t$.

EXAMPLE 3 $(3 + b) + 8 = 3 + (b + 8)$ for every number b
$(2b)y = 2(by)$ for all numbers b and y

Subtraction and division are *not* associative.

EXAMPLE 4 $12 - (4 - 2) \neq (12 - 4) - 2$
$24 \div (8 \div 2) \neq (24 \div 8) \div 2$

Tell which property of addition or multiplication is illustrated.

Sample 1: $3 \cdot 9 = 9 \cdot 3$
What you say: Commutative property of multiplication

Sample 2: $2(3 \cdot 9) = (2 \cdot 3)9$
What you say: Associative property of multiplication

EXERCISES

1. $\frac{1}{2} + \frac{1}{3} = \frac{1}{3} + \frac{1}{2}$ comm. prop. of add.
2. $(2 \cdot 3)4 = 2(3 \cdot 4)$ assoc. prop. of mult.
3. $2 + 7 = 7 + 2$ comm. prop. of add.
4. $\frac{1}{2} + \left(\frac{1}{3} + \frac{1}{4}\right) = \left(\frac{1}{2} + \frac{1}{3}\right) + \frac{1}{4}$ assoc. prop. of add.
5. $\frac{5}{8} + \frac{1}{6} = \frac{1}{6} + \frac{5}{8}$ comm. prop. of add.
6. $\frac{1}{3} \cdot \frac{3}{4} = \frac{3}{4} \cdot \frac{1}{3}$ comm. prop. of mult.
7. $8 \cdot 9 = 9 \cdot 8$ comm. prop. of mult.
8. $5 + (2 + 3) = (5 + 2) + 3$ assoc. prop. of add.
9. $8 + (6 + 2) = 8 + (2 + 6)$ comm. prop. of add.
10. $9(8 \cdot 4) = 9(4 \cdot 8)$ comm. prop. of mult.
11. $2 + (3 + 5) = (2 + 3) + 5$ assoc. prop. of add.
12. $7 \cdot 9 = 9 \cdot 7$ comm. prop. of mult.
13. $8 + 4 = 4 + 8$ comm. prop. of add.

Show that the sentence is true. Name the property illustrated.

Sample: $4(10 \cdot 3) = (4 \cdot 10)3$

Solution: $4(10 \cdot 3) = (4 \cdot 10)3$

$4 \cdot 30$	$40 \cdot 3$
120	120

Associative property of multiplication

A

1. $10 + (8 + 9) = (10 + 8) + 9$ assoc. prop. of add.

2. $2\frac{1}{2} + 3\frac{1}{8} = 3\frac{1}{8} + 2\frac{1}{2}$ comm. prop. of add.

3. $3.6 + (8.2 + 9.4) = 3.6 + (9.4 + 8.2)$ comm. prop. of add.

4. $\dfrac{1}{2}\left(\dfrac{1}{3} \cdot \dfrac{1}{4}\right) = \left(\dfrac{1}{2} \cdot \dfrac{1}{3}\right)\dfrac{1}{4}$ assoc. prop. of mult.

5. $6 + \left(\dfrac{1}{2} \cdot \dfrac{1}{3}\right) = 6 + \left(\dfrac{1}{3} \cdot \dfrac{1}{2}\right)$ comm. prop. of mult.

6. $0.8(6 + 0.5) = 0.8(0.5 + 6)$ comm. prop. of add.

7. $\left(1 + \dfrac{3}{4}\right) + \dfrac{2}{4} = 1 + \left(\dfrac{3}{4} + \dfrac{2}{4}\right)$ assoc. prop. of add.

8. $\dfrac{2}{5} \cdot \dfrac{5}{2} = \dfrac{5}{2} \cdot \dfrac{2}{5}$ comm. prop. of mult.

9. $(0.03 + 0.58)3 = 3(0.03 + 0.58)$ comm. prop. of mult.

10. $(1\frac{1}{3} \cdot 2\frac{3}{4})\dfrac{1}{2} = 1\frac{1}{3}\left(2\frac{3}{4} \cdot \dfrac{1}{2}\right)$ assoc. prop. of mult.

11. comm. prop. of add.   **12.** comm. prop. of mult.

13. assoc. prop. of add.   **14.** comm. prop. of add.

Let $r = 1.2$, $s = 4$ and $t = 0.03$. Show that the statement is true. Name the property illustrated.

Sample: $(rs)t = r(st)$ *Solution:*

15. comm. prop. of add.

16. comm. prop. of add.

17. assoc. prop. of add.

18. comm. prop. of add.

$(rs)t = r(st)$	
$(4.8)(0.03)$	$(1.2)(0.12)$
0.144	0.144

Associative property of multiplication.

B

11. $r + t = t + r$

12. $s \cdot t = t \cdot s$

13. $r + (s + t) = (r + s) + t$

14. $(r + s)t = (s + r)t$

15. $(r + s)t = t(r + s)$

16. $r(s + t) = r(t + s)$

17. $s + (r + t) = (s + r) + t$

18. $s + (r + t) = s + (t + r)$

19. $s(tr) = (tr)s$

20. $t(rs) = (tr)s$

19. comm. prop. of mult.   **20.** assoc. prop. of mult.

Justify each step.

Sample:

$$\frac{1}{2} + \left(2 + \frac{1}{2}\right) = \left(2 + \frac{1}{2}\right) + \frac{1}{2}$$

$$= 2 + \left(\frac{1}{2} + \frac{1}{2}\right)$$

$$= 2 + 1$$

$$= 3$$

Solution:

Commutative property of addition

Associative property of addition

Substitution principle

Substitution principle

21. comm. prop. of add.
$7 + (8 + 3) = 7 + (3 + 8)$
assoc. prop. of add. $= (7 + 3) + 8$
substitution prin. $= 10 + 8$
substitution prin. $= 18$

22. comm. prop. of add.
$16 + (8 + 4) = 16 + (4 + 8)$
assoc. prop. of add. $= (16 + 4) + 8$
substitution prin. $= 20 + 8$
substitution prin. $= 28$

23. comm. prop. of mult.
$5(17 \cdot 6) = 5(6 \cdot 17)$
assoc. prop. of mult. $= (5 \cdot 6)17$
substitution prin. $= 30 \cdot 17$
substitution prin. $= 510$

24. $(9 \cdot 5)\frac{2}{3} = (5 \cdot 9)\frac{2}{3}$ comm. prop. of mult.

assoc. prop. of mult. $= 5\left(9 \cdot \frac{2}{3}\right)$

$= 5 \cdot 6$ substitution prin.

$= 30$ substitution prin.

Replace the __?__ with $=$ or $\neq$ to make a true statement. Use {the numbers of arithmetic} as replacement set for the variables.

25. $16 \cdot 17$ __?__ $17 \cdot 16 =$

26. $4 \div 2$ __?__ $2 \div 4 \neq$

27. $a(b \cdot c)$ __?__ $(a \cdot b)c =$

28. $6 + \frac{1}{2}$ __?__ $\frac{1}{2} + 6 =$

29. $3 \cdot 4$ __?__ $4 + 3 \neq$

30. $6 - 7$ __?__ $7 - 6 \neq$

31. $(3 + 1)4$ __?__ $4(3 + 1) =$

32. $6(5 + 8)$ __?__ $6(8 + 5) =$

33. $d + (ef)$ __?__ $(ef) + d =$

34. $(3 + 1) \div 2$ __?__ $2 \div (3 + 1) \neq$

35. $(4 + 5) \div 3$ __?__ $(5 + 4) \div 3 =$

36. $1 \div a$ __?__ $a \div 1 \neq$

37. $\frac{2 + 1}{3} + 6$ __?__ $6 + \frac{2 + 1}{3} =$

6-4 *The Distributive Property*

OBJECTIVE
Identify and apply the distributive property.

Recall that the formula for the perimeter of a rectangle can be written as $P = 2l + 2w$, or as $P = 2(l + w)$. Since 2 is a common factor of the terms of $2l + 2w$, $2(l + w) = 2l + 2w$. We say that multiplication is distributive over addition.

The Distributive Property ► For every number r, every number s, and every number t, $r(s + t) = rs + rt$.

EXAMPLE 1   Find P, the perimeter of the rectangle shown.

> 15 cm
> 30 cm

$$P = 2l + 2w \qquad\qquad P = 2(l + w)$$
$$= 2(30) + 2(15) \qquad\quad = 2(30 + 15)$$
$$= 60 + 30 = 90 \text{(cm)} \qquad = 2(45) = 90 \text{(cm)}$$

EXAMPLE 2 $4(62) = 4(60 + 2) = 4(60) + 4(2) = 240 + 8 = 248$

EXAMPLE 3 $5 \cdot 3\frac{1}{8} = 5\left(3 + \frac{1}{8}\right) = 5(3) + 5\left(\frac{1}{8}\right) = 15 + \frac{5}{8} = 15\frac{5}{8}$

From the commutative property, we know that $r(s + t) = (s + t)r$ and $rs + rt = sr + tr$. Then we can also state the distributive property as follows.

The Distributive Property ► For every number r, every number s, and every number t, $(s + t)r = sr + tr$.

EXERCISES

Tell how to complete the sentence to illustrate the distributive property.

Sample: $4(7 + 6) = \underline{\ ?\ }$ *What you say:* $4(7 + 6) = 4 \cdot 7 + 4 \cdot 6$

1. $2(3 + 4) = \underline{\ ?\ }$ $2 \cdot 3 + 2 \cdot 4$ **2.** $\underline{\ ?\ } = 2 \cdot 6 + 2 \cdot 8$ $2(6 + 8)$

3. $\underline{\ ?\ } = 5(6 + 1)$ $5 \cdot 6 + 5 \cdot 1$ **4.** $8(9 + 3) = \underline{\ ?\ }$ $8 \cdot 9 + 8 \cdot 3$

5. $2(7 + 7) = \underline{\ ?\ }$ $2 \cdot 7 + 2 \cdot 7$ **6.** $\underline{\ ?\ } = 6(9 + 1)$ $6 \cdot 9 + 6 \cdot 1$

7. $\underline{\quad?\quad} = 4(5 + 3)$ $4 \cdot 5 + 4 \cdot 3$ 8. $a(b + 2) = \underline{\quad?\quad}$ $a \cdot b + a \cdot 2$

9. $(k + 3)4 = \underline{\quad?\quad}$ $k \cdot 4 + 3 \cdot 4$ 10. $5(n + m) = \underline{\quad?\quad}$ $5 \cdot n + 5 \cdot m$

Show that the statement is true. For Ex. 1–10, only the final step is given.

Sample 1: $6(9 + 2) = 6(9) + 6(2)$
Solution: $6(9 + 2) = 6(9) + 6(2)$

$$\begin{array}{c|c} 6 \cdot 11 & 54 + 12 \\ 66 & 66 \end{array}$$

A

1. $3(4 + 5) = 3 \cdot 4 + 3 \cdot 5$ $27 = 27$
2. $(7 + 3)5 = 7 \cdot 5 + 3 \cdot 5$ $50 = 50$
3. $8 \cdot 9 + 8 \cdot 4 = 8(9 + 4)$ $104 = 104$
4. $6 \cdot 7 + 8 \cdot 7 = (6 + 8)7$ $98 = 98$
5. $10 \cdot 7 + 13 \cdot 7 = (10 + 13)7$ $161 = 161$
6. $0(5 + 9) = 0 \cdot 5 + 0 \cdot 9$ $0 = 0$
7. $1(3 + 9) = 1 \cdot 3 + 1 \cdot 9$ $12 = 12$
8. $\dfrac{1}{2}(3 + 5) = \dfrac{1}{2} \cdot 3 + \dfrac{1}{2} \cdot 5$ $4 = 4$
9. $0.6(5 + 0.3) = (0.6)5 + (0.6)(0.3)$ $3.18 = 3.18$
10. $8 \cdot \dfrac{3}{4} + 6 \cdot \dfrac{3}{4} = (8 + 6)\dfrac{3}{4}$ $10\frac{1}{2} = 10\frac{1}{2}$

Apply the distributive property to complete. Use the substitution principle as shown in the sample.

Sample: $316 \cdot 4$ *Solution:* $316 \cdot 4 = (300 + 10 + 6) \cdot 4$
$$= 300 \cdot 4 + 10 \cdot 4 + 6 \cdot 4$$
$$= 1200 + 40 + 24$$
$$= 1264$$

11. $3 \cdot 612$ 1836
14. $213 \cdot 7$ 1491
17. $419 \cdot 7$ 2933
20. $5 \cdot 6798$ 33,990

12. $5 \cdot 115$ 575
15. $618 \cdot 6$ 3708
18. $6 \cdot 792$ 4752
21. $4113 \cdot 9$ 37,017

13. $2 \cdot 461$
16. $139 \cdot 9$
19. $2 \cdot 6248$
22. $9999 \cdot 8$

922
B
1251
12,496
79,992

Justify each step by naming the principle or property used.

23. $16 \cdot 4 = 4 \cdot 16$ comm. prop. of mult.
  sub. prin. $= 4(10 + 6)$
distrib. prop. $= 4 \cdot 10 + 4 \cdot 6$
  sub. prin. $= 40 + 24$
  sub. prin. $= 64$

24. $2(619) = 2(600 + 10 + 9)$ sub. prin.
distrib. prop. $= 2 \cdot 600 + 2 \cdot 10 + 2 \cdot 9$
  sub. prin. $= 1200 + 20 + 18$
  sub. prin. $= 1238$

25. $7 \cdot 35 = 7(30 + 5)$ sub. prin.
distrib. prop. $= 7 \cdot 30 + 7 \cdot 5$
  sub. prin. $= 35 + 210$
  sub. prin. $= 245$

26. $298 \cdot 6 = (200 + 90 + 8)6$ sub. prin.
distrib. prop. $= 200 \cdot 6 + 8 \cdot 6 + 90 \cdot 6$
  sub. prin. $= 1200 + 48 + 540$
  sub. prin. $= 1788$

Tell whether or not the statement is true for all members of {the numbers of arithmetic}.

C

27. $mn + mp = m(n + p)$ True

28. $ab + ac + ad = a(b + c + d)$ True

29. $(r + s + t)x = rx + sx + tx$ True

30. $mn(x + y) = mnx + mny$ True

calculator corner

There are some multiplication problems which have answers too large for a calculator display. However, you can still use your calculator to solve them if you also use some algebra. To solve $32{,}051 \times 4060$ use the distributive property, $32{,}051 \times 4060 = 32{,}051(4000 + 60) = 32{,}051(4000) + 32{,}051(60) = 32{,}051(4)(1000) + 32{,}051(60)$. You can solve $32{,}051 \times 4$ with a calculator and multiply by 1000 by adding three zeros. You can solve $32{,}051(60)$ with an 8-digit calculator directly or with a 6-digit calculator by solving $32{,}051(6)(10)$. Now add both answers to obtain $32{,}051 \times 4060$. Make up and solve a multiplication problem too large for your calculator's display.

6-5 *More about the Distributive Property*

OBJECTIVE

Apply the distributive property to subtraction and to division.

We usually think of the distributive property as applying to multiplication and addition. However, multiplication is also distributive over subtraction.

$$7(58) = 7(60 - 2) = 7(60) - 7(2)$$
$$= 420 - 14 = 406$$

The Distributive Property ► For every number r, every number s, and every number t, $r(s - t) = rs - rt$, and $(s - t)r = sr - tr$.

EXAMPLE 1

$$\frac{1}{3}(17) = \frac{1}{3}(18 - 1)$$

$$= \frac{1}{3}(18) - \frac{1}{3}(1)$$

$$= 6 - \frac{1}{3} = 5\frac{2}{3}$$

Dividing by a number is the same as multiplying by the reciprocal of the number. Then we can also apply the distributive property to distributing division over either addition or subtraction.

EXAMPLE 2

$$396 \div 3 = 396 \times \frac{1}{3} = (300 + 90 + 6)\frac{1}{3}$$

$$= 100 + 30 + 2 = 132$$

EXAMPLE 3

$$19\frac{1}{2} \div 4 = \left(20 - \frac{1}{2}\right) \div 4 = 20\left(\frac{1}{4}\right) - \left(\frac{1}{2}\right)\left(\frac{1}{4}\right)$$

$$= 5 - \frac{1}{8} = 4\frac{7}{8}$$

Tell how to complete to illustrate the distributive property.

Sample 1: $\dfrac{600 + 60 + 9}{3} = \underline{\ \ ?\ \ }$

What you say: $\dfrac{600 + 60 + 9}{3} = \dfrac{600}{3} + \dfrac{60}{3} + \dfrac{9}{3}$

Sample 2: $5(70 - 8) = \underline{\ \ ?\ \ }$
What you say: $5(70 - 8) = 5 \cdot 70 - 5 \cdot 8$

1. $6(8 - 2) = \underline{\ \ ?\ \ }$ $\ 6 \cdot 8 - 6 \cdot 2$

2. $\dfrac{1}{2}\left(9 - \dfrac{1}{2}\right) = \underline{\ \ ?\ \ }$ $\quad \dfrac{1}{2} \cdot 9 - \dfrac{1}{2} \cdot \dfrac{1}{2}$

3. $\left(7 - \dfrac{2}{3}\right)\dfrac{3}{4} = \underline{\ \ ?\ \ }$ $\ 7 \cdot \dfrac{3}{4} - \dfrac{2}{3} \cdot \dfrac{3}{4}$

4. $(x - y)4 = \underline{\ \ ?\ \ }$ $\quad x \cdot 4 - y \cdot 4$

5. $(z - 3)5 = \underline{\ \ ?\ \ }$ $\ z \cdot 5 - 3 \cdot 5$

6. $\dfrac{800 + 40 + 6}{2} = \underline{\ \ ?\ \ }$

7. $\dfrac{600 + 90 + 3}{3} = \underline{\ \ ?\ \ }$ $\quad \dfrac{600}{3} + \dfrac{90}{3} + \dfrac{3}{3}$

8. $\underline{\ \ ?\ \ } = \dfrac{1000}{5} + \dfrac{50}{5} + \dfrac{5}{5}$

9. $\underline{\ \ ?\ \ } = 3 \cdot 5 - 3 \cdot 2$ $\ 3(5 - 2)$

10. $\underline{\ \ ?\ \ } = \dfrac{1}{4} \cdot 7 - \dfrac{1}{4} \cdot 3$ $\quad \dfrac{1}{4}(7 - 3)$

6. $\dfrac{800}{2} + \dfrac{40}{2} + \dfrac{6}{2}$

8. $\dfrac{1000 + 50 + 5}{5}$

Show that the statement is true.

For Ex. 1–10, only the final step is given.

Sample: $4(30 - 8) = 4 \cdot 30 - 4 \cdot 8$
Solution: $4(30 - 8) = 4 \cdot 30 - 4 \cdot 8$

$$\begin{array}{c|c} 4 \cdot 22 & 120 - 32 \\ 88 & 88 \end{array}$$

A

1. $(10 - 3)4 = 10 \cdot 4 - 3 \cdot 4$ $\quad 28 = 28$

2. $\left(8 - \dfrac{1}{2}\right)\dfrac{1}{2} = 8 \cdot \dfrac{1}{2} - \dfrac{1}{2} \cdot \dfrac{1}{2}$ $\quad 3\tfrac{3}{4} = 3\tfrac{3}{4}$

3. $\dfrac{300 + 60 + 9}{3} = \dfrac{300}{3} + \dfrac{60}{3} + \dfrac{9}{3}$ $\quad 123 = 123$

4. $\dfrac{400 + 40 + 8}{8} = \dfrac{400}{8} + \dfrac{40}{8} + \dfrac{8}{8}$ $\quad 56 = 56$

5. $5(40 - 7) = 5 \cdot 40 - 5 \cdot 7$ $\quad 165 = 165$

6. $\left(9 - \dfrac{1}{4}\right)\dfrac{1}{3} = 9 \cdot \dfrac{1}{3} - \dfrac{1}{4} \cdot \dfrac{1}{3}$ $\quad 2\tfrac{11}{12} = 2\tfrac{11}{12}$

7. $(7 - 0.3)5 = 7 \cdot 5 - 0.3 \cdot 5$ $\quad 33.5 = 33.5$

8. $\left(\dfrac{3}{4} - \dfrac{1}{2}\right)\dfrac{1}{3} = \dfrac{3}{4} \cdot \dfrac{1}{3} - \dfrac{1}{2} \cdot \dfrac{1}{3}$ $\quad \dfrac{1}{12} = \dfrac{1}{12}$

9. $\left(100 - \dfrac{1}{10}\right)\dfrac{1}{5} = 100 \cdot \dfrac{1}{5} - \dfrac{1}{10} \cdot \dfrac{1}{5}$ $\quad 19\frac{49}{50} = 19\frac{49}{50}$

10. $(100 - 0.2)6 = 100 \cdot 6 - 0.2 \cdot 6$ $\quad 598.8 = 598.8$

Complete. Use the substitution principle and the distributive property.

Sample 1: $4 \cdot 67$ *Solution:* $4 \cdot 67 = 4(70 - 3)$
$$= 4 \cdot 70 - 4 \cdot 3$$
$$= 280 - 12 = 268$$

Sample 2: $844 \div 4$ *Solution:* $844 \div 4 = \dfrac{800 + 40 + 4}{4}$
$$= \dfrac{800}{4} + \dfrac{40}{4} + \dfrac{4}{4}$$
$$= 200 + 10 + 1 = 211$$

11. $3 \cdot 39$ 117

12. $\dfrac{135}{5}$ 27

13. $\dfrac{448}{8}$ ^{56} **B**

14. $10 \cdot 11.9$ 119

15. $\dfrac{866}{2}$ 433

16. $99 \cdot 9$ 891

17. $\dfrac{284}{4}$ 71

18. $5 \cdot 69$ 345

19. $\dfrac{245}{5}$ 49

20. $119 \cdot 3$ 357

21. $\dfrac{428}{4}$ 107

22. $7 \cdot 5\frac{6}{7}$ 41

23. $8.9 \cdot 6$ 53.4

24. $3 \cdot 12\frac{2}{3}$ 38

25. $9 \cdot 8\frac{5}{6}$ $79\frac{1}{2}$

Replace the _?_ with $=$ or $\neq$ to make a true statement.

Sample: $(10 \cdot 5) + 6 \underline{\ ?\ } 10 \cdot 5 + 10 \cdot 6$

Solution: $(10 \cdot 5) + 6 \neq 10 \cdot 5 + 10 \cdot 6$

26. $6(30 + 7) \underline{\ ?\ } 6 \cdot 30 + 6 \cdot 7$ $=$

27. $5 \cdot \dfrac{1}{2} + 6 \cdot \dfrac{1}{2} \underline{\ ?\ } (5 + 6)\dfrac{1}{2}$ $=$

28. $2 \cdot 3 + 2 \cdot 4 + 2 \cdot 5 \underline{\ ?\ } (3 + 4 + 5)2$ $=$

29. $(4 + 5)8 \underline{\ ?\ } (4 + 8)(5 + 8)$ $\neq$

30. $3 + (4 \cdot 6) \underline{\ ?\ } (3 + 4)(3 + 6)$ $\neq$

SELF-TEST 2

Be sure that you understand these terms.

Commutative property (p. 162) Associative property (p. 162)
Distributive property (p. 166)

Show that the statement is true. Name the property illustrated.

Section 6-3, p. 162

1. $3 \cdot 4 = 4 \cdot 3$ comm. prop. of mult.

2. $2 + (3 + 4) = (3 + 4) + 2$ comm. prop. of add.

3. $2\left(4 \cdot \dfrac{1}{2}\right) = \left(4 \cdot \dfrac{1}{2}\right)2$ comm. prop. of mult.

4. $6(10 \cdot 8) = (6 \cdot 10)8$ assoc. prop. of mult.

Section 6-4, p. 166

5. $6(10 + 4) = (6 \cdot 10) + (6 \cdot 4)$ dist. prop.

6. $(10 - 5)3 = (10 \cdot 3) - (5 \cdot 3)$ dist. prop.

Use the distributive property and the substitution principle to complete.

7. $4 \cdot 316 = (4 \cdot 300) + (4 \cdot 10) + (4 \cdot 6) = \underline{\ ?\ }$ 1264

Section 6-5, p. 169

8. $6 \cdot 89 = (6 \cdot 90) - (6 \cdot 1) = \underline{\ ?\ }$ 534

Complete. Use the distributive property.

9. $\dfrac{488}{2} = \dfrac{400}{2} + \dfrac{80}{2} + \dfrac{8}{2} = \underline{\ ?\ }$ 244   10. $\dfrac{369}{3} = \underline{\ ?\ }$
$100 + 20 + 3 = 123$

Check your answers with those printed at the back of the book.

Time out

Two jet planes, A and B, are flying in the same direction. Plane B is 800 kilometers ahead of plane A.

If plane A flies at a speed of 880 kilometers per hour and plane B flies at 800 kilometers per hour, how long will it take plane A to catch plane B? How far will plane A have flown by the time it catches up with B? A will catch B in 10 hours. A will have flown 8800 km.

6-6 *The Properties of Zero and One*

OBJECTIVES

Apply the additive property of 0.

Apply the multiplicative properties of 0 and 1.

When you add 0 to any number, the sum is the number that was added to 0. We call 0 the additive identity element.

The Additive Property of 0 ▶ For every number r, $r + 0 = 0 + r = r$.

EXAMPLES $0 + 9 = 9$ $3\frac{1}{2} + 0 = 3\frac{1}{2}$ $2.3 + 0 = 2.3$

When any number is multiplied by 0, the product is 0.

The Multiplicative Property of 0 ▶ For every number r, $r \cdot 0 = 0 \cdot r = 0$.

EXAMPLES $0 \cdot 17 = 0$ $4 \cdot 3 \cdot 0 = 0$ $0 \cdot 0 = 0$

There is no similar division property of 0. Division by 0 is not possible.

When you multiply any number by 1, the product is the number that was multiplied by 1. We call 1 the multiplicative identity element.

The Multiplicative Property of 1 ▶ For every number r, $r \cdot 1 = 1 \cdot r = r$.

There is no similar division property of 1. Note that $12 \div 1 = 12$ is true, but $1 \div 12 = 12$ is not true.

EXAMPLES $1 \cdot 75 = 75$ $\frac{3}{4} \cdot 1 = \frac{3}{4}$ $m \cdot 1 = m$

Tell whether the expression names the number 1 or the number 0.

Oral EXERCISES

1. $1 \div 1$ ₁

2. $0 \div 1$ ₀

3. $1 - 0$ ₁

4. $0 - 0$ ₀

5. $0 + 0$ ₀

6. $1 + 0$ ₁

Solve.

7. $m \div 12 = 1$ {12} **8.** $6 + r = 6$ {0} **9.** $\dfrac{3}{4} \cdot \dfrac{4}{3} = s$ {1}

10. $\dfrac{1}{2} = t \cdot 1$ $\left\{\dfrac{1}{2}\right\}$ **11.** $k(12 - 1) = 1$ $\left\{\dfrac{1}{11}\right\}$ **12.** $0 + k = 0$ {0}

Written EXERCISES

Solve.

A

1. $4 \cdot m = 4$ 1
2. $\dfrac{6}{6} \cdot r = 7$ 7

3. $2 \cdot 3 + m = 6$ 0
4. $2(5 + m) = 10$ 0

5. $(8 + 4)0 = n$ 0
6. $p\left(\dfrac{3}{3} + 4\right) = 10$ 2

7. $q\left(\dfrac{2}{2} + 7\right) = 0$ 0
8. $4\left(\dfrac{3}{3} + \dfrac{2}{2}\right) = a$ 8

9. $5\left(\dfrac{7}{7} + b\right) = 10$ 1
10. $8\left(\dfrac{3}{3} + c\right) = 8$ 0

11. $(4 \div 4)d = 2$ 2
12. $f + 0.125 = 0.125$ 0
13. $(0 + 1.8)k = 0$ 0
14. $2^2 \cdot m = 4$ 1
15. $n^2 \cdot 10 = 0$ 0
16. $12^2 \cdot 0 = w$ 0

B

17. $b + b + b = 0$ 0
18. $a + a + a = 3$ 1
19. $0 - y = y$ 0
20. $0 + r = r + 0$ Every number
21. $t - 0 = t$ Every number
22. $m \cdot 1 = m$ Every number
23. $n + 0 = n$ Every number
24. $(8 + 1) = y(8 + 1)$ 1
25. $r \cdot 0 = 0$ Every number
26. $\dfrac{s}{s} = 1$ Every number except 0

27. $m \div m = 1$ Every number except 0
28. $k^2 = (0 + 1)^2$ 1

C

29. $\dfrac{n}{n} \cdot 1 = 3$ No solution
30. $m^2 \cdot \dfrac{6}{6} = 36$ 6

31. $\dfrac{2}{2}(0 + 2) = v$ 2
32. $\dfrac{y}{y} + 0 = 1$ Every number except 0

33. $5 + z^3 = 6$ 1
34. $3^2 \cdot m^2 = 9$ 1

35. $k \cdot 1 = k + 0$ Every number
36. $\dfrac{x}{x} \cdot 0 = 1$ No solution

37. $n^2 \cdot 5^2 = 0$ 0
38. $1 + x^3 = 2$ 1

6-7 *Function Equations*

> **OBJECTIVE**
>
> **Use a function equation and re-placement set to write and graph number pairs.**

Earlier we used a "function machine" and a rule such as $x - 4$ to develop a set of ordered number pairs called a function. Number pairs in a function table are in the form (input, output). **Input values** are specified numbers used to replace x in the rule. The resulting values from the rule are the output numbers. We call each **output number** a **value of the function for x**, which we indicate by the symbol $f(x)$.

EXAMPLE 1 $f(x) = x - 4$; replacement set: $\{5, 10, 15, 20\}$

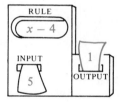

Input	Output	Ordered pair
x	$f(x)$	$(x, f(x))$
5	1	$(5, 1)$
10	6	$(10, 6)$
15	11	$(15, 11)$
20	16	$(20, 16)$

Function: $\{(5, 1), (10, 6), (15, 11), (20, 16)\}$

To graph ordered pairs from a function equation, we label the horizontal axis x and the vertical axis $f(x)$.

EXAMPLE 2 $f(x) = 2x - 1$; replacement set: $\{1, 2, 3\}$

x	$f(x)$	$(x, f(x))$
1	1	$(1, 1)$
2	3	$(2, 3)$
3	5	$(3, 5)$

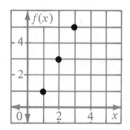

Function: $\{(1, 1), (2, 3), (3, 5)\}$

Use the function machine and tell how to complete the table.

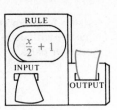

Rule: $\dfrac{x}{2} + 1$

	x	$f(x)$	$(x, f(x))$
1.	0	1; ?	(0, ? 1)
2.	2	2; ?	(2, ? 2)
3.	4	3; ?	(4, ? 3)
4.	8	5; ?	(8, ? 5)
5.	12	7; ?	(12, ? 7)
6.	16	9; ?	(16, ? 9)

Complete the table.

A

	x	$f(x) = \dfrac{1}{2}x$	$(x, f(x))$
1.	0	0	(0, 0)?
2.	1	$\dfrac{1}{2}$	$(1, \frac{1}{2})$?
3.	2	1	(2, 1)?
4.	3	? $\frac{3}{2}$	$(3, \frac{3}{2})$?
5.	4	? 2	(4, 2)?
6.	5	? $\frac{5}{2}$	$(5, \frac{5}{2})$?

	m	$f(m) = m + \dfrac{1}{2}$	$(m, f(m))$
7.	0	$\dfrac{1}{2}$	$(0, ? \frac{1}{2})$
8.	3	$3\frac{1}{2}$	$(3, ? 3\frac{1}{2})$
9.	6	? $6\frac{1}{2}$	$(6, ? 6\frac{1}{2})$
10.	9	? $9\frac{1}{2}$	$(9, ? 9\frac{1}{2})$
11.	12	? $12\frac{1}{2}$	$(12, ? 12\frac{1}{2})$
12.	15	? $15\frac{1}{2}$	$(15, ? 15\frac{1}{2})$

	n	$f(n) = 3n + 1$	$(n, f(n))$
13.	0	1	(0, ? 1)
14.	1	4	(1, ? 4)
15.	4	? 13	(4, ? 13)
16.	9	? 28	(9, ? 28)
17.	10	? 31	(10, ? 31)
18.	20	? 61	(20, ? 61)

	s	$f(s) = s^2 - s$	$(s, f(s))$
19.	0	? 0	(0, ? 0)
20.	1	? 0	(1, ? 0)
21.	3	? 6	(3, ? 6)
22.	4	12	(4, ? 12)
23.	6	? 30	(6, ? 30)
24.	9	? 72	(9, ? 72)

Write the set of ordered pairs that represents the function. Draw axes and graph the ordered pairs.

Sample: Rule: $3t$

Replacement set:

$$\left\{0, \frac{1}{2}, 1\right\}$$

Solution:

$$\left\{(0, 0), \left(\frac{1}{2}, 1\tfrac{1}{2}\right), (1, 3)\right\}$$

B

25. Rule: $m^2 + 1$
Replacement set:
$\{0, 1, 2\}$ {(0, 1), (1, 2), (2, 5)}

26. Rule: $2n - 1$
Replacement set:
$\{1, 2, 3\}$ {(1, 1), (2, 3), (3, 5)}

27. $\frac{1}{3}s + \frac{1}{2}$
Replacement set:
$\{0, 3, 6\}$ $\left\{\left(0, \frac{1}{2}\right), (3, 1\tfrac{1}{2}), (6, 2\tfrac{1}{2})\right\}$

28. Rule: $2a \cdot \dfrac{1}{a}$
Replacement set:
$\{1, 2, 3, 4\}$ {(1, 2), (2, 2), (3, 2), (4, 2)}

29. $\dfrac{3}{4} + \dfrac{b}{4}$
Replacement set:
$\{0, 1, 2, 3\}$ $\left\{\left(0, \frac{3}{4}\right), (1, 1), \left(2, \frac{5}{4}\right),\right.$ $\left.\left(3, \frac{3}{2}\right)\right\}$

30. $1 + \dfrac{c}{c}$
Replacement set:
$\{1, 2, 3, 4, 5\}$
{(1, 2), (2, 2), (3, 2), (4, 2), (5, 2)}

31. $p^2 + p$
Replacement set:
$\{0, 1, 2\}$ {(0, 0), (1, 2), (2, 6)}

32. $3r - 3$
Replacement set:
$\{1, 2, 3\}$ {(1, 0), (2, 3), (3, 6)}

Use the given equation and replacement set to make a table of values of the variable, function of the variable, and the resulting number pairs.

Sample: $f(s) = 2s^2 - 1$ *Solution:*
$\{1, 3, 5, 7\}$

s	$f(s) = 2s^2 - 1$	$(s, f(s))$
1	1	(1, 1)
3	17	(3, 17)
5	49	(5, 49)
7	97	(7, 97)

33. $f(a) = 2a + 1$
$\{0, 3, 6, 9\}$
{(0, 1), (3, 7), (6, 13), (9, 19)}

34. $f(b) = (b + 1)^2$
$\{0, 1, 2, 3\}$ {(0, 1), (1, 4), (2, 9), (3, 16)}

35. $f(c) = c^2 - 4$
$\{2, 5, 8, 11\}$ {(2, 0), (5, 21), (8, 60), (11, 117)}

36. $f(d) = \dfrac{1}{2}(d + 3)$
$\{1, 5, 9, 13\}$ {(1, 2), (5, 4), (9, 6), (13, 8)}

37. $f(g) = 29 - g^3$
$\{0, 1, 2, 3\}$ {(0, 29), (1, 28), (2, 21), (3, 2)}

38. $f(r) = \dfrac{1}{3}(r^3 + 1)$
$\{0, 2, 3, 5\}$ $\left\{\left(0, \frac{1}{3}\right), (2, 3),\right.$ $\left.(3, 9\tfrac{1}{3}), (5, 42)\right\}$

Use the given equation to complete the set of ordered pairs.

Sample: $f(z) = 6z - 5$; $\{(1, 1), (2, 7), (3, \underline{\;?\;}), (4, \underline{\;?\;}), (5, \underline{\;?\;})\}$
Solution: $\{(1, 1), (2, 7), (3, 13), (4, 19), (5, 25)\}$

C 39. $f(n) = n^2 + n$; $\{(0, 0), (1, 2), (2, \underset{\underline{\;?\;}}{6}), (3, \underset{\underline{\;?\;}}{12}), (4, \underset{\underline{\;?\;}}{20})\}$

40. $f(r) = r^2 + r + 1$;
$\{(0, 1), (1, 3), (2, \underset{\underline{\;?\;}}{7}), (3, \underset{\underline{\;?\;}}{13}), (4, \underset{\underline{\;?\;}}{21}), (5, \underset{\underline{\;?\;}}{31})\}$

41. $f(y) = y^2 - y + 2$;
$\{(0, 2), (1, 2), (2, \underset{\underline{\;?\;}}{4}), (3, \underset{\underline{\;?\;}}{8}), (4, \underset{\underline{\;?\;}}{14}), (5, \underset{\underline{\;?\;}}{22})\}$

42. $f(x) = \dfrac{1}{2}(x - 2)$; $\{(2, 0), (4, \underset{\underline{\;?\;}}{1}), (6, \underset{\underline{\;?\;}}{2}), (8, \underset{\underline{\;?\;}}{3}), (10, \underset{\underline{\;?\;}}{4})\}$

43. $f(m) = m^2 + \dfrac{1}{m}$; $\{(1, 2), (2, 4\tfrac{1}{2}), (3, \underset{\underline{\;?\;}}{9\tfrac{1}{3}}), (4, \underset{\underline{\;?\;}}{16\tfrac{1}{4}}), (5, \underset{\underline{\;?\;}}{25\tfrac{1}{5}})\}$

44. $f(a) = \pi - a$ for $\pi = \dfrac{22}{7}$;
$\left\{ \left(0, \dfrac{22}{7}\right), \left(\dfrac{1}{7}, \underset{\underline{\;?\;}}{3}\right), \left(1, \underset{\underline{\;?\;}}{\dfrac{15}{7}}\right), \left(2, \underset{\underline{\;?\;}}{\dfrac{8}{7}}\right), \left(2\dfrac{3}{7}, \underset{\underline{\;?\;}}{\dfrac{5}{7}}\right) \right\}$

SELF-TEST 3

Be sure that you understand these terms.

Additive property of zero (p. 173)
Multiplicative property of zero (p. 173)
Multiplicative property of one (p. 173) $f(x)$ (p. 175)

Section 6-6, p. 173   Solve.

1. $5 \cdot b = 0$ 0

2. $\left(\dfrac{7}{7}\right) \cdot r = r$ Every number is a solution.

3. $a \cdot 0 = 1$ no solution

4. $t \div t = 1$ Every number except 0 is a solution.

Section 6-7, p. 175   Write the set of ordered pairs that represents the function.
Draw axes and graph the ordered pairs.

5. $f(x) = x + 3$; replacement set: $\left\{0, \dfrac{1}{2}, 1, 1\tfrac{1}{2}\right\}$ $\left\{(0, 3), \left(\dfrac{1}{2}, 3\tfrac{1}{2}\right), (1, 4), (1\tfrac{1}{2}, 4\tfrac{1}{2})\right\}$

6. $f(x) = x - 2$; replacement set: $\{2, 3, 4\}$ $\{(2, 0), (3, 1), (4, 2)\}$

Check your answers with those printed at the back of the book.

chapter summary

1. The **reflexive property:** $r = r$
   The **symmetric property:** If $r = s$, then $s = r$.
   The **transitive property:** If $r = s$ and $s = t$, then $r = t$.

2. A set of numbers is **closed** under an operation performed on its members if the operation always gives a result that is a member of that set. This is called the **closure** property.

3. The order in which two numbers are added or multiplied does not affect the result.

   The **commutative property of addition:** $r + s = s + r$
   The **commutative property of multiplication:** $rs = sr$

4. The way in which three or more numbers are grouped for addition or multiplication does not affect the result.

   The **associative property of addition:** $r + (s + t) = (r + s) + t$
   The **associative property of multiplication:** $r \cdot (st) = (rs) \cdot t$

5. Multiplication is distributive over addition.
   The **distributive property:** $r(s + t) = rs + rt$.

6. The **additive property of 0:** $r + 0 = 0 + r = r$
   The **multiplicative property of 1:** $r \cdot 1 = 1 \cdot r = r$
   The **multiplicative property of 0:** $r \cdot 0 = 0 \cdot r = 0$
   Division by 0 is not possible.

Time out

Draw a circle with a compass. Mark off the circumference into 16 equal parts. Label the marks clockwise from 1 to 16. With straight lines, connect 1 with 6, 2 with 7, 3 with 8 and so on. Color in your design. Experiment by dividing the circle into a different number of equal parts. What happens if the parts are unequal?

The design has no symmetry if the parts are unequal.

chapter test

Match each statement or equation in Column 1 with the appropriate property in Column 2. You may use a property more than once.

COLUMN 1

COLUMN 2

1. $1 \cdot 14 = 14$ N
2. $2(3 + x) = 2 \cdot 3 + 2 \cdot x$ K
3. $k(j + 4) = k(j + 4)$ A
4. $7 + x = x + 7$ G
5. $8 + (2 + 7) = (8 + 2) + 7$ I
6. $pr^2 + 0 = pr^2$ L
7. $3 \cdot \dfrac{2}{2} = 3 \cdot 1$ F
8. If $2x = 3 + b$, then B
   $3 + b = 2x$.
9. $(x + 1)(x + 2) = (x + 2)(x + 1)$ H
10. $(7 - 2)6 = 7 \cdot 6 - 2 \cdot 6$ K
11. Since 3 and 4 are whole numbers, $3 \cdot 4$ is a whole number. E
12. $x(yz) = (xy)z$ J
13. $r(xyz) = (xyz)r$ H
14. $3 + (a + 4) = (a + 4) + 3$ G
15. $\dfrac{1}{2} = \dfrac{2}{4}$ and $\dfrac{2}{4} = \dfrac{3}{6}$, so $\dfrac{1}{2} = \dfrac{3}{6}$. C
16. $(x^2y^2z^2) \cdot 0 = 0$ M
17. $6(x + 9) = 6(9 + x)$ G
18. $3 + 4 = 5 + 2$ and $5 + 2 = 6 + 1$, so $3 + 4 = 6 + 1$. C
19. $8 + (2 - 2) = 8 + 0$ F
20. Since 7 and 10 are whole numbers, $7 + 10$ is a whole number. D

A. Reflexive property of equality
B. Symmetric property of equality
C. Transitive property of equality
D. Closure under addition
E. Closure under multiplication
F. Substitution principle
G. Commutative property of addition
H. Commutative property of multiplication
I. Associative property of addition
J. Associative property of multiplication
K. Distributive property
L. Additive property of zero
M. Multiplicative property of zero
N. Multiplicative property of one

Complete.

21. $f(p) = p^2 - 4$; replacement set: $\{2, 5, 8, 10\}$

p	$f(p) = p^2 - 4$	$(p, f(p))$	
2	? 0	?	(2, 0)
5	? 21	?	(5, 21)
8	? 60	?	(8, 60)
10	? 96	?	(10, 96)

challenge topics

Symmetry in Three Dimensions

A cube-shaped box just fits into a larger packing box. One face has "THIS SIDE UP" written on it. In how many different ways can the smaller box be placed properly inside the larger one? 4

A child's cube-shaped play block has the letters A, B, C, D, E, F on the faces. The play block just fits into a box. In how many different ways can the block be placed in the box? 24

In how many ways can each object be placed in the "box" below it?

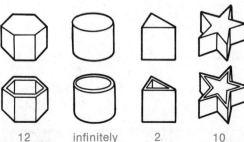

| 12 | infinitely many | 2 | 10 |

REVIEW OF SKILLS

Multiply.

1. $10(4)$ $_{40}$ 2. $(3)(8)$ $_{24}$ 3. $8(2x + 4)$ $^{6x + 32}$

4. $3(5 + n)$ $_{15 + 3n}$ 5. $(2a + 3)6$ $_{15a + 18}$ 6. $4(4h + 2)$ $^{16h + 8}$

7. $5(3n + 2)$ $_{15n + 10}$ 8. $(4s + 6)2$ $_{8s + 12}$ 9. $(8 + 2)m$ $_{10m}$

10. $x(4 + 1)$ $_{5x}$ 11. $y(2 + 3)$ $_{5y}$ 12. $(5 + 6)n$ $_{11n}$

Simplify.

13. $10 + \dfrac{5}{8}$ $_{10\frac{5}{8}}$ 14. $\dfrac{4}{5} - \dfrac{2}{5}$ $_{\frac{2}{5}}$ 15. $\dfrac{9}{10} - \dfrac{3}{10}$ $_{\frac{3}{5}}$

16. $0.77 - 0.34$ $_{0.43}$ 17. $1.07 - 0.25$ $_{0.82}$ 18. $\dfrac{3}{4} - \dfrac{3}{4}$ $_{0}$

19. $0.45 - 0.45$ $_{0}$ 20. $6 + \dfrac{2}{3}$ $_{6\frac{2}{3}}$ 21. $\dfrac{750}{25}$ $_{30}$

22. $\dfrac{120}{15}$ $_{8}$ 23. $\dfrac{42}{14}$ $_{3}$ 24. $8 \div 20$ $_{\frac{2}{5}}$ or 0.4

25. $\dfrac{1}{2} \cdot 2$ $_{1}$ 26. $4 \cdot 2\frac{1}{2}$ $_{10}$ 27. $6 \cdot \dfrac{1}{3}$ $_{2}$

28. $4 \cdot \dfrac{1}{5}$ $_{\frac{4}{5}}$ 29. $\dfrac{1}{7} \cdot 7$ $_{1}$ 30. $5 \cdot \dfrac{1}{5}$ $_{1}$

Tell what number makes each statement true.

31. $9 + 3 = 9 + \underline{\ ?\ }$ $_{3}$ 32. $16 - \underline{\ ?\ } = 16 - 10$ $_{10}$

33. $8 + 2 = \underline{\ ?\ } + 2$ $_{8}$ 34. $\dfrac{1}{2} \cdot \underline{\ ?\ } = 1$ $_{2}$

35. $\dfrac{1}{5} \cdot \underline{\ ?\ } = 1$ $_{5}$ 36. $7 - 3 = \underline{\ ?\ } - 3$ $_{7}$

37–39. Every number is a solution.

37. $(7 + 1) + \underline{\ ?\ } = (4 + 4) + \underline{\ ?\ }$ 38. $(9 + 2) + \underline{\ ?\ } = 11 + \underline{\ ?\ }$

39. $(20 - 6) + \underline{\ ?\ } = 14 + \underline{\ ?\ }$ 40. $10 + \underline{\ ?\ } = (12 - 2) + 3$ $_{3}$

41. $\dfrac{6 \cdot 3}{2} = \dfrac{18}{?}$ $_{2}$ 42. $\dfrac{2 \cdot ?}{2} = 8$ $_{8}$

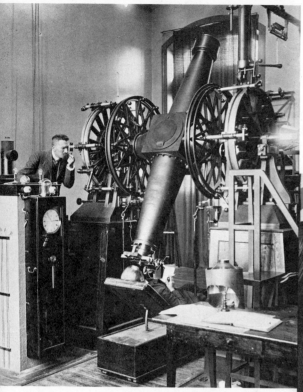

Left: Astronomers using telescope.

Right: Astronomer using solar-image display system to analyze space flight data.

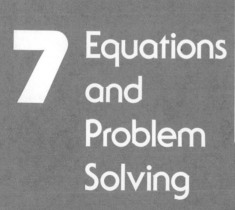

7 Equations and Problem Solving

Basic Properties

7-1 *Combining Similar Terms*

> **OBJECTIVE**
> Simplify expressions by combining similar terms.

We can simplify an expression like $4t + 3t$. We justify the idea that $4t + 3t = 7t$ by using the distributive property: $4t + 3t = (4 + 3)t = 7t$.

Terms such as $4t$ and $3t$ which contain the same variables are called similar terms or like terms. Terms which contain no variables are also called similar terms.

$$5x + 7x \qquad 8rs + 7 - 1 \qquad a^2 + 3a - a$$

similar terms   similar terms   similar terms

These examples further illustrate how we use the distributive property to combine similar terms.

EXAMPLE 1 $8mn + 6mn = (8 + 6)mn$ ◄ The Distributive Property
$\qquad\qquad\qquad\quad = 14mn$

EXAMPLE 2 $5a - a = 5a - 1a$ ◄ $a = 1 \cdot a$
$\qquad\qquad\quad = (5 - 1)a$ ◄ The Distributive Property
$\qquad\qquad\quad = 4a$

EXAMPLE 3 $5t + 9s$ ◄ The terms are unlike. The expression cannot be simplified any further.

EXAMPLE 4 $5x^2 + 4x + 2x = 5x^2 + (4 + 2)x$
$\qquad\qquad\qquad\qquad = 5x^2 + 6x$

 **EXERCISES**

Simplify.

Sample: $6m - 2m$ *What you say:* $(6 - 2)m = 4m$

1. $5a + 2a$ 7a
2. $3b + 5b$ 8b
3. $10k - 4k$ 6k
4. $2m + 3m + 4m$ 9m
5. $5r + r$ 6r
6. $10s + 10s$ 20s
7. $k + k$ 2k
8. $6xy + 4xy$ 10xy
9. $6y - y - y$ 4y

For Extra Practice, see page 419.

Combine similar terms to simplify.

Sample 1: $2b + 6b + 4c$

Solution: $8b + 4c$

Sample 2: $2(6 + 5m) - 4m$

Solution: $12 + 10m - 4m = 12 + 6m$

A

1. $5r + 6r$ $11r$
2. $2t + 7t + 4t$ $13t$
3. $16s - 9s$ $7s$
4. $3k + k + 9k$ $13k$
5. $14 + 3p + 2p$ $14 + 5p$
6. $5ab + ab + 3ab$ $9ab$
7. $cd + 5cd - 2cd$ $4cd$
8. $12rs + rs + rs$ $14rs$
9. $2(2t + 2) + 1$ $4t + 5$
10. $2(2t + 2) + t$ $5t + 4$
11. $6k + 4t + 3k + 8t$ $9k + 12t$
12. $5rs + 2b + 3b + 4rs$ $5b + 9rs$
13. $2m + 6m - 3m$ $5m$
14. $3(6 + b) + 4b$ $18 + 7b$
15. $3mn + 2(4 + 2mn)$ $7mn + 8$
16. $8 + 3(2d + 5)$ $23 + 6d$

B

17. $(5k - 2k) + 8k$ $11k$
18. $(14m - 2m) - m$ $11m$
19. $6a + 2b + 3b - 2a$ $4a + 5b$
20. $8r + 3r + 5t + 7$ $11r + 5b + 7$
21. $16s + 5 - 9s - 2$ $7s + 3$
22. $3r + 5t - r + 6r - 4t$ $8r + t$
23. $100 - 80 + 16q - 4q - 20$ $12q$
24. $r + s + s + r + 3$ $2r + 2s + 3$
25. $14ab - 3ab + 14a + ab$ $12ab + 14a$
26. $6s + 5t - s - t + 2$ $4s + 4t + 2$

Sample: $4(r + s) + 3(2r + 3s)$

Solution: $4(r + s) + 3(2r + 3s) = 4r + 4s + 6r + 9s$
$$= 10r + 13s$$

27. $5(a + b) + 6(a + b)$ $11a + 11b$
28. $3(r + 4) + 5(r + 6)$ $8r + 42$
29. $7(k + m) + 2(k + m)$ $9k + 9m$
30. $2(2k + 4) + 3(3k - 2)$ $13k + 2$
31. $8(4x + 3) + 7(2x - 3)$ $46x + 3$
32. $9(2z + 6) + 4(3z - 8)$ $30z + 22$
33. $8(5f + 4g) + 2(f - 9g)$
34. $(5n + 4)6 + 2(9n - 8)$ $48n + 8$
35. $2(3x + 2y + 4) + 3(4x + 8)$ $18x + 4y + 32$ **33.** $42f + 14g$
36. $7(5a + 5b + 2) + 4(2a - 3b)$ $43a + 23b + 14$

C

37. $4[2(3k + m) + 2k] + 3m$ $32k + 11m$
38. $7s + 5[(6r + 2s)4 + 2r]$ $47s + 130r$
39. $[(3 + 2c + 4d)5 + 2d]3$
40. $4t + 2[3(m + r) + 5m]$ $4t + 16m + 6r$
41. $2m[4(m + 3) + 5n] + [2(m + n)4]$ $8m^2 + 10mn + 32m + 8n$
42. $[3(y + x) + 4(y + x)]2x + 2(x^2 + xy)$ $16x^2 + 16xy$

39. $45 + 30c + 66d$

7-2 *Addition and Subtraction Properties of Equality*

OBJECTIVES

Use the addition property of equality to solve equations.

Use the subtraction property of equality to solve equations.

Recall that if the same number is added to both members of an equation, the truth of the equation is unchanged.

If $x - 3 = 9$, then $x - 3 + 3 = 9 + 3$, and $x = 12$.

3 is added to both members.

The Addition Property ▸ For every number r, every number s, and every
of Equality   number t, if $r = s$, then $r + t = s + t$.

EXAMPLE 1 Solve: $x - 10 = 8$

$x - 10 + 10 = 8 + 10$ ◂ Add 10 to both members.

$x = 18$ ◂ Solution

We check the work by replacing ▸
x with 18 in the given equation.

$$x - 10 = 8$$

$18 - 10$	8
8	8	✓

If we subtract the same number from both members of an equation, the truth of the equation is unchanged.

If $m + 7 = 19$, then $m + 7 - 7 = 19 - 7$, and $m = 12$.

7 is subtracted from both members.

The Subtraction Property ▸ For every number r, every number s, and every
of Equality   number t, if $r = s$, then $r - t = s - t$.

EXAMPLE 2 Solve: $a + 3 = 15$

$a + 3 - 3 = 15 - 3$ ◂ Subtract 3 from both members.

$a = 12$ ◂ Solution

The addition property and the subtraction property are part of our basic equation solving strategy. We change (or *transform*) the equation so the variable stands alone as one member. The successive changes are called transformations.

Give the simplest name.

Sample 1: $6 + 4 - 4$ *What you say:* 6

Sample 2: $m - 7 + 7$ *What you say:* m

1. $a - 3 + 3$ a
2. $b + 8 - 8$ b
3. $14 + 2 - 2$ 14
4. $k + 9 - 9$ k
5. $m + \dfrac{2}{3} - \dfrac{2}{3}$ m
6. $r - \dfrac{1}{2} + \dfrac{1}{2}$ r

Tell how to change the expression so the variable will stand alone.

Sample: $r + 4$ *What you say:* Subtract 4.

7. $r + 8$ subtract 8
8. $z - 10$ add 10
9. $t - 0.5$ add 0.5
10. $s + 4$ subtract 4
11. $a + \dfrac{1}{2}$ subtract $\dfrac{1}{2}$
12. $k + 0.24$ subtract 0.24

For Extra Practice, see page 420.

Complete. Then check the solution.

Sample:

$$t + 4 = 11$$
$$t + 4 - 4 = 11 - 4$$
$$t = \underline{\;?\;}$$

Solution:

$$t + 4 = 11$$
$$t + 4 - 4 = 11 - 4$$
$$t = 7$$

Check: $t + 4 = 11$

$$\begin{array}{c|c} 7 + 4 & 11 \\ \hline 11 & 11 \end{array} \checkmark$$

A

1.
$$m - 6 = 3$$
$$m - 6 + 6 = 3 + 6$$
$$m = \underline{\;?\;} \; 9$$

2.
$$n - 1 = 24$$
$$n - 1 + 1 = 24 + 1$$
$$n = \underline{\;?\;} \; 25$$

3.
$$p + 7 = 13$$
$$p + 7 - 7 = 13 - 7$$
$$p = \underline{\;?\;} \; 6$$

4.
$$s + 9 = 13$$
$$s + 9 - 9 = 13 - 9$$
$$s = \underline{\;?\;} \; 4$$

5.
$$n + \dfrac{2}{3} = 6$$
$$n + \dfrac{2}{3} - \dfrac{2}{3} = 6 - \dfrac{2}{3}$$
$$n = \underline{\;?\;} \; 5\tfrac{1}{3}$$

6.
$$a + 0.1 = 4$$
$$a + 0.1 - 0.1 = 4 - 0.1$$
$$a = \underline{\;?\;} \; 3.9$$

7.
$$36 = k + 10$$
$$36 - 10 = k + 10 - 10$$
$$26 \; \underline{\;?\;} = k$$

8.
$$0.62 = r + 0.38$$
$$0.62 - 0.38 = r + 0.38 - 0.38$$
$$0.24 \; \underline{\;?\;} = r$$

Solve. Begin by stating what number should be added to or subtracted from both members.

Sample: $m - 2 = 8$

Solution: Add 2 to both members.   Check: $m - 2 = 8$

$$m - 2 + 2 = 8 + 2$$
$$m = 10$$

$$\begin{array}{c|c} 10 - 2 & 8 \\ \hline 8 & 8 \end{array} \checkmark$$

9. $r + 3 = 11$ 8

10. $s - 2 = 14$ 16

11. $t + 8 = 13$ 5

12. $p - 4 = 7$ 11

13. $q + \dfrac{1}{2} = 4$ $3\frac{1}{2}$

14. $a + \dfrac{2}{3} = 1$ $\frac{1}{3}$

15. $b - \dfrac{1}{5} = \dfrac{3}{5}$ $\frac{4}{5}$

16. $c - \dfrac{1}{3} = 4$ $4\frac{1}{3}$

17. $100 + r = 167$ 67

18. $r + 0.3 = 6$ 5.7

B

19. $\dfrac{5}{6} = a - 1$ $1\frac{5}{6}$

20. $\dfrac{13}{9} = b + \dfrac{5}{9}$ $\frac{8}{9}$

21. $\dfrac{5}{4} = r - \dfrac{1}{4}$ $\frac{3}{2}$

22. $\dfrac{3}{7} = s - \dfrac{3}{7}$ $\frac{6}{7}$

23. $\dfrac{7}{8} = y + \dfrac{1}{8}$ $\frac{3}{4}$

24. $0.05 = k + 0.01$ 0.04

Solve and check.

Sample: $x - 7 = 9$

Solution: $x - 7 = 9$
$$x - 7 + 7 = 9 + 7$$
$$x = 16$$

Check: $x - 7 = 9$

$$\begin{array}{c|c} 16 - 7 & 9 \\ \hline 9 & 9 \end{array} \checkmark$$

25. $a - 2 = 37$ 39

26. $b - 19 = 61$ 80

27. $25 = c - 17$ 42

28. $92 = r - 47$ 139

29. $s + 14 = 88$ 74

30. $t + 19 = 101$ 82

31. $r - 26 = 26$ 52

32. $s - 14 = 0$ 14

33. $k + 2.7 = 3.6$ 0.9

34. $m - 1.8 = 5.9$ 7.7

35. $p + 1.08 = 1.09$ 0.01

36. $q - 1.08 = 0.071$ 1.151

C

37. $\dfrac{3}{4} + a = \dfrac{5}{4}$ $\frac{1}{2}$

38. $\dfrac{2}{3} + b = 1$ $1\frac{1}{3}$

39. $a - \dfrac{2}{5} = \dfrac{1}{5}$ $\frac{3}{5}$

40. $k - \dfrac{7}{8} = \dfrac{7}{8}$ $\frac{7}{4}$

7-3 *The Division Property of Equality*

OBJECTIVE

Use the division property of equality to solve equations.

If both members of an equation are divided by the same number, the truth of the equation is unchanged.

$$\text{If } 7x = 21, \text{ then } \frac{7x}{7} = \frac{21}{7}, \text{ and } x = 3.$$

Both members are divided by 7.

The Division Property ▶ — For every number r, every number s, and every **of Equality** number t except 0, if $r = s$, then $\frac{r}{t} = \frac{s}{t}$.

EXAMPLE 1 Solve: $9n = 12 + 6$

$$\frac{9n}{9} = \frac{12 + 6}{9} \quad \blacktriangleleft \text{ Divide both members by 9.}$$

$$n = \frac{18}{9} = 2 \quad \blacktriangleleft \text{ Solution}$$

Check:

$$\begin{array}{c|c} 9n &= 12 + 6 \\ \hline 9(2) & 12 + 6 \\ 18 & 18 \quad \checkmark \end{array}$$

Sometimes solving an equation requires the use of more than one property of equality.

EXAMPLE 2 Solve:

$$2k - 5 = 19$$
$$2k - 5 + 5 = 19 + 5 \quad \blacktriangleleft \text{ Add 5 to both members.}$$
$$2k = 24$$
$$\frac{2k}{2} = \frac{24}{2} \quad \blacktriangleleft \text{ Divide both members by 2.}$$
$$k = 12 \quad \blacktriangleleft \text{ Solution}$$

Check:

$$\begin{array}{c|c} 2k - 5 &= 19 \\ \hline 2(12) - 5 & 19 \\ 24 - 5 & 19 \\ 19 & 19 \quad \checkmark \end{array}$$

Tell how to change the equation so that the variable will stand alone. Do not solve the equation.

Sample: $8k = 40$ *What you say:* Divide both members by 8.

1. $3t = 21$ ÷ both by 3
2. $7r = 49$ ÷ both by 7
3. $36 = 6b$ ÷ both by 6
4. $29m = 290$ ÷ both by 29
5. $5p = 60$ ÷ both by 5
6. $15y = 2$ ÷ both by 15
7. $3r = 1$ ÷ both by 3
8. $0.2t = 1$ ÷ both by 0.2
9. $24 = 5n$ ÷ both by 5

For Extra Practice, see page 420.

Solve. Check your answer.

Sample:

$$3s + 2 = 14$$
$$3s + 2 - 2 = 14 - 2$$
$$3s = 12$$
$$\frac{3s}{3} = \frac{12}{3}$$
$$s = 4$$

Check:

$3s + 2 = 14$	
$3(4) + 2$	14
$12 + 2$	14
14	14	√

A

1. $8m = 56$ 7
2. $9n = 72$ 8
3. $13a = 52$ 4
4. $12b = 156$ 13
5. $8r = 92$ $11\frac{1}{2}$
6. $35s = 105$ 3
7. $92 = 14t$ t
8. $16a = 12$ $\frac{3}{4}$
9. $1 = 20b$ $\frac{1}{20}$
10. $36 = 5c$ $\frac{36}{5}$
11. $\frac{1}{2} = 4x$ $\frac{1}{8}$
12. $6m = 6.66$ 1.11
13. $5n = 0$ 0
14. $3a + 1 = 10$ 3
15. $5b - 1 = 29$ 6
16. $2x + 7 = 31$ 12
17. $7c - 4 = 52$ 8
18. $4c + 5c = 81$ 9

B

19. $18 = 4a + 2$ 4
20. $6b - 6 = 0$ 1
21. $5 + 3x = 38$ 11
22. $7z + 8 = 64$ 8
23. $14 + 18k = 14$ 0
24. $9m - 1 = 14 + 12$ 3
25. $0.06p - 0.2 = 1$ 17
26. $21 + 3q = 30$ 3
27. $0 = 4t - 22$ $\frac{11}{2}$
28. $13 = 9d + 13$ 0

C

29. $\frac{1}{2} + 4n = 6\frac{1}{2}$ $1\frac{1}{2}$
30. $0.7 + 2r = 1.8$ 0.55
31. $5s - 0.01 = 1.44$ 0.29
32. $0.2t - 6.21 = 18.95$ 125.8
33. $6.1285 = 2m + 0.0085$ 3.06
34. $0.12 + 7k = 1.07 + 0.03$ $\frac{0.98}{7}$

7-4 *The Multiplication Property of Equality*

An equation like $\dfrac{t}{5} = 3$ can be solved by multiplying both members by the same number.

$$\text{If } \frac{t}{5} = 3, \text{ then } 5 \cdot \frac{t}{5} = 5 \cdot 3, \text{ and } t = 15.$$

The Multiplication Property of Equality ▶ For every number r, every number s, and every number t, if $r = s$, then $rt = st$.

EXAMPLE 1 Solve: $\dfrac{w}{7} = 2$

$$7 \cdot \frac{w}{7} = 7 \cdot 2 \quad \blacktriangleleft \text{ Multiply both members by 7.}$$

$$w = 14 \quad \blacktriangleleft \text{ Solution}$$

Check: $\dfrac{w}{7} = 2$

$$\begin{array}{c|c} \dfrac{14}{7} & 2 \\ \hline 2 & 2 \end{array} \quad \checkmark$$

EXAMPLE 2 Solve: $\dfrac{2y}{3} = 6$

$$3\left(\frac{2y}{3}\right) = 3(6) \quad \blacktriangleleft \text{ Multiply both members by 3.}$$

$$2y = 18$$

$$\frac{2y}{2} = \frac{18}{2} \quad \blacktriangleleft \text{ Divide both members by 2.}$$

$$y = 9 \quad \blacktriangleleft \text{ Solution}$$

The transformations in Example 2 could have been done in the opposite order. We could have divided by 2 first, and then multiplied by 3. The final result would have been the same: $y = 9$.

Tell how to change the equation so that the variable will stand alone. Do not solve the equation.

Sample: $\dfrac{k}{4} = 5$ *What you say:* Multiply both members by 4.

1. $\dfrac{c}{5} = 2$ **2.** $\dfrac{d}{4} = 1$ **3.** $\dfrac{m}{2} = 0.5$

4. $\dfrac{n}{19} = 1$ **5.** $\dfrac{a}{6} = 0.214$ **6.** $\dfrac{b}{41} = 1$

Solve. Begin by telling the number by which both members are to be multiplied.

Sample: $\dfrac{f}{6} = 7$ *What you say:* Multiply both members by 6.

$$f = 42$$

7. $\dfrac{a}{2} = 10$ $a = 20$ **8.** $\dfrac{b}{3} = 6$ $b = 18$ **9.** $\dfrac{m}{2} = 26$ $m = 52$

10. $6 = \dfrac{n}{10}$ $n = 60$ **11.** $\dfrac{1}{3}y = 4$ $y = 12$ **12.** $\dfrac{1}{5}z = \dfrac{1}{2}$ $z = 2\frac{1}{2}$

For Extra Practice, see page 420.

Solve. Check your answer.

A

1. $\dfrac{m}{2} = 8$ 16   **2.** $\dfrac{n}{6} = 1.2$ 7.2   **3.** $\dfrac{a}{5} = 0.5$ 2.5

4. $\dfrac{b}{2} = 1.04$ 2.08   **5.** $\dfrac{m}{10} = 0.018$ 0.18   **6.** $\dfrac{3}{4}q = 4$ $5\frac{1}{3}$

7. $\dfrac{3}{4}k = 3$ 4   **8.** $\dfrac{5}{8}x = 0$ 0   **9.** $\dfrac{5}{8}y = 1$ $\frac{8}{5}$

10. $\dfrac{z}{2} = \dfrac{1}{2}$ 1   **11.** $\dfrac{t}{3} = 8 + 9$ 51   **12.** $\dfrac{2x}{3} = 6$ 9

13. $\dfrac{4x}{3} = 6$ $4\frac{1}{2}$ **14.** $\dfrac{2}{1}m = 9$ $\frac{9}{2}$ **15.** $\dfrac{7a}{3} = 14$ 6

B

16. $\dfrac{7}{9}a = 7$ 9   **17.** $0.2 = \dfrac{8}{7}b$ 0.175   **18.** $0 = \dfrac{3}{2}c$ 0

19. $\dfrac{13m}{4} = 0.26$ 0.08   **20.** $\dfrac{4}{3}d = 1\frac{3}{4}$ **21.** $3 = \dfrac{1}{3}w$ 9

22. $\dfrac{2}{3} = \dfrac{3}{4}x$ $\dfrac{8}{9}$ **23.** $\dfrac{1}{7}x = \dfrac{2}{3}$ $\dfrac{14}{3}$ **24.** $\dfrac{7}{8}y = \dfrac{7}{9}$ $\dfrac{8}{9}$

25. $5m \div 2 = 1.5$ $\dfrac{3}{5}$ **26.** $\dfrac{n}{0.7} = 6.3$ 4.41   **27.** $\dfrac{3z}{2} = 2\frac{1}{2}$ $1\frac{2}{3}$

28. $\dfrac{1}{5}n = 1.21$ 6.05   **29.** $\dfrac{5}{4}k = 0.4$ 0.32   **30.** $\dfrac{x+3}{5} = 4$ 17   **C**

31. $\dfrac{y-2}{4} = 1$ 6   **32.** $\dfrac{2y}{5} = 3.4$ 8.5   **33.** $\dfrac{7y}{5} = 0.14$ 0.1

34. $\dfrac{2n+1}{3} = 7$ 10   **35.** $\dfrac{3y-2}{4} = 1$ 2   **36.** $30 = 2w \div 7$ 105

37. $v \div 1\frac{1}{3} = 33$ 44   **38.** $\dfrac{2n}{3} + 1 = 7$ 9   **39.** $\dfrac{3y}{4} - 2 = 1$ 4

SELF-TEST 1

Be sure that you understand these terms.

similar terms (p. 184)
like terms (p. 184)
unlike terms (p. 184)
addition property of equality (p. 186)
subtraction property of equality (p. 186)
division property of equality (p. 189)
multiplication property of equality (p. 191)

Combine similar terms to simplify.   Section 7-1, p. 184

1. $3a + 4a$ 7a   **2.** $2r + 2s - r$ r + 2s

3. $5(t + 1) - t$ 4t + 5   **4.** $3mn + 4m + mn$ 4mn + 4m

Solve.

5. $t - 7 = 18$ t = 25   **6.** $1.7 + s = 10$ s = 8.3   Section 7-2, p. 186

7. $6x = 48$ x = 8   **8.** $0.4z = 0.36$ z = 0.9   Section 7-3, p. 189

9. $\dfrac{1}{3}a = 5$ a = 15   **10.** $\dfrac{b}{2} = 0.8$ b = 1.6   Section 7-4, p. 191

Check your answers with those printed at the back of the book.

Transformation Strategies

7-5 *More about Equation Solving Strategy*

OBJECTIVE
Solve equations by combining terms and making transformations.

We are now ready to solve more complicated equations, such as $4x + 2x - 3 = 10 + 5$. We need to use a combination of transformations, but it's easy if we proceed logically, step by step.

Step 1 Where possible, combine like terms.
Step 2 Transform the equation so that the term that contains the variable stands alone as one member.
Step 3 Transform the equation so that the coefficient of the variable is 1.

EXAMPLE 1 Solve:
$$4x + 2x - 3 = 10 + 5$$
$$6x - 3 = 15 \quad \blacktriangleleft \text{ Combine terms.}$$
$$6x - 3 + 3 = 15 + 3 \quad \blacktriangleleft \text{ Add 3 to both members.}$$
$$6x = 18$$
$$\frac{6x}{6} = \frac{18}{6} \quad \blacktriangleleft \text{ Divide both members by 6.}$$
$$x = 3 \quad \blacktriangleleft \text{ Solution}$$

Check:

$4x + 2x - 3$	$10 + 5$
$4(3) + 2(3) - 3$	$10 + 5$
$12 + 6 - 3$	15
15	$15 \quad \checkmark$

EXAMPLE 2 Solve:
$$m - \frac{3}{4}m + 3 = 10 - 6$$
$$\frac{1}{4}m + 3 = 4 \quad \blacktriangleleft \text{ Combine terms.}$$
$$\frac{1}{4}m + 3 - 3 = 4 - 3 \quad \blacktriangleleft \text{ Subtract 3 from both members.}$$
$$\frac{1}{4}m = 1$$
$$4\left(\frac{1}{4}m\right) = 4(1) \quad \blacktriangleleft \text{ Multiply both members by 4.}$$
$$m = 4 \quad \blacktriangleleft \text{ Solution}$$

Check: $m - \dfrac{3}{4}m + 3 = 10 - 6$

$4 - \dfrac{3}{4}(4) + 3$	$10 - 6$
$4 - 3 + 3$	4
4	4 ✓

Simplify by combining similar terms. Do not solve the equation.

Oral EXERCISES

Sample: $4z + 5z + 1 = 22 - 8$ *What you say:* $9z + 1 = 14$

1. $4a + 5a = 16$ $9a = 16$ **2.** $4b - 6 = 10 - 8$ $4b - 6 = 2$

3. $6c + 4c + 1 = 18$ $10c + 1 = 18$ **4.** $5m + m = 13 + 2$ $6m = 15$

5. $9n - 5n + 8 = 16$ $4n + 8 = 16$ **6.** $2p + \dfrac{1}{2}p = 6 + \dfrac{1}{2}$ $2\frac{1}{2}p = 6\frac{1}{2}$

7. $2q - \dfrac{1}{2}q = 9 - \dfrac{1}{2}$ $1\frac{1}{2}q = 8\frac{1}{2}$ **8.** $m - \dfrac{1}{2}m = 14 + 1$

$\frac{1}{2}m = 15$

9. $n + \dfrac{1}{3}n = 6 + 1\frac{2}{3}$ $1\frac{1}{3}n = 7\frac{2}{3}$ **10.** $b + b - \dfrac{1}{2}b = (2)(3)(0)$

$1\frac{1}{2}b = 0$

For Extra Practice, see page 421.

Solve.

Written EXERCISES

Sample: $4k - k + 2 = 16 - 8$

Solution: $4k - k + 2 = 16 - 8$ Check: $4k - k + 2 = 16 - 8$

$3k + 2 = 8$

$3k + 2 - 2 = 8 - 2$

$4(2) - 2 + 2$	$16 - 8$
$8 - 2 + 2$	8
8	8 ✓

$3k = 6$

$\dfrac{3k}{3} = \dfrac{6}{3}$

$k = 2$

1. $5p - 2p = 21$ 7 **2.** $6q + 2q = 32$ 4 **A**

3. $4c - c = 36$ 12 **4.** $9d + d = 80$ 8

5. $2a + 3a = 13 + 12$ 5 **6.** $8b - 4b = 7 + 21$ 7

7. $8f + f = 20 - 2$ 2 **8.** $9g - g = 35 + 13$ 6

9. $5z + z - 2 = 28$ 5 **10.** $16v - v + 10 = 85 - 15$ 4

11. $1.2t + 0.3t = 6 - 1.5$ 3 **12.** $6r + 3r - 9 = 0$ 1

13. $1\frac{1}{3}s + 2\frac{1}{2}s = 17 + 7$ 6

14. $4\frac{1}{3}m - 1\frac{1}{3}m + 9 = 21$ 4

15. $0.6n + 5.4n - 1 = 35$ 6

16. $1\frac{1}{3}k + \frac{2}{3}k - 1 = 33$ 17

B

17. $\frac{2}{3}m + 1\frac{1}{3}m = 2\frac{1}{2} + 9\frac{1}{2}$ 6

18. $\frac{1}{2}n + 2\frac{1}{2}n = 18\frac{1}{3} - 2\frac{1}{3}$ $5\frac{1}{3}$

19. $\frac{2r}{3} + \frac{2r}{3} = 1\frac{1}{4} + 2\frac{3}{4}$ 3

20. $\frac{1}{2}s + \frac{3}{4}s - 4 = 1$ 4

21. $\frac{x}{5} + 8 + \frac{4x}{5} = 10$ 2

22. $3(a + 5) + a = 16$ $\frac{1}{4}$

23. $5(2b - 6) - b = 0$ $3\frac{1}{3}$

24. $99 - 62 = 5(2c + 4)$ $1\frac{7}{10}$

25. $\frac{3}{5}t - \frac{3}{4} = 3$ $6\frac{1}{4}$

26. $\frac{x}{3} - 4 + \frac{5}{3} = 0$ 7

27. $15y + 4\frac{1}{3} - 8y = 11\frac{1}{3}$ 1

28. $\frac{8}{5}z + \frac{3}{5}z - 9\frac{1}{3} = 23\frac{2}{3}$ 15

29. $\frac{3}{8}r - 9 + \frac{1}{8} = 0$ $23\frac{2}{3}$

30. $\frac{1}{5}s + \frac{1}{2}s + \frac{2}{4}s = \frac{2}{5}$ $\frac{1}{3}$

31. $6 + 0.2t - 0.1t = 10$ 40

32. $6.2v - 12.8 + 0.3v = 0.2$ 2

C

33. $9.2 + 9m - 4m = 36.7$ 5.5

34. $10n - 4 - n - 5n = 8.4$ 3.1

35. $5r + 6.2 - 0.4r = 10.8$ 1

36. $47 + 18 = 5t - 15 - 3t$ 40

Problems

Write and solve an equation to answer the question.

Sample: The sum of four times a number and 21 is 101. Find the number.

Solution: Let n stand for the number.

$$4n + 21 = 101$$
$$4n + 21 - 21 = 101 - 21$$
$$4n = 80$$
$$\frac{4n}{4} = \frac{80}{4}$$
$$n = 20 \quad \text{The number is 20.}$$

1. The sum of three times a number and 7 is 19. What is the number? $3n + 7 = 19; n = 4$

2. If six times a number is decreased by 5, the result is equal to 61. Find the number. $6n - 5 = 61; n = 11$

3. If four times a number is decreased by 101, the result is equal to 99. Find the number. $4n - 101 = 99; n = 50$

4. If 33 is added to seven times a number, the result is equal to 68. Find the number. $7n + 33 = 68; n = 5$

5. The sum of a whole number and the next greater whole number is 51. Find the numbers. (*Hint:* call the numbers n and $n + 1$.) $n + n + 1 = 51$ $n = 25;$ $n + 1 = 26$

6. When half a number is increased by 9, the result is 32. Find the number. $\frac{1}{2}n + 9 = 33; n = 46$

7. Maria is 5 centimeters taller than Rollie. The sum of their heights is 305 centimeters. How tall is each person? $x + x + 5 = 300$ Maria, 155 cm; Rollie, 150 cm

8. The length of a box lid is three times its width. The perimeter of the box lid is 200 centimeters. How long and how wide is the box lid? (*Hint: P = 2l + 2w*) $200 = 2(3x) + 2(x);$ 25 cm wide 75 cm long

9. The sum of an even number and the next greater even number is 86. Find the numbers. (*Hint:* if n is an even number, the next greater even number is $n + 2$.) $n + n + 2 = 86;$ $n = 42;$ $n + 2 = 44$

10. The lengths of the sides of a triangle are shown in the illustration. The perimeter is 108 meters. Find the length of each side.

    3k 4k 5k

    $3k + 4k + 5k = 108; k = 9$
    The sides are 27 m, 36 m, 45 m.

consumer notes

Listing Ingredients

Find a canned product which is a mixture of separate ingredients, such as fruit cocktail or mixed nuts. On the label of the can, read the order in which the ingredients are listed. They should be listed in order by weight, with the ingredient making up the greatest percentage of weight listed first, and so on.

Open the can and sort its contents into separate containers, all of equal weight when empty. Weigh the containers of ingredients. Which weighs the most? Is this ingredient listed first on the label? Are all the other ingredients listed in the correct order?

7-6 *Equations and Problem Solving*

Equations can be useful in solving many kinds of problems. The basic equation-solving strategies are helpful.

EXAMPLE   Two cement trucks contain a total of 8.1 cubic meters (m^3) of cement. The large truck holds twice as much as the small truck. How much cement is in each truck?

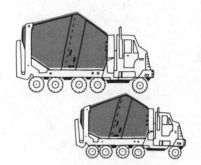

Use a variable to stand for missing information.

▼

Let amount of cement in small truck = t.
Then, amount of cement in large truck = $2t$.
So, total amount of cement = $t + 2t$.

$$\underbrace{\text{cement in small truck}} + \underbrace{\text{cement in large truck}} = \underbrace{\text{total}}$$

$$t \;+\; 2t \;=\; 8.1 \quad \blacktriangleleft \text{Equation}$$
$$3t \;=\; 8.1 \quad \blacktriangleleft \text{Divide each}$$
$$\text{member by 3.}$$
$$t \;=\; 2.7$$

The small truck contains t cubic meters, or $2.7\,m^3$ of cement.
The large truck contains $2t$ cubic meters, or $5.4\,m^3$ of cement.

Let's check: $t + 2t = 2.7\,m^3 + 5.4\,m^3$
$$= 8.1\,m^3 \quad \checkmark$$

Use the information given to answer the question.

Sample: Swimmer swims m meters Friday; $2m$ meters Saturday;
900 meters total. How far on Friday? On Saturday?

Hint: $\underbrace{\text{Friday meters}}$ + $\underbrace{\text{Saturday meters}}$ = $\underbrace{900}$
$\quad\quad\quad\quad m \quad\quad + \quad\quad 2m \quad\quad = 900$

What you say: $3m = 900$; $m = 300$

300 meters on Friday and 600 meters on Saturday.

The equations may vary. One equation is given for each exercise.

1. Basketball team record: lost x games; won $3x$ games; played 24
   games. How many lost? won? $x = 6$; lost 6 games,
   Hint: $\underbrace{\text{games lost}}$ + $\underbrace{\text{games won}}$ = $\underbrace{24}$ won 18 games
   $\quad\quad\quad\quad x \quad\quad + \quad\quad 3x \quad\quad = 24$

2. Sports field: width, w meters; length, $2w$ meters; perimeter,
   450 meters. What is the width? length? $w = 75$; 75 m wide,
   Hint: $\underbrace{2 \cdot \text{width}}$ + $\underbrace{2 \cdot \text{length}}$ = $\underbrace{\text{perimeter}}$ 150 m long
   $\quad\quad\quad 2 \cdot w \quad + \quad 2 \cdot 2w \quad = 450$

3. Two consecutive odd numbers: first number is n; second number is $n + 2$; sum is 36. What are the numbers?
   Hint: $\underbrace{\text{first number}}$ + $\underbrace{\text{second number}}$ = $\underbrace{36}$ $n = 17, 19$
   $\quad\quad\quad\quad n \quad\quad + \quad\quad n + 2 \quad\quad = 36$

4. Three consecutive numbers: first number is m; second number
   is $m + 1$; third number is $m + 2$; sum is 72. What are the
   numbers? $m = 23$; 23, 24, 25
   Hint: $\underbrace{\text{first number}}$ + $\underbrace{\text{second number}}$ + $\underbrace{\text{third number}}$ = $\underbrace{72}$
   $\quad\quad\quad\quad m \quad\quad + \quad\quad m + 1 \quad\quad + \quad\quad m + 2 \quad\quad = 72$

5. Two consecutive multiples of 5: first multiple is z; second multiple is $z + 5$; their sum is 85. What are the multiples?
   Hint: $\underbrace{\text{first multiple}}$ + $\underbrace{\text{second multiple}}$ = $\underbrace{\text{total}}$ $z = 40$;
   $\quad\quad\quad\quad z \quad\quad + \quad\quad z + 5 \quad\quad = \quad\quad 85$ 40, 45

For Extra Practice, see page 421.

Write an equation. Then solve it to answer the question.
$x + 6 = 21$; $x = 15$

1. Alan earned $6 more on his paper route than Freddie. If
   Alan earned $21, how much did Freddie earn?

2. When a swimming pool heater was turned off, the water
   cooled 3.5° Celsius in one hour. If the cooled water was 22°
   Celsius, what was the original water temperature?
   $x - 3.5 = 22$; $x = 25.5$

3. The White Sox beat the Red Sox by 6 runs. If the Red Sox scored 2 runs, how many runs did the White Sox score? $x - 2 = 6$; 8 runs

4. LuAnn and Karen have part-time jobs. LuAnn earns 15¢ per hour more than Karen. If Karen earns $1.85 per hour, how much does LuAnn earn? $x - 0.15 = 1.85$; $x = 2.00$

5. Brand X costs 19¢ more than Brand Y. If Brand X costs 98¢, what is the cost of Brand Y? $x + 0.19 \stackrel{?}{=} 0.98$; $x = 0.79$

6. Betty has 25 new customers on her paper route this month. If she now has 158 customers, how many did she have last month? $x + 25 = 158$; $x = 133$

B 7. The class had 11 pieces of chicken remaining at the end of the party. If this was one-fifth of the original number of pieces of chicken, how many pieces were eaten? $\frac{1}{5}x = 11$; $x = 55$

8. Listo has 4 more record albums than Nathan. Together they have 20 albums. How many does Listo have? $x + x + 4 = 20$; $x = 8$ Listo has 12

9. The sum of three consecutive numbers is 156. What are the three numbers? $x + x + 1 + x + 2 = 156$; $x = 51$; The numbers are 51, 52, and 53

10. The sum of a number and 18.3 is 99. What is the number?

11. The sum of two consecutive even numbers is 74. What are the two numbers? $x + 1 + x + 3 = 74$; $x = 35$; The numbers are 36 and 38

12. The sum of two consecutive odd numbers is 40. What are the two numbers? $x + x + 2 = 40$; $x = 19$; The numbers are 19 and 21.

C 13. The length of a rectangle is 6 meters more than the width. If half the perimeter is 48 meters, what are the length and width? $2 \cdot 48 = 2(x + 6) + 2(x)$; $x = 21$ 21 m wide and 27 m long

14. The width of a piece of fabric is 25 centimeters less than the length. If the perimeter is 190 centimeters, what are the length and width? $190 = 2(x) + 2(x - 25)$; $60 = x$ 35 m wide and 60 m long

15. Sam is 7 years older than his brother. If the sum of their ages is 11 years, how old is each? $x + x + 7 = 11$; $x = 2$ Sam is 9; his brother is 2

16. Louella is three times as old as her sister. If the difference between their ages is 14 years, how old is each? $3x - x = 14$; $x = 7$ Louella: 21; sister: 7

17. Priscilla rode her bike half as far this week as last week. If she rode 63 kilometers altogether, how far did she ride last week? $x + \frac{1}{2}x = 63$; $x = 42$

18. The average of 18 and a number is 24. What is the number?

19. The average of two consecutive even numbers is 27. What are the two numbers?

10. $x + 18.3 = 99$; $x = 80.7$

18. $\frac{x + 18}{2} = 24$; $x = 30$ 19. $\frac{n + 1 + n + 3}{2} = 27$; 26 and 28

7-7 *Equations with the Variable in Both Members*

OBJECTIVE

Solve equations in which the variable appears in both members.

In the equation $4n = 15 - n$, the variable, n, appears in both members. The properties of equality can be used to make transformations leading to the solution.

EXAMPLE 1 Solve: $4n = 15 - n$

$$4n = 15 - n$$
$$4n + n = 15 - n + n \quad \blacktriangleleft \text{ Add } n \text{ to both members.}$$
$$5n = 15$$
$$\frac{5n}{5} = \frac{15}{5} \qquad \blacktriangleleft \text{ Divide both members by 5.}$$
$$n = 3 \qquad \blacktriangleleft \text{ Solution}$$

Check:

$4n$	$= 15 - n$
$4(3)$	$15 - 3$
12	$12 \;\checkmark$

EXAMPLE 2 Solve: $2x + 3x + 5 = 4 + 9 + x$

$$2x + 3x + 5 = 4 + 9 + x$$
$$5x + 5 = 13 + x \quad \blacktriangleleft \text{ Combine terms.}$$
$$5x + 5 - x = 13 + x - x \quad \blacktriangleleft \text{ Subtract } x \text{ from both members.}$$
$$4x + 5 = 13$$
$$4x + 5 - 5 = 13 - 5 \qquad \blacktriangleleft \text{ Subtract 5 from both members.}$$
$$4x = 8$$
$$\frac{4x}{4} = \frac{8}{4} \qquad \blacktriangleleft \text{ Divide both members by 4.}$$
$$x = 2 \qquad \blacktriangleleft \text{ Solution}$$

Check:

$2x + 3x + 5$	$= 4 + 9 + x$
$2(2) + 3(2) + 5$	$4 + 9 + 2$
$4 + 6 + 5$	15
15	$15 \;\checkmark$

For Extra Practice, see page 421.

Written
EXERCISES

Solve. Check your answer.

Sample:

$$6k = 28 + 2k$$
$$6k - 2k = 28 + 2k - 2k$$
$$4k = 28$$
$$\frac{4k}{4} = \frac{28}{4}$$
$$k = 7$$

Check:

$$6k = 28 + 2k$$

$6(7)$	$28 + 2(7)$
42	$28 + 14$
42	42 ✓

A

1. $3a = 6 + a$ 3
3. $7c = 42 + c$ 7
5. $13n = 19 + 3n$ 1.9
7. $15s = 9s + 54$ 9
9. $14v + 72 = 17v$ 24
11. $13w = 8 - 3w$ $\frac{1}{2}$
13. $6k - 9 = 3k$ 3
15. $7g - 21 = 4g$ 7
17. $5r = 43 - 5r$ $4\frac{3}{10}$
19. $b + 6 = 10 - b$ 2
21. $4m - 9 = m + 6$ 5

2. $4b = 35 - b$ 7
4. $8m = 49 + m$ 7
6. $8r = 16 - 2r$ $\frac{8}{5}$
8. $6t + 10 = 7t$ 10
10. $3w + 6 = 12w$ $\frac{2}{3}$
12. $11z + 8 = 15z$ 2
14. $f = 2f - 7$ 7
16. $t = 44 - t$ 22
18. $2 + 3m = 4 + 2m$ 2
20. $16 - c = 13 + c$ $\frac{3}{2}$
22. $7n + 5 = 9n - 13$ 9

B

23. $a + 10 = 14a - 16$ 2
25. $15 - 2m = 16 - 5m$ $\frac{1}{3}$
27. $0.25 + c = 3c - 0.75$ $\frac{1}{2}$

24. $14g - g = 10g + 20$ $\frac{20}{3}$
26. $\frac{1}{2}b + 8 = 1\frac{1}{2}b - 2$ 10
28. $16m = 10m$ 0

C

29. $9m - 8 = 6m - 10 + 4m$ 2
31. $c(8 + 1) + c = 8 + 2c$ 1
33. $\frac{1}{4}(p + 6) = \frac{3}{4}p - 4$ 11

30. $3(n - 3) = 4n - 13$ 4
32. $5(2n - 4) = 8 + 7n$ $9\frac{1}{3}$
34. $5q - 4 + 5q = 7q + 18$ $\frac{22}{3}$

Problems

1. If a number is increased by 15, the result is the same as two times the number. Find the number. 15

2. If a number is increased by 36, the result is seven times the number. What is the number? 6

3. If four times a number is increased by 35, the result is nine times the number. Find the number. 7

4. If 41 is decreased by three times a number, the result is the same as the number increased by five. Find the number. 9

5. The length of a rectangle is one meter less than twice the width. The perimeter is 46 meters. Find the length and width. 8 m wide; 15 m long

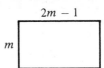
$2m - 1$

m

6. Four times a number, decreased by ten, is the same as three times the same number increased by three. What is the number? 13

7. Tony has one more nickel than he has pennies. He has 15 coins. How many nickels does he have? 7 pennies; 8 nickels

8. Jolene has two more nickels than pennies, and five more dimes than nickels. She has 21 coins. How many of each coin does she have? 4 pennies, 6 nickels, 11 dimes

SELF-TEST 2

Solve.

1. $5c + 6c = 22$ 2

2. $4r - 4 = 2.4$ 1.6

3. $2m = 6 + 10$ 8

4. $3s + 8 + s = 40$ 8

Section 7-5, p. 194

Write and solve an equation to answer the question.

5. The sum of a number and 9 is 25. Find the number. 16

6. When twice a number is decreased by 1, the result is 21. Find the number. 11

7. The sum of two consecutive odd numbers is 76. What are the numbers? 37 and 39

Section 7-6, p. 198

8. Rover and Grover are beagles. Rover is 3 years older than Grover. The sum of their ages is 13. How old is Rover? 8

Solve.

Section 7-7, p. 201

9. $4m + 2 = m + 11$ 3

10. $3n - 6 = 2n + 2$ 8

11. Three times a number, increased by 4, is 8 less than four times the number. Find the number. 12

Check your answers with those printed at the back of the book.

chapter summary

1. Expressions may be simplified by combining similar terms.

$$3x + 12x = 15x$$

similar terms

2. The addition property of equality: If $r = s$, then $r + t = s + t$.

3. The subtraction property of equality: If $r = s$, then $r - t = s - t$.

4. The division property of equality: If $r = s$, then $\dfrac{r}{t} = \dfrac{s}{t}$. $(t \neq 0)$

5. The multiplication property of equality: If $r = s$, then $rt = st$.

6. The process of changing the form of an equation by using a property of equality is called **transforming**. The successive changes are called **transformations**.

chapter test

Solve. Check your answer.

1. $n + 13 = 20$ 7

2. $m - 2 = 19$ 21

3. $6p = 60$ 10

4. $\dfrac{1}{3}y = 4$ 12

5. $4a - 2 = 10$ 3

6. $b + \dfrac{1}{2} = 20$ $19\frac{1}{2}$

7. $100 = 10x + 30$ 7

8. $30 - \dfrac{1}{2}x = 20$ 20

9. $16k - 1 = 0$ $\frac{1}{16}$

10. $\dfrac{c}{5} = \dfrac{1}{10}$ $\frac{1}{2}$

11. $\dfrac{3}{4}w = 45$ 60

12. $42 - v = 5v$ 7

13. $26 - 3m = 2m - 14$ 8

14. $\dfrac{n}{4} + 5 = 7$ 8

Write an equation. Explain what number the variable represents. Do not solve the equation.

15. The sum of a number and 13 is 24. Find the number. 11

16. If 12 is subtracted from three times a number, the result is equal to the number increased by six. Find the number. 9

17. The sum of an even number and the next greater even number is 46. Find the number. 22 and 24

18. Each of two sides of a triangle is two-thirds as long as the third side. The perimeter is 168 centimeters. Find the length of the third side. 72 cm

2. Nine trips:
2 children over; 2 children over;
2 children over; 1 child back;
1 child back; 1 adult over;
1 adult over; 1 child back;
1 child back; 2 children over.

challenge topics

Puzzles

1. Three discs are arranged on peg *A*. The object of the puzzle is to move the discs one at a time so they end up in the same arrangement on peg *B*. You may use all three pegs but a larger disc may *not* be placed on a smaller disc. What is the least number of moves required? 7

2. A group of campers consists of two children and two adults. They want to cross a river but their canoe will hold only one adult or two children at a time. How can they cross the river? (Hint: two children cross first)

3. A row of ten coins is arranged as illustrated. In this puzzle you are to move any coin over the *two* next to it and onto the coin beyond to make a stack of two coins. You are to finish with five equally spaced stacks of two coins each. (Hint: move only those coins in odd numbered positions or those in even numbered positions.)

cumulative review

Simplify.

1. $\dfrac{15 + 5}{4}$ 5

2. $3^2 - 3$ 6

3. $16 - (4 \times 4)$ 0

True or false?

4. $\dfrac{30}{5} = 2 \times 3$ true

5. $58 = 13 + 35$ false

6. $5^2 - 5 = 4 \times 5$ true

Name the value. Let $b = 2$, $l = 6$, $u = 10$.

7. $l + u + b$ 18

8. $\dfrac{b + u}{l}$ 2

9. $b^2 + u^2$ 104

Write the equation in symbols.

10. The quotient of 42 and 7 is equal to some number. $42 \div 7 = n$

Solve.

11. $20 = 3 + x$ 17

12. $27 - 5 = y$ 22

13. $13 \times z = 39$ 3

14. $\dfrac{22}{c} = 2$ 11

15. $28 = 4 \times d$ 7

16. $n^3 = 8$ 2

Name the coordinate.

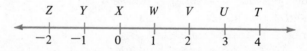

17. U 3

18. W 1

19. Y ⁻1

20. Z ⁻2

Name the opposite.

21. ⁻17 ⁺17

22. 2 ⁻2

23. 4 ⁻4

24. $\dfrac{^-3}{8}$ $\dfrac{^+3}{8}$

Complete the pattern.

25. 0, __?__, __?__, 3, __?__, 5 1, 2, 4

26. ⁻20, ⁻19, __?__, ⁻17, __?__, __?__ ⁻18, ⁻16, ⁻15

Graph. Use the number line. Check students' graphs.

27. The integers less than 0.

28. The numbers between ⁻3 and 2.

Complete. Use $<$ or $>$.

29. $4 \underline{\ ?\ } \ ^-3 \ >$

30. $^-4 \underline{\ ?\ } \ 0 \ <$

Draw axes and graph. Check students' graphs.

31. $(2, 4)$, $(1, \ ^-2)$, $(0, 3)$, $(^-1, 2)$, and $(^-2, \ ^-4)$

Name the LCM and GCF.

32. 2 and 3 $_{6;\ 1}$

33. 3 and 12 $_{12;\ 3}$

Write each number in the form $2 \times n$ or $(2 \times n) + 1$.

34. 14 $_{2\ \times\ 7}$

35. 21 $_{(2\ \times\ 10)\ +\ 1}$

36. 31 $_{(2\ \times\ 15)\ +\ 1}$

Complete.

37. $\dfrac{3}{4} = \dfrac{?}{16}$ $_{12}$

38. $\dfrac{2}{7} = \dfrac{?}{28}$ $_{8}$

Write as a fraction in lowest terms.

39. 5 out of 20 $\frac{1}{4}$

40. 7 out of 21 $\frac{1}{3}$

Write as a decimal and as a percent.

41. $\dfrac{3}{20}$ 0.15; 15%

42. $\dfrac{2}{50}$ 0.04; 4%

43. $\dfrac{99}{100}$ 0.99; 99%

Complete.

44. 10 cm $\leftrightarrow \underline{\ ?\ }$ m $_{0.1\ \text{or}\ \frac{1}{10}}$

45. 10 m $\leftrightarrow \underline{\ ?\ }$ cm 1000

Add, subtract, multiply, or divide.

46. $\dfrac{3}{5} + \dfrac{1}{5}$ $\frac{4}{5}$

47. $\dfrac{7}{22} - \dfrac{4}{22}$ $\frac{3}{22}$

48. $6\frac{2}{3} - 2\frac{1}{3}$ $4\frac{1}{3}$

49. $\dfrac{4}{7} \div \dfrac{2}{3}$ $\frac{6}{7}$

50. $1\frac{1}{3} \times \dfrac{3}{4}$ 1

51. $\dfrac{6}{11} \div \dfrac{2}{22}$ 6

Find the value when $z = 2$, $y = 3$, and $w = 1$.

52. $\dfrac{z \cdot y}{w}$ 6

53. $\dfrac{2}{y} + \dfrac{w}{y}$ 1

54. $\dfrac{y}{w} + z$ 5

Solve.

55. $q = 2.3 + 5.4$ 7.7

56. $r = 11.1 \times 2.34$ 25.974

57. $s = 54.2 \div 27.1$ 2

58. $t = 15\% \times 50$ 7.5

Simplify.

59. $(4 + 3)2$ 14

60. $(2 + 5)(4 + 1)$ 35

Find the values of the expression. The replacement set for m is $\{1, 2, 5\}$.

61. $m + 13$ 14; 15; 18

62. $3m - 3$ 0; 3; 12

Complete the factorization of the product.

63. Product: $14\,lm$; $14(\underline{\ ?\ })$, $7(\underline{\ ?\ })$, $2l(\underline{\ ?\ })$ *lm*; 2*lm*; 7*m*

Complete the set of ordered pairs by using the given rule.

64. $2(t + 3)$ $\{(1, 8), (2, \underset{10}{\underline{\ ?\ }}), (3, 12), (4, \underset{14}{\underline{\ ?\ }}), (5, \underset{16}{\underline{\ ?\ }})\}$

Write a number sentence.

65. The product of 2.7 and some number is 7.1. $(2.7)(n) = 7.1$

Find the solution set. The replacement set is given.

66. $3n + 1 = 7$; $\{0, 2, 4, 6\}$ {2}

67. $6.07 + t < 10.07$; {the numbers of arithmetic} {the numbers of arithmetic less than 4}

To solve, perform the operation as indicated.

68. $6t = 24$; divide by 6 4

69. $13 + q = 27$; subtract 13 14

70. $\dfrac{m}{3} = 21$; multiply by 3 63

71. $n - 5 = 21$; add 5 26

Write an equation. Then solve it to answer the question.

72. The sum of a number and 23 is 57. What is the number? $n + 23 = 57$; $n = 34$

73. A rectangle has width 2 cm and length 6 cm. What is the area? $A = 6(2) = 12$; the area is 12 cm²

Solve and graph the solution set. The replacement set is {the numbers of arithmetic}. Check students' graphs.

74. $g + 2 > 3$ **75.** $7 - h \leq 4$

77. assoc. prop. of add.   **78.** reflex. prop.   **79.** common. prop. of add.
80. symmetric prop.   **81.** trans. prop.   **82.** closure prop.

Write an inequality. Then write the solution set if the replacement set is {the numbers of arithmetic}.

76. The product of 5 and some number is greater than 20. What can the number be? $5x > 20$; $x > 4$; {the numbers of arithmetic greater than 4}

83. closure prop.   **84.** sub. prin.   **85.** comm. prop. of mult.
86. assoc. prop. of mult.

Tell which property is illustrated.

77. $(3 + 2) + 5 = 3 + (2 + 5)$ **78.** $52 = 52$

79. $3 + 2 = 2 + 3$ **80.** If $ab = c$, then $c = ab$.

81. If $4 + 2 = 6$ and $6 = 3 \cdot 2$, then $4 + 2 = 3 \cdot 2$.

82. Since 2 and 5 are whole numbers, $2 + 5$ is a whole number.

83. Since 2 and 4 are whole numbers, $2 \cdot 4$ is a whole number.

84. $2 + 4 + 5 = 6 + 5$ **85.** $2 \cdot 4 = 4 \cdot 2$

86. $a(bc) = (ab)c$ **87.** $2(3 + 5) = 2(3) + 2(5)$

88. $3(5 - 1) = 3(5) - 3(1)$ **89.** $154 + 0 = 154$

90. $x^2y \cdot 0 = 0$ **91.** $b \cdot \dfrac{5}{5} = b$

87. distrib. prop.   **88.** distrib. prop.   **89.** add. prop. of 0
90. mult. prop. of 0   **91.** mult. prop. of 1

Write the set of ordered pairs that represents the function. Draw axes and graph the ordered pairs.

92. Rule: $a + 1$
Replacement set: {0, 2, 4}
{(0, 1), (2, 3), (4, 5)}

93. Rule: $2b$
Replacement set: {1, 2, 3}
{(1, 2), (2, 4), (3, 6)}

Solve. Check your answer.

94. $h - 3 = 7$ 10   **95.** $3 + f = 41$ 38

96. $14t = 56$ 4   **97.** $7k + 2 = 9$ 1

98. $\dfrac{4l}{3} = 12$ 9   **99.** $\dfrac{7m}{2} - 2 = 12$ 4

100. $3n + 1 = 4n$ 1   **101.** $2p - 14 = 4$ 9

Tell what letter on the number line below names the point which represents the number described.

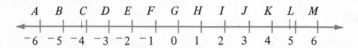

1. One kilometer above sea level H   2. One kilometer below sea level F
3. Four degrees above zero K   4. Five degrees below zero B
5. Six points scored M   6. A loss of three dollars D

Answer the question. Assume that 2 points won is written 2 and 3 points lost is written ⁻3.

7. Mario's score was 8, and then he lost 3 points. What was his new score? 5
8. When Mario's score was ⁻1, he won 3 points. What was his new score? 2
9. If Mario's score was 2, and then he lost 4 points, what was his score then? ⁻2

Answer the question. Assume that 2 kilometers east is written 2 and 3 kilometers west is written ⁻3.

10. Susan walked 3 kilometers west, and then 4 kilometers east. Where was she in relation to her starting point? 1 km east
11. The next day Susan walked 5 kilometers east, then 8 kilometers west, then 4 kilometers east. Where was she in relation to her starting position? 1 km east

Answer the question. Assume that a rise of 3 floors is written 3 and a movement down of 2 floors is written ⁻2.

12. The elevator rose 3 floors (3) and then moved down 4 floors (⁻4). What was the result of the two moves? down 1 floor or ⁻1
13. The elevator moved down 2 floors and then down 3 more floors. What was the result of the two moves? down 5 floors or ⁻5

Left: Physician performing laboratory tests.

Right: Medical technician operat-
ing computer-assisted blood ana-
lyzer.

8 Working with Directed Numbers

8-1 *Directed Numbers and the Number Line*

OBJECTIVES

Name the magnitudes of directed numbers.

Use directed numbers in practical situations.

Recall that when we refer to the position of a number on the number line, we mean the position of its graph. Then we can say the following about a horizontal number line:

Numbers to the left of 0 are **negative.**

Numbers to the right of 0 are **positive.**

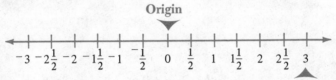

We write 3 instead of ⁺3.
⁺3 is the same as 3.

On a vertical number line, numbers above 0 are **positive.** Numbers below 0 are **negative.**

Pairs of numbers like 3 and ⁻3 are opposites. Note that the point 3 is just as far from 0 as is the point ⁻3. To locate 3 on the horizontal number line, we move 3 units to the right of 0. To locate ⁻3, we move 3 units to the left of 0. The distance between 0 and any number is called the magnitude of the number.

EXAMPLE 1

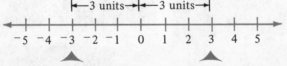

The magnitude of ⁻3 is 3.   The magnitude of 3 is 3.

Since direction is important with respect to positive and negative numbers, we often use the term directed numbers. Although 0 is neither positive nor negative, it is called a directed number.

Directed numbers occur in many practical situations.

EXAMPLE 2 To preserve the food, the temperature in the freezer must be kept at 18°C below 0. We write ⁻18°C.

Tell the meaning in words.

Sample: ⁻1⅓   *What you say:* Negative one and one-third

1. ⁻5 negative 5
2. ⁻2.36 negative 2.36
3. ⁻2⅔ negative 2⅔
4. $\dfrac{^{-}1}{10}$ negative $\frac{1}{10}$

Name the directed number described.

Sample: 2 units to the left of zero   *What you say:* Negative two

5. 3 units to the right of 0 positive 3
6. 2 units to the left of 0 negative 2
7. 5⅓ units to the right of 0 positive 5⅓
8. The negative number of magnitude 5 negative 5
9. The positive number of magnitude 3¾ positive 3¾
10. The positive number of magnitude 0.03 positive 0.03

Use a directed number to express the thermometer reading.

11. 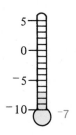 ⁻7

12. 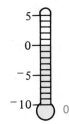 0

13. 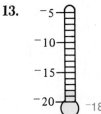 ⁻18

Written
EXERCISES

Name the coordinate of the point.

$$A \quad C \; N \quad K \quad\quad I \; L \; J \quad\quad M \; F \quad\quad D$$

-2		-1	0	1	2

Sample: A *Solution:* $^-2$

A

1. D 2

2. K $\frac{^-3}{4}$

3. N $^-1\frac{1}{4}$

4. F $1\frac{1}{2}$

5. M $1\frac{1}{4}$

6. I 0

7. C $^-1\frac{1}{2}$

8. J $\frac{1}{2}$

9. L $\frac{1}{4}$

Use a directed number to express the following.

Sample: 20 degrees below zero   *Solution:* $^-20$

10. 15 degrees above zero 15

11. 12 degrees below zero $^-12$

12. 30 meters above sea level 30

13. 100 meters below sea level $^-100$

14. 6.8 degrees below zero $^-6.8$

15. 22.8 degrees above zero 22.8

16. 5.5 degrees above zero 5.5

17. 0.1 degree below zero $^-0.1$

18. 2.7 degrees below zero $^-2.7$

19. 0 degrees 0

Name the directed number.

Sample: The positive number 4 units from zero   *Solution:* 4

20. The negative number 6 units from 0 $^-6$

21. The positive number 5 units from 0 5

22. The negative number 3.8 units from 0 $^-3.8$

23. The positive number 20.9 units from 0 20.9

B

24. The number that lies the same distance from 0 as $^-4$ 4

25. The number that lies the same distance from 0 as 15 $^-15$

26. The number that lies the same distance from 0 as $^-6\frac{2}{3}$ $6\frac{2}{3}$

27. The number that lies the same distance from 0 as $45\frac{7}{9}$ $^-45\frac{7}{9}$

28. The number that lies the same distance from 0 as $^-0.075$ 0.075

29. Two directed numbers, each 11 units from 0 11; $^-11$

30. Two directed numbers, each 23 units from 0 23; $^-23$

31. Two directed numbers, each $5\frac{1}{9}$ units from 0 $5\frac{1}{9}$; $^-5\frac{1}{9}$

32. Two directed numbers, each 9.3 units from 0 9.3; $^-9.3$

C

33. Two directed numbers, each 4 units from 5 1; 9

34. Two directed numbers, each 13 units from $^-3$ 10; $^-16$

35. Two directed numbers, each 13 units from 3 16; $^-10$

36. Two directed numbers, each $4\frac{1}{4}$ units from $^-5\frac{3}{4}$ $^-1\frac{1}{2}$; $^-10$

career
capsule   *Telephone Installer*

Telephone installers deliver, install, and remove telephones in homes and offices. Installers begin their jobs by first inspecting the work area and planning installation procedures. They attach outside wires to a pole or cable and then connect the cable terminals with inside wiring. After installing related inside wiring, they test the telephones to ensure proper working order.

About seven months of classroom and on-the-job training are required to be a telephone installer. Courses in mathematics, physics, and shopwork are helpful. Other qualifications include physical fitness, the ability to work in cramped areas, manual dexterity, good eyesight, normal color vision, and a friendly and patient disposition.

8-2 *Arrows to Represent Directed Numbers*

OBJECTIVE

Use arrows on the number line to represent directed numbers.

We can use positive and negative directed numbers to show gains and losses in football.

EXAMPLE 1   First play ▶ Gain of 4 yards: 4
Second play ▶ Loss of 7 yards: ⁻7

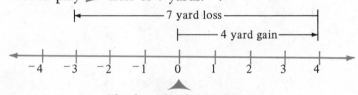

The first play begins here.
Call this point the origin.

Total after two plays ▶ Loss of 3 yards: ⁻3

In general, we can represent a directed number on the number line by an arrow. The arrow begins at a specified point. Its length corresponds to the magnitude of the directed number. For a positive number, the arrow points to the right. For a negative number, the arrow points to the left.

EXAMPLE 2   5 (Start at the origin.)

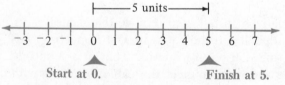

Start at 0.   Finish at 5.

EXAMPLE 3   ⁻2 (Start at the origin.)

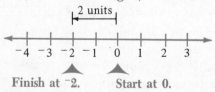

Finish at ⁻2.   Start at 0.

EXAMPLE 4   3 (Start at ⁻1.)

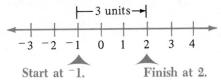

Start at ⁻1.   Finish at 2.

EXAMPLE 5 $1\frac{1}{2}$ (Start at 3.)

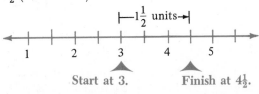

Start at 3.   Finish at $4\frac{1}{2}$.

Locate the directed number with respect to 0.

Oral
EXERCISES

Sample: 30   *What you say:*   Thirty units in the positive direc-
    pos = positive direction   tion.
    neg = negative direction

1.   6 6 units; pos.   **2.**   ⁻16 16 units; neg.   **3.**   0.5 0.5 units, pos.

4.   ⁻$2\frac{1}{2}$ $2\frac{1}{2}$ units, neg.   **5.**   8 8 units; pos.   **6.**   ⁻0.7 0.7 units, neg.

Use the number line to complete.

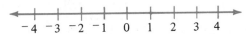

Sample: A move from 0 to 2 is __?__ units to the __?__.

What you say: A move from 0 to 2 is two units to the right.

7.   A move from 0 to 1 is __?__ units to the __?__. 1; right

8.   A move from 0 to ⁻3 is __?__ units to the __?__. 3; left

9.   A move from 1 to ⁻3 is __?__ units to the __?__. 4; left

10.   A move from ⁻3 to 1 is __?__ units to the __?__. 4; right

11.   A move from ⁻$2\frac{1}{2}$ to 0 is __?__ units to the __?__. $2\frac{1}{2}$; right

12.   A move from ⁻4 to 0 is __?__ units to the __?__. 4; right

Written
EXERCISES

Make a number line sketch to show the moves. Tell where you finish. Check students' drawings.

Sample: Start at 0. Move 2 units in the positive direction. Then move 3 units in the negative direction.

Solution:

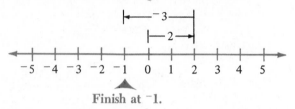

Finish at ⁻1.

A

1. Start at 0. Move 4 units in the positive direction. Then move 2 units in the negative direction. Finish at 2.

2. Start at 0. Move 5 units in the positive direction. Then move 4 units in the negative direction. Finish at 1.

3. Start at 0. Move 5 units in the negative direction. Then move 4 units in the positive direction. Finish at ⁻1.

4. Start at 0. Move 4 units in the negative direction. Then move 3 units in the positive direction. Finish at ⁻1.

5. Start at 0. Move $4\frac{1}{2}$ units in the positive direction. Then move $4\frac{1}{2}$ units in the negative direction. Finish at 0.

6. Start at 0. Move $2\frac{1}{3}$ units in the negative direction. Then move $2\frac{1}{3}$ units in the positive direction. Finish at 0.

Name the directed number suggested by the arrow.

Sample:

![number line from ⁻4 to 4 with arrow]

Solution: ⁻4

7.

![number line from ⁻4 to 4]   3

8.

![number line from ⁻4 to 4]   ⁻6

9.

![number line from ⁻4 to 4]   ⁻2

10.  3

11.  2

12. 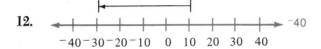 ⁻40

The statement refers to moves on the number line. Make a number line sketch and tell where you finish.

Sample: Start at 4 and move 6 units in the negative direction.

Solution:

```
        ├────── 6 ──────┤
 ◄──┼──┼──┼──┼──┼──┼──┼──┼──┼──┼──┼──►
   ⁻4  ⁻3 ⁻2 ⁻1  0  1  2  3  4  5  6
           ▲
      Finish at ⁻2.
```

13. Start at 3 and move 4 units in the negative direction.   Finish at ⁻1

14. Start at ⁻2 and move 5 units in the positive direction.   Finish at 3

15. Start at ⁻3 and move 4 units in the positive direction.   Finish at 1

16. Start at 2 and move 5 units in the negative direction.   Finish at ⁻3

17. Start at $2\frac{1}{2}$ and move $2\frac{1}{2}$ units in the negative direction.   Finish at 0

18. Start at ⁻5 and move 5 units in the positive direction.   Finish at 0

19. Start at ⁻3. Move 4 units in the positive direction. Then move 3 units in the negative direction.

B
Finish at ⁻2

20. Start at 2. Move 3 units in the negative direction. Then move 4 units in the positive direction.   Finish at 3

Complete. Use *positive* or *negative*.

21. A move from 4 to 3 is in the _?_ direction. negative

22. A move from 3 to 5 is in the _?_ direction. positive

23. A move from ⁻3 to ⁻6 is in the _?_ direction. negative

24. A move from ⁻5 to ⁻1 is in the _?_ direction. positive

25. A move from 0 to ⁻4 is in the _?_ direction. negative

26. A move from 0 to 4 is in the _?_ direction. positive

C | Name the point described.

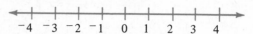

27. The point half the distance from 2 to 4. ₃
28. The point half the distance from ⁻4 to 0. ⁻₂
29. The point half the distance from ⁻3 to 3. ₀
30. The point one-fifth the distance from ⁻1 to 4. ₀
31. The point one-third the distance from ⁻2 to 4. ₀
32. The point one-fourth the distance from 4 to ⁻4. ₂
33. The point half the distance from 2 to ⁻2. ₀
34. The point one-third the distance from 0 to ⁻1. $\frac{^-1}{3}$
35. The point one-fourth the distance from ⁻4 to 0. ⁻₃
36. The point half the distance from ⁻1 to ⁻2. ⁻1½

Time out

Do you know how a camera works? Cut off one end of a cracker box and cover it tightly with tissue paper. Center a pinhole in the closed end of the box. Move the box into a dark room. Place a lighted candle one meter before the pinhole. Observe the tissue paper. What image do you see? Can you explain the image with a diagram? The image on the tissue page is an inverted image of the candle flame.

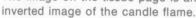

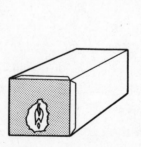

8-3 *Comparing Directed Numbers*

OBJECTIVE

Classify inequality statements as true or false.

We use the symbols $<$, $>$, and $=$, as well as $\leq$, $\geq$, and $\neq$ to compare numbers.

Symbol	Meaning	Symbol	Meaning
$=$	is equal to	$\neq$	is not equal to
$<$	is less than	$\leq$	is less than or equal to
$>$	is greater than	$\geq$	is greater than or equal to

We can use the number line to check the truth of statements that contain these symbols.

EXAMPLE 1   Is $^-2 < 4$ a true statement?

$^-2$ **is to the left** of 4. ▶ $^-2$ **is less than** 4.
$^-2 < 4$ is a true statement.

EXAMPLE 2   Is $^-1 \geq\, ^-3$ a true statement?

$^-1$ **is to the right** of $^-3$. ▶ $^-1$ **is greater than** $^-3$.
$^-1 \geq\, ^-3$ is a true statement.

EXAMPLE 3   Is $^-3 > 2$ a true statement?

$^-3$ **is to the left** of 2 ▶ $^-3$ **is less than** 2.
$^-3 > 2$ is a false statement. But these statements are true:

$$^-3 < 2 \qquad ^-3 \leq 2 \qquad ^-3 \neq 2 \qquad 2 > \,^-3$$

Complete to make a true statement. Use *right* or *left* and $>$ or $<$.

Sample: 4 is to the __?__ of ⁻2; 4 __?__ ⁻2.

What you say: 4 is to the right of ⁻2; 4 $>$ ⁻2.

1. $1\frac{2}{3}$ is to the __?__ of 0; $1\frac{2}{3}$ __?__ 0. right; $>$

2.   ⁻8.5 is to the __?__ of ⁻8; ⁻8.5 __?__ ⁻8. left; $<$

3.   ⁻$5\frac{1}{3}$ is to the __?__ of ⁻6; ⁻$5\frac{1}{3}$ __?__ ⁻6. right; $>$

4.   10 is to the __?__ of 0; 10 __?__ 0. right; $>$

5.   15 is to the __?__ of ⁻15; 15 __?__ ⁻15. right; $>$

6.   ⁻12 is to the __?__ of 2; ⁻12 __?__ 2. left; $<$

7.   5.6 is to the __?__ of 5; 5.6 __?__ 5. right; $>$

8.   ⁻2 is to the __?__ of 3; ⁻2 __?__ 3. left; $<$

9.   ⁻2 is to the __?__ of ⁻3; ⁻2 __?__ ⁻3. right; $>$

10.   ⁻9.4 is to the __?__ of 9; ⁻9.4 __?__ 9. left; $<$

Complete to make a true statement. Use right or left and $>$ or $<$.

Sample: ⁻5 is to the __?__ of ⁻4; so ⁻5 __?__ ⁻4.

Solution: ⁻5 is to the left of ⁻4, so ⁻5 $<$ ⁻4.

A

1.   3 is to the __?__ of ⁻3, so 3 __?__ ⁻3. right; $>$

2.   ⁻5 is to the __?__ of ⁻6, so ⁻5 __?__ ⁻6. right; $>$

3.   ⁻1 is to the __?__ of 2; so ⁻1 __?__ 2. left; $<$

4.   2 is to the __?__ of 4, so 2 __?__ 4. left; $<$

5. $\dfrac{⁻1}{2}$ is to the __?__ of 0, so $\dfrac{⁻1}{2}$ __?__ 0. left; $<$

6. $\dfrac{1}{2}$ is to the __?__ of ⁻$10\frac{1}{2}$, so $\dfrac{1}{2}$ __?__ ⁻$10\frac{1}{2}$. right; $>$

Complete to make a true statement. Use $>$ or $<$.

7.   1 __?__ ⁻1 $>$ 8.   ⁻6 __?__ 6 $<$

9.   ⁻6 __?__ ⁻5 $<$ 10.   4 __?__ 8 $<$

11.   ⁻8 __?__ ⁻10 $>$ 12.   3 __?__ ⁻4 $>$

13.   0 __?__ ⁻7 $>$ 14. $\dfrac{1}{2}$ __?__ 0 $>$

15. $0 \underline{\ ?\ } \frac{-1}{5}$ $>$

16. $^-0.02 \underline{\ ?\ } 0.002$ $<$

17. $4.67 \underline{\ ?\ } ^-4.68$ $>$

18. $^-2.32 \underline{\ ?\ } ^-2.31$ $<$

19. $\frac{1}{3} \underline{\ ?\ } \frac{-2}{3}$ $>$

20. $\frac{-3}{4} \underline{\ ?\ } \frac{1}{4}$ $<$

21. $\frac{-5}{6} \underline{\ ?\ } 0$ $<$

True or false?

22. $6 > 12$ false

23. $^-2 > ^-5$ true

24. $^-6 \le 6$ true

25. $^-0.5 \ge 1.0$ false

26. $^-5 \ne 5$ true

27. $0 > ^-20$ true

28. $\frac{-5}{6} \ge \frac{10}{12}$ false

29. $5 \le 6$ true

30. $1.5 \le 1.0$ false

31. $^-1.5 \le ^-1.0$ true

32. $0.0001 < ^-0.0006$ false

33. $^-5.003 > ^-5.004$ true

34. $0 \le 0.0007$ true

35. $\pi \le ^-\pi$ false

36. $2 < ^-3$ false

Tell which statements are true and which are false, if A, B, C, W, X, Y, and Z are directed numbers. Use the number line below.

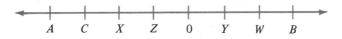

37. Y is a positive number true

38. A is a positive number. false

39. $Y < B$ true

40. Z is a negative number. true

41. $B \ge A$ true

42. $Y \ge 0$ true

B

C

calculator corner

   Did you know that you can find out someone's age and the amount of change up to $1 that is in that person's pocket by using a calculator? Hand a calculator to a friend whom you've either just met or haven't seen for a while. Claim that you have a magic calculator that can electronically scan to find unknown facts about people. Give these directions: "Take the number that is double your age. Add 5. Multiply by 50. Add the whole number that represents the amount of change in your pocket, up to $1. Subtract the number of days in a year. Add 115. Divide by 100." The display will show your friend's age plus the amount of change.

SELF-TEST 1

Be sure that you understand these terms.

origin (p. 212)   opposites (p. 212)
magnitude (p. 212)   directed numbers (p. 212)

Section 8-1, p. 212 Use a directed number to express the following.

1. 2 degrees above zero 2
2. 5 degrees below zero ⁻5
3. The positive number 3 units from zero 3
4. The negative number 5 units from zero ⁻5

Section 8-2, p. 216 Make a number line sketch to show the moves. Tell where you finish.

5. Start at 0. Move 3 units in the negative direction. Then move 5 units in the positive direction. Finish at 2
6. Start at ⁻4 and move 3 units in the positive direction. Finish at ⁻1

Section 8-3, p. 221 Complete to make a true statement. Use *right* or *left* and > or <.

7. 5 is to the __?__ of ⁻1, so 5 __?__ ⁻1. right; >
8. ⁻3 is to the __?__ of ⁻4, so ⁻3 __?__ ⁻4. right; >

True or false?

9. ⁻2 < 2 true   10. 0 < ⁻3 false   11. ⁻1 < ⁻2 false

Check your answers with those printed at the back of the book.

Directed Numbers and Inequalities

8-4 *Integers as Solutions of Inequalities*

OBJECTIVES

Solve an inequality that has a set of integers as its replacement set.

Graph the solution set of an inequality for which the solutions are integers.

Recall that the integers are directed numbers that are either whole numbers or their opposites. Using members of {the integers}, we can now solve an inequality like $x < 1$.

EXAMPLE 1 $x < 1$; replacement set: $\{^-2, ^-1, 0, 1, 2\}$

Replacement	$x < 1$	True/False	Solution?
$^-2$	$^-2 < 1$	True	YES
$^-1$	$^-1 < 1$	True	YES
0	$0 < 1$	True	YES
1	$1 < 1$	False	No
2	$2 < 1$	False	No

The solutions are $^-2$, $^-1$, and 0. Solution set: $\{^-2, ^-1, 0\}$

Graph:

$^-4 \quad ^-3 \quad ^-2 \quad ^-1 \quad 0 \quad 1 \quad 2$

EXAMPLE 2 $x \leq 2$; replacement set: $\{^-2, 0, 2, 4\}$

Replacement	$x \leq 2$	True/False	Solution?
$^-2$	$^-2 \leq 2$	True	YES
0	$0 \leq 2$	True	YES
2	$2 \leq 2$	True	YES
4	$4 \leq 2$	False	No

The solutions are $^-2$, 0, and 2. Solution set: $\{^-2, 0, 2\}$

Graph:

$^-4 \quad ^-3 \quad ^-2 \quad ^-1 \quad 0 \quad 1 \quad 2 \quad 3 \quad 4$

Name the set of directed numbers graphed.

Sample: 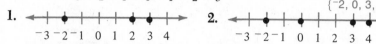 What you say: {⁻1, 0, 1}

{⁻2, 2, 3}
1. ‹—+—●—+—+—+—●—●—+—+—›
‾3 ‾2 ‾1 0 1 2 3 4

{⁻2, 0, 3, 4}
2. ‹—+—●—+—+—+—+—●—●—›
‾3 ‾2 ‾1 0 1 2 3 4

{0, 1, 2}
3. ‹—+—+—+—●—●—●—+—+—›
‾3 ‾2 ‾1 0 1 2 3 4

4. ‹—●—●—+—+—+—+—●—+—›
‾3 ‾2 ‾1 0 1 2 3 4
{⁻3, ⁻2, 2, 3}

{⁻3, ⁻2, ⁻1}
5. ‹—●—●—●—+—+—+—+—+—›
‾3 ‾2 ‾1 0 1 2 3 4

6. ‹—+—+—+—●—+—+—+—●—+—›
‾3 ‾2 ‾1 0 1 2 3 4 {0, 3}

True, false, or neither?

7. $6 > ^-3$ true   **8.** $^-6 < ^-4$ true   **9.** $^-6 > x$ neither

10. $5 \geq ^-5$ true   **11.** $x < ^-5$ neither   **12.** $k \leq 7$ neither

---

Written
EXERCISES

Show whether a true or a false statement results when the variable in the sentence is replaced by each member of the replacement set. Then state the solution set.

Sample: $n \geq ^-4$: {⁻6, ⁻8, 0, 2}

Solution: $^-6 \geq ^-4$, false; $^-8 \geq ^-4$, false;
$0 \geq ^-4$, true; $2 \geq ^-4$, true
Solution set: {0, 2}

A

1. $a > ^-2$: {⁻3, ⁻1, 0} {⁻1, 0}   **2.** $z < 3$: {0, 1, 3, 4} {0, 1}

3. $^-4 < x$: {⁻6, 0, 5} {0, 5}   **4.** $y \geq 5$: {⁻5, 0, 5, 6} {5, 6}

5. $w \leq 0$: {⁻3, ⁻2, 0, 1} {⁻3, ⁻2, 0}   **6.** $^-7 \geq z$: {⁻8, ⁻7, 6} {⁻8, ⁻7}

7. $k \geq \frac{^-1}{2}$: {⁻1, ⁻2, 0} {0}   **8.** $m \leq ^-2\frac{1}{3}$: {⁻4, ⁻3, 0} {⁻4, ⁻3}

9. $\frac{1}{2} \geq n$: {1, 2, 3} ∅   **10.** $7 \geq t$: {⁻3, 0, 9, 12} {⁻3, 0}

11. $0 \leq s$: {⁻1, 0, 1, 2} {0, 1, 2}   **12.** $c \leq 1$: {⁻4, ⁻2, 0, 2, 4}
{⁻4, ⁻2, 0}

Write and graph the solution set. The replacement set is {⁻3, ⁻2, ⁻1, 0, 1, 2, 3}.

Sample: $x \geq ^-2$ Solution: {⁻1, 0, 1, 2, 3}

13. $s < 2$ **14.** $y \leq 3$ **15.** $2.5 > s$
{⁻3, ⁻2, ⁻1, 0, 1}   {⁻3, ⁻2, ⁻1, 0, 1, 2, 3}   {⁻3, ⁻2, ⁻1, 0, 1, 2}

16. $n \le {}^-3 \{{}^-3\}$ **17.** $r \ge 1 \{1, 2, 3\}$ **18.** $\frac{1}{2} > a$ $\{{}^-3, {}^-2, {}^-1, 0\}$

B

19. ${}^-2 \ge t \{{}^-3, {}^-2\}$ **20.** ${}^-3.2 < m$ **21.** $0 \ge w$

22. $0 \le d \{0, 1, 2, 3\}$ **23.** $s \ge 0.4 \{1, 2, 3\}$ **24.** $2 \le v \{2, 3\}$

20. $\{{}^-3, {}^-2, {}^-1, 0, 1, 2, 3\}$ **21.** $\{{}^-3, {}^-2, {}^-1, 0\}$

Write the solution set. Use the designated replacement set.

25. $3.5 > x$: $\{4, 3.6, 3.4, 2.0\}$ $\{3.4, 2.0\}$

26. $y > {}^-0.005$: $\{0.002, 0.004, {}^-0.002\}$ $\{0.002, 0.004, {}^-0.002\}$

27. ${}^-2.3 > t$: $\{{}^-2.1, {}^-2.0, {}^-3.6\}$ $\{{}^-3.6\}$

28. $\frac{{}^-2}{3} \le p$: $\left\{\frac{2}{3}, \frac{{}^-2}{3}, \frac{1}{3}\right\}$ $\left\{\frac{2}{3}, \frac{{}^-2}{3}, \frac{1}{3}\right\}$

C

29. $w \ge {}^-4.8$: $\{5.0, {}^-5.3, {}^-3.2, 2.0\}$ $\{5.0, {}^-3.2, 2.0\}$

30. $4.6 < s$: $\{0, 2.4, 4.7, 4.9\}$ $\{4.7, 4.9\}$

31. $\frac{{}^-5}{6} < n$: $\left\{{}^-1, \frac{{}^-4}{6}, \frac{5}{6}, 0\right\}$ $\left\{\frac{{}^-4}{6}, \frac{5}{6}, 0\right\}$

32. $\frac{{}^-3}{4} \le r$: $\left\{{}^-1\frac{1}{2}, {}^-1, \frac{{}^-1}{4}, 0, \frac{1}{4}, \frac{1}{2}\right\}$ $\left\{\frac{{}^-1}{4}, 0, \frac{1}{4}, \frac{1}{2}\right\}$

Charlotte Angas Scott   1858–1931

Charlotte Scott studied mathematics at Cambridge University in England. As a woman, she was admitted informally to the examinations, in which she attained the equivalent of eighth place in mathematics. She received a doctorate of science from the University of London in 1885. She was later called to Bryn Mawr, Pennsylvania to help establish the mathematics programs at Bryn Mawr College. She continued to direct these programs for forty years. Charlotte Scott published a mathematical text as well as thirty papers in American and European journals and contributed greatly to the development of algebraic geometry.

8-5 *Graphing Inequalities*

OBJECTIVE

Graph an inequality for which the replacement set is {the directed numbers}

Now let's consider inequalities for which the replacement set is {the directed numbers}. We can graph the solution sets on the number line. The graph of an inequality is the graph of its solution set.

EXAMPLE 1 $x > \dfrac{1}{2}$; replacement set: {the directed numbers}

The inequality is true when x is any number greater than $\dfrac{1}{2}$.

Solution set: $\left\{\text{the directed numbers greater than } \dfrac{1}{2}\right\}$

Graph:

```
◄─┼─┼─┼─┼─┼─○─┼─┼─┼─┼─►
     ⁻2  ⁻1  0  1  2  3
             ▲
```

Note the hollow dot.
$\dfrac{1}{2}$ is not included.

EXAMPLE 2 $m \leq {}^-1$; replacement set: {the directed numbers}
The inequality is true when m is $^-1$ or any number less than $^-1$.

Solution set: $\{^-1 \text{ and the directed numbers less than } ^-1\}$
Graph:

```
◄───┼─┼─●─┼─┼─┼─►
   ⁻3  ⁻2  ⁻1  0  1  2
             ▲
```

Note the solid dot.
$^-1$ is included.

The double inequality $^-2 < x < 3$ means that x can be any number between $^-2$ and 3. On the number line:

```
◄─┼─○─┼─┼─┼─┼─○─┼─►
  ⁻3  ⁻2  ⁻1  0  1  2  3  4
      ▲              ▲
```

Note the hollow dots at $^-2$ and 3.

EXAMPLE 3 $^-1 < n < 2\frac{1}{2}$; replacement set: {the directed numbers}

The inequality is true when n is any number that is greater than $^-1$ and less than $2\frac{1}{2}$.

Solution set: {the directed numbers between $^-1$ and $2\frac{1}{2}$}.

Graph:

$\begin{array}{ccccccc} ^-2 & ^-1 & 0 & 1 & 2 & 3 & 4 \end{array}$

1. *a* is any number greater than 2. 2. *c* is any number less than $^-4$.

3. *s* is any number greater than $\dfrac{^-2}{3}$.

Tell what the inequality means.

Sample 1: $t > ^-6$ *What you say:* t is any number greater than $^-6$.

Sample 2: $^-6 \le x < ^-4$

What you say: x is either $^-6$ or a number between $^-6$ and $^-4$.

4. *t* is either $\dfrac{1}{8}$ or a number less than $\dfrac{1}{8}$.

1. $a > 2$ **2.** $c < ^-4$ **3.** $s > \dfrac{^-2}{3}$

4. $t \le \dfrac{1}{8}$ **5.** $^-7 \ge z$ **6.** $4 \le b$

7. $^-4 \le z < ^-2$ **8.** $\dfrac{^-3}{4} < a \le 0$ **9.** $^-3 \le x < 0$

5. *z* is either $^-7$ or a number less than $^-7$.

6. *b* is either 4 or a number greater than 4.

Match each inequality with its graph. The replacement set is {the directed numbers}.

7. *z* is either $^-4$ or a number between $^-4$ and $^-2$.

10. $a > \dfrac{^-1}{4}$ D

11. $x \le 1$ A

12. $y > \dfrac{^-2}{3}$ B

13. $^-2 < r < 2$ C

A.
$\begin{array}{ccccc} ^-2 & ^-1 & 0 & 1 & 2 \end{array}$

B.
$\begin{array}{ccccc} ^-2 & ^-1 & 0 & 1 & 2 \end{array}$

C.
$\begin{array}{ccccc} ^-2 & ^-1 & 0 & 1 & 2 \end{array}$

D.
$\begin{array}{ccccc} ^-2 & ^-1 & 0 & 1 & 2 \end{array}$

8. *a* is either 0 or a number between $\dfrac{^-3}{4}$ and 0.

9. *x* is either $^-3$ or a number between $^-3$ and 0.

Graph. Check students' graphs.

Sample: {the directed numbers less than 3}

Solution:
$\begin{array}{ccccccc} ^-3 & ^-2 & ^-1 & 0 & 1 & 2 & 3 \end{array}$

1. {the directed numbers greater than $^-3$}

2. {the directed numbers greater than 0}

3. {the directed numbers less than $^-2$}

4. {the directed numbers less than 4}

5. {the directed numbers greater than or equal to ⁻3}

6. {the directed numbers between ⁻2 and 0}

Name the set of directed numbers graphed.

Sample 1:

Solution: $\left\{\text{the directed numbers less than } \dfrac{1}{2}\right\}$

Sample 2:

Solution: {the directed numbers between ⁻1½ and 2½}

See page A1 at the back of the book for Ex. 7–14 and Ex. 23–32.

7.

8.

9.

10.

11.

12.

13.

14.

Name and graph the solution set. The replacement set is {the directed numbers}. Check students' graphs.

B

15. $y > {}^{-}2$

16. $3 < x$

17. $t \le 0$

18. $m \ge {}^{-}1$

19. $b \le 3$

20. $^{-}2 < z < 0$

21. $0 \le x < 1$

22. $^{-}3 < n \le 4$

Name the solution set. The replacement set is {the directed numbers}.

C

23. $^{-}3 > s \ge {}^{-}8$

24. $0 \le n \le 2\frac{1}{3}$

25. $^{-}5 \ge x > {}^{-}10$

26. $k \ge \left(\dfrac{1}{3} + 1\frac{2}{3}\right)$

27. $^{-}5.6 < p + 1 < {}^{-}3.2$

28. $^{-}2.6 < a < 1.2$

29. $y \ne 6\left(\dfrac{1}{3} + \dfrac{1}{2}\right)$

30. $^{-}0.0005 \le t \le 0$

31. $0 < z + 2 \le {}^{-}6$

32. $^{-}5(5 - 2) \ge s$

SELF-TEST 2

Be sure that you understand these terms.

integers (p. 225)   graph of an inequality (p. 228)

Write the solution set. The replacement set is $\{^-1, 0, 1\}$.   Section 8-4, p. 225

1. $m > 0$ {1}   **2.** $n \le 3$ **3.** $^-2 \le t$ **4.** $^-3 < z$
                  $\{^-1, 0, 1\}$ $\{^-1, 0, 1\}$ $\{^-1, 0, 1\}$

Name and graph the solution set. The replacement set is {the set   Section 8-5, p. 228
of directed numbers}. **5.** {the directed numbers less than 2}
                        6. {the directed numbers greater than 1}

5. $r < 2$ **6.** $b > 1$ **7.** $x \ge ^-3$ **8.** $^-2 < c < ^-1$

7. {$^-3$ and the directed numbers greater than $^-3$}

Check your answers with those printed at the back of the book.

8. {the directed numbers between $^-2$ and $^-1$}

chapter summary

1. The point to which 0 is assigned on the number line is called the **origin.**

2. The distance between 0 and any number is called the **magnitude** of the number.

3. Arrows may be used with the number line to represent directed numbers.

   For $^-3$, the arrow points to the left and is three units in length.

   For 2, the arrow points to the right and is two units in length.

4. One directed number **is greater than** another if it lies to the **right** of it on the number line. One directed number **is less than** another if it lies to the **left** of it on the number line.

5. The graph of a set of directed numbers like the following can be shown on the number line.

   Graph of $\{^-1, 0, 1, 2\}$:

   Graph of {directed numbers less than $1\frac{1}{2}$}:

   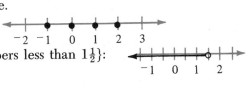

chapter test

Name the directed number described.

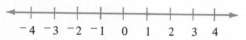

$$-4 \quad -3 \quad -2 \quad -1 \quad 0 \quad 1 \quad 2 \quad 3 \quad 4$$

1. 1 unit to the right of 0 ¹
2. 2 units to the left of 0 ⁻2
3. the negative number of magnitude 4 ⁻4
4. the positive number 8 units from 0 8

The statement refers to moves on the number line. Make a sketch and tell where you finish.

5. Start at 0. Move 1 unit in the positive direction. Then move 4 units in the negative direction. Finish at ⁻3.
6. Start at 0. Move 3 units in the negative direction. Then move 2 units in the positive direction. Finish at ⁻1.
7. Start at ⁻4 and move 5 units in the positive direction. Finish at 1.

Complete. Use right or left and $>$ or $<$.

8. 3 is to the __?__ of 0, so 3 __?__ 0. right; $>$
9. ⁻3 is to the __?__ of ⁻4, so ⁻3 __?__ ⁻4. right; $>$

Show whether a true or a false statement results when the variable is replaced by each member of the replacement set. Then state the solution set.

10. $y < {}^-2; \{{}^-3, {}^-2, {}^-1, 0\}$ ⁻3 $<$ ⁻2, true; ⁻2 $<$ ⁻2, false; ⁻1 $<$ ⁻2, false; 0 $<$ ⁻2, false; solution set: {⁻3}
11. $x > 5; \{4, 5, 6, 7\}$ 4 $>$ 5, false; 5 $>$ 5, false; 6 $>$ 5, true; 7 $>$ 5; true; solution set: {6, 7}

Graph. Check students' graphs.

12. {the directed numbers less than 2}
13. {the directed numbers between ⁻3 and 1}

Name the solution set. The replacement set is {the directed numbers}.

14. $x < {}^-1$ less than ⁻1
15. $y \geq {}^-2$ greater than or equal to ⁻2
16. $t > 3$ greater than 3
17. $0 < n < 3$ between 0 and 3

challenge topics

Logic and Inductive Reasoning

Suppose you observe a pattern that occurs consistently, or see some event take place repeatedly. If you draw some general conclusion from what you have experienced, we say you have used **inductive reasoning.**

Inductive reasoning helps us make "educated guesses" about many things. For example, you might observe mosquitoes biting people. If you could tell that all the mosquitoes were female, you might generalize that "only female mosquitoes bite." Although this generalization seems sensible, we would need to make further observations to be certain. However, if even one male mosquito bites a person, the conclusion is contradicted. Such a contradiction is called a **counterexample.**

Consider the illustration and conclusion. Then either agree with the conclusion or cite at least one counterexample.

1.
Conclusion: All vehicles have at least four wheels.
disagree—bicycles have two wheels

2.
$$\begin{array}{ccc} 15 & 101 & 17 \\ + 7 & + 35 & 19 \\ \hline 22 & 136 & 36 \text{ agree} \end{array}$$

Conclusion: The sum of any two odd numbers is an even number.

3. $\frac{1}{9} = 0.1111\ldots$

   $\frac{2}{9} = 0.2222\ldots$

   $\frac{3}{9} = 0.3333\ldots$ agree

Conclusion: $\frac{7}{9} = 0.7777\ldots$

4. $1 = 1^2$
   $1 + 3 = 2^2$
   $1 + 3 + 5 = 3^2$
   $1 + 3 + 5 + 7 = 4^2$ agree

Conclusion: The sum of the first ten odd numbers is 10^2.

Name the opposite of the number or expression.

1. rise fall
2. forward backward
3. north south
4. ⁻6 6
5. 0 0
6. 4 ⁻4

Think of moves along a number line. Tell which move you would make to have the indicated result.

7. What move would you make after a move of ⁻3 to have as an end result no change (0)? move of 3

8. What move would you make after a move of 6, to have as an end result no change? move of ⁻6

9. What move would you make after a move of ⁻1, to have as an end result no change? move of 1

Simplify.

10. $3 + 6$ 9
11. $16 - 6 + 5$ 15
12. $x + (5 - 5)$ x
13. $x + (3 - 2)$ x + 1
14. $(x + 3) - 2$ x + 1
15. $r + 14 - 6$ r + 8
16. $(6 + 8) - 5$ 9
17. $6 + (8 - 5)$ 9
18. $6 + 7 - 4 + 5$ 14

Try to determine a value of k that will make the statement true.

19. $k = 3 + 8$ 11
20. $5 + k = 3$ ⁻2
21. $18 - k = 5$ 13

Complete the set of number pairs according to the given function rule.

22. $f(z) = z + 3$: $\{(1, 4), (0, 3), (4, \underline{?}), (\underline{?}, 6), (15, \underline{?})\}$
23. $f(y) = y - 5$: $\{(5, 0), (9, 4), (7, \underline{?}), (\underline{?}, 6), (\underline{?}, 8)\}$

22. $\{(1, 4), (0, 3), (4, 7), (3, 6), (15, 18)\}$
23. $\{(5, 0), (9, 4), (7, 2), (11, 6), (13, 8)\}$

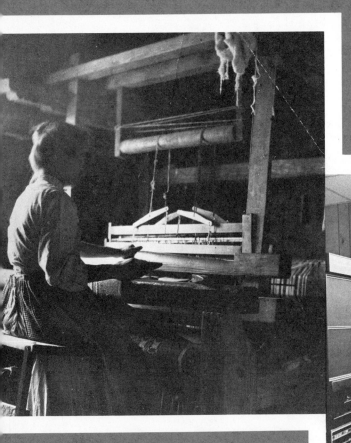

Left: Weaver using hand loom, 1914.

Right: Technician developing pattern by computer. The computer controls the knitting machine shown.

9 Addition and Subtraction of Directed Numbers

9-1 *Adding Directed Numbers on the Number Line*

> **OBJECTIVE**
>
> **Use the number line to find sums of directed numbers.**

Addition of directed numbers is easy to understand on the number line.

EXAMPLE 1 $4 + 3 = \underline{\ ?\ }$

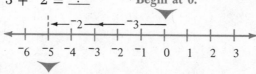

Finish at 7. ► $4 + 3 = 7$

EXAMPLE 2 $^-3 + {}^-2 = \underline{\ ?\ }$

Finish at $^-5$. ► $^-3 + {}^-2 = {}^-5$

EXAMPLE 3 $5 + {}^-3 = \underline{\ ?\ }$

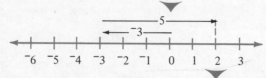

Finish at 2. ► $5 + {}^-3 = 2$

EXAMPLE 4 $^-3 + 5 = \underline{\ ?\ }$

Finish at 2. ► $^-3 + 5 = 2$

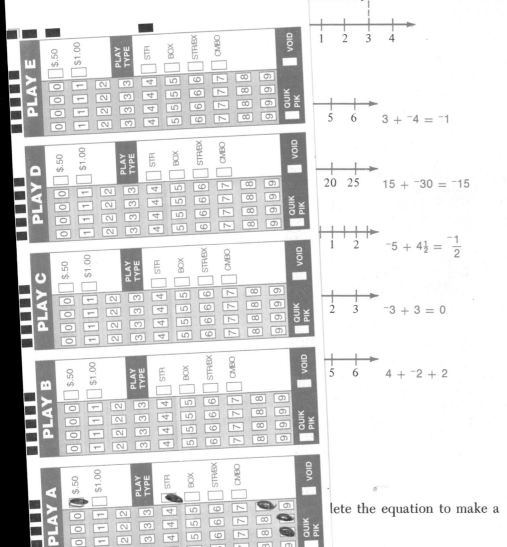

$3 + {}^-4 = {}^-1$

$15 + {}^-30 = {}^-15$

$-5 + 4\frac{1}{2} = \frac{{}^-1}{2}$

$-3 + 3 = 0$

$4 + {}^-2 + 2$

...lete the equation to make a

$3 + 2 = \underline{\ ?\ }\ 5$

$\underline{\ ?\ } = 4\frac{1}{2} + {}^-2\frac{1}{2}\ \ 2$

$7\frac{1}{2} + 3\frac{1}{2} = \underline{\ ?\ }\ \ 11$

A

ADDITION AND SUBTRACTION OF DIRECTED NUMBERS / 237

Tell whether the expression names a positive number, a negative number, or 0. Then simplify. Use the number line for help if necessary. p = positive; n = negative

7. $2 + 3$ p; $2 + 3 = 5$ **8.** $1 + 6$ p; $1 + 6 = 7$

9. $^-16 + 6$ n; $^-16 + 6 = ^-10$ **10.** $15 + ^-9$ p; $15 + ^-9 = 6$

11. $15 + ^-25$ n; $15 + ^-25 = ^-10$ **12.** $^-8 + 10$ p; $^-8 + 10 = 2$

13. $^-12 + 12$ 0; $^-12 + 12 = 0$ **14.** $8 + ^-4$ p; $8 + ^-4 = 4$

15. $^-8 + 4$ n; $^-8 + 4 = ^-4$ **16.** $^-3 + ^-8$ n; $^-3 + ^-8 = ^-11$

17. $16 + ^-6$ p; $16 + ^-6 = 10$ **18.** $^-20 + 21$ p; $^-20 + 21 = 1$

Copy and complete the addition table.

19.

+	2	$^-4$	6	$^-8$
2	? 4	? $^-2$	? 8	? $^-6$
$^-4$	? $^-2$	? $^-8$	? 2	? 12
6	? 8	? 2	? 12	? $^-2$
$^-8$	? $^-6$	? 12	? $^-2$	? $^-16$

20.

+	3	$^-3$	5	$^-7$
3	? 6	? 0	? 8	? $^-4$
$^-3$	? 0	? $^-6$	? 2	? $^-10$
5	? 8	? 2	? 10	? $^-2$
$^-7$	? $^-4$	? $^-10$	? $^-2$	? $^-14$

Find the sum.

B

21. $^-2\frac{1}{3} + 2$ $^{-1}\!\!\!\;\frac{}{3}$ **22.** $5\frac{1}{4} + \frac{3}{4}$ 6

23. $^-6\frac{1}{2} + 7\frac{1}{2}$ 1   **24.** $8 + ^-4\frac{1}{2}$ $3\frac{1}{2}$

25. $^-8 + 4\frac{1}{2}$ $^-3\frac{1}{2}$ **26.** $104\frac{5}{8} + ^-104\frac{5}{8}$ 0

Find the sum.

27.
$$\begin{array}{r} 10 \\ ^-7 \\ \hline 3 \end{array}$$

28.
$$\begin{array}{r} ^-20 \\ 18 \\ \hline ^-2 \end{array}$$

29.
$$\begin{array}{r} ^-17 \\ 20 \\ \hline 3 \end{array}$$

30.
$$\begin{array}{r} 12 \\ ^-11 \\ \hline 1 \end{array}$$

C

31.
$$\begin{array}{r} ^-58 \\ 16 \\ \hline ^-42 \end{array}$$

32.
$$\begin{array}{r} ^-63 \\ 98 \\ \hline 35 \end{array}$$

33.
$$\begin{array}{r} 143 \\ ^-482 \\ \hline ^-339 \end{array}$$

34.
$$\begin{array}{r} ^-897 \\ 897 \\ \hline 0 \end{array}$$

Complete to make a true statement.

Sample: $8 = 10 + \underline{\ ?\ }$ *Solution:* $8 = 10 + ^-2$

35. $16 = 11 + \underline{\ ?\ }$ 5   **36.** $^-8 = 8 + \underline{\ ?\ }$ $^-16$

37. $12 = ^-16 + \underline{\ ?\ }$ 28   **38.** $7 + \underline{\ ?\ } = ^-2$ $^-9$

39. $6 = 12 + \underline{\ ?\ }$ $^-6$ **40.** $^-9 + \underline{\ ?\ } = 18$ 27

Express the problem as the sum of two directed numbers. Find the sum and answer the question.

Sample: Twelve students signed up for a cooking class. The first dish they learned to make was coconut-spinach cream puffs. Ten students quit the course. How many are still in the class?

Solution: $12 + {}^-10 = \underline{\quad?\quad}$
$12 + {}^-10 = 2$ ▸ Two students are still in the class.

1. A truck was carrying 31 bales of hay. It hit a pothole in the road and 5 bales fell out. How many bales of hay were left in the truck? $31 + {}^-5 = 26$

2. Rita had eighty-five cents in change. She put the money in her pocket, which had a small hole in it. A quarter fell out. How much money did she have left? $85 + {}^-25 = 60$

3. Three hikers are at the top of a cliff 975 meters above sea level. They can see a lake 60 meters directly below them. How many meters above sea level is the lake? $975 + {}^-60 = 915$

4. Joe won $50 in a public speaking contest. He paid $25 to rent a formal outfit for the award presentation. How much did he have left? $50 + {}^-25 = 25$

5. An elevator stopped at an insurance company on the twenty-first floor. It then went down sixteen floors, where it stopped at a lawyer's office. On what floor is the lawyer's office? $21 + {}^-16 = 5$

6. Jack baked 36 chocolate-chip cookies for Marcie's birthday. Marcie ate 5 cookies. How many cookies were left? $36 + {}^-5 = 31$

7. A pair of denim pants was 90 cm long. It shrank 2 cm when it was washed. How long is it now? $90 + {}^-2 = 88$

8. A train travels 45 kilometers north. Then it travels 23 kilometers south. How far is the train from its original position? $45 + {}^-23 = 22$

9. An airplane takes off and climbs to an altitude of 5630 meters. Then it descends 2360 meters. What is the new altitude?
$5630 + {}^-2360 = 3270$

9-2 *Additive Inverses and the Identity Element for Addition*

Every directed number has an opposite. Both are the same distance from 0 on the number line, but on opposite sides of 0. Zero is its own opposite. The distance a number is from 0 is called its magnitude.

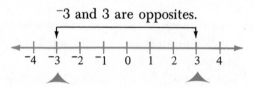

$^-3$ and 3 are opposites.

The magnitude of $^-3$ is 3. The magnitude of 3 is 3.

You know that the symbol "−" in $9 - 7$ means "minus." The same symbol can mean "the opposite of." $- 3$ means "the opposite of 3," and $-(^-3)$ means "the opposite of $^-3$."

EXAMPLE 1 $-(^-8)$ means "the opposite of $^-8$."
$-(^-8) = 8$

EXAMPLE 2 $- 0.1$ means "the opposite of 0.1."
$- 0.1 = ^-0.1$

► In general, for every directed number m:

If m is positive, then $- m$ is negative;
If m is negative, then $-m$ is positive;
If m is 0, $-m$ is also 0.

The sum of 0 and any directed number is that number.

$$0 + 8 = 8 \qquad ^-3 + 0 = ^-3 \qquad 0 + 0 = 0$$

0 is called the identity element for addition.
The sum of any number and its opposite is the identity element, 0.

$$^-8 + 8 = 0 \qquad -(^-1) + ^-1 = 0$$

The opposite of a number is also called its additive inverse.

EXAMPLE 3 $8 + {}^-8 = 0$; 8 is the additive inverse of ${}^-8$.
${}^-8$ is the additive inverse of 8.

Complete to make a true statement.

Sample: 4 is the additive inverse of _?_.

What you say: 4 is the additive inverse of ${}^-4$.

1. ${}^-3$ is the additive inverse of _?_. 3
2. 3 is the additive inverse of _?_. ${}^-3$
3. $10\frac{2}{3}$ is the additive inverse of _?_. ${}^-10\frac{2}{3}$
4. 3.4 is the opposite of _?_. ${}^-3.4$
5. _?_ is the additive inverse of ${}^-6\frac{2}{3}$. $6\frac{2}{3}$

Simplify.

Sample: $6 + {}^-6$ *What you say:* Zero

6. $40 + {}^-40$ 0
7. ${}^-32 + 32$ 0
8. $\dfrac{{}^-2}{3} + \dfrac{2}{3}$ 0

9. $152 + 0$ 152
10. $0 + {}^-9$ ${}^-9$
11. $\dfrac{1}{4} + \dfrac{{}^-1}{4}$ 0

Show two ways to express in symbols.

Sample: The opposite of ${}^-2$ *Solution:* $-({}^-2)$; 2

1. The opposite of ${}^-4$ $-({}^-4)$; 4
2. The opposite of ${}^-8$ $-({}^-8)$; 8
3. The opposite of 3 $-(3)$; ${}^-3$
4. The opposite of 7 $-(7)$; ${}^-7$
5. The opposite of ${}^-16$ $-({}^-16)$; 16
6. The opposite of $\dfrac{1}{2}$ $-\left(\dfrac{1}{2}\right)$; $\dfrac{{}^-1}{2}$

7. The opposite of ${}^-3.4$ $-({}^-3.4)$; 3.4
8. The opposite of $\dfrac{{}^-3}{4}$ $-\left(\dfrac{{}^-3}{4}\right)$; $\dfrac{3}{4}$

9. The opposite of -6.08 $-({}^-6.08)$; 6.08
10. The opposite of $-({}^-3)$ $-(-({}^-3))$; ${}^-3$

ADDITION AND SUBTRACTION OF DIRECTED NUMBERS / 241

Name the two numbers described.

Sample: 2 units from zero on the number line   *Solution:* 2, ⁻2

11. 5 units from zero on the number line ⁻5; 5
12. 8 units from zero on the number line ⁻8; 8
13. 20 units from zero on the number line ⁻20; 20
14. $3\frac{1}{2}$ units from zero on the number line ⁻$3\frac{1}{2}$; $3\frac{1}{2}$
15. 5 units from ⁻5 on the number line ⁻10; 0
16. 5 units from 5 on the number line 0; 10

Tell whether or not the numbers are additive inverses.

17. 8, ⁻8 yes   18. ⁻6, 6 yes   19. 0, ⁻4 no
20. 8, 0 no   21. −(⁻4), 4 no   22. −10, ⁻10 no
23. −(⁻7), ⁻7 yes 24. −3.2, 3.2 yes 25. −n, n yes

Solve.

Sample: $x = 4 + {}^{-}4$ *Solution:* $x = 4 + {}^{-}4$
$$x = 0$$

26. $m = {}^{-}12 + 12$ 0   27. $t = 6 + 0$ 6   28. $n = {}^{-}3.4 + 3.4$ 0

B 29. $a + {}^{-}3 = 0$ 3   30. $4 + b = 0$ -4   31. ${}^{-}6 + y = {}^{-}6$ 0

32. $8 + x = 8$ 0   33. $2\frac{2}{3} + y = 0$ -$2\frac{2}{3}$ 34. $s + {}^{-}10 = 0$ 10

35. $r + \dfrac{{}^{-}2}{3} = 0$ $\frac{2}{3}$ 36. $t + 0 = \dfrac{1}{4}$ $\frac{1}{4}$ 37. $0 = u + {}^{-}4$ 4

38. $c = {}^{-}25 + 25$ 0   39. $d = {}^{-}18 + 18$ 0   40. ${}^{-}4.8 + m = 0$ 4.8

Give the meaning. Then write in simplest form.

Sample: -9 *Solution:* The opposite of 9; ⁻9

41. $-({}^{-}16)$ 16   42. -8 -8   43. -1.42 -1.42

44. $-\dfrac{3}{4}$ -$\frac{3}{4}$ 45. $-({}^{-}0.63)$ 0.63   46. -3.4 -3.4

Simplify.

Sample: $-(-({}^{-}1))$ *Solution:* $-(-({}^{-}1)) = -(1) = {}^{-}1$

C 47. $-({}^{-}3)$ 3   48. $-(-4)$ 4   49. $-(-({}^{-}3))$ -3
50. $-({}^{-}4)$ 4   51. $-(-1.3)$ 1.3   52. $-(-({}^{-}10))$ -10

9-3 *Simplifying Expressions*

OBJECTIVES

Simplify expressions like $-(3 + 7)$ **and** $-(-5 + 8)$.

Solve equations like $5 + t = 2$.

The symbol -6 means "the opposite of 6." However, we have seen that $-6 = {}^-6$. From now on we will use the symbol -6 for both **negative 6** and **the opposite of 6.** Lowered minus signs will now be used to represent all negative numbers.

Now let's consider the meaning of an expression like $-(3 + 7)$. We will see that the opposite of a sum is the sum of the opposites.

EXAMPLE 1 $\quad -(3 + 7) \overset{?}{=} -3 + (-7)$ ◄ $\overset{?}{=}$ means "does it equal?"

$$
\begin{array}{c|c}
-(10) & -10 \\
-10 & -10 \quad \text{Yes}
\end{array}
$$

EXAMPLE 2 $\quad -(-5 + 8) \overset{?}{=} 5 + (-8)$

$$
\begin{array}{c|c}
-(3) & -3 \\
-3 & -3 \quad \text{Yes}
\end{array}
$$

EXAMPLE 3 $\quad -[-2 + (-8)] \overset{?}{=} 2 + 8$

$$
\begin{array}{c|c}
-(-10) & 10 \\
10 & 10 \quad \text{Yes}
\end{array}
$$

A number line sketch can help us find a solution for an equation like $5 + t = 2$.

EXAMPLE 4 $\quad 5 + t = 2$

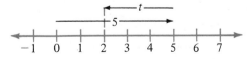

Solution: $\quad t = -3$

State the expression in two ways.

Sample 1: $-(3 + 6)$ *What you say:*   The opposite of 9; the opposite of 3 plus the opposite of 6.

1. the opposite of 6; the opposite of 4 plus the opposite of 2.

Sample 2: $-(-6 + 3)$ *What you say:*   The opposite of -3; the opposite of -6 plus the opposite of 3.

2. the opposite of 7; the opposite of 6 plus the opposite of 1.

1. $-(4 + 2)$ 2. $-(6 + 1)$ 3. $-[4 + (-2)]$

4. $-[5 + (-7)]$ 5. $-(-8 + 11)$ 6. $-\left(\dfrac{1}{2} + \dfrac{1}{2}\right)$

3. the opposite of 2; the opposite of 4 plus the opposite of -2.
4. the opposite of -2; the opposite of 5 plus the opposite of -7.

Name the sum.

7. $\begin{array}{r} -4 \\ 5 \\ \hline 1 \end{array}$ 8. $\begin{array}{r} 8 \\ 4 \\ \hline 12 \end{array}$ 9. $\begin{array}{r} 18 \\ -6 \\ \hline 12 \end{array}$ 10. $\begin{array}{r} 20 \\ 18 \\ \hline 38 \end{array}$

11. $\begin{array}{r} -8 \\ -6 \\ \hline -14 \end{array}$ 12. $\begin{array}{r} -3 \\ 17 \\ \hline 14 \end{array}$ 13. $\begin{array}{r} -40 \\ 12 \\ \hline -28 \end{array}$ 14. $\begin{array}{r} -30 \\ -20 \\ \hline -50 \end{array}$

5. the opposite of 3; the opposite of -8 plus the opposite of 11.

6. the opposite of 1; the opposite of $\dfrac{1}{2}$ plus the opposite of $\dfrac{1}{2}$.

For Extra Practice, see page 422.

Give a simpler name for the expression.

Sample 1: $-(6 + 1)$ *Solution:* $-(6 + 1) = -7$

Sample 2: $-4 + (-2)$ *Solution:* $-4 + (-2) = -6$

A

1. $-(4 + 3)$ $_{-7}$ 2. $-(12 + 6)$ $_{-18}$ 3. $-(22 + 17)$ $_{-39}$

4. $-4 + (-8)$ $_{-12}$ 5. $-9 + (-7)$ $_{-16}$ 6. $-20 + (-1)$ $_{-21}$

7. $-3.6 + (-4.4)$ $_{-8.0}$ 8. $-\dfrac{1}{8} + (-2\frac{1}{4})$ $_{-2\frac{3}{8}}$ 9. $-(-6 + 7)$ $_{-1}$

Solve.   Use a number line sketch.

Sample: $-4 + t = 1$

Solution:

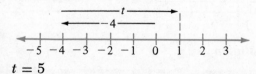

$t = 5$

10. $6 + x = 4$ -2 11. $-7 + m = -4$ 3   12. $6 + y = 2$ -4

13. $5 + r = 4$ -1 14. $-5 + n = 6$ 11   15. $4 = -5 + s$ 9

Show whether the statement is true or false.

Sample 1: $-(6 + 5) = -6 + (-5)$

Solution: $-(6 + 5) \overset{?}{=} -6 + (-5)$

$$
\begin{array}{c|c}
-(11) & -11 \\
-11 & -11 \quad \text{True}
\end{array}
$$

Sample 2: $-(-8 + 2) = -8 + (-2)$

Solution: $-(-8 + 2) \overset{?}{=} -8 + (-2)$

$$
\begin{array}{c|c}
-(-6) & -10 \\
6 & -10 \quad \text{False}
\end{array}
$$

16. $-(6 + 8) = -6 + (-8)$ true

17. $-(8 + 2) = -8 + (-2)$ true

18. $-16 + 5 = -(16 + 5)$ false

19. $-7 + (-12) = -(7 + 12)$ true

20. $-(-5 + 8) = -(-5) + (-8)$ true

21. $-[20 + (-6)] = -20 + (-6)$ false

22. $-(2\frac{2}{3} + 1\frac{1}{3}) = -2\frac{2}{3} + (-1\frac{1}{3})$ true

23. $-5\frac{1}{4} + 2\frac{1}{8} = -[5\frac{1}{4} + (-2\frac{1}{8})]$ true

24. $-[6 + (-9)] = -6 + 9$ true

25. $-(-0.15 + 1.2) = 0.15 + (-1.2)$ true

B

Solve.

26. $-6 = 5 + c$ -11 27. $5 = -3 + p$ 8

28. $a = -4 + (-4)$ -8 29. $w + 6 = 7$ 1

30. $-5 = -4 + t$ -1 31. $11 = t + 7\frac{1}{4}$ $3\frac{3}{4}$

32. $-1 = u + 5$ -6 33. $s = -7 + (-1)$ -8

34. $n + 3.5 = 7$ 3.5   35. $-(5 + 9) = s$ -14

36. $-(15 + 3) = t$ -18 37. $-[7 + (-13)] = r$ 6

38. $-(-15 + 9) = a$ 6   39. $-(-3 + 7) = b$ -4

40. $-(-7 + 19) = c$ -12 41. $x = -[15 + (-20)]$ 5

42. $y = -(-25 + 15)$ 10

9-4 *Addition Properties for Directed Numbers*

OBJECTIVES

Apply the properties of addition to directed numbers.

Add directed numbers.

Addition of directed numbers involves use of many of the same properties as addition of positive numbers. Let's restate them and see how we use them to add directed numbers.

The Commutative Property ▶— For all directed numbers r and s,
of Addition $r + s = s + r$.

The Associative Property ▶— For all directed numbers r, s, and t,
of Addition $(r + s) + t = r + (s + t)$.

The Additive Property ▶— For every directed number r,
of Zero $r + 0 = 0 + r = r$.

EXAMPLE 1 $(-2 + 4) + 2 = [4 + (-2)] + 2$ ◀ Commutative Property
                $= 4 + (-2 + 2)$ ◀ Associative Property
                $= 4 + 0$
                $= 4$ ◀ Additive Property of 0

We have seen two more addition properties in this chapter.

▶— For every directed number r, $-r + r = r + (-r) = 0$.

▶— For all directed numbers r and s, $-(r + s) = -r + (-s)$.

These properties suggest a method for adding two directed numbers without using the number line. We rename the number that has greater magnitude.

EXAMPLE 2 $-8 + 3 = [-5 + (-3)] + 3$ ◀ Rename -8 as a sum, with one addend the opposite of 3.
             $= -5 + (-3 + 3)$ ◀ Associative Property
             $= -5 + 0$ ◀ Additive Property of Inverses
             $= -5$ ◀ Additive Property of 0

Tell which number has the greater magnitude. Then tell how that number should be renamed to simplify the expression.

Sample: $-15 + 3$ *What you say:* -15 has the greater magnitude. Rename -15 as $-12 + (-3)$.

1. $\underline{-23} + 5$ $-18 + (-5)$ 2. $\underline{-7} + 3$ $-4 + (-3)$ 3. $\underline{9} + (-5)$ $4 + 5$

4. $\underline{19} + (-9)$ $10 + 9$ 5. $9 + (\underline{-15})$ $-9 + (-6)$ 6. $\underline{23} + (-9)$ $14 + 9$

7. $-7 + \underline{19}$ $7 + 12$ 8. $-11 + \underline{33}$ $11 + 22$ 9. $7 + (\underline{-20})$ $-7 + (-13)$

10. $15 + (\underline{-29})$ $-15 + (-14)$ 11. $19 + (\underline{-23})$ $-19 + (-4)$ 12. $7\frac{7}{8} + (-7)$ $-\frac{7}{8} + 7$

For Extra Practice, see page 422.

Name the property illustrated.

Sample 1: $5 + (-4) = -4 + 5$ Comm = Commutative Prop
Solution: Commutative Property   Assoc = Associative Prop
                                      Add = Additive Prop

Sample 2: $6 + (-6 + 9) = [6 + (-6)] + 9$

Solution: Associative Property

A

1. $8 + (-6) = -6 + 8$ Comm.   2. $-11 + 9 = 9 + (-11)$ Comm.
3. $-4 + 4 = 0$ Add. of opposites   4. $-2 + (7 + 4) = (-2 + 7) + 4$ Assoc.
5. $0 + (-13) = -13$ Add. of 0   6. $(-9 + 9) + 4 = -9 + (9 + 4)$ Assoc.

Complete to make a true statement.

Sample 1: $-7 + \underline{\ ?\ } = -7$ *Solution:* $-7 + 0 = -7$

Sample 2: $11 + (-5 + \underline{\ ?\ }) = 11$

Solution: $11 + (-5 + 5) = 11$

7. $\underline{\ ?\ } + (-16) = 0$ 16   8. $\underline{\ ?\ } + (-18) = -18 + 9$ 9
9. $7 + \underline{\ ?\ } = 7$ 0   10. $20 + (-20) = \underline{\ ?\ }$ 0
11. $22 + \underline{\ ?\ } = 22$ 0   12. $33 + (-13) = \underline{\ ?\ } + 33$ -13
13. $(-3 + 3) + \underline{\ ?\ } = 7$ 7   14. $(-15 + 15) + 5 = \underline{\ ?\ }$ 5
15. $9 + (-17 + \underline{\ ?\ }) = 9$ 17   16. $-9 = (-2 + 2) + \underline{\ ?\ }$ -9
17. $(-15 + 17) + 5 = 5 + (17 + \underline{\ ?\ })$ (-15)
18. $(-7 + 7) + \underline{\ ?\ } = 15$ 15

Add. Use the additive property of inverses.

19. $7 + 13 + (-13)$ 7   20. $-7 + 7 + 16$ 16

ADDITION AND SUBTRACTION OF DIRECTED NUMBERS / 247

21. $3 + (-3) + (-27)$ -27 **22.** $-15 + (-17) + 17$ -15

23. $23 + 11 + (-11) + 17 + (-17)$ 23 **24.** $-4 + 20 + (-20)$ -4

25. $3\frac{1}{4} + 5\frac{2}{3} + (-5\frac{2}{3})$ $3\frac{1}{4}$ **26.** $-3\frac{2}{5} + 5 + 19 + 3\frac{2}{5}$ 24

Add.

27.	**28.**	**29.**	**30.**
-5	27	13	43
3	-9	-13	-27
-3	9	33	-43
$\overline{-5}$	$\overline{27}$	$\overline{33}$	$\overline{-27}$

Show that the statement is true.

Sample: $-7 + 15 = 28 + (-20)$

Solution: $-7 + 15 \overset{?}{=} 28 + (-20)$

$-7 + (7 + 8)$	$(8 + 20) + (-20)$
$(-7 + 7) + 8$	$8 + [20 + (-20)]$
$0 + 8$	$8 + 0$
8	8

B

31. $-5 + 15 = -7 + 17$

32. $-11 + 21 = -9 + 19$

33. $-13\frac{1}{3} + 5 = 3\frac{1}{3} + (-11\frac{2}{3})$

34. $25 + (-15) = (-37) + 47$

35. $-13 + 21 + 13 = 35 + (-14)$

36. $-12 + (-3) + 7 = -16\frac{1}{3} + 8\frac{1}{3}$

Simplify.

Sample: $-13 + 5 + 3 + (-17)$

Solution: $-13 + 5 + 3 + (-17) = [-13 + (-17)] + (5 + 3)$

$$= -30 + 8$$
$$= -22 + (-8) + 8$$
$$= -22 + 0$$
$$= -22$$

37. $5 + 11 + 3 + (-17)$ 2 **38.** $-11 + (-5) + 5 + 19$ 8

39. $-11 + (-7) + 25$ 7 **40.** $19 + (-19) + 3 + 31$ 34

C

41. $-31 + 69 + 25 + 11$ 74

42. $19 + (-15) + (-11) + 5 + (-3)$ -5

43. $-43 + (-33) + 15 + (-17) + 7$ -71

44. $-5.8 + 8.8 + 1.4 + (-4.6)$ −0.2

45. $-6.4 + 16.6 + (-12.9) + 8.4$ 5.7

46. $48 + (-43) + 18 + 34 + 28 + (-2)$ 83

SELF-TEST 1

Be sure that you understand these terms.

magnitude (p. 240)   identity element (p. 240)
additive inverse (p. 241) $\stackrel{?}{=}$ (p. 243)

Sketch a number line solution. Check students' drawings.   **Section 9-1, p. 236**

1. $2 + {}^-1 = \underline{\ ?\ }$ 1   **2.** $3 + {}^-5 = \underline{\ ?\ }$ ⁻2

3. $4 + {}^-3 = \underline{\ ?\ }$ 1   **4.** ${}^-1 + 4 = \underline{\ ?\ }$ 3

Give the meaning. Then write in simplest form.   **Section 9-2, p. 240**

5. ${}^-5$ negative 5; ⁻5   **6.** $-({}^-6)$ the opposite of ⁻6; 6

Solve.

7. $z = {}^-17 + 17$ 0   **8.** $12 + {}^-12 = n$ 0

9. $0 + {}^-7 = b$ ⁻7   **10.** $y + {}^-4 = 0$ 4

Give a simpler name for the expression.   **Section 9-3, p. 243**

11. $-(7 + 2)$ −9   **12.** $-4 + (-3)$ −7

Solve. Use the number line for help if necessary.

13. $-2 + x = 1$ 3   **14.** $3 + t = -2$ −5

Show that the statement is true. Name the property illustrated.   **Section 9-4, p. 246**

15. $-2 + 2 = 0$ add. prop. of inverses

16. $(-3 + 7) + 4 = -3 + (7 + 4)$ assoc. prop.

17. $-(-9 + 2) = 9 + (-2)$ prop. of opp. of a sum

18. $-12 + 10 = 10 + (-12)$ commutative prop.

Check your answers with those printed at the back of the book.

Subtraction; Functions

9-5 Subtracting Directed Numbers

OBJECTIVE

Subtract a directed number by adding the opposite of the number.

We can use the number line to subtract directed numbers, such as $7 - 5$. It seems reasonable to show "Subtract 5" by an arrow much like the one for "Add 5." However, the arrow should point in the opposite direction.

EXAMPLE 1 $7 - 5 = \underline{\ ?\ }$

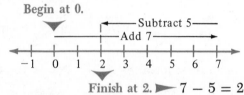

Begin at 0.

Subtract 5

Add 7

$-1 \quad 0 \quad 1 \quad 2 \quad 3 \quad 4 \quad 5 \quad 6 \quad 7$

Finish at 2. ► $7 - 5 = 2$

Look at the sketch in Example 1. It is just like the number line sketch for the equation $7 + (-5) = 2$. In fact, subtracting a directed number gives the same result as adding its opposite.

$$7 - 5 = 2 \qquad 7 + (-5) = 2$$

The equations $7 - 5 = 2$ and $7 + (-5) = 2$ are **equivalent equations**.

EXAMPLE 2 $10 - (-6) = \underline{\ ?\ }$
$10 - (-6) = 10 + 6$ ◄ To subtract -6, add its opposite.
$\qquad\qquad\quad = 16$

EXAMPLE 3 $2 - 8 = \underline{\ ?\ }$
$2 - 8 = 2 + (-8)$ ◄ To subtract 8, add its opposite.
$\qquad\quad = -6$

EXAMPLE 4 $-5 - (-4) = \underline{\ ?\ }$
$-5 - (-4) = -5 + 4$ ◄ To subtract -4, add its opposite.
$\qquad\qquad\quad = -1$

► For all directed numbers a and b, $a - b = a + (-b)$.

State the subtraction equation indicated and the equivalent addition equation.

Sample:

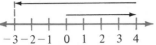

What you say: $4 - 7 = -3$
$4 + (-7) = -3$

1. 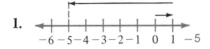 **2.**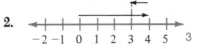

3. 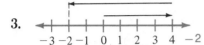 **4.**

State the equivalent addition equation.

Sample 1: $6 - (-3) = 9$ *What you say:* $6 + 3 = 9$

Sample 2: $-8 - (-3) = -5$ *What you say:* $-8 + 3 = -5$

5. $5 - (-2) = 7$ **6.** $12 - (-6) = 18$ **7.** $5 - (-9) = 14$

8. $16 - (-8) = 24$ **9.** $26 - (-6) = 32$ **10.** $33 = 24 - (-9)$

11. $-8 - 2 = -10$

12. $-12 - (-4) = -8$

13. $-3 - (-9) = 6$

5. $5 + 2 = 7$ **6.** $12 + 6 = 18$
7. $5 + 9 = 14$ **8.** $16 + 8 = 24$
9. $26 + 6 = 32$ **10.** $33 = 24 + 9$
11. $-8 + (-2) = -10$ **12.** $-12 + 4 = -8$ **13.** $-3 + 9 = 6$

For Extra Practice, see page 422.

Solve. Begin by writing the equivalent addition equation.

Sample 1: $14 - 6 = m$ *Solution:* $14 + (-6) = m$
$8 = m$

Sample 2: $2 - 9 = r$ *Solution:* $2 + (-9) = r$
$-7 = r$

A

1. $14 - 2 = b$ 12   **2.** $22 - 6 = x$ 16   **3.** $1 - 12 = s$ −11

4. $4 - 17 = x$ −13   **5.** $15 - (-2) = c$ 17   **6.** $19 - (-3) = y$ 22

7. $9 - (-24) = p$ 33   **8.** $6 - 13 = t$ −7   **9.** $21 - 7 = d$ 14

10. $6 - 16 = k$ −10   **11.** $n = 2 - (-12)$ 14   **12.** $s = 6 - (-4)$ 10

13. $-1 - (-9) = k$ 8   **14.** $-20 - (-10) = z$ -10

15. $r = 17 - 8$ 9   **16.** $p = 2 - 13$ -11

17. $w = 11 - 16$ -5 **18.** $-16 - 4 = m$ -20

19. $-4 - 21 = r$ -25 **20.** $t = -5\frac{1}{2} - 1$ $-6\frac{1}{2}$

21. $-6\frac{3}{4} - (-2\frac{1}{4}) = k$ $-4\frac{1}{2}$

Subtract. Add to check.

Sample: $\begin{array}{r} 16 \\ -8 \\ \hline \end{array}$ *Solution:* $\begin{array}{r} 16 \\ -8 \\ \hline 24 \end{array}$ Check: (Add) $\begin{array}{r} -8 \\ 24 \\ \hline 16 \end{array}$

22. $\begin{array}{r} 18 \\ -8 \\ \hline 26 \end{array}$ **23.** $\begin{array}{r} -2 \\ 8 \\ \hline -10 \end{array}$ **24.** $\begin{array}{r} 57 \\ -14 \\ \hline 71 \end{array}$ **25.** $\begin{array}{r} -32 \\ -15 \\ \hline -17 \end{array}$

26. $\begin{array}{r} 14 \\ -7 \\ \hline 21 \end{array}$ **27.** $\begin{array}{r} -2 \\ -9 \\ \hline 7 \end{array}$ **28.** $\begin{array}{r} 20 \\ -3 \\ \hline 23 \end{array}$ **29.** $\begin{array}{r} -5 \\ -19 \\ \hline 14 \end{array}$

Subtract.

Sample 1: $-4 - 7$ *Solution:* $-4 - 7 = -4 + (-7)$
$$= -11$$

Sample 2: $-15 - (-19)$

Solution: $-15 - (-19) = -15 + 19$
$$= 4$$

B

30. $7 - 19$ -12 **31.** $-22 - 4$ -26 **32.** $15 - (-16)$ 31

33. $2 - (-9)$ 11   **34.** $-4 - 12$ -16 **35.** $-10 - 32$ -42

36. $-10 - (-40)$ 30   **37.** $-0.6 - 0.4$ -1.0 **38.** $-0.50 - (-0.20)$ -0.30

Sample 3: $12 - [-(5 + 3)]$ *Solution:* $12 - [-(5 + 3)] =$
$$12 - (-8) = 12 + 8 = 20$$

C

39. $4 - [-(15 + 1)]$ 20   **40.** $11 - (1 + 4)$ 6

41. $-20 - (2 + 3)$ -25 **42.** $(6 + 1) - (-5)$ 12

43. $-(3 + 13) - 6$ -22 **44.** $(2 + 10) - 16$ -4

45. $-5 - (12 - 2)$ -15 **46.** $-17 - [-(4 - 1)]$ -14

47. $-(2 + 4) - [-(8 + 4)]$ 6

48. $-(3 - 5) - (2 + 3)$ -3

49. $2 - [-(1 - 3)]$ 0

50. $[-1 - (-1)] - [1 + (-1)]$ 0

career capsule

Plumber

Plumbers install water, gas, and waste disposal systems in homes, factories, schools and other buildings. They lay out pipe systems as a building is being built. In the final stages of construction, they install air conditioning units and connect systems of radiators, water heaters, plumbing fixtures and sprinkler systems.

The usual training for a plumber is a five year apprenticeship program. Courses in mathematics, physics and chemistry are helpful.

9-6 *Functions and Directed Numbers*

OBJECTIVE

Solve function equations that involve adding and subtracting directed numbers.

A function "machine" may accept directed numbers as inputs and give them as outputs. Recall that inputs are values for x. The outputs are values for $f(x)$.

EXAMPLE 1 $f(x) = x + 4$; replacement set: $\{-3, -2, -1, 0, 1\}$

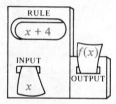

x	$f(x)$	$(x, f(x))$
-3	1	$(-3, 1)$
-2	2	$(-2, 2)$
-1	3	$(-1, 3)$
0	4	$(0, 4)$
1	5	$(1, 5)$

Function: $\{(-3, 1), (-2, 2), (-1, 3), (0, 4), (1, 5)\}$

EXAMPLE 2 $f(x) = x - 5$; replacement set: $\{2, 1, 0, -1, -2\}$

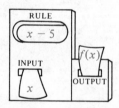

x	$f(x)$	$(x, f(x))$
2	-3	$(2, -3)$
1	-4	$(1, -4)$
0	-5	$(0, -5)$
-1	-6	$(-1, -6)$
-2	-7	$(-2, -7)$

Function: $\{(2, -3), (1, -4), (0, -5), (-1, -6), (-2, -7)\}$
Graph:

EXAMPLE 3 $f(x) = 3 - x$; replacement set: $\{4, 3, 2, 1, 0, -1, -2, \ldots\}$

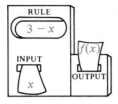

x	$f(x)$	$(x, f(x))$
4	-1	$(4, -1)$
3	0	$(3, 0)$
2	1	$(2, 1)$
1	2	$(1, 2)$
0	3	$(0, 3)$
-1	4	$(-1, 4)$
-2	5	$(-2, 5)$
.	.	.
.	.	.
.	.	.

Function: $\{(4, -1), (3, 0), (2, 1), (1, 2), (0, 3), (-1, 4),$
$(-2, 5), \ldots\}$

Tell how to complete.

$$f(x) = x + 5$$

	x	$f(x)$	$(x, f(x))$	
Sample:	1	6	$(1, 6)$	
1.	2	?	?	$7; (2, 7)$
2.	3	?	?	$8; (3, 8)$
3.	-1	?	?	$4; (-1, 4)$
4.	-2	?	?	$3; (-2, 3)$
5.	-3	?	?	$2; (-3, 2)$

$$f(x) = x - 2$$

	x	$f(x)$	$(x, f(x))$	
Sample:	-1	-3	$(-1, -3)$	
6.	-2	?	?	$-4; (-2, -4)$
7.	-3	?	?	$-5; (-3, -5)$
8.	1	?	?	$-1; (1, -1)$
9.	2	?	?	$0; (2, 0)$
10.	3	?	?	$1; (3, 1)$

Written EXERCISES

Complete the table according to the function machine. Then graph the function. Can you give a simpler rule that gives the same result?

A

RULE

$x - (-x)$

$f(x$

INPUT

OUTPUT

x

	x	$f(x)$	$(x, f(x))$	
Sample:	3	6	(3, 6)	
1.	2	?	?	4; (2, 4)
2.	1	?	?	2; (1, 2)
3.	0	?	?	0; (0, 0)
4.	−1	?	?	−2; (−1, −2)
5.	−2	?	?	−4; (−2, −4)
6.	−3	?	?	−6; (−3, −6)

8. $\{(-1, -3), (-2, -4), (-3, -5), (-4, -6)\}$
9. $\{(0, -1), (2, 1), (4, 3), (6, 5), (8, 7)\}$
10. $\{(12, 17), (10, 15), (8, 13), (6, 11), (4, 9)\}$

Complete according to the given function equation.

Sample: $f(t) = t + (-1)$: $\{(0, \underline{?}), (-1, \underline{?}), (2, \underline{?})\}$
Solution: $\{(0, -1), (-1, -2), (2, 1)\}$

7. $f(y) = y + 3$: $\{(-6, -3), (-3, \underline{?}), (0, \underline{?}), (3, \underline{?}), (6, \underline{?}),$
   $(9, \underline{?})\}$ $\{(-6, -3), (-3, 0), (0, 3), (3, 6), (6, 9), (9, 12)\}$

8. $f(m) = m + (-2)$: $\{(-1, -3), (-2, -4), (-3, \underline{?}), (-4, \underline{?})\}$

9. $f(k) = k - 1$: $\{(0, -1), (2, \underline{?}), (4, \underline{?}), (6, \underline{?}), (8, \underline{?})\}$

10. $f(t) = t - (-5)$: $\{(12, 17), (10, \underline{?}), (8, \underline{?}), (6, \underline{?}), (4, \underline{?})\}$

11. $f(n) = n + (-10)$: $\{(10, 0), (8, -2), (6, \underline{?}), (4, \underline{?}), (2, \underline{?})\}$

12. $f(k) = -9 + k$: $\{(-1, -10), (-3, \underline{?}), (-5, \underline{?}), (1, \underline{?}),$
    $(3, \underline{?})\}$ $\{(-1, -10), (-3, -12), (-5, -14), (1, -8), (3, -6)\}$

13. $f(t) = t - 0$: $\{(4, 4), (\underline{?}, 6), (\underline{?}, 8), (\underline{?}, -4), (\underline{?}, -6)\}$
    $\{(4, 4), (6, 6), (8, 8), (-4, -4), (-6, -6)\}$

11. $\{(10, 0), (8, -2), (6, -4), (4, -6), (2, -8)\}$

Tell whether or not the set of number pairs is a function.

Sample: $\{(1, -1), (2, -2), (2, -3), (4, -3)\}$
Solution: The set is not a function. Two different pairs have the same first member.

14. $\{(10, 2), (9, 2), (7, 2), (5, 2), (3, 2), (1, 2)\}$ function

15. $\{(2, 5), (3, 7), (1, 4), (2, 6), (8, 8), (3, 1)\}$ not a function

16. $\{(1, 0), (-1, 4), (2, 4), (-2, 5), (3, -1), (-3, 4)\}$ function

17. $\{(3, 3), (4, 4), (-1, 2), (2, 5), (-3, 6), (0, 0)\}$ function

18. $\{(-50, 2), (-40, 2), (-30, 2), (-20, 2), (-10, 2), (0, 2), \ldots\}$
    function

Match each set of number pairs in Column 1 with its function equation in Column 2.

COLUMN 1

19. $\{(2, -1), (3, 0), (4, 1), (5, 2), (6, 3)\}$ B   **B**

20. $\{(6, 4), (4, 2), (2, 0), (0, -2), (-2, -4)\}$ D

21. $\left\{(7, 7), (-4, -4), \left(\dfrac{1}{2}, \dfrac{1}{2}\right), (0, 0), (5, 5)\right\}$ A

22. $\{(-1, 4), (-2, 3), (-3, 2), (-4, 1), (-5, 0), (0, 5)\}$ E

23. $\{(-6, -3), (-2, 1), (2, 5), (6, 9), (10, 13)\}$ C

COLUMN 2

A. $f(t) = t + 0$

B. $f(x) = x - 3$

C. $f(s) = s - (-3)$

D. $f(y) = y + (-2)$

E. $f(n) = n - (-5)$ **26.** $\{(6, 14), (3, 11), (0, 8), (-3, 5), (-6, 2), (-8, 0)\}$

         27. $\{(-21, -28), (-14, -21), (-7, -14), (7, 0), (14, 7), (21, 14)\}$

Use the function equation and the given replacement set to write a function.   **28.** $\{(-3, 2), (-2, 3), (-1, 4), (0, 5), (1, 6), (2, 7), (3, 8)\}$

Sample: $y - (-3) = f(y)$; $\{-2, 0, 2, 4, 6\}$

Solution: $\{(-2, 1), (0, 3), (2, 5), (4, 7), (6, 9)\}$

24. $f(k) = k - 4$; $\{12, 14, 16, 18, 20\}$ $\{(12, 8), (14, 10), (16, 12), (18, 14),$ $(20, 16)\}$

25. $t + 2 = f(t)$; $\{5, 10, 15, 20, 25\}$ $\{(5, 7), (10, 12), (15, 17), (20, 22),$ $(25, 27)\}$

26. $f(h) = h - (-8)$; $\{6, 3, 0, -3, -6, -8\}$

27. $f(m) = m + (-7)$; $\{-21, -14, -7, 0, 7, 14, 21\}$

28. $f(z) = 5 - (-z)$; $\{-3, -2, -1, 0, 1, 2, 3\}$

29. $f(n) = -n$; $\left\{\dfrac{1}{2}, 1, 1\dfrac{1}{2}, 2\dfrac{3}{4}\right\}$ $\left\{\left(\dfrac{1}{2}, -\dfrac{1}{2}\right), (1, -1), (1\dfrac{1}{2}, -1\dfrac{1}{2}), (2\dfrac{3}{4}, -2\dfrac{3}{4})\right\}$ **C**

30. $f(x) = x - 1.5$; $\{-1.5, 3, 3.5, 7\}$ $\{(-1.5, -3.0), (3, 1.5), (3.5, 2.0),$ $(7, 5.5)\}$

31. $f(t) = t - (-1)$; $\left\{-\dfrac{1}{2}, -\dfrac{1}{4}, 0, \dfrac{1}{4}\right\}$ $\left\{\left(-\dfrac{1}{2}, \dfrac{1}{2}\right), \left(-\dfrac{1}{4}, \dfrac{3}{4}\right), (0, 1)\right.$ $\left.\left(\dfrac{1}{4}, 1\dfrac{1}{4}\right)\right\}$

32. $f(z) = z + 0.2$; $\{-1, 0, 1, 2\}$ $\{(-1, -0.8), (0, 0.2), (1, 1.2), (2, 2.2)\}$

33. $b - 2 = f(b)$; $\left\{-1, -\dfrac{1}{2}, \dfrac{1}{2}, 1\right\}$ $\left\{(-1, -3), \left(-\dfrac{1}{2}, -2\dfrac{1}{2}\right), \left(\dfrac{1}{2}, -1\dfrac{1}{2}\right)\right.$ $\left.(1, -1)\right\}$

34. $0.5 + (-n) = f(n)$; $\{-0.5, 0.5, 1, 1.5\}$ $\{(-0.5, 1.0), (0.5, 0)(1, -0.5)$ $(1.5, -1.0)\}$

SELF-TEST 2

Solve. Begin by writing the equivalent addition equation.

Section 9-5, p. 250

1. $8 - 3 = x$ $8 + (-3) = x; 5 = x$ **2.** $5 - 7 = y$ $5 + (-7) = y; -2 = y$

3. $m = -2 - 7$ $m = -2 + (-7);$ **4.** $n = -3 - (-4)$ $n = -3 + 4;$
$\quad m = -9$ $\quad\qquad n = 1$

Subtract.

5. $\begin{array}{r} 20 \\ -12 \\ \hline 32 \end{array}$
    6. $\begin{array}{r} 4 \\ -5 \\ \hline 9 \end{array}$

7. $\begin{array}{r} 10 \\ -10 \\ \hline 20 \end{array}$
    8. $\begin{array}{r} 7 \\ -3 \\ \hline 10 \end{array}$

9. $\{(3, 1), (2, 0), (-2, -4), (0, -2)\}$

Section 9-6, p. 254 Complete according to the given function equation.

9. $f(x) = x + (-2)$: $\{(3, \underline{\ ?\ }), (2, \underline{\ ?\ }), (-2, \underline{\ ?\ }), (0, \underline{\ ?\ })\}$

10. $f(x) = x - 4$: $\{(4, \underline{\ ?\ }), (3, \underline{\ ?\ }), (-1, \underline{\ ?\ }), (\underline{\ ?\ }, 0)\}$

10. $\{(4, 0), (3, -1), (-1, -5), (4, 0)\}$

Check your answers with those printed at the back of the book.

calculator corner

When you use your calculator, you must remember to follow the rules of algebra. To solve a problem such as $(65 \times 2) + (75 \times 3)$ you must first multiply 65×2 and write down the answer, multiply 75×3 and write down the answer, and finally add the two answers. What is $(65 \times 2) + (75 \times 3)$?

Here is a formula which shows a quicker way to calculate a problem like $(65 \times 2) + (75 \times 3)$ without any writing: $(A \times B) + (C \times D) = [((A \times B) \div D) + C] \times D$. Let $A = 65$, $B = 2$, $C = 75$

and $D = 3$. Find $[((A \times B) \div D) + C] \times D$ with your calculator. Does $(65 \times 2) + (75 \times 3) = [((65 \times 2) \div 3) + 75] \times 3$? Multiply $13 \times 2 + 14 \times 5$ with your calculator, using the formula and $A = 13, B = 2, C = 14, D = 5$. Have a friend multiply $(13 \times 2) + (14 \times 5)$ with a calculator, without any formulas. Which method is faster?

Of course, your calculator may have a feature which enables it to calculate $(65 \times 2) + (75 \times 3)$ directly. If so, try both methods. You'll see what a time-saver that feature is. 355

chapter summary

1. Addition of directed numbers can be shown on the number line.

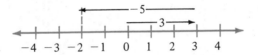

2. Every directed number has an **opposite**. Its opposite is also called its **additive inverse.**

3. The sum of any directed number and its opposite is 0.

4. The opposite of every positive number is negative.
   The opposite of every negative number is positive.
   The opposite of 0 is 0.

5. The sum of any directed number and 0 is that number. That is, 0 is the **identity element** for addition.

6. Subtraction of directed numbers can be shown on the number line.

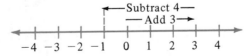

7. Subtracting one directed number from another is the same as adding the opposite. In general, we say $r - s = r + (-s)$.

chapter test

Name the opposite.

1. 5 –5

2. -3 3 or $-(-3)$

3. $^-2$ 2 or $-(^-2)$

Solve.

4. $n + {}^-4 = 0$ 4

5. $-7 + 7 = t$ 0

6. $-12 + x = 0$ 12

7. $6 + n = 5$ –1

8. $3 = -2 + m$ 5

9. $4 + a = -2$ –6

Name the property illustrated.

10. $-12 + 12 = 0$ add. prop. of opposites
11. $2 + [-3 + (-2)] = [2 + (-3)] + (-2)$ assoc. prop.
12. $7 + (-12) = -12 + 7$ comm. prop.
13. $-5 + [-(-5)] = 0$ add. prop. of opposites

Solve. Begin by writing an equivalent addition equation.

14. $x = 4 - 5$ –1

15. $n = 8 - 12$ –4

16. $-2 - 5 = x$ –7

17. $2 - (-9) = b$ 11

18. $z = -5 - (-6)$ 1

19. $4 - (-2) = c$ 6

Complete according to the function equation.

20. $f(t) = t + (-2)$; $\{(0, \underline{\ ?\ }), (-1, \underline{\ ?\ })\}$ $\{(0, -2), (-1, -3)\}$
21. $f(z) = z - (-8)$; $\{(-5, \underline{\ ?\ }), (2, \underline{\ ?\ })\}$ $\{(-5, 3), (2, 10)\}$
22. $f(c) = -1 - c$; $\{(8, \underline{\ ?\ }), (-6, \underline{\ ?\ })\}$ $\{(8, -9), (-6, 5)\}$

Complete the table according to the given function rule. Then write the set of ordered pairs and graph the function.

$$f(a) = -a + 1$$

	a	$f(a)$	$(a, f(a))$	
Sample:	-1	2	$(-1, 2)$	
23.	-2	? 3	?	$(-2, 3)$
	-3	? 4	?	$(-3, 4)$
	0	? 1	?	$(0, 1)$
	1	? 0	?	$(1, 0)$
	2	? –1	?	$(2, -1)$
	3	? –2	?	$(3, -2)$

challenge topics

Absolute Value

Study the number line pictured below. How many units from 0 is 4? How many units from 0 is −4? Do you see that each number is 4 units from 0? The distance from 0 of a number on the number line is called the absolute value of the number.

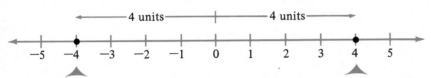

The absolute value of −4 is 4.   The absolute value of 4 is 4.

The symbol $|x|$ stands for the absolute value of the directed number x. The symbol is read "the absolute value of x."

$$|10| = 10; \qquad \left|\frac{2}{3}\right| = \frac{2}{3}; \qquad |35| = 35$$

$$|-10| = 10; \qquad \left|-\frac{2}{3}\right| = \frac{2}{3}; \qquad |-35| = 35$$

We define absolute value for directed numbers as follows:

When $r \geq 0$, $|r| = r$. When $r < 0$, $|r| = -r$.

Note that the absolute value of *any* directed number is *always* a **positive** number or **zero.**

True or false?

1. $|-3| = 3$ true
2. $|2| = |-2|$ true
3. $|-1| > |0|$ true
4. $|-1| \neq |1|$ false
5. $|0| < |-1|$ true
6. $|4| > |-4|$ false

Solve.

7. $|x| = 2$ {2, −2}
8. $-|y| = -2$ {2, −2}
9. $-|t| + 2 = 0$ {2, −2}
10. $|t| = 0$ {0}
11. $|a| + 1 = 3$ {2, −2}
12. $|y| - 2 = 4$ {6, −6}

Find the value.

13. $|2 + 3|$ 5
14. $|8 - 7|$ 1
15. $-|3| + |3|$ 0
16. $|8 - 7| - |1|$ 0
17. $|6 - 9|$ 3
18. $|-3 + 2| - |1|$ 0

Review of Skills

Perform the operation indicated.

1. $1\frac{1}{4} \times \frac{1}{2}$ $\frac{5}{8}$

2. $1\frac{5}{9} \div 4\frac{2}{3}$ $\frac{1}{3}$

3. $3\frac{1}{8} \div 1\frac{1}{4}$ $\frac{5}{2}$ or $2\frac{1}{2}$

4. 0.02×7.1 0.142

5. $\dfrac{12.3}{0.3}$ 41

6. 0.66×0.5 0.33

Complete.

7. $-4 + (-4) + (-4) = \underline{\ ?\ }$ -12
   $3 \cdot (-4) = \underline{\ ?\ }$ -12

8. $-2 + (-2) + (-2) + (-2) = \underline{\ ?\ }$ -8
   $4 \cdot (-2) = \underline{\ ?\ }$ -8

Complete the pattern.

9.
$3 \cdot 4 = 12$
$2 \cdot 4 = 8$
$1 \cdot 4 = 4$
$0 \cdot 4 = \underline{\ ?\ }$ 0
$-1 \cdot 4 = \underline{\ ?\ }$ -4
$-2 \cdot 4 = \underline{\ ?\ }$ -8

10.
$3 \cdot 7 = 21$
$2 \cdot 7 = 14$
$1 \cdot 7 = \underline{\ ?\ }$ 7
$0 \cdot 7 = \underline{\ ?\ }$ 0
$-1 \cdot 7 = \underline{\ ?\ }$ -7
$-2 \cdot 7 = \underline{\ ?\ }$ -14

True or false? (*a* and *b* are directed numbers.)

11. $a \cdot b = b \cdot a$ true

12. $a + b = b + a$ true

13. $a + (b + c) = (a + b) + c$ true

14. $a \cdot (b \cdot c) = (a \cdot b) \cdot c$ true

15. $a \cdot (b + c) = (a + b) + (a \cdot c)$ false

16. $a \cdot (b + c) = a \cdot b + a \cdot c$ true

Simplify. Use the distributive property.

17. $7y + 9y$ 16y

18. $2z + 3z$ 5z

19. $4t + 2n + 3n + t$ 5t + 5n

20. $14m + m + 2b$ 15m + 2b

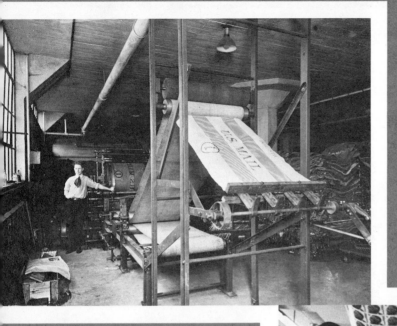

Left: Machine which produced eight mail bags per minute.

Right: Press which can print one hundred million stamps per day.

10
Multiplication and Division of Directed Numbers

Multiplication

10-1 *Multiplication by a Positive Number or by Zero*

OBJECTIVES

Multiply a directed number by a positive number.

Apply the multiplication properties of 0 and 1.

Multiplication by a positive integer can be thought of as repeated addition.

EXAMPLE 1 $3 \cdot 5 = 5 + 5 + 5 = 15$
$3 \cdot 5 = 15$

EXAMPLE 2 $2(-7) = -7 + (-7) = -14$
$2(-7) = -14$

It is not clear how to interpret $-9 \cdot 3$ as repeated addition. The commutative property, which we assume to be true for directed numbers, makes it easier.

EXAMPLE 3 $-9 \cdot 3 = 3(-9) = -9 + (-9) + (-9)$
$-9 \cdot 3 = -27$

We also assume the multiplicative properties of 0 and 1 for directed numbers.

EXAMPLE 4 $5 \cdot 0 = 0 \cdot 5 = 0$ $-4 \cdot 0 = 0(-4) = 0$

EXAMPLE 5 $1 \cdot 8 = 8 \cdot 1 = 8$ $1(-7) = -7 \cdot 1 = -7$

Look closely at Examples 1–5. Notice the following patterns.

▶ The product of two **positive** numbers is **positive**.
The product of a **positive** number and a **negative** number is **negative**.
The product of 0 and any directed number is 0.
The product of 1 and any directed number is that directed number.

Tell how to complete each of the following to make a true statement.

Sample: $-7 + (-7) + (-7) + (-7) = \underline{\ ?\ }$
$4(-7) = \underline{\ ?\ }$

What you say: $-7 + (-7) + (-7) + (-7) = -28$
$4(-7) = -28$

1. $5 + 5 + 5 = \underline{\ ?\ }$ 15
$3 \cdot 5 = \underline{\ ?\ }$ 15

2. $3 + 3 + 3 + 3 = \underline{\ ?\ }$ | 12
$4 \cdot 3 = \underline{\ ?\ }$ | 12

3. $-8 + (-8) + (-8) = \underline{\ ?\ }$ -24
$3(-8) = \underline{\ ?\ }$ -24

4. $-12 + (-12) = \underline{\ ?\ }$ | -24
$2(-12) = \underline{\ ?\ }$ | -24

5. $0 + 0 + 0 + 0 = \underline{\ ?\ }$ 0
$4 \cdot 0 = \underline{\ ?\ }$ 0

6. $-11 + (-11) = \underline{\ ?\ }$ | -22
$2(-11) = \underline{\ ?\ }$ | -22

For Extra Practice, see page 423.

Simplify. Assume that the variable represents a positive number.

Written EXERCISES

Sample 1: $-16 \cdot 3$ Solution: $-16 \cdot 3 = -48$

Sample 2: $n\left(-\dfrac{2}{3}\right)$ Solution: $n\left(-\dfrac{2}{3}\right) = -\dfrac{2}{3} \cdot n = -\dfrac{2n}{3}$

1. $4(-6)$ -24 **2.** $5(-8)$ -40 **3.** $11 \cdot 13$ 143   **A**

4. $3 \cdot 7m$ 21m   **5.** $4(-9)$ -36 **6.** $6(-7)$ -42

7. $-5 \cdot w$ $-5w$ **8.** $-8 \cdot 7$ -56 **9.** $p\left(-\dfrac{1}{5}\right)$ $-\dfrac{p}{5}$

10. $-4 \cdot \dfrac{1}{3}$ $-\dfrac{4}{3}$ **11.** $-0.15(2)$ -0.30 **12.** $\left(-\dfrac{1}{5}\right)\left(\dfrac{3}{4}\right)$ $-\dfrac{3}{20}$

13. $-3(0.321)$ -0.963 **14.** $\dfrac{2}{7}\left(-\dfrac{2}{5}\right)$ $-\dfrac{4}{35}$ **15.** $\left(\dfrac{1}{6}\right)(5)$ $\dfrac{5}{6}$

Complete.

16.

×	0	1	3	6	9
0	?0	?0	?0	?0	?0
−1	?0	?−1	?−3	?−6	?−9
−3	?0	?−3	?−9	−18	? −27
−6	?0	?−6	?−18	?−36	? −54
−9	?0	?−9	−27	?−54	? −81

17.

×	−1	−3	−5	−7
1	?−1	?−3	?−5	?−7
3	?−3	?−9	?−15	−21
5	?−5	−15	?−25	?−35
7	?−7	?−21	?−35	?−49
9	?−9	?−27	?−45	?−63

Simplify.

Sample: $2(-6) + 2(-4)$

Solution: $2(-6) + 2(-4) = -12 + (-8) = -20$

18. $3(-9) + 3 \cdot 4$ −15   **19.** $(-2 \cdot 7) + (-2 \cdot 5)$ −24

20. $2(-6) + 2(-4)$ −20   **21.** $8 + (-2 \cdot 7)$ −6

22. $(-1 \cdot 8) + 19$ 11   **23.** $(-7 \cdot 1) + 8$ 1

24. $(-3 \cdot 5) + (-3 \cdot 5)$ −30   **25.** $9(-3) + 9(-3)$ −54

Solve.

Sample: $3 \cdot m = 21$ *Solution:* $3 \cdot m = 21$

$$3 \cdot 7 = 21$$

So, $m = 7$.

B

26. $4(-8) = h$ −32   **27.** $2(m) = -36$ −18

28. $6(-11) = h$ −66   **29.** $3 \cdot m = -45$ −15

30. $-8 \cdot \dfrac{3}{4} = x$ −6   **31.** $-4\frac{1}{5}x = -4\frac{1}{5}$ 1

32. $p(-6) = -48$ 8   **33.** $5(-12) = a$ −60

34. $-12 \cdot 0 = n$ 0   **35.** $-1 \cdot 17 = x$ −17

36. $-1 \cdot 39 = a$ −39   **37.** $-5\frac{1}{3} = 5\frac{1}{3}m$ −1

C

38. $a + 2(-3) = -10$ −4   **39.** $-13 \cdot b = -3 \cdot 13$ 3

40. $5 + 3(-4) = m$ −7   **41.** $(-6 \cdot 3) + p = 15$ 33

42. $-5 \cdot 11 = -5 \cdot m$ 11   **43.** $(-7 \cdot 4) + m = -31$ −3

Kotaro Honda   *1870–1954*

Kotaro Honda was one of Japan's leading metallurgists. (A metallurgist is a scientist who experiments with metals.) In 1916 he found that the addition of cobalt to tungsten steel produced a more powerful magnet than steel. Nothing more advanced in the field of magnetics was discovered until the mid 20th century. In 1937 Honda was awarded the Cultural Order of the Rising Sun, Japan's equivalent of the Nobel Prize.

10-2 *Multiplication Properties for Directed Numbers*

OBJECTIVES

Apply the associative property of multiplication to directed numbers.

Apply the distributive property to directed numbers.

We assume that multiplication of directed numbers is **associative.** That is, the way that factors are grouped does not affect the product.

EXAMPLE 1 $(4 \cdot 2)(-3) = 8(-3) = -24;$

$4 \cdot [2(-3)] = 4(-6) = -24$

We also assume that the **distributive** property holds for multiplication of directed numbers.

EXAMPLE 2 $\underline{4[5 + (-2)] = 4 \cdot 5 + 4(-2)}$

$4 \cdot 3$	$20 + (-8)$
12	12 $\checkmark$

EXAMPLE 3 $\underline{-3(1 + 6) = -3 \cdot 1 + (-3 \cdot 6)}$

$-3 \cdot 7$	$-3 + (-18)$
-21	-21 $\checkmark$

Let's review the properties of multiplication which we will assume to be true for directed numbers. For all directed numbers r, s, and t:

1. $r \cdot s = s \cdot r$ ◀ The Commutative Property
2. $(r \cdot s) \cdot t = r \cdot (s \cdot t)$ ◀ The Associative Property
3. $r(s + t) = r \cdot s + r \cdot t$ ◀ The Distributive Property
4. $r \cdot 0 = 0 \cdot r = 0$ ◀ The Multiplicative Property of Zero
5. $r \cdot 1 = 1 \cdot r = r$ ◀ The Multiplicative Property of One

Oral EXERCISES

Name the property illustrated.

Sample: $5(-8) = -8 \cdot 5$ *What you say:* Commutative property.

1. $8(-5 + 4) = 8(-5) + 8 \cdot 4$ distributive
2. $17(-14) = -14 \cdot 17$ commutative
3. $12 \cdot 0 = 0$ multiplicative property of 0
4. $-1 \cdot 25 = 25(-1)$ commutative
5. $2(-9) + 5(-9) = (2 + 5)(-9)$ distributive
6. $-2\frac{1}{5} = -2\frac{1}{5} \cdot 1$ multiplicative property of 1
7. $(5 \cdot 4)3 = 3(5 \cdot 4)$ commutative

True or false?

8. $-8(3) = -8[5 + (-2)]$ true
9. $7(-2) > 6(-2)$ false
10. $8(-4)3 = 8 \cdot 4(-3)$ true
11. $[-7 + (-5)]2 = (-7 \cdot 2) + (-5 \cdot 2)$ true
12. $0(-3) > 2 \cdot 5$ false
13. $[-5 + (-4)]3 = -5 \cdot 3 + (-4)3$ true
14. $-4 \cdot 5 \neq -5 \cdot 4$ false
15. $(-2\frac{3}{5})(4) = (4)(-2\frac{3}{5})$ true

Written EXERCISES

Write the expression as a sum of two products. Then simplify the expressions to show that they are equal.

Sample 1: $7[6 + (-9)]$

Solution: $7[6 + (-9)] = 7(6) + 7(-9)$

$7(-3)$	$42 + (-63)$
-21	-21

A

1. $3(4 + 1)$ 15
2. $4(-5 + 2)$ -12
3. $7[-8 + (-1)]$ -63
4. $-1(3 + 5)$ -8
5. $-8(1 + 6)$ -56
6. $10[-10 + (-10)]$ -200
7. $[3 + (-6)]8$ -24
8. $[-4 + (-2)]3$ -18
9. $(-8 + 6)\frac{1}{2}$ -1
10. $[9 + (-12)]\frac{1}{3}$ -1
11. $-\frac{1}{4}(12 + 16)$ -7
12. $-0.8(6 + 0.1)$ -4.88

Show that the statement is true.

For Ex. 13–24, the final step is given.

Sample: $-4(3 + 5) = (-4 \cdot 3) + (-4 \cdot 5)$

Solution: $-4(3 + 5) = (-4 \cdot 3) + (-4 \cdot 5)$

$$\begin{array}{c|c} -4(8) & -12 + (-20) \\ -32 & -32 \ \checkmark \end{array}$$

13. $(4)[5(-2)] = (4 \cdot 5)(-2) \ {\scriptstyle -40 \ = \ -40}$

14. $(4 - 7)3 = (4 \cdot 3) - (7 \cdot 3) \ {\scriptstyle -9 \ = \ -9}$

15. $[2 + (-5)]3 = (2 \cdot 3) + (-5 \cdot 3) \ {\scriptstyle -9 \ = \ -9}$

16. $-4(6 + 4) = (-4 \cdot 6) + (-4 \cdot 4) \ {\scriptstyle -40 \ = \ -40}$

17. $-3(8 + 2) = (-3 \cdot 8) + (-3 \cdot 2) \ {\scriptstyle -30 \ = \ -30}$

18. $-4(2 + 3 + 1) = (-4 \cdot 2) + (-4 \cdot 3) + (-4 \cdot 1) \ {\scriptstyle -24 \ = \ -24}$

19. $(5 - 8)(3) = 5 \cdot 3 - 8 \cdot 3 \ {\scriptstyle -9 \ = \ -9}$

20. $(-7 + 5)4 = (-7 \cdot 4) + (5 \cdot 4) \ {\scriptstyle -8 \ = \ -8}$

21. $[-4 + 1 + (-2)]3 = (-4 \cdot 3) + (1 \cdot 3) + (-2 \cdot 3) \ {\scriptstyle -15 \ = \ -15}$

22. $-3\left(\dfrac{1}{4} + \dfrac{1}{2}\right) = \left(-3 \cdot \dfrac{1}{4}\right) + (-3)\left(\dfrac{1}{2}\right) \ {\scriptstyle -\frac{9}{4} \ = \ -\frac{9}{4}}$

23. $5[-3 + (-4)] = 5(-3) + 5(-4) \ {\scriptstyle -35 \ = \ -35}$

24. $(-7 \cdot 2) + (-6 \cdot 2) = [-7 + (-6)]2 \ {\scriptstyle -26 \ = \ -26}$

Let $w = 3$, $x = 5$, $y = \dfrac{1}{4}$, $z = -8$. Show that the resulting statement is true.

25. $w(x + z) = wx + wz \ {\scriptstyle -9}$

26. $x \cdot z = z \cdot x \ {\scriptstyle -40}$ **B**

27. $y(w + z) = (w + z)y \ {\scriptstyle -\frac{5}{4}}$

28. $(xz)y = x(zy) \ {\scriptstyle -10}$

29. $-1 \cdot x = x(-1) \ {\scriptstyle -5}$

30. $2(w + z) = 2w + 2z \ {\scriptstyle -10}$

31. $-3(x - w) = (-3x) - (-3w) \ {\scriptstyle -6}$

32. $4z + 4x = 4(z + x) \ {\scriptstyle -12}$

33. $w \cdot 0 = 0 \cdot w \ {\scriptstyle 0}$

34. $w(x + z) = wz + wx \ {\scriptstyle -9}$

Multiply.

Sample: $3(-47)$ *Solution:*

$$\begin{array}{r} -50 + 3 \\ 3 \\ \hline -150 + 9 = -141 \end{array}$$

35. $3(-216) \ {\scriptstyle -648}$

36. $-2(321) \ {\scriptstyle -642}$

37. $4(-28) \ {\scriptstyle -112}$ **C**

38. $10(-327) \ {\scriptstyle -3270}$

39. $-5(149) \ {\scriptstyle -745}$

40. $2(-607) \ {\scriptstyle -1214}$

41. $-4(48) \ {\scriptstyle -192}$

42. $9(-118) \ {\scriptstyle -1062}$

43. $6(-248) \ {\scriptstyle -1488}$

10-3 *Multiplication of Negative Numbers*

We have seen the following pattern for the product of two directed numbers:

$$\text{positive} \times \text{positive} = \text{positive}$$
$$\text{positive} \times \text{negative} = \text{negative}$$
$$\text{negative} \times \text{positive} = \text{negative}$$

We can use the distributive property to show:

$$\text{negative} \times \text{negative} = \text{positive}$$

EXAMPLE 1
$$-2(-3 + 7) = -2(-3) + (-2 \cdot 7)$$
$$-2 \cdot 4 = -2(-3) + (-2 \cdot 7)$$
$$-8 = -2(-3) + (-14)$$
$$-8 = 6 + (-14)$$
Then it must be true that $-2(-3) = 6$.

EXAMPLE 2
$$-5[10 + (-6)] = -5 \cdot 10 + [-5(-6)]$$
$$-5 \cdot 4 = -50 + [-5(-6)]$$
$$-20 = -50 + 30 \quad \blacktriangleleft -5(-6) \text{ must equal } 30.$$
$$-20 = -20$$

EXAMPLE 3
$$-1(-2 + 9) = -1(-2) + (-1 \cdot 9)$$
$$-1 \cdot 7 = 2 + (-9) \quad \blacktriangleleft -1(-2) = 2$$
$$-7 = -7$$

Here's a simple way to remember the sign of the product of two directed numbers.

▶ If the two signs are **alike**, the product is **positive**.
If the two signs are **unlike**, the product is **negative**.

We use these same rules to determine the sign of a product when more than two factors are multiplied.

$$3 \cdot 4(-1) \blacktriangleright (3 \cdot 4) \cdot (-1) = -12$$

positive   negative

▶ The product of two or more directed numbers is positive if there is an even number of negative factors. The product is negative if there is an odd number of negative factors.

Tell whether the expression represents a positive number or a negative number. Do not simplify.

1. $-4 \cdot 2$ neg.

2. $-4(-5)(-7)$ neg.

3. $\frac{1}{2}(-2)5$ neg.

4. $-2(-3)4$ pos.

5. $\frac{1}{2}(-2)\left(-\frac{1}{3}\right)$ pos.

6. $-3 \cdot 9(-1)(-2)$ neg.

7. $\frac{1}{5}\left(-\frac{1}{8}\right)(-2)$ pos.

8. $-7\left(-\frac{1}{2}\right)(-2)$ neg.

9. $-4(-4)$ pos.

10. $-1 \cdot 3\left(-\frac{1}{2}\right)$ pos.

11. $-6(-6)(-6)$ neg.

12. $-5 \cdot 5(-5)$ pos.

For Extra Practice, see page 423.

Tell whether the expression represented is positive, negative, or zero. Then simplify.

EXERCISES

Sample: $-4 \cdot 3 \cdot 2$ *Solution:* negative; -24

1. $-3 \cdot 6$ -18

2. $-5 \cdot 1(-4)$ 20

3. $-3(-2)(-5)$ -30 A

4. $-\frac{1}{2} \cdot 0(-5)$ 0

5. $-3(-5)2$ 30

6. $\frac{1}{3}(-2)$ $-\frac{2}{3}$

7. $-3(-2)2(-1)$ -12

8. $(-2)^3$ -8

9. $-7(-3)1$ 21

10. $-5(-2)2$ 20

11. $(-5)^2$ 25

12. $-6 \cdot 2 \cdot 0 \cdot \frac{1}{2}$ 0

Rewrite the expression as a sum of two products. Then simplify both expressions to show that they are equal.

Sample 1: $5[2 + (-4)]$

Solution: $5[2 + (-4)] = 5(2) + 5(-4)$

$$\frac{5(-2)\ \big|\ 10 + (-20)}{-10\ \big|\ -10\ \checkmark}$$

Sample 2: $-3(-2 + 4)$

Solution: $-3(-2 + 4) = -3(-2) + (-3)(4)$

$$\frac{-3(2)\ \big|\ 6 + (-12)}{-6\ \big|\ -6\ \checkmark}$$

13. $-3[5 + (-3)]$ −6   14. $5[-3 + (-2)]$ −25 15. $-2(7 + 5)$ −24

16. $3(-8 + 5)$ −9   17. $(-6 + 3)(-2)$ 6   18. $(-8 + 2)(-2)$ 12

19. $-\dfrac{1}{2}(5 + 0)$ $-\dfrac{5}{2}$ 20. $2[8 - (-2)]$ 20

21. $-5[-3 + (-4)]$ 22. $-2(-5 + 5)$ 21. 35   22. 0

23. $-6\left[\dfrac{1}{2} + (-1)\right]$ 24. $-3[-2 + (-1)]$ 23. 3   24. 9

Evaluate. Let $a = 3$, $b = -2$, $c = 1$, $d = 5$, $f = -3$.

Sample: $a \cdot b \cdot f$

Solution: $3(-2)(-3) = 18$

B

25. $-a \cdot b$ 6   26. $b \cdot c(-2)$ 4   27. $(a)(-d)(-1)$ 15

28. $-3(a)(-c)$ 9   29. $f \cdot a - 2c$ −11   30. $3 \cdot a \cdot c \cdot f$ −27

31. $ab + cf$ −9   32. $(-c) + df$ −16   33. $(-2)f + a$ 9

Simplify.

C

34. $-3 + [-6 + (2 \cdot 4)]$ −1

35. $[-3 + (-2)] + [-4(-1 + 5)]$ −21

36. $-3[4(-6 + 4)]$ −24

37. $5[-1(2 \cdot 3) + (-2 \cdot 0)]$ −30

Evaluate.

38. $(-1)^2$ 1   39. $(-1)^3$ −1   40. $(-3)^2$ 9

41. $(-3)^3$ −27   42. $(-8)^2$ 64   43. $(-8)^3$ −512

10-4 *The Distributive Property in Simplifying Expressions*

OBJECTIVE
Simplify algebraic expressions by combining similar terms.

We use the distributive property to simplify expressions.

EXAMPLE 1 $7x + 3x = (7 + 3)x = 10x$

EXAMPLE 2 $7m - 3a + 2m + a = 7m + 2m + (-3a) + a$ ◀ Group like terms.
$$= (7 + 2)m + (-3 + 1)a \blacktriangleleft \text{Distributive}$$
$$= 9m + (-2)a \qquad\qquad\qquad \text{property}$$
$$= 9m - 2a$$

EXAMPLE 3 $7n + 6n + n - 3 = (7n + 6n + n) - 3$
$$= (7 + 6 + 1)n - 3$$
$$= 14n - 3$$

EXAMPLE 4 $15t + 9s + 3t - 7s = 15t + 3t + 9s + (-7s)$
$$= (15 + 3)t + [9 + (-7)]s$$
$$= 18t + 2s$$

EXAMPLE 5 $-5x + y + 3y - x = -5x + (-x) + y + 3y$
$$= -5x + (-1x) + 1y + 3y \quad \blacktriangleleft -x = -1x$$
$$= [-5 + (-1)]x + (1 + 3)y$$
$$= -6x + 4y$$

As soon as you understand this, you will probably do much of it in your head.

Simplify by combining similar terms.

Sample: $5w - 2w + 4$ *What you say:* $3w + 4$

1. $3p + 4p + 2p$ $9p$
2. $10a - 6 + 4a$ $14a - 6$
3. $8n + (-3n)$ $5n$
4. $4x - 2 + (-2x)$ $2x - 2$
5. $3w - 3n + 5w$ $8w - 3n$
6. $6x + 3y + (-4x) + 3$

EXERCISES

$2x + 3y + 3$

7. $\frac{2}{3}x + 3y + \frac{1}{3}x$ $x + 3y$ 8. $-7a + 3b + 5a$ $-2a + 3b$

9. $3m + 2n + (-5m) + (-2n)$ 10. $3c + 4a + (-5c) + (-2)$
$-2m$ $-2c + 4a - 2$

For Extra Practice, see page 424.   6. $\frac{1}{3}y + \frac{1}{6}x$ 8. $2w - 2$

Written EXERCISES

Simplify by combining similar terms.

10. $2a^2 + 2a + 1$ 12. $7r^2 + 1$

Sample 1: $5x + 3m - 2x + m$

Solution: $5x + 3m - 2x + m = [5x + (-2x)] + (3m + m)$
$$= 3x + 4m$$

Sample 2: $3a + 2(-4a - 6b)$

Solution: $3a + 2(-4a - 6b) = 3a + (-8a) + (-12b)$
$$= -5a + (-12b) = -5a - 12b$$

A
$-3a + 4b$
1. $3b + (-4a) + b + a$ 2. $-a + 5b + 2a - 3b$ $a + 2b$

$-9xy - 3p$ 3. $-2xy - 3p + (-7xy)$ 4. $2r - 5s - 6r + 3s$ $-4r - 2s$

$3m - 6x - 2$ 5. $-m - 7x - 2 + 4m + x$ 6. $\frac{2}{3}y - \frac{1}{2}x + \left(-\frac{1}{3}y\right) + \frac{2}{3}x$

$2a + 2c + 6$ 7. $4a + 3c + (-2a) + 6 - c$ 8. $3w + y + (-w) - 2 + (-y)$

$ab - 5$ 9. $2ab + (-6) + (-ab) + 1$ 10. $-3a^2 + 2a + a^2 - 1$

$2c^2 + 7c$ 11. $4c^2 + 7c + (-6c^2)$ 12. $-2r^2 + 3 + (-5r^2) + (-2)$

B   13. $x(3x - 2) + (x^2 + 5)$ 14. $-5(a + 4) + (-3a)$ $-8a - 20$

$ab + 3c + 1$ 15. $3(ab + c) + (-2ab) + 1$ 16. $-3(mn + 2) - mn + 4$

$-13s + 6$ 17. $-7s + 3(-2s + 2)$ 18. $13a + [-7(a - 1)]$ $6a + 7$

19. $\frac{2}{5}m + \frac{1}{3}m^2 + \frac{1}{5}m + \left(-\frac{2}{3}m^2\right)$ $\frac{3}{5}m - \frac{1}{3}m^2$ 13. $4x^2 - 2x + 5$

16. $-4mn - 2$

20. $10y - 12w + 2(-8y + 3w)$ $-6y - 6w$

C   21. $2[x(3x - 2)] + (x^2 + 5)$ $7x^2 - 4x + 5$

22. $-3y[(2y - 1)2 + (-y)]$ $-9y^2 + 6y$

23. $-5a(6a + 4) + [-4(20a^2 + 15a)]$ $110a^2 - 80a$

24. $[-5 + (-3c) + (-4c^2)]10 + c^2$ $-39c^2 - 30c - 50$

Find the value of the expression, if $a = -2$, $b = 7$, $c = -1$, and $d = 5$.

25. $6 + (b + d)^2$ $15c$ 26. $5(a + b) + d$ 30

27. $\frac{1}{2}(ab) + 1$ $_{-7}$

28. $a(3c - d)$ 16

29. $a^2 \cdot c$ $_{-4}$

30. $-2(cd + a) - 1$ $_{-13}$

31. $(d + c)^2 + a$ 14

32. $(a + c)^2 + cd$ 4

33. $\left(\frac{1}{5}d + c\right)\frac{1}{2}$ 0

34. $cab - 2c^2$ 12

35. $0.3cd + (7a)^2$ 194.5

36. $0.2c + 0.3ac^2$ $_{-0.8}$

SELF-TEST 1

Multiply.

1. $3(-4)$ $_{-12}$ **2.** $-8 \cdot 5$ $_{-40}$ **3.** $-9 \cdot 0$ 0   Section 10-1, p. 264

Solve.

4. $-4 \cdot n = -4$ 1   **5.** $-2 \cdot 5 = m$ $_{-10}$ **6.** $3x = -6$ $_{-2}$

Name the property illustrated.   Section 10-2, p. 267

7. $-3 \cdot 4 = 4(-3)$ commutative

8. $(-6 \cdot 4)8 = 8(-6 \cdot 4)$ commutative

9. $4[-6 + (-2)] = 4(-6) + 4(-2)$ distributive

Tell whether the product is positive, negative, or zero.   Section 10-3, p. 270
Then simplify.

10. $-2(-3) \cdot 4$ positive; 24   **11.** $-4(-6)(-5)$ negative; -120

Use the distributive property to rewrite the expression as a sum of
two products. Then simplify both expressions to show they are equal.

12. $-4(-6 + 2)$ 16   **13.** $-8[-9 + (-14)]$ 184

Simplify by combining similar terms.   Section 10-4, p. 273

14. $-3m + 4n - 5m$ $4n - 8m$ **15.** $-13k - 3 + 5 - 8k$ $_{-21k + 2}$

Check your answers with those printed at the back of the book.

10-5 *Division of Directed Numbers*

> **OBJECTIVES**
>
> **Divide directed numbers.**
>
> **Write the decimal equivalent of a fraction like** $-\dfrac{2}{5}$.

Multiplication and division are **inverse operations.** Given a division sentence we can write a related multiplication sentence.

$$32 \div 8 = 4 \blacktriangleright 8 \cdot 4 = 32 \text{ (or } 4 \cdot 8 = 32)$$

We can use this relationship to form a pattern of signs for dividing directed numbers.

EXAMPLE 1 $-32 \div 8 = \underline{\ ?\ } \blacktriangleright$ Since $-4 \cdot 8 = -32$, $-32 \div 8 = -4$.

EXAMPLE 2 $\dfrac{-35}{-5} = \underline{\ ?\ } \blacktriangleright$ Since $7(-5) = -35$, $\dfrac{-35}{-5} = 7$.

EXAMPLE 3 $\dfrac{18}{-9} = \underline{\ ?\ } \blacktriangleright$ Since $-2(-9) = 18$, $\dfrac{18}{-9} = -2$.

If you look closely at the examples, you'll note that the pattern is the same as for multiplication. When dividing directed numbers:

$\blacktriangleright$ If the two signs are **alike,** the answer is **positive.**
  If the two signs are **unlike,** the answer is **negative.**

Example 3 names the fraction $\dfrac{18}{-9}$. $\dfrac{-18}{9}$ names the same number.

Look at these other examples:

$$\dfrac{-4}{2} = \dfrac{4}{-2} = -\dfrac{4}{2} = -2 \qquad\qquad \dfrac{-15}{-3} = \dfrac{15}{3} = 5$$

A number expressed as a fraction can be changed to decimal form by doing the indicated division.

EXAMPLE 4 $\dfrac{-3}{5} \blacktriangleright \begin{array}{r} -0.6 \\ 5)\overline{-3.0} \end{array} \blacktriangleright \dfrac{-3}{5} = -0.6$

Complete to make a true statement.

1. Because $6 \cdot 8 = 48$, we know $\dfrac{48}{6} = \underline{\ ?\ }$. 8

2. Because $9(-4) = -36$, we know $-36 \div 9 = \underline{\ ?\ }$. -4

3. Because $-7 \cdot 8 = -56$, we know $\dfrac{-56}{-7} = \underline{\ ?\ }$. 8

4. Because $-5(-6) = 30$, we know $\dfrac{30}{-5} = \underline{\ ?\ }$. -6

5. Because $\underline{\ ?\ }(-10) = -60$, we know $-60 \div 6 = -10$. 6

For Extra Practice, see page 424.

Simplify.

1. $-45 \div 9$ -5 2. $50 \div 5$ 10   3. $-60 \div 5$ -12

A

4. $-48 \div 2$ -24 5. $-24 \div 4$ -6 6. $-5\overline{)55}$ -11

7. $\dfrac{-36}{5+4}$ -4 8. $\dfrac{-8+(-2)}{5}$ -2 9. $\dfrac{-56}{-8}$ 7

10. $\dfrac{-54}{6}$ -9 11. $\dfrac{-32}{-6+(-2)}$ 4   12. $\dfrac{40}{-8}$ -5

13. $\dfrac{-12}{7+(-3)}$ -3 14. $\dfrac{-25}{8+(-3)}$ -5 15. $\dfrac{-7+(-8)}{-3}$ 5

16. $\dfrac{-35}{-2+(-5)}$ 5   17. $\dfrac{-28}{8+(-4)}$ -7 18. $\dfrac{-36}{-12}$ 3

Write the decimal equivalent.

19. $-\dfrac{3}{5}$ -0.6 20. $\dfrac{15}{-100}$ -0.15 21. $\dfrac{6}{-10}$ -0.6 22. $\dfrac{4}{5}$ 0.8

23. $\dfrac{-4}{-20}$ 0.2   24. $\dfrac{-9}{20}$ -0.45 25. $\dfrac{3}{25}$ 0.12   26. $\dfrac{-7}{8}$ -0.875

27. $\dfrac{-6}{-8}$ 0.75   28. $\dfrac{-7}{5}$ -1.4 29. $\dfrac{-3}{4}$ -0.75 30. $\dfrac{-1}{-5}$ 0.2

True or false?

31. $-\dfrac{3}{4} = \dfrac{-3}{-4}$ false   32. $\dfrac{-4}{10} = -0.4$ true   33. $\dfrac{-4}{-7} = \dfrac{4}{7}$ true

B

34. $\dfrac{7}{8} = \dfrac{-7}{-8}$ true   35. $\dfrac{2}{-4} = \dfrac{-2}{4}$ true   36. $-\dfrac{1}{8} = \dfrac{8}{-1}$ false

10-6 *Reciprocals of Directed Numbers*

OBJECTIVES

Name the reciprocal of a directed number.

Divide by a number by multiplying by its reciprocal.

Use reciprocals to solve equations like $3c = -18$.

Two numbers are reciprocals of each other if their product is 1. A number and its reciprocal have like signs.

EXAMPLE 1 $\frac{1}{2} \cdot 2 = 1$ ▶ $\frac{1}{2}$ and 2 are reciprocals.

$-\frac{1}{2}(-2) = 1$ ▶ $-\frac{1}{2}$ and -2 are reciprocals.

$-0.25(-4) = 1$ ▶ -0.25 and -4 are reciprocals.

The Reciprocal Property ▶ **For every directed number r except 0,**

$$r \cdot \frac{1}{r} = \frac{1}{r} \cdot r = 1.$$

When we are dividing by a number, it is often easier to multiply by its reciprocal.

EXAMPLE 2 $-8 \div 4 = -2$ $-8 \cdot \frac{1}{4} = -2$ ◀ 4 and $\frac{1}{4}$ are reciprocals.

$3 \div \frac{1}{5} = 15$ $3 \times 5 = 15$ ◀ $\frac{1}{5}$ and 5 are reciprocals.

▶ **For every directed number r, and every directed number s except 0,**

$$r \div s = r \cdot \frac{1}{s}.$$

We can use reciprocals to solve equations. First we identify the coefficient of the variable. Then we multiply both members of the equation by the reciprocal of the coefficient of the variable.

EXAMPLE 3 Solve $\dfrac{3n}{5} = -6$.

$$\dfrac{3}{5} \cdot n = -6 \quad \blacktriangleleft \text{ The coefficient of } n \text{ is } \dfrac{3}{5}.$$

$$\dfrac{5}{3} \cdot \dfrac{3}{5} \cdot n = \dfrac{5}{3}(-6) \blacktriangleleft \text{ The reciprocal of } \dfrac{3}{5} \text{ is } \dfrac{5}{3}.$$

$$1 \cdot n = -10$$

$$n = -10$$

EXAMPLE 4 Solve $-6s = 42$.

$$-6 \cdot s = 42 \quad \blacktriangleleft \text{ The coefficient of } s \text{ is } -6.$$

$$-\dfrac{1}{6} \cdot -6 \cdot s = -\dfrac{1}{6} \cdot 42 \blacktriangleleft \text{ The reciprocal of } -6 \text{ is } -\dfrac{1}{6}.$$

$$1 \cdot s = -7$$

$$s = -7$$

Name the reciprocal.

Oral
EXERCISES

Sample 1: $\dfrac{2}{-3}$ *What you say:* $\dfrac{-3}{2}$

Sample 2: -0.5 *What you say:* $-2 \left(\text{Note: } -0.5 = -\dfrac{1}{2}\right)$

1. $\dfrac{3}{4}$ $\dfrac{4}{3}$ 2. $\dfrac{1}{-3}$ -3 3. $-\dfrac{3}{5}$ $-\dfrac{5}{3}$

4. $\dfrac{5}{-8}$ $-\dfrac{8}{5}$ 5. $1\frac{1}{2}$ $\dfrac{2}{3}$ 6. $-1\frac{3}{4}$ $-\dfrac{4}{7}$

7. -0.2 -5 8. 0.25 4 9. 8 $\dfrac{1}{8}$

Tell how to complete to make a true statement.

10. $\dfrac{2}{5} \div \dfrac{1}{4} = \dfrac{2}{5} \cdot \underline{\ ?\ }$ 4 11. $\dfrac{1}{-5} \div 4 = \dfrac{1}{-5} \cdot \underline{\ ?\ }$ $\dfrac{1}{4}$

12. $-\dfrac{2}{3} \div \underline{\ ?\ } = -\dfrac{2}{3} \cdot \dfrac{4}{5}$ $\dfrac{5}{4}$ 13. $\dfrac{1}{-4} \div \underline{\ ?\ } = \dfrac{1}{-4} \cdot 5$ $\dfrac{1}{5}$

14. $\dfrac{-7}{8} \div 1\frac{1}{3} = \dfrac{-7}{8} \cdot \underline{\ ?\ }$ $\dfrac{3}{4}$ 15. $0.5 \div 0.2 = 0.5 \cdot \underline{\ ?\ }$ 5

For Extra Practice, see page 424.

Written EXERCISES

Complete to make a true statement.

Sample: $\dfrac{3}{-10} \cdot \dfrac{-10}{3} = \underline{\;?\;}$. *Solution:* $\dfrac{3}{-10} \cdot \dfrac{-10}{3} = \dfrac{-30}{-30} = 1$

A

1. $\dfrac{7}{8} \cdot \dfrac{8}{7} = \underline{\;?\;}$ 1 2. $-\dfrac{3}{5}\left(-\dfrac{5}{3}\right) = \underline{\;?\;}$ 1 3. $\dfrac{1}{-3} \cdot \underline{\;?\;} = 1$ -3

4. $-\dfrac{4}{9} \cdot \underline{\;?\;} = 1$ $-\dfrac{9}{4}$ 5. $\underline{\;?\;}(-1) = 1$ -1 6. $\dfrac{9}{-5}\left(-\dfrac{5}{9}\right) = \underline{\;?\;}$ 1

Simplify. Use reciprocals.

Sample 1: $\dfrac{-2}{3} \div 4$ *Solution:* $\dfrac{-2}{3} \div 4 = \dfrac{-2}{3} \cdot \dfrac{1}{4} = \dfrac{-2}{12}$

$$= \dfrac{-1}{6} = -\dfrac{1}{6}$$

Sample 2: $\dfrac{1}{-6} \div \dfrac{2}{-3}$ *Solution:* $\dfrac{1}{-6} \div \dfrac{2}{-3} = \dfrac{1}{-6} \cdot \dfrac{-3}{2}$

$$= \dfrac{-3}{-12} = \dfrac{1}{4}$$

7. $\dfrac{1}{-3} \div \dfrac{2}{3}$ $-\dfrac{1}{2}$ 8. $\dfrac{-5}{6} \div 2$ $-\dfrac{5}{12}$ 9. $\dfrac{1}{5} \div \dfrac{2}{-5}$ $-\dfrac{1}{2}$

10. $10 \div \dfrac{1}{-3}$ -30 11. $7 \div \dfrac{1}{-4}$ -28 12. $\dfrac{7}{-10} \div 3$ $-\dfrac{7}{30}$

13. $\dfrac{2}{-5} \div \dfrac{2}{-5}$ 1 14. $\dfrac{1}{-8} \div \dfrac{1}{-2}$ $\dfrac{1}{4}$ 15. $-14 \div \dfrac{2}{7}$ -49

16. $\dfrac{1}{2} \div \dfrac{1}{-3}$ $-\dfrac{3}{2}$ 17. $\dfrac{5}{-6} \div \dfrac{6}{-5}$ $\dfrac{25}{36}$ 18. $\dfrac{1}{7} \div \dfrac{1}{-6}$ $-\dfrac{6}{7}$

Solve.

Sample: $-3w = 21$ *Solution:* $-3w = 21$

$$-\dfrac{1}{3}(-3)w = -\dfrac{1}{3} \cdot 21$$

$$1 \cdot w = -\dfrac{21}{3}$$

$$w = -7$$

19. $8b = -16$ -2 20. $-x = -9$ 9 21. $-4a = 28$ -7

22. $10 = -4a$ $-\dfrac{5}{2}$ 23. $-3x = 15$ -5 24. $-24 = 2x$ -12

25. $-11x = -11$ 1 26. $-12y = -4$ $\dfrac{1}{3}$ 27. $-2a = -20$ 10

28. $-9m = 72$ _−8_ **29.** $-6y = -3\frac{1}{2}$ **30.** $-21 = -3y$ _7_

Solve. No variable is 0.

Sample: $\dfrac{2}{m} = 6$ *Solution:* $\dfrac{2}{m} = 6$

$$m \cdot \frac{2}{m} = m \cdot 6$$

$$2 = 6m$$

$$\frac{1}{6} \cdot 2 = \frac{1}{6} \cdot 6m$$

$$\frac{2}{6} = m \text{ or } m = \frac{1}{3}$$

31. $\dfrac{1}{x} = -6$ _−$\frac{1}{6}$_ **32.** $-\dfrac{3}{m} = 5$ _−$\frac{3}{5}$_ **33.** $\dfrac{2}{5}a = -10$ _−25_ **B**

34. $\dfrac{5}{-4c} = 8$ _−$\frac{5}{32}$_ **35.** $\dfrac{1}{3c} = 7$ _$\frac{1}{21}$_ **36.** $\dfrac{5}{-a} = 10$ _−$\frac{1}{2}$_

Divide.

Sample: $4a \div \dfrac{1}{3}$ *Solution:* $4a \div \dfrac{1}{3} = 4a \cdot \dfrac{3}{1} = 12a$

37. $-3a \div \dfrac{1}{5}$ _−15a_ **38.** $(-2d + 5d) \div \dfrac{1}{2}$ _6d_ **39.** $\dfrac{m}{5} \div 3$ _$\frac{m}{15}$_

40. $-6w \div 5$ _$\frac{6w}{5}$_ **41.** $\dfrac{-t}{10} \div \dfrac{-1}{5}$ _$\frac{1}{2}t$_ **42.** $4k \div (-3)$ _$\frac{4k}{3}$_

Find the unknown number. _−$\frac{1}{2}$_

1. 7 divided by some number is equal to -14. What is the number?

2. The product of -15 and some number is -6. Find the number. _$\frac{2}{5}$_

3. The product of $\dfrac{5}{-7}$ and some number is 1. What is the number? _−$\frac{7}{5}$_

4. The reciprocal of some number is -9. Find the number. _−$\frac{1}{9}$_

5. The reciprocal of some number is $-\dfrac{5}{7}$. Find the number. _−$\frac{7}{5}$_

6. The reciprocal of some number is -8. Find the number. _−$\frac{1}{8}$_

7. When -16 is divided by a number the result is -2. What is the number? _8_

career capsule

Television and Radio Service Technician

Television and radio service technicians repair electronic products, television sets, radios, stereo components, tape recorders, intercoms and public address systems. Using voltmeters and signal generators they check suspected circuits for loose or broken connections and other probable causes of trouble. Technicians refer to wiring diagrams which show connections and contain information on repair.

Becoming a qualified service technician requires two to four years on-the-job experience. High school training should include courses in mathematics, electronics and physics.

10-7 *Functions and Directed Numbers*

OBJECTIVE

Use function equations that involve multiplying and dividing directed numbers.

The work of a function machine may involve multiplying or dividing directed numbers.

EXAMPLE 1 $f(t) = \dfrac{2t}{3}$; replacement set: $\{-3, -2, -1, 0, 1\}$

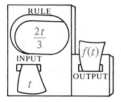

t	$f(t)$	$(t, f(t))$
-3	-2	$(-3, -2)$
-2	$-1\frac{1}{3}$	$(-2, -1\frac{1}{3})$
-1	$-\frac{2}{3}$	$\left(-1, -\frac{2}{3}\right)$
0	0	$(0, 0)$
1	$\frac{2}{3}$	$\left(1, \frac{2}{3}\right)$

Function: $\left\{(-3, -2), (-2, -1\frac{1}{3}), \left(-1, -\dfrac{2}{3}\right), (0, 0), \left(1, \dfrac{2}{3}\right)\right\}$

EXAMPLE 2 $f(x) = -\dfrac{1}{2}x$; replacement set: $\{-2, -1, 0, 1, 2\}$

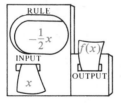

x	$f(x)$	$(x, f(x))$
-2	1	$(-2, 1)$
-1	$\frac{1}{2}$	$\left(-1, \frac{1}{2}\right)$
0	0	$(0, 0)$
1	$-\frac{1}{2}$	$\left(1, -\frac{1}{2}\right)$
2	-1	$(2, -1)$

Function: $\left\{(-2, 1), \left(-1, \dfrac{1}{2}\right), (0, 0), \left(1, -\dfrac{1}{2}\right), (2, -1)\right\}$

Graph:

Match each function equation in Column 1 with the correct function from Column 2. The replacement set for each is $\{-2, -1\}$.

COLUMN 1   COLUMN 2

1. $f(a) = -a(-a)$ **D**   **A.** $\{(-2, 3), (-1, 6)\}$

2. $f(b) = -\dfrac{1}{2} \cdot b$ **F**   **B.** $\{(-2, -10), (-1, -5)\}$

                                   C. $\{(-2, 4), (-1, 0)\}$

3. $f(y) = -3 \cdot y$ **E**   **D.** $\{(-2, 4), (-1, 1)\}$

4. $f(c) = \dfrac{6}{-c}$ **A**   **E.** $\{(-2, 6), (-1, 3)\}$

5. $f(m) = m \div \dfrac{1}{5}$ **B**   **F.** $\left\{(-2, 1), \left(-1, \dfrac{1}{2}\right)\right\}$

Tell whether or not the set of ordered number pairs is a function.

6. $\{(-1, 2), (-2, 4), (3, -6)\}$ function

7. $\{(-3, 4), (5, 2), (-3, 1), (7, -1)\}$ not a function

8. $\{(-2, -3), (1, 3), (0, 1), (5, 9)\}$ function

9. $\left\{(3, -1), \left(1, -\dfrac{1}{3}\right), \left(-2, \dfrac{2}{3}\right), \left(-1, \dfrac{1}{3}\right)\right\}$ function

10. $\{(6, -3), (5, 2), (-2, 1), (-2, 3)\}$ not a function

Written

EXERCISES

Complete the table according to the rule shown on the function machine. The replacement set is $\{-2, -1, 0, 2, 3\}$. Draw the graph of the function.

A

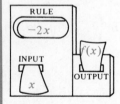

	x	$f(x)$	$(x, f(x))$	
Sample:	-2	4	$(-2, 4)$	
1.	-1	2 ?	?	$(-1, 2)$
2.	0	0 ?	?	$(0, 0)$
3.	2	-4 ?	?	$(2, -4)$
4.	3	-6 ?	?	$(3, -6)$

Use the replacement set $\{2, 1, 0, -1, -2, -3\}$ to complete the table for each function machine. Compare the completed tables and comment.

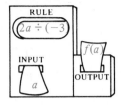

	a	$f(a)$	$(a, f(a))$
Sample:	2	$-\dfrac{4}{3}$	$\left(2, -\dfrac{4}{3}\right)$
5.	1	$-\frac{2}{3}$?	?
6.	0	0?	?
7.	-1	$\frac{2}{3}$?	?
8.	-2	$\frac{4}{3}$?	?
9.	-3	2?	?

$\left(1, -\dfrac{2}{3}\right)$

$(0, 0)$

$\left(-1, \dfrac{2}{3}\right)$

$\left(-2, \dfrac{4}{3}\right)$

$(-3, 2)$

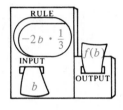

	b	$f(b)$	$(b, f(b))$
Sample:	2	$-\dfrac{4}{3}$	$\left(2, -\dfrac{4}{3}\right)$
10.	1	$-\frac{2}{3}$?	?
11.	0	0 ?	?
12.	-1	$\frac{2}{3}$?	?
13.	-2	$\frac{4}{3}$?	?
14.	-3	2 ?	?

$\left(1, -\dfrac{2}{3}\right)$

$(0, 0)$

$\left(-1, \dfrac{2}{3}\right)$

$\left(-2, \dfrac{4}{3}\right)$

$(-3, 2)$

Complete the set of number pairs by using the given function equation. The replacement set is $\{-3, -2, 0, 1\}$.

Sample: $f(x) = -\dfrac{x}{4}$: $\left\{\left(-3, \dfrac{3}{4}\right), \left(-2, \dfrac{1}{2}\right), (0, \underline{\ ?\ }), (1, \underline{\ ?\ })\right\}$

Solution: $\left\{\left(-3, \dfrac{3}{4}\right), \left(-2, \dfrac{1}{2}\right), (0, 0), \left(1, -\dfrac{1}{4}\right)\right\}$

15. $f(a) = \dfrac{a}{3}$: $\left\{(-3, -1), \left(-2, -\dfrac{2}{3}\right), (0, \underline{\ ?\ }), (1, \underline{\ ?\ })\right\}$ $0, \frac{1}{3}$

16. $f(x) = -2x + 1$: $\{(-3, 7), (-2, 5), (0, \underline{\ ?\ }), (1, \underline{\ ?\ })\}$ $1, -1$

17. $f(w) = -w \cdot \dfrac{1}{5}$: $\left\{\left(-3, \dfrac{3}{5}\right), \left(-2, \dfrac{2}{5}\right), (0, \underline{\ ?\ }), (1, \underline{\ ?\ })\right\}$ $0, -\frac{1}{5}$

18. $f(y) = 3y - 2$: $\{(-3, -11), (-2, \underline{\ ?\ }), (0, -2), (1, \underline{\ ?\ })\}$ $-8, 1$

19. $f(t) = \dfrac{-2t}{5}$: $\left\{\left(-3, \dfrac{6}{5}\right), (-2, \underline{\ ?\ }), (0, \underline{\ ?\ }), (1, \underline{\ ?\ })\right\}$ $\frac{1}{5}, 0, -\frac{2}{5}$

20. $f(c) = 4c \div 2$: $\{(-3, \underline{\ ?\ }), (-2, \underline{\ ?\ }), (0, 0), (1, \underline{\ ?\ })\}$ $-6, -4, 2$

25. $\left\{\left(-5, -\frac{6}{5}\right), \left(-4, -\frac{3}{2}\right), (-3, -2), (-2, -3), (-1, -6)\right\}$

Write the number pairs indicated by the given function equation and replacement set.

$\{(-2, 8), (-1, 5), (0, 2), (2, -4), (3, -7), (4, -10)\}$

B **21.** $f(n) = -3n + 2$: $\{-2, -1, 0, 2, 3, 4\}$

22. $f(a) = \frac{3}{4}(a) - 1$: $\{-4, -2, 0, 2, 8\}$

$\left\{(-4, -4), (-2, -2\frac{1}{2}), (0, -1), \left(2, \frac{1}{2}\right), (8, 5)\right\}$

23. $f(y) = \frac{y}{2} + 1$: $\{-2, -1, 0, 6\}$ $\left\{(-2, 0), \left(-1, \frac{1}{2}\right), (0, 1), (6, 4)\right\}$

24. $f(b) = (-b)(-b)$: $\{-3, -2, 4, 5\}$ $\{(-3, 9), (-2, 4), (4, 16), (5, 25)\}$

25. $f(m) = \frac{6}{m}$: $\{-5, -4, -3, -2, -1\}$

26. $f(w) = -2 \cdot 3w$: $\left\{-\frac{1}{2}, -\frac{1}{3}, -2, 3\right\}$ $\left\{\left(-\frac{1}{2}, 3\right), \left(-\frac{1}{3}, 2\right),\right.$ $\left.(-2, 12), (3, -13)\right\}$

27. $f(c) = \frac{7}{-c}$: $\{-4, -3, -2, 3\}$

28. $f(h) = -3 \div h$: $\left\{-9, -3, 6, \frac{1}{3}, \frac{1}{2}\right\}$

29. $f(n) = 5n^2$: $\{-2, -1, 0, 2\}$ $\{(-2, 20), (-1, 5), (0, 0), (2, 20)\}$

30. $f(d) = d \cdot \frac{2}{d}$: $\{-5, -3, -1, 2, 4\}$

27. $\left\{\left(-4, \frac{7}{4}\right), \left(-3, \frac{7}{3}\right), \left(-2, \frac{7}{2}\right), \left(3, -\frac{7}{3}\right)\right\}$ **28.** $\left\{\left(-9, \frac{1}{3}\right), (-3, 1), \left(6, -\frac{1}{2}\right), \left(\frac{1}{3}, -9\right), \left(\frac{1}{2}, -6\right)\right\}$

SELF-TEST 2

30. $\{(-5, 2), (-3, 2), (-1, 2), (2, 2), (4, 2)\}$

Be sure that you understand the term *reciprocal* (p. 278).

Simplify if necessary. Then divide.

Section 10-5, p. 276

1. $\frac{-36}{4}$ -9

2. $\frac{-15}{-5}$ 3

3. $\frac{-7 + (-5)}{3}$ -4

4. $\frac{-1 + (-6)}{-3}$ $\frac{7}{3}$

Section 10-6, p. 278 Simplify. Use reciprocals.

5. $\frac{-9}{16} \div \frac{-9}{16}$ 1

6. $-21 \div \frac{-3}{7}$ 49

7. $-18 \div \frac{3}{4}$ -24

8. $-\frac{4}{3} \div \frac{3}{4}$ $-\frac{16}{9}$

Solve.

9. $2x = -18$ -9

10. $-3y = -24$ 8

11. $4z = 28$ 7

12. $8m = -16$ -2

Complete. Section 10-7, p. 283

13. $f(x) = -\dfrac{1}{2}x;\ \{(0, \underset{0}{\underline{\ ?\ }}),\ (-4, \underset{2}{\underline{\ ?\ }}),\ (4, \underset{-2}{\underline{\ ?\ }}),\ (-5, \underset{2\frac{1}{2}}{\underline{\ ?\ }})\}$

Check your answers with those printed at the back of the book.

---

chapter summary

1. The product of any directed number and 0 is 0.

2. The product of any directed number and 1 is that directed number.

3. Multiplication of directed numbers is commutative and associative. Also, the distributive property holds true for directed numbers.

$$r \cdot s = s \cdot r \qquad \blacktriangleleft \text{ The commutative property}$$
$$r \cdot (st) = (rs) \cdot t \qquad \blacktriangleleft \text{ The associative property}$$
$$r(s + t) = rs + rt \ \blacktriangleleft \text{ The distributive property}$$

4. In multiplying two directed numbers, if the two signs are **alike,** the product is **positive.** If the two signs are **unlike,** the product is **negative.**

5. In dividing two directed numbers, if the two signs are **alike,** the quotient is **positive.** If the two signs are **unlike,** the quotient is **negative.**

6. The distributive property can be used to simplify expressions which involve directed numbers.

7. Two numbers are **reciprocals** of each other if their product is 1.

8. Dividing by a directed number is the same as multiplying by its reciprocal. That is, for every directed number r, and every directed number s except 0, $r \div s = r \cdot \dfrac{1}{s}$.

challenge topics *Probability*

Suppose you toss a penny and a nickel at the same time. What combinations of heads and tails might come up? Study the chart.

Nickel

		H	T
Penny	h	h,H	h,T
	t	t,H	t,T

The chart shows that there are four possible outcomes, that is, four ways for the coins to come up heads and tails. A coin is honest if after many tosses it comes up heads very nearly the same number of times it comes up tails. Assuming the two coins are honest, each of the four outcomes is said to be equally likely. From the chart we see that the probability of an outcome of two heads (h,H) is 1 out of 4 or $\frac{1}{4}$. We write: $P(h,H) = \frac{1}{4}$.

What are the possible outcomes when two dice are tossed? The sum of the number of dots on the upper faces is considered the outcome.

	1	2	3	4	5	6
1	2	3	4	5	6	7
2	3	4	5	6	7	8
3	4	5	6	7	8	9
4	5	6	7	8	9	10
5	6	7	8	9	10	11
6	7	8	9	10	11	12

◀ $1 + 6 = 7$

The table shows that there are 36 possible outcomes. If the dice are honest, each of the outcomes is equally likely. The probability of any one is $\frac{1}{36}$. The table shows that an outcome of 6 may occur in 5 different ways. $P(6) = 5 \times \frac{1}{36} = \frac{5}{36}$

1. Two coins are tossed. What is the probability of getting two tails? $\frac{1}{4}$

2. Two coins are tossed. What is the probability of getting a head and a tail? $\frac{1}{4}$

3. A single coin is tossed 100 times. It comes up heads 52 times and tails 48 times. Is the coin honest? Explain.

4. Two dice are tossed. Assuming equally likely outcomes, what is the probability of an outcome of 4? of 6? of 8? $\frac{1}{12}$; $\frac{5}{36}$; $\frac{5}{36}$

5. Two dice are tossed. What is the probability of getting an 8 by having both dice come up with a 4? $\frac{1}{36}$

3. Yes, because the number of times it came up heads is about the same number of times it came up tails.

chapter test

Simplify.

1. $6(-8)$ -48

2. $-9 \cdot 5$ -45

3. $\frac{36}{-4}$ -9

4. $\frac{-54}{6}$ -9

5. $-3(-3)(-3)$ -27

6. $-2(-2)(-2)$ -8

7. $-1 \cdot 1(-8)$ 8

8. $-2 \cdot 4(-5)$ 40

9. $\frac{5 + (-17)}{4}$ -3

10. $\frac{-28}{-4}$ 7

11. $2m + 3t - 6m$ $-4m + 3t$

12. $4z - 3b + 2b - 6z$ $-2z - b$

Name the reciprocal.

13. $-\frac{1}{2}$ -2

14. $1\frac{2}{3}$ $\frac{3}{5}$

15. -0.75 $-1\frac{1}{3}$

Solve.

16. $4y + (-2) = 22$ 6

17. $\frac{x}{3} = -7$ -21

18. $5k + (-k) = 20$ 5

19. $5a - 2 = 13$ 3

20. $-4m = -3 \cdot 8$ 6

21. $\frac{2}{3}x = -8$ -12

Complete using the given function equation.

22. $f(m) = 5m + 2$: $\{(0, \underline{\ ?\ }), (1, \underline{\ ?\ }), (3, \underline{\ ?\ })\}$ $2; 7; 17$

23. $f(a) = 3a - 4$: $\{(0, \underline{\ ?\ }), (-1, \underline{\ ?\ }), (5, \underline{\ ?\ }), (-2, \underline{\ ?\ })\}$ $-4; -7; 11; -10$

Review of Skills

Simplify.

1. $3 + (-8)$ -5

2. $\dfrac{3}{4} \cdot \dfrac{2}{5}$ $\dfrac{3}{10}$

3. $-4 \cdot 3$ -12

4. $\dfrac{2}{3} \cdot \dfrac{1}{2}$ $\dfrac{1}{3}$

5. $-7 \cdot 2$ -14

6. $-6 + (-2)$ -8

7. $-6 + 6$ 0

8. $-1.8 \cdot \dfrac{1}{2}$ -0.9

9. $5(-3)$ -15

Name the additive inverse.

10. -7 7

11. 10 -10

12. $-\dfrac{5}{7}$ $\dfrac{5}{7}$

13. 5 -5

Solve.

14. $x + 4 = 17$ 13

15. $-3x = 27$ -9

16. $5x = 21$ $4\frac{1}{5}$

17. $21 = 6 + x$ 15

18. $\dfrac{2}{5} \cdot x = -6$ -15

19. $32 = 4x$ 8

20. $x + 9 = 6$ -3

21. $\dfrac{1}{3} \cdot x = 11$ 33

22. $7x = -56$ -8

Use the distributive property to simplify.

23. $8m - 3m$ $5m$

24. $\dfrac{5}{6}x + \dfrac{2}{3}x$ $1\frac{1}{2}x$

25. $5y + 4y$ $9y$

Complete to make a true statement. Use $<$ or $>$.

26. $7 \underline{\ ?\ } 9$ $<$

27. $-8 \underline{\ ?\ } -5$ $<$

28. $-7 \underline{\ ?\ } 4$ $<$

29. $3 \cdot 2 \underline{\ ?\ } 5$ $>$

30. $-\dfrac{1}{3} \underline{\ ?\ } -\dfrac{4}{5}$ $>$

31. $4(-1) \underline{\ ?\ } -3$ $<$

32. $-6 \underline{\ ?\ } -9$ $>$

33. $10 \underline{\ ?\ } 12$ $<$

34. $3 + 4 \underline{\ ?\ } 9$ $<$

True or false?

35. $6 > 5$ and $5 > 6$. false

36. $4 < 7$ and $4 + 3 < 7 + 3$. true

37. $5 > 4$ and $5 \cdot 3 > 4 \cdot 3$. true

38. $4 < 6$ and $4(-2) < 6(-2)$. false

Left: Early aerial weather observers.

Right: Console of all-electronic weather forecasting system.

Solving Equations and Inequalities

Types of Equations

11-1 *Equations of Type* $x + a = b$

OBJECTIVE
Solve equations like $x + 7 = 15$.

The equation $x + 7 = 15$ is of the type $x + a = b$, where x is the variable, and a and b are directed numbers.

These equations are also of type $x + a = b$, although they may look different.

$$m - 3 = 7 \qquad 15 = x + 3.5 \qquad -6 = 2 + h$$

$$m + (-3) = 7 \qquad x + 3.5 = 15 \qquad h + 2 = -6$$

We solve equations of the type $x + a = b$ by using the addition property of equality.

EXAMPLE 1

$$x + 7 = 15$$
$$x + 7 + (-7) = 15 + (-7) \quad \blacktriangleleft \text{Add } -7 \text{ to both members.}$$
$$x + 0 = 15 + (-7) \quad \blacktriangleleft \ 7 + (-7) = 0$$
$$x = 8$$

EXAMPLE 2

$$3 = -10 + n$$
$$10 + 3 = 10 + (-10) + n \quad \blacktriangleleft \text{Add 10 to both members.}$$
$$10 + 3 = 0 + n \quad \blacktriangleleft \ 10 + (-10) = 0$$
$$13 = n$$

EXAMPLE 3

$$t - 7 = -10$$
$$t + (-7) = -10 \quad \blacktriangleleft \text{To subtract 7, add its opposite.}$$
$$t + (-7) + 7 = -10 + 7 \quad \blacktriangleleft \text{Add 7 to both members.}$$
$$t + 0 = -10 + 7 \quad \blacktriangleleft \ -7 + 7 = 0$$
$$t = -3$$

 **Oral EXERCISES**

State in the form $x + a = b$.

Sample: $m - 3 = -5$ *What you say:* $m + (-3) = -5$

1. $-13 + k = 20$ $k + (-13) = 20$ 2. $-4 + y = -3$ $y + (-4) = -3$

3. $-6 = x + 3$ $x + 3 = -6$ 4. $w - (-9) = 20$ $w + 9 = 20$

5. $z + (-2) = -7$ $z + (-2) = -7$ **6.** $u - 10 = 52$ $u + (-10) = 52$

7. $-\dfrac{3}{4} + t = 1$ $t + \left(-\dfrac{3}{4}\right) = 1$ **8.** $-\dfrac{1}{3} + z = -6$ $z + \left(-\dfrac{1}{3}\right) = -6$

9. $(-3 + 2) + y = 10$ $y + (-1) = 10$ **10.** $q - (3 + 6) = -9$ $9 + (-9) = -9$

11. $14 = r + 2$ $r + 2 = 14$ **12.** $16 = -4 + y$ $y + (-4) = 16$

For Extra Practice, see page 425.

Complete.

Sample:
$$y + 3 = 12$$
$$y + 3 + (-3) = 12 + (-3)$$
$$y + \underline{} = 12 + (-3)$$
$$y = \underline{}$$

Solution:
$$y + 3 = 12$$
$$y + 3 + (-3) = 12 + (-3)$$
$$y + 0 = 12 + (-3)$$
$$y = 9$$

A

1.
$$m - 6 = 19$$
$$m + (-6) + 6 = 19 + 6$$
$$m + \underline{} = 19 + 6 \quad 0$$
$$m = \underline{} \quad 25$$

2.
$$k + 14 = 7$$
$$k + 14 + \underline{} = 7 + (-14) \quad -14$$
$$k + \underline{} = 7 + (-14) \quad 0$$
$$k = \underline{} \quad -7$$

3.
$$5\tfrac{1}{4} + v = 8$$
$$-5\tfrac{1}{4} + 5\tfrac{1}{4} + v = 8 + \underline{} \quad -5\tfrac{1}{4}$$
$$\underline{} + v = 8 + (-5\tfrac{1}{4}) \quad 0$$
$$v = \underline{} \quad 2\tfrac{3}{4}$$

4.
$$-6 + x = 12$$
$$6 + (-6) + x = 12 + \underline{} \quad 6$$
$$\underline{} + x = 12 + 6 \quad 0$$
$$x = \underline{} \quad 18$$

Solve and check.

Sample: $t + (-8) = 24$

Solution:
$$t + (-8) = 24$$
$$t + (-8) + 8 = 24 + 8$$
$$t = 32$$

Check: $t + (-8) \overset{?}{=} 24$

$32 + (-8)$	24
24	24	✓

5. $-2 + r = 20$ 22 **6.** $x + 4 = -13$ -17

7. $z + (-13) = 9$ 22 **8.** $-5 + w = 15$ 20

9. $-16 + n = 25$ 41 **10.** $-44 = u + 5$ -49

11. $b + \dfrac{2}{3} = -5$ $-5\tfrac{2}{3}$ **12.** $s + \dfrac{1}{5} = 2\tfrac{3}{10}$ $2\tfrac{1}{10}$

13. $-6 = m + (-9)$ 3 **14.** $n + (-51) = 0$ 51

15. $x + 14 = 0$ -14

16. $\dfrac{5}{6} + z = 5$ $4\frac{1}{6}$

17. $m + 1.8 = 6.9$ 5.1

18. $-18.5 + k = 26.3$ 44.8

19. $y - 4 = -2\frac{1}{2}$ $1\frac{1}{2}$

20. $a + \left(-\dfrac{2}{3}\right) = 6$ $6\frac{2}{3}$

B **21.** $n - \left(-\dfrac{1}{3}\right) = \dfrac{1}{6}$ $-\dfrac{1}{6}$

22. $y + \dfrac{3}{8} = \dfrac{3}{4}$ $\dfrac{3}{8}$

23. $-\dfrac{1}{6} + t = -\dfrac{1}{12}$ $\dfrac{1}{12}$

24. $-\dfrac{3}{5} = -\dfrac{10}{2} + p$ $4\frac{2}{5}$

25. $-0.6 = -3.2 + v$ 2.6

26. $-0.06 = -6.2 + s$ 6.14

Solve for y.

27. $y + s = z$ $z - s$

28. $-n + y = h$ $h + n$

29. $d = b + y$ $d - b$

30. $k = y - w$ $k + w$

31. $r + y = -t$ $-t - r$

32. $y + (-a) = -b$ $a - b$

Solve for z.

C **33.** $z - r = s$ $s + r$

34. $z + (-p) = q$ $q + p$

35. $m = z + q$ $m - q$

36. $z - (-x) = y$ $y - x$

37. $z - g = -h$ $g - h$

38. $d + z = f$ $f - d$

Williamina Fleming 1857–1911

Williamina Fleming entered the field of astronomy as a clerk at the Harvard College Observatory in 1881. Demonstrating an obvious talent for astronomy, she became assistant to Professor Edward Pickering and soon excelled in the field. Her chief work dealt with classification of stars based on patterns photographed when their light was scattered through a prism. In 1890 Fleming published the *Draper Catalogue of Stella Spectra*, classifying 10,351 stars. She was admitted to the Royal Astronomical Society in 1906.

11-2 Equations of Type $ax = b$

The equation $3x = 45$ is of the type $ax = b$. x is a variable. a and b are directed numbers.

Here are some other equations of this type.

$$0.3 = -2n \blacktriangleright -2n = 0.3 \qquad \frac{3}{7} = -3b \blacktriangleright -3b = \frac{3}{7}$$

We solve equations of the type $ax = b$ by using the multiplication property of equality. We multiply by the reciprocal of a, the coefficient of x.

EXAMPLE 1 $4m = 3$

$\frac{1}{4} \cdot 4m = \frac{1}{4} \cdot 3$ ◀ Multiply both members by the reciprocal of 4.

$1 \cdot m = \frac{1}{4} \cdot 3$ ◀ $\frac{1}{4} \cdot 4 = 1$

$m = \frac{3}{4}$

EXAMPLE 2 $-\frac{3t}{2} = 9$

$-\frac{3}{2} \cdot t = 9$ ◀ The coefficient of t is $-\frac{3}{2}$.

$-\frac{2}{3}\left(-\frac{3}{2}\right)t = -\frac{2}{3} \cdot 9$ ◀ Multiply both members by $-\frac{2}{3}$.

$1 \cdot t = -\frac{2}{3} \cdot 9$ ◀ $-\frac{2}{3}\left(-\frac{3}{2}\right) = 1$

$t = -6$

EXAMPLE 3   Solve $\frac{x}{m} = t$ for x. ◀ $m \neq 0$. Otherwise, $\frac{1}{m}$ has no meaning.

$m \cdot \frac{1}{m} \cdot x = m \cdot t$ ◀ Multiply both members by m.

$1 \cdot x = m \cdot t$ ◀ $m \cdot \frac{1}{m} = 1$ if $m \neq 0$.

$x = mt$

Oral
EXERCISES

Name the reciprocal of the coefficient of the variable.

Sample: $4x = 20$ *What you say:* $\dfrac{1}{4}$

1. $3y = 12$ $\frac{1}{3}$
2. $-6m = 36$ $-\frac{1}{6}$
3. $-7t = 49$ $-\frac{1}{7}$
4. $8y = \dfrac{1}{4}$ $\frac{1}{8}$
5. $\dfrac{k}{6} = 3$ 6
6. $\dfrac{s}{-5} = -2$ -5
7. $\dfrac{1}{4}q = 6\frac{1}{4}$ 4
8. $\dfrac{2}{5}b = -10$ $\frac{5}{2}$
9. $\dfrac{-z}{4} = 7$ -4

Tell what number should replace the question mark to complete the equation.

Sample: $\underline{\quad ? \quad} \cdot \dfrac{3}{2}y = y$

What you say: $\dfrac{2}{3}$, since $\dfrac{2}{3} \cdot \dfrac{3}{2} \cdot y = 1y = y.$

10. $\underline{\quad ? \quad} \cdot 5s = s$ $\frac{1}{5}$
11. $\underline{\quad ? \quad}(-8z) = z$ $-\frac{1}{8}$
12. $-\dfrac{1}{2}p \cdot \underline{\quad ? \quad} = p$ -2
13. $\underline{\quad ? \quad}\left(-\dfrac{7g}{9}\right) = g$ $-\frac{9}{7}$
14. $\underline{\quad ? \quad} \cdot \dfrac{2}{3}x = x$ $\frac{3}{2}$
15. $\dfrac{-6m}{-5} \cdot \underline{\quad ? \quad} = m$ $\frac{5}{6}$

For Extra Practice, see page 425.

Written
EXERCISES

Solve and check.

Sample: $\dfrac{4}{5}c = 20$

Solution: $\dfrac{4}{5}c = 20$ *Check:* $\dfrac{4}{5}c \overset{?}{=} 20$

$\dfrac{5}{4} \cdot \dfrac{4}{5} \cdot c = 20 \cdot \dfrac{5}{4}$ $\dfrac{4}{5} \cdot 25 \,\Big|\, 20$

$c = 25$ $20 \,\Big|\, 20$ ✓

A

1. $3x = 39$ 13
2. $-y = 56$ -56
3. $-18 = 5q$ $-3\frac{3}{5}$
4. $-7a = 63$ -9
5. $\dfrac{-3u}{5} = -21$ 35
6. $\dfrac{1}{4}b = 3$ 12
7. $w(-4) = -23$ $5\frac{3}{4}$
8. $16s = 64$ 4
9. $7d = -9$ $-\frac{9}{7}$
10. $\dfrac{-1}{6}v = 10$ -60

Solve for the underlined variable. Assume that no divisor has the value 0.

11. $g\underline{d} = v$ $\frac{v}{g}$
12. $\underline{g}d = v$ $\frac{v}{d}$
13. $a = l\underline{w}$ $\frac{a}{l}$
14. $a = l\underline{w}$ $\frac{a}{w}$
15. $i = p\underline{r}$ $\frac{i}{p}$
16. $i = p\underline{r}$ $\frac{i}{r}$
17. $\pi\underline{d} = c$ $\frac{c}{\pi}$
18. $\pi\underline{c} = s$ $\frac{s}{\pi}$
19. $s = \underline{c}t$ $\frac{s}{t}$

Solve.

20. $15 = 7v$ $\frac{15}{7}$
21. $8y = -40$ -5
22. $-3j = -5$ $\frac{5}{3}$
23. $\frac{n}{7} = 2$ 14
24. $\frac{2z}{3} = 4$ 6
25. $\frac{7}{-5}q = 25$ $-17\frac{6}{7}$
26. $0.3s = 18$ 60
27. $-0.02 = -0.05x$ 0.4
28. $-2.3a = 46$ -20
29. $2.5h = -0.5$ -0.2

Solve the formula for the indicated variable.

30. Distance:
$d = rt$
Solve for t. $\frac{d}{r}$

31. Circumference of a circle:
$c = 2\pi r$
Solve for r. $\frac{c}{\pi}$

32. Volume of a cone:
$v = \frac{1}{3}bh$
Solve for b. $\frac{3v}{h}$

33. Volume of a cylinder:
$v = bh$
Solve for h. $\frac{v}{b}$

Solve for the underlined variable. Then let $r = -\frac{4}{5}$, $t = \frac{1}{5}$, $s = 4$, and $u = 8$, and find the value of that variable.

Sample: $s\underline{x} = r$

Solution: $\frac{1}{s} \cdot sx = \frac{1}{s} \cdot r \blacktriangleright x = \frac{r}{s}$

If $r = -\frac{4}{5}$ and $s = 4$, $x = -\frac{4}{5} \div 4 = -\frac{4}{5} \cdot \frac{1}{4} = -\frac{1}{5}$

34. $b\underline{r} = t$ $-\frac{1}{4}$
35. $u\underline{n} = s$ $\frac{1}{2}$
36. $\underline{a}t = u$ 40
37. $-r = u\underline{m}$ $\frac{1}{10}$
38. $-st = -r\underline{k}$ -1
39. $-t\underline{p} = t$ -1
40. $\frac{1}{s} \cdot \underline{b} = u$ 32
41. $t\underline{c} = -s$ -20
42. $ut = \underline{d}s$ $\frac{2}{5}$

B

C

11-3 *Equations of Type* $ax + bx = c$

OBJECTIVE

Solve equations like
$5x + 6x = 14$ and
$-w = -8w + 14$.

The equation $5x + 6x = 14$ is of the type $ax + bx = c$.

$ax + bx = c$ ◀ x is a variable.

            a, b, and c are directed numbers.

            ax and bx are similar terms.

These equations are also of this type.

$$\frac{1}{4}t + \frac{1}{2}t = 5 \qquad y = 12 - 3y \qquad 7 + 5n = 3n$$

$$\blacktriangledown \qquad\qquad\qquad \blacktriangledown$$

$$1y + 3y = 12 \qquad 3n + (-5n) = 7$$

To solve equations in which the variable appears in more than one term, we first combine similar terms. Then we use the properties of equality.

EXAMPLE 1 $3r + 7r = 40$

$$10r = 40 \qquad\qquad ◀ \text{Add } 3r + 7r.$$

$$\frac{1}{10} \cdot 10r = \frac{1}{10} \cdot 40 \qquad ◀ \text{Multiply both members by } \frac{1}{10}.$$

$$1 \cdot r = \frac{1}{10} \cdot 40 \qquad ◀ \frac{1}{10} \cdot 10 = 1$$

$$r = 4$$

EXAMPLE 2 $\dfrac{3}{5}x + \dfrac{4}{5}x = -21$

$$\frac{7}{5}x = -21 \qquad ◀ \text{Add } \frac{3}{5}x + \frac{4}{5}x.$$

$$\frac{5}{7} \cdot \frac{7}{5}x = \frac{5}{7}(-21) \qquad ◀ \text{Multiply both members by } \frac{5}{7}.$$

$$1 \cdot x = \frac{5}{7}(-21) \qquad ◀ \frac{5}{7} \cdot \frac{7}{5} = 1$$

$$x = -15$$

EXAMPLE 3

$$-w = -8w + 14$$

$$8w + (-w) = 8w + (-8w) + 14 \quad \blacktriangleleft \text{ Add } 8w \text{ to both members.}$$

$$8w + (-w) = 0 + 14 \quad \blacktriangleleft \; -8w + 8w = 0$$

$$7w = 14 \quad \blacktriangleleft \text{ Combine similar terms.}$$

$$\frac{1}{7} \cdot 7w = \frac{1}{7} \cdot 14 \quad \blacktriangleleft \text{ Multiply both members by } \frac{1}{7}.$$

$$1 \cdot w = \frac{1}{7} \cdot 14 \quad \blacktriangleleft \; \frac{1}{7} \cdot 7 = 1$$

$$w = 2$$

EXAMPLE 4

$$3x + 6 = 2x - 4$$

$$-2x + 3x + 6 = -2x + 2x - 4 \quad \blacktriangleleft \text{ Add } -2x \text{ to both members.}$$

$$x + 6 = 0 - 4$$

$$x + 6 + (-6) = 0 - 4 + (-6) \quad \blacktriangleleft \text{ Add } -6 \text{ to both members.}$$

$$x + 0 = -10 \quad \blacktriangleleft \; 6 + (-6) = 0$$

$$x = -10$$

EXAMPLE 5

$$rx + tx = k, \text{ and } r + t \neq 0$$

$$x(r + t) = k \quad \blacktriangleleft \text{ Distributive property}$$

$$x(r + t)\frac{1}{r + t} = k \cdot \frac{1}{r + t} \quad \blacktriangleleft \text{ Multiply both members by } \frac{1}{r + t}.$$

$$x \cdot 1 = k \cdot \frac{1}{r + t} \quad \blacktriangleleft \; (r + t) \cdot \frac{1}{r + t} = 1$$

$$x = \frac{k}{r + t}$$

Combine similar terms to give a simpler name.

Oral EXERCISES

1. $3x + 5x$ 8x

2. $2y + (-3y)$ $-y$

3. $\frac{1}{7}a + \frac{2}{7}a$ $\frac{3}{7}a$

4. $\frac{3}{2}q + \left(-\frac{1}{2}q\right)$ q

5. $3k + 7k + 4k$ 14k

6. $2s + (-3s) + 5s$ 4s

7. $10m + 2n - 3m$ 7m + 2n

8. $-25p + 12g + 13p$ 12g − 12p

9. $-7t + (-3t)$ $-10t$

10. $15r + r + (-1)$ 16r − 1

True or false?

11. $\frac{2z}{5} + \frac{z}{5}$ is the same as $\frac{3z}{5}$. true

12. $\frac{c}{8} + \frac{c}{4}$ is the same as $\frac{3c}{4}$. false

13. $5f + \frac{1}{5}f$ is the same as $\frac{6}{5}f$. false

14. $-17b + 3b$ is the same as $-14b$. true

For Extra Practice, see page 425.

Written EXERCISES

Complete.

Sample: $3n + (-5n) = 12$ Solution: $3n + (-5n) = 12$
$\underline{\quad?\quad} \cdot n = 12$ $-2n = 12$
$\underline{\quad?\quad}(-2n) = 12 \cdot \underline{\quad?\quad}$ $-\frac{1}{2}(-2n) = 12\left(-\frac{1}{2}\right)$
$n = \underline{\quad?\quad}$ $n = -6$

A

1. $6a - 2a = 12$
$\overset{4}{\underline{\quad?\quad}} \cdot a = 12$
$\overset{\frac{1}{4}}{\underline{\quad?\quad}} \cdot 4a = 12 \cdot \underline{\overset{\frac{1}{4}}{\quad?\quad}}$
$a = \underline{\overset{3}{\quad?\quad}}$

2. $-7b + (-b) = -16$
$\overset{-8}{\underline{\quad?\quad}} \cdot b = -16$
$\overset{-\frac{1}{8}}{\underline{\quad?\quad}}(-8b) = -16 \cdot \underline{\overset{-\frac{1}{8}}{\quad?\quad}}$
$b = \underline{\overset{2}{\quad?\quad}}$

3. $\frac{1}{4}x = \frac{3}{4}x + 5$

$-\frac{3}{4}x + \frac{1}{4}x = \frac{3}{4}x - \frac{3}{4}x + 5$

$-\frac{2}{4}x \underline{\overset{}{\quad?\quad}} \cdot x = \underline{\quad?\quad} + 5 \quad 0$

$x = \underline{\quad?\quad} \quad -10$

4. $\frac{n}{2} + \frac{2n}{5} = 18$

$\overset{\frac{9}{10}}{\underline{\quad?\quad}} \cdot n = 18$

$\overset{\frac{10}{9}}{\underline{\quad?\quad}} \cdot \frac{9}{10} \cdot n = 18 \cdot \underline{\overset{\frac{10}{9}}{\quad?\quad}}$

$n = \underline{\quad?\quad} \quad 20$

5. $-13m + (-5m) = 36$
$\overset{-18}{\underline{\quad?\quad}} \cdot m = 36$
$-\frac{1}{18} \underline{\quad?\quad}(-18)m = 36 \cdot \underline{\overset{-\frac{1}{18}}{\quad?\quad}}$
$m = \underline{\quad?\quad} \quad -2$

6. $-\frac{2}{3}s = \frac{1}{3}s + (-8)$

$-\frac{1}{3}s + \left(-\frac{2}{3}s\right) = -\frac{1}{3}s + \frac{1}{3}s + (-8)$

$-\frac{3}{3} \underline{\quad?\quad} \cdot s = -8$

$s = \underline{\quad?\quad} \quad 8$

300 / CHAPTER ELEVEN

Solve for the variable indicated. Assume that no divisor has the value 0.

Sample: Solve $tm + km = s$ for m.　*Solution:*　$tm + km = s$

$$m(t + k) = s$$

$$m = \frac{s}{t + k}$$

7. $\dfrac{b}{c + d}$　　**9.** $\dfrac{g}{v + r}$　　**11.** $\dfrac{-3}{3 + j}$

8. $\dfrac{u}{2 - k}$

7. Solve $cz + dz = b$ for z.　　**8.** Solve $2d - kd = u$ for d.

9. Solve $g = vq + rq$ for q.　　**10.** Solve $-wt - ft = 20$ for t.

10. $\dfrac{20}{-w - t}$

11. Solve $3h + jh = -3$ for h.　　**12.** Solve $ds + ps = -a$ for s.

12. $\dfrac{-a}{d + r}$

Solve and check.

13. $2x + 5x = 13$ $\frac{13}{7}$　　　　　　**14.** $-2z + (-4z) = -18$ 3

15. $3y + (-2y) = -7$ -7　　　　**16.** $4 = -5w + (-4w)$ $-\frac{4}{9}$

17. $-38 = 15b + (-34b)$ 2　　　**18.** $2.22 = 0.4m - 0.03m$ 6

19. $-0.25 = -1.3q - 1.2q$ 0.1　　**20.** $44h - 9h = 70$ 2

21. $\dfrac{7m}{9} + \dfrac{2m}{9} = -\dfrac{1}{9}$ $-\frac{1}{9}$　　　**22.** $\dfrac{r}{6} + \dfrac{5r}{6} = 2$ 2

Sample: $7r = 2r + 10$

Solution:　　　$7r = 2r + 10$　　　*Check:*　$7r \overset{?}{=} 2r + 10$

$$7r - 2r = 2r - 2r + 10$$

$7(2)$	$2(2) + 10$
14	$4 + 10$
14	14 ✓

$$5r = 10$$

$$\frac{1}{5} \cdot 5r = 10 \cdot \frac{1}{5}$$

$$r = 2$$

23. $12a = 19a - 35$ 5　　　　　**24.** $-8t = 2t + 5$ $-\frac{1}{2}$

25. $3x = -2x + 30$ 6　　　　　**26.** $-10k = 3k + 39$ -3

27. $-1 + (-5n) = -6n$ 1　　　　**28.** $6c = 5c + 5$ 5

29. $9y = 73 + y - 1$ 7.2　　　　**30.** $-8 + 3x + (-2) = 2x + 8$ 18 **B**

31. $3p + 5 = 2p + 8$ 3　　　　　**32.** $c = 6.3 - 0.05c$ 6

33. $-\dfrac{5}{8}x + 3 = -\dfrac{3}{8}x + 7$ -16　**34.** $\dfrac{2m}{3} - \dfrac{5m}{6} = 24$ -144

35. $0.6z = 0.2z + 1.6$ 4　　　　**36.** $11a - 4a = -2a + 45$ 5

37. $-3[x + (-2)] = 11 - 5x$ $\frac{5}{2}$　**38.** $3(z + 1) = 2(z - 2)$ -7　　**C**

39. $7y - 2(y - 5) = 25$ 3　　　**40.** $4(h + 5) + h = 35$ 3

41. $p - \dfrac{1}{2}(p + 4p) + 4 = 30$ $-17\frac{1}{3}$　**42.** $\dfrac{1}{3}(9w - 18) + 2w = 14$ 4

11-4 *Applying Formulas*

OBJECTIVE

Use basic formulas to solve practical problems.

EXAMPLE 1 Find the volume.

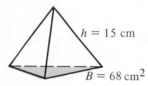

Formula: $V = \dfrac{1}{3}Bh$

$$B = 68, \; h = 15$$

$$V = \dfrac{1}{3} \cdot 68 \cdot 15$$

$$= 340$$

The volume is 340 cm^3.

EXAMPLE 2 Find the height.

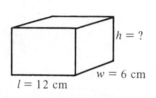

Volume: 576 cm³

Formula: $V = lwh$

$$\dfrac{1}{lw} \cdot V = \dfrac{1}{lw} \cdot lwh$$

$$\dfrac{V}{lw} = h$$

Formula for h: $h = \dfrac{V}{lw}$

$$h = \dfrac{576}{12 \times 6} = 8$$

The height of the box is 8 cm.

EXAMPLE 3 A bank makes a one-year loan of \$25,000 to a business. If the interest for the year is \$1750, what is the rate of interest?
Formula: Interest = principal × rate × time or, $I = prt$.

$$\dfrac{1}{pt} \cdot I = \dfrac{1}{pt} \cdot prt$$

$$\dfrac{I}{pt} = r$$

► Formula for r: $r = \dfrac{I}{pt}$.

$$r = \dfrac{1750}{25,000} = 0.07 = 7\%$$

The rate of interest is 7%.

  In Examples 1–3 we simplified the arithmetic by first solving for the variable. Of course, we could have substituted first, and then solved the resulting equation.

For Extra Practice, see page 426.

Solve. Use the formula $A = lw$ (Area = length $\times$ width).

1. The length of a rectangular swimming pool is 19 meters. The area of the pool is 133 square meters. Find the width. 7m

2. The dimensions of a rectangular picture are 12 cm by 18 cm. What is the area of the picture? 216 cm²

Solve. Use the formula $I = prt$ (Interest = principal $\times$ rate $\times$ time).

3. Sheila borrowed $2700.00 from the bank at a rate of $9\frac{1}{4}\%$. She paid $499.50 in interest. How long did she keep the loan? 2 years

4. Rudy borrowed $1000 for a period of 9 months $\left(\dfrac{3}{4} \text{ year}\right)$. He paid $30.00 interest. What was the rate of interest? 4%

5. A homeowner borrowed money at the rate of $8\frac{1}{2}\%$ for a period of 2 years. The interest charged was $255. What was the amount of the loan? $1500

6. A couple borrowed $4000 for 2 years at 7% interest. How much interest did they pay? $560

Solve. Use the formula Area $= \dfrac{1}{2} \times$ altitude $\times$ sum of bases,

$A = \dfrac{1}{2}as.$

7. The area of a trapezoid is 640 square centimeters. The sum of the bases is 80 cm. Find the altitude. 16 cm

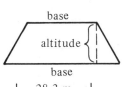

8. What is the area of this trapezoid? 202 m²

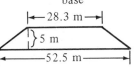

9. The area of this trapezoid is 18 square centimeters. What is the altitude? 3 cm

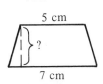

10. The area of this trapezoid is 540 square meters. What is the sum of the lengths of the bases? 72 m

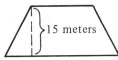

SELF-TEST 1

Solve and check.

Section 11-1, p. 292 1. $r + 3 = 8$ 5

2. $-7 = t - 2$ -5

Section 11-2, p. 295 Solve and check.

3. $2t = -10$ -5

4. $-7a = 21$ -3

Section 11-3, p. 298 Solve and check.

5. $8k - 4k = -2$ $-\frac{1}{2}$

6. $2b = 3b - 4$ 4

Section 11-4, p. 302 Solve. Use $A = lw$.

7. The area of a rectangle is 26 square meters. The width is 4 meters. What is the length? $6\frac{1}{2}$ m

8. A plot of land is 13 meters long and has an area of 143 square meters. What is the width? 11 m

Check your answers with those printed at the back of the book.

consumer notes *Energy Guide*

The Energy Efficiency Ratio, EER, is used to determine the efficiency of room air conditioners. The EER is computed by using the air conditioner's cooling capacity, which is measured in British thermal units (Btu's) and the electrical power it requires, which is measured in watts. To obtain the EER, divide $\dfrac{\text{Btu's}}{\text{watts}}$. The greater the EER, the more efficient the machine is.

Sometimes the air conditioner which is more expensive to buy is cheaper to operate. Would you rather buy an 8,000 Btu air conditioner which requires 900 watts or one which requires 1300 watts? Find the EER of each.

Properties of Inequality

11-5 *The Addition Property of Inequality*

OBJECTIVES

Solve inequalities like
$x + 5 > -1$ and $x + 6 \leq 3$.

Graph solution sets of inequalities.

Let's see if it seems reasonable to assume an addition property for inequality like the one for equality.

Add 3 to both members.

$5 < 9$ ◀ True
$8 < 12$ ◀ True

Add -2 to both members.

$-8 \geq -10$ ◀ True
$-10 \geq -12$ ◀ True

The Addition Property ▶ For all directed numbers r, s, and t,
of Inequality   if $r < s$, then $r + t < s + t$;
                     if $r > s$, then $r + t > s + t$.

Recall that subtracting a directed number is the same as adding its opposite. Then if $r < s$, $r - t < s - t$ because $r + (-t) < s + (-t)$.

Now let's use the addition property to solve inequalities.

EXAMPLE 1

$$x + 5 < 8$$
$$x + 5 + (-5) < 8 + (-5) \quad ◀ \text{Add } -5 \text{ to both members.}$$
$$x + 0 < 8 + (-5) \quad ◀ 5 + (-5) = 0$$
$$x < 3$$

Graph:

3 is not included.

EXAMPLE 2

$$3 + m \leq 2$$
$$-3 + 3 + m \leq -3 + 2 \quad ◀ \text{Add } -3 \text{ to both members.}$$
$$0 + m \leq -3 + 2 \quad ◀ -3 + 3 = 0$$
$$m \leq -1$$

Graph:

-1 is included

Oral

EXERCISES

Tell whether the statement is true or false. Explain your answer in terms of the number line.

Sample: $-5 < 3$ *What you say:* True; -5 is to the left of 3 on the number line.

1. $7 > -6$ T
2. $0 > -2$ T
3. $0 < -5$ F
4. $-6 > -7$ T
5. $\frac{2}{3} > \frac{5}{6}$ F
6. $-\frac{1}{4} < \frac{1}{5}$ T

Tell why the statement is true.

Sample: If $7 < b$, then $7 + (-3) < b + (-3)$.

What you say: -3 is added to both members.

7. If $2 > m$, then $2 - 3 > m - 3$. -3 is added to both members.
8. If $d > -6$, then $5 + d > 5 - 6$. 5 is added to both members.
9. If $20 > y$, then $20 + 4 > y + 4$. 4 is added to both members.
10. If $13 < r$, then $-9 + 13 < -9 + r$. -9 is added to both members.

Written

EXERCISES

Match each inequality in Column 1 with an equivalent inequality in Column 2.

A

COLUMN 1

1. $3 + x \leq 5$ E
2. $x + 7 > 16$ C
3. $x - 9 \leq 12$ A
4. $x + 4 \leq -10$ F
5. $x - 2 < 3$ D
6. $x + (-5) \geq -8$ B

COLUMN 2

A. $x \leq 21$
B. $x \geq -3$
C. $x > 9$
D. $x < 5$
E. $x \leq 2$
F. $x \leq -14$

Complete. Then graph the solution set.

Sample:
$$y + (-5) < -2$$
$$y + (-5) + \underline{\ ?\ } < -2 + \underline{\ ?\ }$$
$$y < \underline{\ ?\ }$$

Solution:
$$y + (-5) < -2$$
$$y + (-5) + 5 < -2 + 5$$
$$y < 3$$

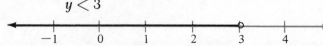

7.
$$(-11)\ x + 11 > 3$$
$$x + 11 + \underline{?} > 3 + \underline{?} \quad (-11)$$
$$x > \underline{?}\ _{-8}$$

8.
$$(-8)\ k + 8 \leq 7$$
$$k + 8 + \underline{?} \leq 7 + \underline{?}(-8)$$
$$k \leq \underline{?}\ _{-1}$$

9.
$$_4\ -6 \geq t + (-4)$$
$$-6 + \underline{?} \geq t + (-4) + \underline{?}\ _4$$
$$-2\ \underline{?} \geq t$$

10.
$$_{-1}\ a + 1 \geq -5$$
$$a + 1 + \underline{?} \geq -5 + \underline{?}\ ^{-1}$$
$$a \geq \underline{?}\ _{-6}$$

11.
$$9 + p < 4$$
$$_{-9}\ \underline{?} + 9 + p < \underline{?} + 4\ _{-9}$$
$$p < \underline{?}\ _{-5}$$

12.
$$s - 10 < -5$$
$$s - 10 + \underline{?} < -5 + \underline{?}\ ^{10}$$
$$_{10}s < \underline{?}\ _5$$

Solve the inequality and write its solution set.

Sample: $m + 2 > 5$

Solution:
$$m + 2 > 5$$
$$m + 2 + (-2) > 5 + (-2)$$
$$m > 3$$
$$\{\text{the directed numbers greater than 3}\}$$

13. $z + (-7) \geq 3$ $_{\geq 10}$

14. $v + 4 < -5$ $_{< -9}$

15. $14 < n + 2$ $_{> 12}$

16. $x - 15 > -22$ $_{> -7}$

17. $5 < g + 18$ $_{> -13}$

18. $10 + q > 5$ $_{> -5}$

19. $-4 + w \geq -9$ $_{> -5}$

20. $m + 6 < 8\frac{1}{2}$ $_{< 2\frac{1}{2}}$

21. $2 + u < -2$ $_{< -4}$

22. $10 + r \geq -20$ $_{> -30}$

23. $-5 \leq k + (-5)$ $_{\geq 0}$

24. $s + 16 > -9$ $_{> -25}$

25. $-32 \geq -16 + b$ $_{\leq -16}$

26. $0 > -12 + d$ $_{< 12}$

27. $4\left(-1 + \dfrac{a}{4}\right) \leq 0$ $_{\leq 4}$

28. $\dfrac{1}{3}(3z - 12) < -3$ $_{< 1}$

29. $\dfrac{5m - 10}{5} \geq -15$ $_{\geq 13}$

30. $6\left(2 + \dfrac{b}{6}\right) < -10$ $_{< -22}$

31. $(-24 + 2r)\dfrac{1}{2} \geq -20$ $_{\geq -8}$

32. $0.1(10w - 30) < 5$ $_{< 8}$

33. $5y - 12 > 4y$ $_{> 12}$

34. $0.5(8 + 2k) < -2$ $_{< -6}$

35. $3(x - 1) \geq 2(x + 2)$ $_{\geq 7}$

36. $\dfrac{4}{5}\left(\dfrac{5}{4}j + 15\right) > 9$ $_{> -3}$

37. $8(-2n + n) + 9n \leq -\dfrac{1}{4}$ $_{\leq -\frac{1}{4}}$

38. $\dfrac{18 + 4h}{3} < \dfrac{h}{3} + (-1)$ $_{< -7}$

39. $7(m - 2) < 6m + (-10)$ $_{< 4}$

40. $\dfrac{-18 + 6a}{3} \geq -5a + (-3.2)$ $_{> 0.4}$

B

C

11-6 *The Multiplication Property of Inequality*

OBJECTIVES
Solve inequalities like $4y > 12$ and $-5x + (-2) \le 8$, and graph the solution sets.

Let's look at some inequalities where both members are multiplied by the same directed number.

The multiplier can be positive:

$4 < 9$ ◄ True
$2(4) < 2(9)$
$8 < 18$ ◄ True

The multiplier can be negative:

$6 > 4$ ◄ True
$-3(6) > -3(4)$
$-18 > -12$ ◄ False. $-18 < -12$

When the multiplier is negative, the sense of the inequality is reversed. The multiplier can also be 0. Then, of course, both members would be 0.

The Multiplication Property ► For all directed numbers, r, s, and t:
of Inequality

(1) If t is positive and $r < s$, $rt < st$.
If t is positive and $r > s$, $rt > st$.

(2) If t is negative and $r < s$, $rt > st$.
If t is negative and $r > s$, $rt < st$.

(3) If $t = 0$ and $r < s$ or $r > s$, then $rt = st = 0$.

EXAMPLE 1

$4y > 12$

$\frac{1}{4} \cdot 4y > \frac{1}{4} \cdot 12$ ◄ Multiply both members by $\frac{1}{4}$.

$y > 3$

Graph:

EXAMPLE 2

$-5x + (-2) \le 8$

$-5x + (-2) + 2 \le 8 + 2$ ◄ Add 2 to both members.

$-5x \le 10$

$-\frac{1}{5}(-5x) \ge -\frac{1}{5} \cdot 10$ ◄ Multiply both members by $-\frac{1}{5}$.
Change $\le$ to $\ge$.

$x \ge -2$

Graph:

EXAMPLE 3

$$-3x + 1 < 5$$
$$-3x + 1 + (-1) < 5 + (-1)$$ ◀ Add -1 to both members.
$$-3x < 4$$
$$-\frac{1}{3}(-3)x > -\frac{1}{3} \cdot 4$$ ◀ Multiply both members by $-\frac{1}{3}$. Change $<$ to $>$.
$$x > -\frac{4}{3}$$

Graph:

Tell whether the symbol $>$, $<$, or $=$ should replace each question mark to make a true statement.

1. $15 \underline{\ ?\ } 10$ and $15 \cdot 2 \underline{\ ?\ } 10 \cdot 2$ $>; >$
2. $-8 \underline{\ ?\ } 5$ and $-8(-3) \underline{\ ?\ } 5(-3)$ $<; >$
3. $2 \underline{\ ?\ } 3$ and $2 \cdot 0 \underline{\ ?\ } 3 \cdot 0$ $<; =$
4. $-14 \underline{\ ?\ } 7$ and $-14 \cdot 3 \underline{\ ?\ } 7 \cdot 3$ $<; <$
5. $-4 \underline{\ ?\ } -7$ and $-4(-1) \underline{\ ?\ } -7(-1)$ $>; <$
6. $5 \underline{\ ?\ } 0$ and $5 \cdot 2 \underline{\ ?\ } 0 \cdot 2$ $>; >$
7. $-6 \underline{\ ?\ } 15$ and $-6 \cdot 0 \underline{\ ?\ } 15 \cdot 0$ $<; =$

Use the multiplication property of inequality to change the first inequality into the second.

Written EXERCISES

Sample: $12 < 25$; $48 < 100$ *Solution:* $12 < 25$
1. $-6(-5) > -5(-5)$ 2. $3 \cdot 9 > 3(-3)$ $12 \cdot 4 < 25 \cdot 4$
4. $-\frac{1}{5}(-30) \geq -\frac{1}{5}(45)$ 5. $4(-4) > 4(-10)$ $48 < 100$

1. $-6 < -5$; $30 > 25$ 2. $9 > -3$; $27 > -9$ A
3. $26 \geq 13$; $2 \geq 1\frac{1}{13}(26) \geq \frac{1}{13}(13)$ 4. $-30 \leq 45$; $6 \geq -9$
5. $-4 > -10$; $-16 > -40$ 6. $8 < 9$; $16 < 18$ $2(8) < 2(9)$
7. $-2 \geq -5$; $6 \leq 15$ 8. $3m > 24$; $m > 8$ $\frac{1}{3} \cdot 3m > \frac{1}{3} \cdot 24$
9. $-\frac{d}{3} \geq 7$; $d \leq -21$ 10. $5z \leq 10$; $z \leq 2$ $\frac{1}{5}(5z) \leq \frac{1}{5}(10)$

7. $-3(-2) \leq -3(-5)$ 9. $-3\left(-\frac{d}{3}\right) \leq -3(7)$

Solve. Then graph the solution set.

Sample: $3x > 9$ Solution: $3x > 9$

$$\frac{1}{3} \cdot 3x > \frac{1}{3} \cdot 9$$

$$x > 3$$

11. $14m \geq 42$ $m \geq 3$ 12. $18r < 12$ $r < \frac{2}{3}$ 13. $\frac{4}{5}w \leq 16$ $w \leq 20$

14. $-5 < 10n$ $-\frac{1}{2} < n$ 15. $8d \leq -16$ $d \leq -2$ 16. $-18 < 9v$ $-2 < v$

17. $\frac{1}{5}s > -1$ $s > -5$ 18. $100 < 10p$ $10 < p$ 19. $-15 < 5b$ $-3 < b$

Solve each inequality and write the solution set. Remember that multiplying both members of an inequality by a negative number reverses the sense of the inequality.

Sample: $-2b < 10$

Solution: $-2b < 10$

$$-\frac{1}{2}(-2b) > 10\left(-\frac{1}{2}\right)$$

$$b > -5$$

Solution set: {the directed numbers greater than -5}.

20. $-t > 29$ < -29 21. $-7y \geq -9$ $\leq \frac{9}{7}$ 22. $27 < -3m$ < -9

$< -\frac{1}{2}$ 23. $-\frac{1}{6} > \frac{1}{3}u$ 24. $-14 > -z$ > 14 25. $-\frac{n}{5} \leq 3$ ≥ -15

> 2

< 6 26. $2m - 1 < 11$ 27. $3 - 2s \geq 13$ ≤ -5 28. $-8b + 1 < -15$

B   29. $\frac{n}{4} < 7$ < 28 30. $-\frac{4}{5}z \geq 40$ ≤ -50 31. $24 > -\frac{3}{4}x$ > -32

≤ -6 32. $\frac{m}{3} + 2 \leq 0$ 33. $-(5r + 2) \leq 18$ > -4 34. $-0.50x < -2$ > 4

35. $-7a + 4 - a < -5 + a$ > 1 36. $-2[3t + (-6)] \geq 2t$ $\leq \frac{3}{2}$

37. $c + 3[-9 + (-c)] < -c + 3$ > -30 38. $\frac{1}{5}u - 8 > -2$ > 30

39. $5d - 3(4 - d) \geq 20$ ≥ 4 40. $-w + 5 < -4w - 7$ < -4

C   41. $10\left(\frac{9}{2} - \frac{1}{5}\right) > 2q$ $< 21\frac{1}{2}$ 42. $12\left(\frac{1}{4} - \frac{w}{6}\right) \leq -3w$ ≤ -3

SELF-TEST 2

Be sure that you understand these terms.

Addition Property of Inequality (p. 305)
Multiplication Property of Inequality (p. 308)

Solve. Graph the solution set.   Section 11-5, p. 305

1. $x + 4 > 7$ $x > 3$ 2. $t - 3 \leq -4$ $t \leq -1$

3. $-2 \leq t + 1$ $-3 \leq t$ 4. $-2 - m > -5$ $m < 3$

5. $7n \leq 21$ $n \leq 3$ 6. $-\dfrac{1}{2} y < 8$ $y > -16$ Section 11-6, p. 308

7. $-3b + 3 > 12$ $b < -3$ 8. $2c - 2 \geq -2$ $c \geq 0$

Check your answers with those printed at the back of the book.

chapter summary

1. Equations of the type $x + a = b$ are solved for x by applying the addition property of equality.

2. Equations of the type $ax = b$ are solved for x by applying the multiplication property of equality.

3. Equations of the type $ax + bx = c$ are solved for x by first combining similar terms and then applying the properties of equality.

4. The **addition property of inequality** can be stated as follows. For all directed numbers r, s, and t: If $r < s$, then $r + t < s + t$;
   If $r > s$, then $r + t > s + t$.

5. The **multiplication property of inequality** can be stated as follows. For all directed numbers r, s, and t:
   (1) If t is a positive number, if $r < s$, then $rt < st$;
        if $r > s$, then $rt > st$.
   (2) If t is a negative number, if $r < s$, then $rt > st$;
        if $r > s$, then $rt < st$.
   (3) If t is 0 and $r < s$ or $r > s$, then $rt = st = 0$.

challenge topics

Exponential Notation

As we have seen, exponential notation is very useful in expressing, in a simple way, a product consisting of repeating factors. For example:

$$8 \cdot 8 \cdot 8 \cdot 8 \cdot 8 \cdot 8 = 8^6$$

Study the pattern below:

$$3 \cdot 3 \cdot 3 \cdot 3 = 3^4$$
$$3 \cdot 3 \cdot 3 = 3^3$$
$$3 \cdot 3 = 3^2$$
$$3 = 3^1$$

Do you see why $3 = 3^1$? In a similar manner $y^1 = y$ and $k^1 = k$. We read k^1 as "k to the first power."

Often a large number can be expressed as the product of two numbers of which one is a power of **10**.

For example, here are two ways of writing **3800** as a product involving a power of 10:

$$3800 = 38 \times 100 \qquad 3800 = 3.8 \times 1000$$
$$= 38 \times 10^2 \qquad \qquad = 3.8 \times 10^3$$

A number is said to be in **scientific notation** when it is written as the product of a number between 1 and 10 and a power of 10 expressed in exponential notation. Which of the methods shown above expresses **3800** in **scientific notation**? To write **21,500** in scientific notation, we note that **2.15** is a number between 1 and 10.

$$21,500 = 2.15 \times 10,000$$
$$= 2.15 \times 10^4$$

What power of 10 should replace n to make the statement true?

1. $2000 = 2 \times n$ 10^3 **2.** $94000 = 9.4 \times n$ 10^4

3. $1000 = 1 \times n$ 10^3 **4.** $19,800 = 1.98 \times n$ 10^4

5. $30,500 = 3.05 \times n$ 10^4 **6.** $2,190,000 = 2.19 \times n$ 10^6

Express as a regular decimal numeral.

7. 1.02×10^2 102   **8.** 6.6×10^2 660

9. 2.07×10^3 2070   **10.** 6.02×10^5 602,000

Write in scientific notation.

11. 1200 1.2×10^3 **12.** 6000 6.0×10^3 **13.** 250 2.5×10^2

14. 75,000 7.5×10^4 **15.** 1,900,000 1.9×10^6 **16.** 2,870 2.87×10^3

17. two thousand **18.** five hundred **19.** two million 2.0×10^6

20. nine million **21.** five thousand **22.** four billion

9.0×10^6 5.0×10^3 4.0×10^9

17. 2.0×10^3 **18.** 5.0×10^2

chapter test

Solve.

1. $x - 5 = -7$ $_{-2}$

2. $6t + 2t = 4$ $\frac{1}{2}$

3. $-8k = 56$ $_{-7}$

4. $25 = -14 + x$ $_{39}$

5. $5x = 30$ $_6$

6. $2x - 5x = 15$ $_{-5}$

7. $\frac{x}{6} + 5 = 7$ $_{12}$

8. $-\frac{2}{7}t + \frac{1}{7}t = 5$ $_{-35}$

9. $-\frac{3x}{4} = 9$ $_{-12}$

10. $-1.5x + 0.5 = -1$ $_1$

Solve for y. Assume that no divisor has the value 0.

11. $y + c = d$ $_{d-c}$

12. $yb = c$ $\frac{c}{b}$

13. $cy - d = -b$ $\frac{d-b}{c}$

14. $dy - by = c$ $\frac{c}{d-b}$

Solve. Graph the solution set.

15. $4x < -16$ $_{x < -4}$

16. $-\frac{1}{2}x > 2$ $_{x < -4}$

17. $-4 + x > -1$ $_{x > 3}$

18. $3x + 4x \le 49$ $_{x \le 7}$

19. $x + 5 > -1$ $_{x > -6}$

20. $-3x + 6 \ge 0$ $_{x \le 2}$

21. The figure has volume 78 cubic centimeters. Use the formula $V = lwh$ and solve for l. Then substitute and find the value of l. 13 cm

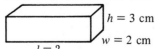

$h = 3$ cm

$w = 2$ cm

$l = ?$

22. A garden plot is to be 15 meters wide. How long will you make it if you want its total area to be 75 square meters? 5 m

Review of Skills

Write as a power of 10.

Sample: 100   *Solution:* $100 = 10 \cdot 10 = 10^2$

1. 1000 10^3 **2.** 10 10^1 **3.** 10,000,000 10^6 **4.** 100,000 10^5

5. $6 \times 10^3 + 3 \times 10^2 + 2 \times 10 + 5$

6. $1 \times 10^2 + 5 \times 10 + 7$

Write in expanded form.

Sample: 543

Solution: $500 + 40 + 3 = 5 \cdot 10^2 + 4 \cdot 10 + 3$

5. 6325   **6.** 157   **7.** 80,491   **8.** 674

7. $8 \times 10^4 + 4 \times 10^2 + 9 \times 10 + 1$

8. $6 \times 10^2 + 7 \times 10 + 4$

Simplify.

9. $9x + 2x$ $11x$ **10.** $10y - 8y$ $2y$ **11.** $-6q - 8q$ $-14q$

12. $4 \cdot 5 + 4 \cdot 2$ 28   **13.** $4(5 + 2)$ 28   **14.** $11 \cdot 3 + 11 \cdot 7$ 110

15. $31 \cdot 2 + 31 \cdot 8$ 310   **16.** $14 \cdot 14 + 14 \cdot 6$ 280   **17.** $12 + (-5)$ 7

18. $12 - 5$ 7   **19.** $18 + (-18)$ 0   **20.** $18 - 18$ 0

21. $53 + 32$ 85   **22.** $42 + 34$ 76   **23.** $34 + 42$ 76

24. $(19 + 3) + 7$ 29   **25.** $19 + (3 + 7)$ 29   **26.** $29 + (1 + 15)$ 45

27. $11(3 + 7)$ 110   **28.** $31(2 + 8)$ 310   **29.** $-14 + 14$ 0

Name the additive inverse.

30. 72 -72 **31.** -10 10   **32.** -61 61

33. $\frac{1}{5}$ $-\frac{1}{5}$ **34.** 2 -2 **35.** $-\frac{4}{3}$ $\frac{4}{3}$

Solve.

36. $-6 = -16 + 2m$ $m = 5$ **37.** $2m + 6 = 6 - 3m$ $m = 0$

38. $3m + 7 = -2$ $m = -3$ **39.** $5m - 13m = -24$ $m = 3$

40. $4m = 2m - 4$ $m = -2$ **41.** $-21 + m = -8m + 12$ $m = \frac{11}{3}$

Left: Public transportation in Seattle, about 1889.

Right: Modern public transportation in San Francisco.

12
Addition and Subtraction
of Polynomials

Adding Polynomials

12-1 *Polynomials*

> **OBJECTIVE**
> Identify types of polynomials.

Expressions like $12t$, $m + 3$, and $x^2y^2 + 3x + y + 2$ are called **polynomials.** A polynomial in one variable takes the form

$$ax^m + bx^n + cx^p + \cdots + d$$

where a, b, c, and d are directed numbers and m, n, and p are positive integers.

EXAMPLE 1 $3x + 2$ ◄ $a = 3, m = 1$
$d = 2$

EXAMPLE 2 $5x^2 + x + 3$ ◄ $a = 5, m = 2$
$b = 1, n = 1$
$d = 3$

Some polynomials have names that indicate the number of terms.

$12t$ ► one term ► **monomial**
$m + 3$ ► two terms ► **binomial**
$x^2 + 3x + 2$ ► three terms ► **trinomial**

A polynomial which has more than three terms, such as $x^2y^2 + 3x + y + 2$, has no special name. It is simply called a polynomial.

Here are more examples of polynomials.

Monomials ► $5k^2$ s^3 $\dfrac{2b}{3}$ 7

Binomials ► $n + 8$ $2 + x^2$ $ab - c$

Trinomials ► $2x^2 + 3x + 1$ $a + 2b - c$

Polynomials ► $m^2 + m + n + 3$ $b^3 + c^2 + c + 2$

Tell whether the polynomial is a monomial, a binomial, or a trinomial.

1. $-4a^2$ monomial

2. $2n + 6 + 7m$ trinomial

3. $-4rst - 5rs$ binomial

4. $-2xy$ monomial

5. $-14cd + c + 15$

6. $15rst^2$ monomial

5. trinomial

7. $10x^2yz^3$ monomial

8. $\frac{1}{4}ab^2c^3$ monomial

9. $x - y$ binomial

10. $3^2 + 4^2$ binomial

11. $12wxyz$ monomial

12. $6 + 3x + x^2$

trinomial

Tell whether the right member is a monomial, a binomial, or a trinomial.

1. $V = \frac{1}{3}Bh$ monomial

2. $A = lw$ monomial

3. $A = p + prt$ binomial

4. $a = s^2$ monomial

5. $q = 4D + 5$ binomial

6. $n = 10t + u$ binomial

7. $P = a + b + c$ trinomial

8. $I = 0.03PT$ monomial

9. $V = 6 + c^2 + b$ trinomial

Use the symbols 3, c^2, b, and b^5 to write an expression.

10. A monomial $3b^5c^2$

11. A polynomial with four terms $3 + c^2 - b^5 - b$

12. A binomial $c^2 - 3b$

13. A trinomial $b^5 + b + 3$

14. A polynomial with five terms $3b^5 - c^2 + bc^2 + b - 3$

15. A polynomial with six terms $c^2b - c^2 - 3 + b - b^5 + c^2b^5$

Nikolai Ivanovich Lobachevski

In 1829 Nikolai Ivanovich Lobachevski developed a new geometry which revolutionized mathematics. For centuries geometry had been based on the axioms of Euclid. One of these states that parallel lines never intersect. Lobachevski invented a geometry in which parallel lines *can* intersect. Nearly eighty years later non-Euclidean geometry would have a profound effect on the development of Albert Einstein's theory of relativity.

12-2 *Standard Form*

A polynomial in one variable is in standard form when the terms are ordered so that the variable in the first term has the greatest exponent, the variable in the second term has the next greatest exponent, and so on. The greatest exponent names the degree of the polynomial.

EXAMPLE 1 $3x + 10x^2 + 5 = \mathbf{10x^2 + 3x + 5}$ ◄ standard form

The degree is 2.

EXAMPLE 2 $x + x^4 - 2 = \mathbf{x^4 + x - 2}$ ◄ standard form

The degree is 4.

Notice in Example 2, there are no terms of degree two or three. But we can insert "missing" terms to complete the polynomial.

EXAMPLE 3 $x + x^4 - 2 = x^4 + 0x^3 + 0x^2 + x - 2$

0   0

The meaning of the polynomial is unchanged.

Some polynomials have more than one variable. We write these polynomials in standard form by ordering the terms according to values of exponents of one of the variables.

EXAMPLE 4 $5m^2n - m^3 + 4mn^2 = -m^3 + 5m^2n + 4mn^2$

Terms ordered according to exponents of m.

$$= 4mn^2 + 5m^2n - m^3$$

Terms ordered according to exponents of n.

Name the degree of the polynomial.

1. $x^2 + 2$ 2
2. $b - 7$ 1
3. $3c^2 + c + 1$ 2
4. $x^5 - 2$ 5
5. $2y^2 + y + 1$ 2
6. m^3 3
7. $t^5 - 2$ 5
8. $2n^2 - n + 1$ 2
9. $y + 5$ 1
10. $x^8 - x^2 + 1$ 8

Write in standard form.

Written
EXERCISES
A

Sample: $x^3 - 3 + 7x^5 + x$ *Solution:* $7x^5 + x^3 + x - 3$

1. $c - 7c^5$ $-7c^5 + c$
2. $-4 + 2x^2$ $2x^2 - 4$
3. $9k^2 - 5 + 3k^5 - k^3$
4. $12t^2 + 10 - 9t^4$
5. $3n^5 - 4n + n^2$ $3n^5 + n^2 - 4n$
6. $2a^2 - 4a^3 + a - 1$
7. $14a^5 - 8a^7 + a^3$
8. $d^2 - 5 + d$ $d^2 + d - 5$
9. $5r^4 - 2r^2 + r^9$ $r^9 + 5r^4 - 2r^2$
10. $2x^3 - 3x^5 + 4 - 7x^7$
   $-7x^7 - 3x^5 + 2x^3 + 4$

3. $3k^5 - k^3 + 9k^2 - 5$
4. $-9t^4 + 12t^2 + 10$
6. $4a^3 + 2a^2 + a - 1$
7. $-8a^7 + 14a^5 + a^3$

Write in standard form.

Sample: $2mn - 5m^2n^2 + 6m^3 - 1$

Solution: $6m^3 - 5m^2n^2 + 2mn - 1$
   $(\text{or } -5m^2n^2 + 2mn + 6m^3 - 1)$

13. $7a^6 - 6a^3b + 2a^2b^2$
14. $r^3 - 2rs^3 - 3s^2 + 4$
15. $2a^2 + ab + 2b^2$
16. $3p^4 - 6p^2q^2 + q^4 + 7$
   $m^3 - m^2n + n^2$

11. $4xy - 3x^2$ $-3x^2 + 4xy$
12. $n^2 + m^3 - m^2n$
13. $2a^2b^2 - 6a^3b + 7a^6$
14. $r^3 - 3s^2 - 2rs^3 + 4$
15. $2a^2 + 2b^2 + ab$
16. $7 + 3p^4 + q^4 - 6p^2q^2$
17. $m^3n^2 + n^3 + m^4n^2 + m^2$
   $m^4n^2 + m^3n^2 + m^2 + n^3$
18. $13z^2 + 2x^5 - 7x + 4x^3$
   $2x^5 + 4x^3 - 7x + 13z^2$

B

Write in standard form. Insert any "missing" terms.

21. $s^5 + 0s^4 + 0s^3 - 3s^2 + 0s + 2$
22. $2y^5 + 0y^4 - y^3 + 0y^2 + 0y + 5$

Sample: $2x^5 + x - 2$

Solution: $2x^5 + 0x^4 + 0x^3 + 0x^2 + x - 2$
   $n^6 + 0n^5 + 0n^4 + 0n^3 + 3n^2 + 0n + 8$

19. $3m^4 - 1$ $3m^4 + 0x^3 + 0x^2 + 0x - 1$
20. $3n^2 + 8 + n^6$
21. $s^5 - 3s^2 + 2$
22. $5 - y^3 + 2y^5$
23. $x^7 + x^4$
24. $12 - p^2 + 7p^3$
25. $k - 2k^5 + k^7 - 19$
26. $1 + 6d^5$

$7p^3 - p^2 + 0p + 12$

$k^7 + 0k^6 - 2k^5 + 0k^4 + 0k^3 + 0k^2 + k - 19$ $6d^5 + 0d^4 + 0d^3 + 0d^2 + 0d + 1$

23. $x^7 + 0x^6 + 0x^5 + x^4 + 0x^3 + 0x^2 + 0x + 0$

12-3 *Polynomials and Function Machines*

OBJECTIVE

Evaluate polynomials used as function rules.

As you have seen, the rule for finding values of $f(x)$ with a function machine is often a polynomial.

EXAMPLE 1 $f(x) = 3x - 1$; replacement set: $\{-3, -1, 0, 1, 3, 5\}$

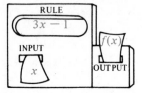

x	$f(x)$	$(x, f(x))$
-3	-10	$(-3, -10)$
-1	-4	$(-1, -4)$
0	-1	$(0, -1)$
1	2	$(1, 2)$
3	8	$(3, 8)$
5	14	$(5, 14)$

Function: $\{(-3, -10), (-1, -4), (0, -1), (1, 2), (3, 8), (5, 14)\}$

EXAMPLE 2 $f(x) = 2x^2 - 2$; replacement set: $\{2, 0, -1, -2, -3\}$

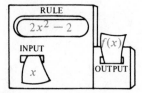

x	$f(x)$	$(x, f(x))$
2	6	$(2, 6)$
0	-2	$(0, -2)$
-1	0	$(-1, 0)$
-2	6	$(-2, 6)$
-3	16	$(-3, 16)$

Function: $\{(2, 6), (0, -2), (-1, 0), (-2, 6), (-3, 16)\}$

calculator corner

You can use your calculator as a function machine. Let the polynomial $x^2 + x + 4$ be the rule. Choose five input values for x which are between $^-5$ and 5. Find the corresponding output values, make a list of ordered pairs, and graph them. Now choose five input values between -10 and -5 and five between 5 and 10. Find the output values and graph the ordered pairs. Can you determine what the graph looks like? The more ordered pairs you find, the more exact your graph will be. Try to graph other polynomials with your function machine. A u-shaped curve

Tell how to complete the table.

$f(x) = x^2 + 1$

x	$f(x)$	$(x, f(x))$
5	26	(5, 26)
1. 0	? 1	? (0, 1)
2. 2	? 5	? (2, 5)
3. 4	?17	? (4, 17)
4. -1	? 2	? (−1, 2)
5. -3	?10	? (−3, 10)

6.
7.
8.
9.
10.

$f(x) = x^2 + x + 1$

x	$f(x)$	$(x, f(x))$
-5	21	(−5, 21)
0	? 1	? (0, 1)
-1	? 1	? (−1, 1)
1	? 3	? (1, 3)
-2	? 3	? (−2, 3)
2	? 7	? (2, 7)

Find the value of $f(a)$. Use the given replacement for a.

Sample: $f(a) = 4 - 3a + 2a^2$
Let $a = -1$.

Solution: $4 - 3(-1) + 2(-1)^2$
$4 + 3 + 2$
9

A

1. $f(a) = a^2 + 7a + 10$
Let $a = -2.$ 0

2. $f(a) = a^2 - 8a - 15$
Let $a = 0.$ −15

3. $f(a) = a^2 - 14a + 33$
Let $a = 3.$ 0

4. $f(a) = 7a^2 - 15a + 2$
Let $a = 1.$ −6

5. $f(a) = 5 + 9a - 18a^2$
Let $a = \dfrac{1}{3}.$ 6

6. $45 - 70a + 25a^2 = f(a)$
Let $a = \dfrac{1}{5}.$ 32

7. $f(a) = 14 - (-6a + 5) - 21$
Let $a = -2.$ −24

8. $f(a) = -(-7a + 12)$
Let $a = 0.3.$ −9.9

9. $f(a) = (a^2 - 12) + 3$
    Let $a = -4$. 7

10. $f(a) = 2a - (a + 3)$
    Let $a = 5$. 2

11. $f(a) = -(5 - 4a)$
    Let $a = 2$. 3

12. $a^2 + 100a + 20 = f(a)$
    Let $a = 10$. 1120

13. $f(a) = (2 - 3a) - 4a$
    Let $a = 2$. -12

14. $f(a) = (3a + 6) - (4a - 7)$
    Let $a = 0.2$. 12.8

Use the function machine and the rule to complete the table.

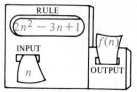

	n	$f(n)$	$(n, f(n))$	
15.	-7	120	?	$(-7, 120)$
16.	-3	28	?	$(-3, 28)$
17.	-1	? 6	?	$(-1, 6)$
18.	0	? 1	?	$(0, 1)$
19.	4	?21	?	$(4, 21)$
20.	6	?55	?	$(6, 55)$
21.	8	?105	?	$(8, 105)$

See page A2 at the back of the book for Ex. 22–32.

Use the function equation and replacement set to write the function.

Sample: $f(c) = 4c^2 - 2c + 4$; $\{-2, -1, 0, 1, 2\}$

Solution: $\{(-2, 24), (-1, 10), (0, 4), (1, 6), (2, 16)\}$

B

22. $f(b) = (2b^2 + 3) + b$; $\{-3, -1, 0, 1, 3\}$

23. $f(d) = 6d - (3d - 5)$; $\{-6, -4, -2, 0, 2, 4, 6\}$

24. $f(h) = 7 - 2h + h^2$; $\left\{-1, -\dfrac{1}{2}, 0, \dfrac{1}{2}, 1\right\}$

25. $f(a) = a^3 - 2a + 4$; $\{-3, -2, 0, 2, 3\}$

26. $f(m) = 5m^2 - 10m - 15$; $\left\{-\dfrac{1}{10}, -\dfrac{1}{5}, 0, \dfrac{1}{5}, \dfrac{1}{10}\right\}$

27. $f(x) = 12 + x - 3x^2$; $\{-4, -2, 0, 2, 4\}$

28. $f(y) = y^2 + y + 3$; $\{-6, -4, -2, 0, 2, 4, 6\}$

29. $f(t) = t^3 - 1$; $\{-2, -1, 0, 1, 2\}$

C

30. $f(x) = (3x + 5) - 4x + 7$; $\{-10, -5, 0, 5, 10\}$

31. $f(y) = 2 + y^2 - (3y - 1)$; $\{-0.1, 0, 0.1\}$

32. $f(s) = 4 - (-5s - 1) - 2s^2 + 3$; $\{-5, -3, -1, 0, 1, 3, 5\}$

12-4 *Addition of Polynomials*

OBJECTIVES

Find the sum of polynomials, such as $4x + 8$ and $x + 6$, or $3x^2 - x + 1$ and $5x + 7x^2$.

To **add** polynomials, we combine **similar** terms. We can arrange the work either vertically or horizontally:

EXAMPLE 1 Add $4x + 1$ and $x + 3$.

Vertical

$$
\begin{array}{r}
4x + 1 \\
x + 3 \\
\hline
5x + 4
\end{array}
$$

Horizontal

$$(4x + 1) + (x + 3) = 4x + x + 1 + 3$$
$$= 5x + 4$$

EXAMPLE 2 Add $5y - 7$ and $3y + 2$.

Vertical

$$
\begin{array}{l}
5y + (-7) \\
3y + 2 \\
\hline
8y + (-5) = 8y - 5
\end{array}
$$

Horizontal

$$[5y + (-7) + (3y + 2)] = 5y + 3y + (-7) + 2$$
$$= 8y + (-5)$$
$$= 8y - 5$$

If the polynomials are not in standard form, we usually express each in standard form before adding.

EXAMPLE 3 Add $5m^2 - 4 + 5m$ and $3 - 5m - m^2$.

$$
\begin{array}{l}
5m^2 - 4 + 5m \blacktriangleright \quad 5m^2 + 5m + (-4) \\
3 - 5m - m^2 \blacktriangleright \quad -1m^2 + (-5m) + 3 \\
\hline
 4m^2 + 0m + (-1) = 4m^2 - 1
\end{array}
$$

We can check the result by substituting for m. Let's use $m = 1$ and check Example 3.

$$
\begin{array}{l}
5m^2 + 5m + (-4) \blacktriangleright \quad 5 + 5 + (-4) \blacktriangleright \quad 6 \\
-1m^2 + (-5m) + 3 \quad\quad -1 + (-5) + 3 \quad\quad \underline{-3} \\
\hline
 3
\end{array}
$$

$$4m^2 - 1 = 4 - 1 = 3 \quad \checkmark$$

Simplify.

Sample: $(4r - 2t) + 3t$ What you say: $4r + t$

1. $(3a + 6) + 2a$ $\,^{5a + 6}$ **2.** $12n + (2m - 6n)$ $\,^{2m + 6n}$

3. $(4c + 3) - 6c$ $\,^{-2c + 3}$ **4.** $(2k^2 - 5h) + 11h$ $\,^{2k^2 + 6h}$

5. $\dfrac{1}{2}b + (c + 4b)$ $\,^{4\frac{1}{2}b + c}$ **6.** $2x + \left(5y + \dfrac{1}{2}x\right)$ $\,^{2\frac{1}{2}x + 5y}$

7. $(2r - 5s) - 2s$ $\,^{2r - 7s}$ **8.** $(2s + t) + (s - t)$ $\,^{3s}$

9. $(2a^3 - 5b^2) - 8b^2$ $\,^{2a^3 - 13b^2}$ **10.** $7.3m + (4.8n + 2.1m)$ $\,^{9.4m\,+}_{4.8n}$

11. $0.7t + (3.1r - 0.9t)$ $\,^{3.1r - 0.2t}$ **12.** $(0.4w + 0.7z) - 0.2w$
$^{0.2w + 0.7z}$

For Extra Practice, see page 427.

Add.

Sample 1: $\begin{aligned} 2t &+ 7 \\ 4t &+ 3 \end{aligned}$ *Sample 2:* $\begin{aligned} r^2 & \quad\;\; + 1 \\ 3r^2 &+ 7s + 5 \end{aligned}$

Solution: $\overline{6t + 10}$ *Solution:* $\overline{4r^2 + 7s + 6}$

A

1. $\begin{aligned} 6w &+ 4 \\ 8w &+ 5 \\ \hline 14w &+ 9 \end{aligned}$ **2.** $\begin{aligned} 3k &+ h \\ 4k &+ 7h \\ \hline 7k &+ 8h \end{aligned}$ **3.** $\begin{aligned} 2a &+ 4a^3 \\ a &+ 5a^3 \\ \hline 3a &+ 9a^3 \end{aligned}$

4. $\begin{aligned} 12x &- 2y \\ 7x & \\ \hline 19x &- 2y \end{aligned}$ **5.** $\begin{aligned} m^3 &+ 5n^2 \\ -m^3 &- 3n^2 \\ \hline &\;\; 2n^2 \end{aligned}$ **6.** $\begin{aligned} 2t &+ 12 \\ 6t &- \;\; 7 \\ \hline 8t &+ 5 \end{aligned}$

7. $\begin{aligned} x^2 &+ 2y \\ x^2 &- 2y \\ \hline 2x^2 & \end{aligned}$ **8.** $\begin{aligned} 3r^2 + 2s^2 +\;\; t^2 \\ 4r^2 + 5s^2 + 3t^2 \\ \hline 7r^2 + 7s^2 + 4t^2 \end{aligned}$ **9.** $\begin{aligned} -7x^2 + 5xy + 4y^2 \\ -3x^2 - 2xy - 6y^2 \\ \hline -10x^2 + 3xy - 2y^2 \end{aligned}$

10. $\begin{aligned} 6a &+ 9 \\ 2a &- 4 \\ \hline 8a &+ 5 \end{aligned}$ **11.** $\begin{aligned} m^2 -\;\; n^2 +\;\; p^2 \\ 4m^2 - 3n^2 - 5p^2 \\ \hline 5m^2 - 4n^2 - 4p^2 \end{aligned}$ **12.** $\begin{aligned} 9z^2 \quad\quad\;\; + 5w^2 \\ 4z + 7w^2 \\ \hline 9z^2 + 4z + 12w^2 \end{aligned}$

13. $\begin{aligned} 7c &+ \;\; 4 \\ 3c &- 12 \\ \hline 10c &- 8 \end{aligned}$ **14.** $\begin{aligned} 5x &- 7 \\ 2x &- 2 \\ \hline 7x &- 9 \end{aligned}$ **15.** $\begin{aligned} 4a &- b \\ -5a &- b \\ \hline -a &- 2b \end{aligned}$

16. $\begin{aligned} 3x &+ y \\ -3x &- y \\ \hline &\;\; 0 \end{aligned}$ **17.** $\begin{aligned} 3x^2 - 4y + \;\; 9 \\ x^2 - \;\; y - 12 \\ \hline 4x^2 - 5y - \;\; 3 \end{aligned}$ **18.** $\begin{aligned} 3r^2 + 8d + 4 \\ r^2 + 3d + 1 \\ \hline 4r^2 + 11d + 5 \end{aligned}$

19. $\begin{aligned} y^2 &+ z \\ y^2 &- z \\ \hline 2y^2 & \end{aligned}$ **20.** $\begin{aligned} 4n^2 - 7n + 9 \\ 2n^2 \quad\quad\; - 1 \\ \hline 6n^2 - 7n + 8 \end{aligned}$ **21.** $\begin{aligned} 7.6a + 2.4b + 5 \\ 1.8a - 1.7b + 3 \\ \hline 9.4a + 0.7b + 8 \end{aligned}$

22. $(12s^2 + 24) + (3s^4 - 12s^2)$ $3s^4 + 24$

23. $(m^2 + 4m - 8) + (2m^2 - 9m + 5)$ $3m^2 - 5m - 3$

24. $(8k - 5h^2) + (13k - h^2)$ $-6h^2 + 21k$

25. $(k^4 - 5k^2 + 4) + (-3k^3 + 6k - 1)$ $k^4 - 3k^3 - 5k^2 + 6k + 3$

26. $(7z^2 - 3z + 6) + (4z^2 + 9z - 11)$ $11z^2 + 6z - 5$

27. $(2x^2 + 4xy + 2y^2) + (2x^2 - 4xy + 2y^2)$ $4x^2 + 4y^2$

Add. Check by using $a = -1$, $b = 2$, $c = 3$, $d = 4$.

Sample:	$3a - 5$	*Check:* $3(-1) - 5 = -8$
	$a + 3$	$1(-1) + 3 = \;\;\;2$
	$2a - 1$	$2(-1) - 1 = -3$
Solution:	$6a - 3$	$6(-1) - 3 = -9$ $\checkmark$

28. $\begin{array}{r} 2d - 8 \\ 4d + 1 \\ \hline 6d - 7 \end{array}$ 　　**29.** $\begin{array}{r} 2b + 5 \\ 5b - 10 \\ \hline 7b - 5 \end{array}$ 　　**30.** $\begin{array}{r} c^2 + 4c - 6 \\ 2c^2 - c + 6 \\ \hline 3c^2 + 3c \end{array}$

B

31. $\begin{array}{r} 8a + 2b - 12 \\ a - b - 7 \\ 4a - 3b + 4 \\ \hline 13a - 2b - 15 \end{array}$ 　**32.** $\begin{array}{r} 3c - d + 5 \\ c + d + 4 \\ 5c - 2d \\ \hline 9c - 2d + 9 \end{array}$ 　**33.** $\begin{array}{r} 2a + 8 - 2d \\ 4a - 13 - d \\ -6a + 2d \\ \hline -5 - d \end{array}$

34. $\begin{array}{r} 5c + 6d \\ 7c - 2d \\ \hline 12c + 4d \end{array}$ 　**35.** $\begin{array}{r} d^3 - 5d^2 + 7d \\ 3d^3 - 3d^2 - 7d \\ \hline 4d^3 - 8d^2 \end{array}$ 　**36.** $\begin{array}{r} a^2 - 4a - 1 \\ -6a^2 - 8 \\ \hline -5a^2 - 4a - 9 \end{array}$

Add.

37. $\begin{array}{r} \frac{1}{2}s^3 + t^3 \\ s^3 - t^3 \\ \hline 1\frac{1}{2}s^3 \end{array}$ 　**38.** $\begin{array}{r} 2\frac{1}{3}k^2 - m^2 \\ \frac{1}{3}k^2 + \frac{1}{2}m^2 \\ \hline 2\frac{2}{3}k^2 - \frac{1}{2}m^2 \end{array}$ 　**39.** $\begin{array}{r} \frac{3}{5}a - 2b \\ \frac{2}{5}a - 2b \\ \hline a - 4b \end{array}$

40. $\begin{array}{r} 0.2rs - 0.61 \\ 0.5rs + 0.26 \\ 4.6rs + 0.54 \\ \hline 5.3rs + 0.19 \end{array}$ 　**41.** $\begin{array}{r} 4.6 + 5ab \\ 7.0 + 3ab \\ -9.2 - 3ab \\ \hline 2.4 + 5ab \end{array}$ 　**42.** $\begin{array}{r} 2c - 15 \\ 6c + 30 \\ -9c - 9 \\ \hline -c + 6 \end{array}$

43. $\begin{array}{r} 13x^4 - 4.2x^2y^2 + 5\;\;y^4 \\ 7x^4 - 8\;\;x^2y^2 - 7.2y^4 \\ \hline 20x^4 - 12.2x^2y^2 - 2.2y^4 \end{array}$ 　**44.** $\begin{array}{r} 4h^3 - 5k^2 - 3 \\ 2h^3 - 3k^2 - 6k \\ \hline 6h^3 - 8k^2 - 6k - 3 \end{array}$

45. $(21x^3 + 4x^2 - 7x + 16) + (7x - 9 - 2x^2 - 8x^3) + (x - 4)$ 　$13x^3 + 2x^2 + x + 3$

46. $(-4k^2) + (2k^2 - 5k^4 - 1 + k) + (7k + 2k^2 + k^5 - 9)$ 　$k^5 - 5k^4 + 8k - 10$

12-5 *Addition Properties*

OBJECTIVE

Use the commutative and associative properties in adding polynomials.

We assume that addition of polynomials is both commutative and associative.

EXAMPLE 1 The Commutative Property of Addition:

$$(3x + 9) + (x - 4) = 3x + 9 + x + (-4)$$
$$= 3x + x + 9 + (-4)$$
$$= 4x + 5$$
$$(x - 4) + (3x + 9) = x + (-4) + 3x + 9$$
$$= x + 3x + (-4) + 9$$
$$= 4x + 5$$

EXAMPLE 2 The Associative Property of Addition:

$$[(4m^2 + 3) + (m^2 - 10)] + 4m^2 = [4m^2 + 3 + m^2 + (-10)] + 4m^2$$
$$= (5m^2 - 7) + 4m^2$$
$$= 9m^2 - 7$$
$$(4m^2 + 3) + [(m^2 - 10) + 4m^2] = (4m^2 + 3) + (m^2 - 10 + 4m^2)$$
$$= 4m^2 + 3 + (5m^2 - 10)$$
$$= 9m^2 - 7$$

If you think of a polynomial as a way of representing a number, it seems logical that polynomials have the same properties as numbers.

EXERCISES

Tell which property of addition justifies the statement.

1. $(7 + x) + 4x^2 = 4x^2 + (7 + x)$ commutative
2. $(y^3 - 6) + (y^3 + 7) = (y^3 + 7) + (y^3 - 6)$ commutative
3. $(2a + 5) + (4a - 6) = (4a - 6) + (2a + 5)$ commutative
4. $[4n + (1 + 3n^2)] + n = 4n + [(1 + 3n^2) + n]$ associative
5. $9s + (2s^2 - 5) = (2s^2 - 5) + 9s$ commutative
6. $[(3d + 2) + d] + 4d^2 = (3d + 2) + [(d + 4d^2)]$ associative

7. $[(x^2 + y) + (xy - 2)] + (x - y) = (x^2 + y) +$
$[(xy - 2) + (x - y)]$ associative

8. $(k + 1) + [(k^2 - 4) + (5k - 2)] = [(k^2 - 4) +$
$(5k - 2)] + (k + 1)$ commutative

9. $(2t + s) + [(t - s) + (s - 3)] = (2t + s) +$
$[(s - 3) + (t - s)]$ commutative

10. $[3r + (r + 5)] + (r + 2) = 3r + [(r + 5) + (r + 2)]$ associative

1-9. The final step of each simplification is given.

Simplify both members to show that they are equal.

Sample: $\underline{(5a - 3) + (4 + a) = (4 + a) + (5a - 3)}$

Solution:

$5a - 3 + 4 + a$	$4 + a + 5a - 3$
$5a + a - 3 + 4$	$a + 5a + 4 - 3$
$6a + 1$	$6a + 1$ $\checkmark$

A

1. $(12 - 4b) + (4 + 2b) = (4 + 2b) + (12 - 4b)$ $-2b + 16$
2. $-5k + (k^2 - 2k + 4) = (k^2 - 2k + 4) + (-5k)$ $k^2 - 7k + 4$
3. $(2b - 7) + (2b - 3) = 2b + [-7 + (2b - 3)]$ $4b - 10$
4. $(6a - b) + (a + b) = (a + b) + (6a - b)$ $7a$
5. $(10z^2 - 8z + 3) + (6z - 5z^2) = (6z - 5z^2) +$
$(10z^2 - 8z + 3)$ $5z^2 - 2z + 3$
6. $[(2x - 4y - 12) + (x + 2y + 9)] + (y - 3) =$
$(2x - 4y - 12) + [(x + 2y + 9) + (y - 3)]$ $3x - y - 6$
7. $(m - 2) + [(2m^2 + m - 7) + (3m + 1)] =$
$[(m - 2) + (3m + 1)] + (2m^2 + m - 7)$ $2m^2 + 5m - 8$
8. $(7x^2 + 10xy + 6y^2) + (-2x^2 - y^2) = (-2x^2 - y^2) +$
$(7x^2 + 10xy + 6y^2)$ $5x^2 + 10xy + 5y^2$
9. $(-a^3 + 2a^2 - 5a) + (a^4 - 4a) = (a^4 - 4a) +$
$(-a^3 + 2a^2 - 5a)$ $a^4 - a^3 + 2a^2 - 9a$

Find both sums. Compare.

10.

$3k - 5$	$k + 8$
$k + 8$	$3k - 5$
$4k + 3$	$4k + 3$

11.

$5y - 12$	$-3y + 10$
$-3y + 10$	$5y - 12$
$2y - 2$	$2y - 2$

12.

$6x - 2$	$-x - 11$
$3x + 10$	$3x + 10$
$-x - 11$	$6x - 2$
$8x - 3$	$8x - 3$

13.

$-3x + 5y$	$-3x + 5y$
$-4x + y$	$- x$
$- x$	$-4x + y$
$-8x + 6y$	$-8x + 6y$

14.

$$2t - 2s - 10 \qquad 5t - 3s + 1$$
$$\underline{5t - 3s + 1} \qquad \underline{2t - 2s - 10}$$
$$7t - 5s - 9 \qquad 7t - 5s - 9$$

15.

$$x^2 + 2y^2 \qquad -x^2$$
$$x^2 - y^2 \qquad x^2 - y^2$$
$$\underline{-x^2} \qquad \underline{x^2 + 2y^2}$$
$$x^2 + y^2 \qquad x^2 + y^2$$

Find the value. Use $a = -2$, $b = 3$, $x = 0$, $y = 1$.

Sample 1: $8b^2 + 3x - 2$

Solution: $8(9) + 3(0) - 2 = 72 - 2 = 70$

Sample 2: $3a^2 + 6a - 10$

Solution: $3(4) + 6(-2) - 10 = 12 - 12 - 10 = -10$

16. $-7x^2 - 4x + 9$ 9

17. $y^3 + 5y^2 - 8y$ -2

18. $2a^2 - 4ax + x^2$ 8

19. $a^2 + b^2 + x^2$ 13

20. $a^3 - a^2 + 2a$ -16

21. $by + ab + bx$ -3

22. $2.1b^2 + 1.3b - 1$ 21.8

23. $xy + bx + ax$ 0

24. $(2a + b) + (-2a - b)$ 0

25. $0.5y + 2.5b + 3$ 11.0

26. $2ab + 3a + 2b$ -12

27. $x^4 + x^3 + x^2 + x + 1$ 1

Show that a true statement results when the variable is replaced by the suggested value.

Sample: $(2a + 5) + (4a - 6) = (4a - 6) + (2a + 5); a = 3$

Solution: $(2a + 5) + (4a - 6) \stackrel{?}{=} (4a - 6) + (2a + 5)$

$(2 \cdot 3 + 5) + (4 \cdot 3 - 6)$	$(4 \cdot 3 - 6) + (2 \cdot 3 + 5)$
$11 + 6$	$6 + 11$
17	17 ✓

B

28. $4r + (3r - 2) = (4r + 3r) - 2; r = 2$ 12

29. $(p^2 - 3p) + 5p = 5p + (p^2 - 3p); p = 4$ 24

30. $(3y + 4) + (6y^2 + 11y - 4) = (6y^2 + 11y - 4) +$ $(3y + 4); y = 5$ 220

31. $2a + [4a + (8a - 3)] = (2a + 4a) + (8a - 3); a = 10$ 137

32. $(y^2 - 2y^3 + y^2) + (-4y + 5) = (-4y + 5) +$ $(y^2 - 2y^3 + y^2); y = 3$ -43

33. $(3x - 2) + (5x - 4) = (5x - 4) + (3x - 2); x = -2$ -22

34. $(k^3 - 2k^2 - 20) + (k^3 - k) = (k^3 - k) +$ $(k^3 - 2k^2 - 20); k = 0$ -20

35. $4r^2 + (5 - 10r + 2r^2) = (4r^2 + 5) + (-10r + 2r^2); r = \dfrac{1}{2}$ $1\frac{1}{2}$

SELF-TEST 1

Be sure that you understand these terms.

polynomial (p. 316)　　monomial (p. 316)
binomial (p. 316)　　　trinomial (p. 316)
standard form (p. 318)　degree (p. 318)

Tell whether the polynomial is a monomial, a binomial, or a trinomial.

Section 12-1, p. 316

1. $6q - r$ binomial

2. $2abc + 4$ binomial

3. $6x + 4ab + 5d$ trinomial

4. $6ab$ monomial

Write in standard form.

Section 12-2, p. 318

5. $2x + 5 + 9x^2$ $9x^2 + 2x + 5$

6. $m^2 + 2 + m^3$ $m^3 + m^2 + 2$

7. $1 + x^2y + xy^2$ $x^2y + xy^2 + 1$

8. $2y + y^5 + 1$ $y^5 + 2y + 1$

Find the value of $f(a)$. Use the given replacement for a.

Section 12-3, p. 320

9. $f(a) = 2a^2 + a + 3;\ a = -1$ $f(a) = 4$

10. $f(a) = a^2 - 4a - 4;\ a = 0$ $f(a) = -4$

Add. Check by using $a = 1$, $b = 2$.

Section 12-4, p. 323

11. $4a + 5$
　　$8a - 14$
　　$\overline{12a - 9}$

12. $-4a^2 \qquad + 6b^2$
　　$\ 9a^2 + ab - 5b^2$
　　$\overline{5a^2 + ab + b^2}$

Tell which property of addition justifies the statement.

Section 12-5, p. 326

13. $[(2y + 3) + (8y^2 + y + 2)] + y^3 = (2y + 3) + [(8y^2 + y + 2) + y^3]$ associative property

14. $(12z^2 + z + 1) + (z^2 - 5) = (z^2 - 5) + (12z^2 + z + 1)$ commutative property

Check your answers with those printed at the back of the book.

Subtracting Polynomials

12-6 *Polynomials and Their Opposites*

OBJECTIVES

Apply the addition property of zero to polynomials.

Give the opposite of a polynomial.

When you add zero to any polynomial, the polynomial is unchanged. Zero is called the **identity element** for addition of polynomials.

EXAMPLE 1 $(3t + 7) + 0 = 0 + (3t + 7) = 3t + 7$

EXAMPLE 2 $(-x^3 + 3x + 1) + 0 = 0 + (-x^3 + 3x + 1) = -x^3 + 3x + 1$

Recall that every number has an **opposite** (or **additive inverse**) and that the sum of a number and its opposite is 0. This is also true for polynomials.

EXAMPLE 3 The opposite of $x + 3$ is written $-(x + 3)$.
$(x + 3) + [-(x + 3)] = 0$

EXAMPLE 4 The opposite of $-x^2 + 1$ is written $-(-x^2 + 1)$.
$(-x^2 + 1) + [-(-x^2 + 1)] = 0$

You will also recall that the opposite of an addition expression such as $x + 3$ is the sum of the opposites of the terms. We use this property to write the opposite of a polynomial without using parentheses.

EXAMPLE 5 $-(x + 3) = -x + (-3) = -x - 3$

EXAMPLE 6 $-(2x^3 - 7x^2 + 22) = -2x^3 + 7x^2 - 22$

Oral EXERCISES

Give the sum of the two polynomials. Justify your answer.

Sample 1: $-(8b - c + 9)$ and $(8b - c + 9)$

What you say: 0; the sum of any polynomial and its opposite is 0.

Sample 2: $(4k - 4k)$ and $(5x^3 + 2x - 1)$

What you say: $5x^3 + 2x - 1$; $4k - 4k = 0$ and 0 is the identity element for addition of polynomials.

1. $(9x^2 - 8x)$ and $-(9x^2 - 8x)$ 0
2. $-(2st - t)$ and $(2st - t)$ 0
3. $(12a^3b^3 - a^2b^2 + b)$ and $(-12a^3b^3 + a^2b^2 - b)$ 0
4. $(20xy^2 - 20xy^2)$ and $(3c^2 - 13cd + 7d^2)$ $3c^2 - 13cd + 7d^2$
5. $-(r^3 - r^2 + r - 5)$ and $(r^3 - r^2 + r^2 - 5)$ $r^2 - r$
6. $(27k^4 - 1)$ and $(7h^3 - 7h^3)$ $27k^4 - 1$

Match each polynomial in Column 1 with its additive inverse in Column 2.

COLUMN 1

7. $5x^2 + 11x + 2$ D
8. $a^2b^2 - 12ab - 20$ C
9. $-a^2b^2 - 12ab + 20$ A
10. $5x^2 - 11x + 2$ B

COLUMN 2

A. $a^2b^2 + 12ab - 20$
B. $-5x^2 + 11x - 2$
C. $-a^2b^2 + 12ab + 20$
D. $-5x^2 - 11x - 2$

3. $-\dfrac{m^2}{10} + \dfrac{2n^2}{7} - \dfrac{4mn}{2}$

5. $-12x^9 - 6x^7 - 3x^5 - x - 1$

8. $3x^4 - 5x^3 + 2x^2 + x + 6$

Give the opposite. Write your answer in standard form.

Sample: $-k^2 + 5k - 6$

Solution: $-(-k^2 + 5k - 6) = k^2 - 5k + 6$

1. $2m^2 - 3m + 15$ $-2m^2 + 3m - 15$
2. $-w^2 + 3w - 40$ $w^2 - 3w + 40$
3. $\dfrac{m^2}{10} - \dfrac{2n^2}{7} + \dfrac{4mn}{2}$
4. $2 + \dfrac{a}{5} - \dfrac{a^2}{12}$ $-2 - \dfrac{a}{5} + \dfrac{a^2}{12}$
5. $12x^9 + 6x^7 + 3x^5 + x + 1$
6. $-2a^2 - a^3 + 5a - 7$ $2a^2 + a^3 - 5a + 7$
7. $-(-5x^2 - 3x + 7)$ $-5x^2 - 3x + 7$
8. $3x^4 + 5x^3 - 2x^2 - x - 6$
9. $-2k + 1 - 7k^2 - 4k^3 + k^6$ $2k - 1 + 7k^2 + 4k^3 - k^6$
10. $-[(-4b^2 + 6b - 1)] + 2b$ $-4b^2 + 6b - 1 - 2b$

Write in standard form without using parentheses.

Sample: $-(3x^6 - 8x^2 + 5x^4 - 1)$

Solution: $-3x^6 - 5x^4 + 8x^2 + 1$

11. $-(3y^2 + 13y + 5)$ $-3y^2 - 13y - 5$
12. $-(y^2 - 8y + 5)$ $-y^2 + 8y - 5$
13. $-(-10 + 4x - x^2)$ $x^2 - 4x + 10$
14. $-\left(-\dfrac{a}{3} - \dfrac{a^2}{4} + 18\right)$ $\dfrac{a^2}{4} - \dfrac{a}{3} - 18$

15. $-(-2x^3 + 25x^2 - 10)$ $2x^3 - 25x^2 + 10$

16. $-(-12a^3 - 4ab^2 + 6b^3 + 12a^2b)$ $12a^3 - 12a^2b + 4ab^2 - 6b^3$

17. $-(1.8x^5 - 6.2x^3 - 0.7x + 15)$ $-1.8x^5 + 6.2x^3 + 0.7x - 15$

18. $-\left(-\dfrac{k^5}{6} - \dfrac{k^4}{2} - \dfrac{k^3}{5} - 1\right)$ $\dfrac{k^5}{6} + \dfrac{k^4}{2} + \dfrac{k^3}{5} + 1$

19. $-(-y^7 + 2y^5 - 6y^3 + y - 2)$ $y^7 - 2y^5 + 6y^3 - y + 2$

20. $-(8.1r^3 + 1.2r^4 - 0.3r - 1.8)$ $-1.2r^4 - 8.1r^3 + 0.3r + 1.8$

Add. First arrange your work in vertical form.

21. $4x^2 + 4x$
22. $-10x$
23. 0
24. $x - 14$
25. 0
26. $-4x^2 + 8x$

Sample: $-(8k^2 - 5k + 10)$ and $-(3k^2 - k + 4)$

Solution: $-(8k^2 - 5k + 10) = -8k^2 + 5k - 10$
$\underline{-(3k^2 - k + 4) = -3k^2 + k - 4}$
$-11k^2 + 6k - 14$

B

21. $(5x^2 + 5x)$ and $-(x^2 + x)$ 22. $-(5x + y)$ and $-(5x - y)$

23. $(3c + 10)$ and $-(3c + 10)$ 24. $-(2x + 4)$ and $(3x - 10)$

25. $(a^2 + b^2)$ and $-(a^2 + b^2)$ 26. $-(2x^2 - 4x)$ and $(4x - 2x^2)$

27. $-(3r^2 - 5s + 2t - 4)$ and $(-5r^2 - 4t - 12)$

28. $-(2.3a^2 + 6.8ab - 3.6b^2)$ and $(3.2a^2 + 4.6ab + 3.6b^2)$

29. $\left(\dfrac{1}{3}x^3 + \dfrac{1}{3}x^2 + \dfrac{3}{2}\right)$ and $-\left(\dfrac{1}{3}x^3 - \dfrac{1}{2}x^2 + \dfrac{1}{2}\right)$ $\dfrac{5}{6}x^2 + 1$

30. $(5x^5 + 7x^3 - x^2 + 4)$ and $-(6x^5 - 2x^3 + 3x^2 + 4)$
$-x^5 + 9x^3 - 4x^2$

27. $-8r^2 + 5s - 6t - 8$ 28. $0.9a^2 - 2.2ab + 7.2b^2$

Write in standard form without using grouping symbols.

Sample: $-[2x^4 - (x + 5x^2 - 6)]$

Solution: $-[2x^4 - (x + 5x^2 - 6)] = -[2x^4 - x - 5x^2 + 6]$
$= -2x^4 + 5x^2 + x - 6$

C

31. $5s^2 - (s^3 + 4s + 8)$ $-s^3 + 5s^2 - 4s - 8$

32. $(x^5 - 7) + (4x^2 + 9)$ $x^5 + 4x^2 + 2$

33. $-[(5a - 10) + (a^5 - 2a^2 + a)]$ $-a^5 + 2a^2 - 6a + 10$

34. $7 - (4x - 2x^2) + 5x^3$ $5x^3 + 2x^2 - 4x + 7$

35. $-(15y^2 - 3y) + (-7 + 2y^3)$ $2y^3 - 15y^2 + 3y - 7$

36. $-[-(2a^4 - 8) + (7a + a^2)]$ $2a^4 - a^2 - 7a - 8$

37. $-[(-x^2 - 5x + 8) - (x^3 + 2x^4)]$ $2x^4 + x^3 + x^2 + 5x - 8$

38. $-[-(2k^5) + (k^3 - 1) - (k^4 + k)]$ $2k^5 + k^4 - k^3 + k + 1$

12-7 *Subtraction with Polynomials*

OBJECTIVE
Subtract a polynomial by adding its opposite.

We can subtract polynomials by applying the same ideas used earlier to do subtraction.

EXAMPLE 1 $10 - 4 = 10 + (-4) = 6$
$5a - (-4b) = 5a + 4b$

EXAMPLE 2 $(8t - 3) - (t + 3) = (8t - 3) + \underbrace{(-t - 3)}_{\text{opposite of } t + 3} = 7t - 6$

EXAMPLE 3   Subtract:

	Add:
$10a^2 + 8a + 5$	$10a^2 + 8a + 5$
$\underline{3a^2 - a + 1}$	$\underline{-3a^2 + a - 1}$
▶	$7a^2 + 9a + 4$

Recall that we can add to check subtraction. We add the answer to the number subtracted.

EXAMPLE 4   Subtract:

	Add:	Check:
$12s + 7$	$12s + 7$	$7s + 10$
$\underline{5s - 3}$	$\underline{-5s + 3}$	$\underline{5s - 3}$
▶	$7s + 10$	$12s + 7$ ✓

Name the opposite.

Sample: $4 - 2t$ *What you say:* $-4 + 2t$

EXERCISES

1. $-2k$ $2k$

2. $4x^5$ $-4x^5$

3. $z + 2$ $-z - 2$

4. $4c - d$ $-4c + d$

5. $-5 - 6x^2$ $5 + 6x^2$

6. $2 - 4a + b$ $-2 + 4a - b$

7. $-3a^2 + 2a - 6$ $3a^2 - 2a + 6$

8. $-\dfrac{1}{2}a - \dfrac{1}{3}b - \dfrac{1}{4}c$ $\dfrac{1}{2}a + \dfrac{1}{3}b + \dfrac{1}{4}c$

Subtract the second polynomial from the first.

Sample: $\begin{array}{r} 2a \\ -5a \\ \hline \end{array}$ *What you say:* 2a plus 5a, or 7a

9. $\begin{array}{r} -15c \\ 10c \\ \hline -25c \end{array}$ **10.** $\begin{array}{r} 3ab \\ 8ab \\ \hline -5ab \end{array}$ **11.** $\begin{array}{r} \dfrac{1}{2}k \\ -\dfrac{1}{4}k \\ \hline \dfrac{3}{4}k \end{array}$ **12.** $\begin{array}{r} 12b \\ 5b \\ \hline 7b \end{array}$

For Extra Practice, see page 427.

Written EXERCISES

Subtract the second polynomial from the first. Check by addition.

Sample: $\begin{array}{l} 3x + 4y \\ 2x - 3y - 1 \\ \hline \end{array}$

Solution: $\begin{array}{l} 3x + 4y \\ 2x - 3y - 1 \\ \hline x + 7y + 1 \end{array}$ *Check:* $\begin{array}{l} x + 7y + 1 \\ 2x - 3y - 1 \\ \hline 3x + 4y \end{array}$ $\checkmark$

A

1. $\begin{array}{l} 4r + 7s \\ 3r - 2s \\ \hline r + 9s \end{array}$ **2.** $\begin{array}{l} 8 + \ c \\ 3 - 2c \\ \hline 5 + 3c \end{array}$ **3.** $\begin{array}{r} 5p + 2 \\ -2p + 2 \\ \hline 7p \end{array}$

4. $\begin{array}{l} 10k^2 + 3 \\ 5k^2 + 8 \\ \hline 5k^2 - 5 \end{array}$ **5.** $\begin{array}{l} 5x^2 + 20 \\ x^2 + 15 \\ \hline 4x^2 + 5 \end{array}$ **6.** $\begin{array}{l} 5 + 4c \\ 15 - \ c \\ \hline -10 + 5c \end{array}$

7. $\begin{array}{l} 3b^2 + 5c \\ b^2 - 8c + 5 \\ \hline 2b^2 + 13c - 5 \end{array}$ **8.** $\begin{array}{l} 8x - 4y \\ 5x - 2y \\ \hline 3x - 2y \end{array}$ **9.** $\begin{array}{l} 2m^2 \qquad\ + 4 \\ m^2 - 7m \\ \hline m^2 + 7m + 4 \end{array}$

10. $\begin{array}{l} 10x + 4y + 9 \\ 2x - \ y + 8 \\ \hline 8x + 5y + 1 \end{array}$ **11.** $\begin{array}{l} -2k + 5h^2 \\ \quad\ \ - 4h^2 \\ \hline -2k + 9h^2 \end{array}$ **12.** $\begin{array}{l} 2x^3 + 13x^2 \\ -x^3 \qquad\quad - 10 \\ \hline 3x^3 + 13x^2 + 10 \end{array}$

Write the expression without parentheses. Do not combine similar terms.

Sample: $(8x - 9) - (6 - x)$ *Solution:* $8x - 9 - 6 + x$

13. $(k - 10) - (2k + 15)$ $k - 10 - 2k - 15$

14. $(4 - 2s) - (-6 + 9s)$ $4 - 2s + 6 - 9s$

15. $(-2y^2 - 4y) - (y^2 - 3y)$ $-2y^2 - 4y - y^2 + 3y$

16. $(3y^2 + 5y) - (-y^2 - 8y)$ $3y^2 + 5y + y^2 + 8y$

17. $(2x - y + 8) - (13x + 4y + 9)$ $2x - y + 8 - 13x - 4y - 9$

18. $(-10 - 3a - 5b) - (-7b + 2a - 8)$

$\qquad\qquad\qquad\qquad -10 - 3a - 5b + 7b - 2a + 8$

Simplify. Write the answer in standard form.

Sample: $(2 + 7b) - (-3 + 5b)$

Solution: $(2 + 7b) - (-3 + 5b) = 2 + 7b + 3 - 5b = 2b + 5$

19. $(3w - 5) - (4w + 7)$ $-w - 12$

20. $(a + 3b) - (a - 8b)$ $11b$

21. $(2t^2 + 5) - (9 - 4t^2)$ $6t^2 - 4$

22. $(3 - 7d) - (1 - d)$ $-6d + 2$

23. $(2r + 3s) - (2r + 3s)$ 0

24. $(4x - 9y) - (-5x + 10)$ $9x - 9y - 10$

25. $(1 - 2w) - (1 - 3w)$ w

26. $(3.5a + 0.5) - (2.1a + 0.2)$ $1.4a + 0.3$

27. $(-2d + 4c) - (-5d + 7c)$ $-3c + 3d$

28. $(2a - b) - (2a - b)$ 0

29. $(3 - 10d + 7d^2) - (4 + 3d^2)$ $4d^2 - 10d - 1$

30. $(2.7s^3 - 0.7s^2) - (-1.3s^2 - 0.9)$ $2.7s^3 + 0.6s^2 + 0.9$

B

Subtract the second polynomial from the first. Check by addition.

Sample: $\begin{aligned}&x^2 + 14x - 49\\&\underline{6x^2 + 5x - 1}\end{aligned}$

Solution: $\begin{aligned}&x^2 + 14x - 49\\&\underline{6x^2 + 5x - 1}\\&-5x^2 + 9x - 48\end{aligned}$ *Check:* $\begin{aligned}&6x^2 + 5x - 1\\&\underline{-5x^2 + 9x - 48}\\&x^2 + 14x - 49 \quad \checkmark\end{aligned}$

31. $\begin{aligned}&6r^2 + 13r + 7\\&\underline{2r^2 + 10r - 4}\\&4r^2 + 3r + 11\end{aligned}$

32. $\begin{aligned}&12w^2 + w - 2\\&\underline{4w^2 + 7w - 3}\\&8w^2 - 6w + 1\end{aligned}$

33. $\begin{aligned}&2h^2 - 15h - 8\\&\underline{h^2 + 6h + 5}\\&h^2 - 21h - 13\end{aligned}$

34. $\begin{aligned}&5x^2 - 11x + 2\\&\underline{-2x^2 - 3x - 1}\\&7x^2 - 8x + 3\end{aligned}$

35. $\begin{aligned}&-3b^2 + 6b + 1\\&\underline{-4b^2 + 3}\\&b^2 + 6b - 2\end{aligned}$

36. $\begin{aligned}&x^2 + 9\\&\underline{-3x^2 - 4x + 13}\\&4x^2 + 4x - 4\end{aligned}$

37.
$$7x^2 - 12x$$
$$\underline{ 20x - 7}$$
$$7x^2 - 32x + 7$$

38.
$$8b^3 - 2b^2 + 5b$$
$$\underline{6b^3 - 4b^2 + b}$$
$$2b^3 + 2b^2 + 4b$$

39.
$$k^2 + 2k$$
$$\underline{ - 3k + 5}$$
$$k^2 + 5k - 5$$

Simplify. Remove all grouping symbols and combine similar terms. Write the answer in standard form.

C

40. $2c - [c - (4c + 6) + 10 - 3c^2] + 6c^2$ $9c^2 + 5c - 4$

41. $(6k - 4) - [(3k + 1) + 7 - (-2k - 6)]$ $k - 18$

42. $-(15b + 10) - [-3b - (2 + b) + 8]$ $-11b - 16$

43. $[(8 + 4x) - (3x - 2)] - [2 - (x - 3 + x^2)]$ $x^2 + 2x + 5$

44. $-[y - (3 + y)] - [-8y + (9y - y^2) + 4] + y$ $y^2 - 1$

SELF-TEST 2

Be sure you understand the term *identity element for addition of polynomials.* (p. 330)

Section 12-6, p. 330 Add. Justify your answer.

1. $-(2c^2 + c + 7) + (2c^2 + c + 7)$ 0

2. $(2x - 2x) + (7x^2 - 1)$ $7x^2 - 1$

Give the opposite. Write your answer in standard form.

3. $-(2y^3 - y + 3y^2)$
  $2y^3 + 3y^2 - y$

4. $12z + z^3 - z$ $-z^3 - 11z$

Section 12-7, p. 333 Subtract the second polynomial from the first.

5.
$$4m - 3$$
$$\underline{-2m - 2}$$
$$6m - 1$$

6.
$$2t + 4$$
$$\underline{t + 1}$$
$$t + 3$$

Simplify.

7. $(2x^2 + 3x) - (4x^2 + 5x)$
  $-2x^2 - 2x$

8. $(3n^2 + 5) - (2n^2 - 8)$
  $n^2 + 13$

Check your answers with those printed at the back of the book.

12-8 *Polynomials and Problem Solving*

The information given in a word problem may be expressed in polynomial form.

EXAMPLE 1 Find the area of the shaded part of the figure. The area of the square is $4m^2 + 4m + 1$. The area of the circle is $4m + 6$.

Subtract:

$$\begin{array}{ll} 4m^2 + 4m + 1 & \blacktriangleleft \text{ Area of square} \\ \underline{ 4m + 6} & \blacktriangleleft \text{ Area of circle} \\ 4m^2 - 5 & \blacktriangleleft \text{ Area of shaded region} \end{array}$$

$4m^2 + 4m + 1$

EXAMPLE 2 Find the distance from R to T.

Add:

$$\begin{array}{ll} a^2 + 3ab - 2 & \blacktriangleleft \text{distance from } R \text{ to } S \\ \underline{4a^2 - ab} & \blacktriangleleft \text{distance from } S \text{ to } T \\ 5a^2 + 2ab - 2 & \blacktriangleleft \text{distance from } R \text{ to } T \end{array}$$

For Extra Practice, see page 427.

Write the answer as a polynomial in standard form.

Problems

1. Find the perimeter of the figure at the right. $a^2 + 6a - 10$

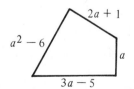

2. A square has sides of length $7 - 12t$. Find the perimeter.
    $28 - 48t$

3. One plot of land has area $\frac{1}{2}a - 9$. A connecting plot has area $a^2 + \frac{1}{3}a + 5$. Find the combined area. $a^2 + \frac{5}{6}a - 4$

4. The length XZ is $2x^3 + 6x^2 + 7$. The length XY is $4x^2 + 6$. Find the length YZ. $2x^3 + 2x^2 + 1$

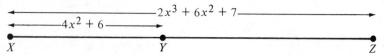

5. The area of $ABDE$ is $4z^2 - 3z + 8$. The area of $ABCE$ is $3z^2 + 5$. Find the area of CDE. $z^2 - 3z + 3$

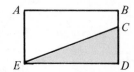

6. The area of the square is $4r^4 + 8r^2 + 16$. The area of the shaded region is $3r^3 + 2r^2$. Find the area of the circle. $4r^4 - 3r^3 + 6r^2 + 16$

7. A board had length $1 - 5a - 6a^2$. A carpenter sawed a piece of length $4 - 2a^2$ from the end. How long is the remaining piece? $-4a^2 - 5a - 3$

8. Two planes flew different routes from London to San Francisco. Plane 1 flew a distance of $\frac{1}{2}x^2 + \frac{2}{3}x - 4$. Plane 2 flew a distance of $\frac{3}{4}x^2 - \frac{1}{6}x + 9$. How much farther did Plane 2 travel? $\frac{1}{4}x^2 - \frac{5}{6}x + 13$

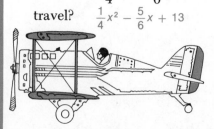

9. Fairfield Farms owned land with combined area $\frac{2}{3}x^3 + \frac{1}{4}x^2 + 5$. They purchased another piece of land with area $\frac{1}{4}x^3 + 2x^2 + x$. Find the total area of their land.

$\frac{11}{12}x^3 + 2\frac{1}{4}x^2 + x + 5$

12-9 *Polynomials and Solving Equations*

OBJECTIVE

Solve and check equations like
$2t + (4 - t) = 10$ and
$5y - (2y + 4) = 12$.

We may need to simplify polynomial expressions as the first step in solving an equation.

EXAMPLE 1   Solve:
$$3x + (2x + 4) = 14$$
$$3x + 2x + 4 = 14$$
$$5x + 4 = 14$$
$$5x = 10$$
$$x = 2$$

Check:
$$3(2) + [2(2) + 4] \overset{?}{=} 14$$
$$6 + (4 + 4) \overset{?}{=} 14$$
$$6 + 8 \overset{?}{=} 14$$
$$14 = 14 \ \checkmark$$

EXAMPLE 2   Solve:
$$3m - (5 - 2m) = 40$$
$$3m - 5 + 2m = 40$$
$$5m - 5 = 40$$
$$5m = 45$$
$$m = 9$$

Check:
$$3(9) - [5 - 2(9)] \overset{?}{=} 40$$
$$27 - (5 - 18) \overset{?}{=} 40$$
$$27 - (-13) \overset{?}{=} 40$$
$$40 = 40 \ \checkmark$$

EXAMPLE 3   Solve:
$$14 = (3n + 4) - (n + 2)$$
$$14 = 3n + 4 - n - 2$$
$$14 = 2n + 2$$
$$12 = 2n$$
$$6 = n$$

Check:
$$14 \overset{?}{=} [3(6) + 4] - (6 + 2)$$
$$14 \overset{?}{=} (18 + 4) - 8$$
$$14 \overset{?}{=} 22 - 8$$
$$14 = 14 \ \checkmark$$

For Extra Practice, see page 428.

Solve. Check your solution.

Written EXERCISES

A

1. $(2n + 4) + (3n - 7) = 2$ $n = 1$
2. $k + (k - 3) + (2k - 4) = 1$ $k = 2$
3. $(9x - 42) - 3x = -6$ $x = 6$
4. $-4z + (11z + 3) = 24$ $z = 3$
5. $8 + (2x - 10) = 14$ $x = 8$
6. $(2y - 28) + 5y = 42$ $y = 10$
7. $(3n + 6) - n = 40$ $n = 17$
8. $6y - (2y - 16) = 0$ $y = -4$
9. $20 + (16 - 8k) = 44$ $k = -1$
10. $(2z + 1) - (12z + 6) = 30$ $z = -3.5$
11. $(0.2x + 1) - 4 = 6$ $x = 45$
12. $-37 = -(5n + 1) + 11n$ $-6 = n$
13. $24 = (4s + 1) - (2s + 1)$ $12 = s$
14. $-(x^2 - 5) + x^2 + 3x = 35$ $x = 10$

Solve.

B

15. $-3n - (4n^2 - 2n + 5) = 12 - 4n^2$ $n = -17$

16. $(4k - 5) + 7 = (9k - 6) + 3$ $k = 1$

17. $2 + (3x - 5) + (4x + 6) = (4x + 1) - 15$ $x = -5\frac{2}{3}$

18. $(4x - 1.2) - 8 = (2x + 5) - (1 + 2x)$ $x = 3.3$

19. $(4k + 8) - (2k - 14) = (8 - 6k) + (6k + 2)$ $k = -6$

20. $(3t - 1) + (4t + 5) - 12 = 7 - (2t + 3)$ $t = \dfrac{4}{3}$

21. $(a^2 + 5) - (a^2 - 6) - 3a = 5a - 21$ $4 = a$

22. $(x - 3) - (4x + 2) = -x - 12$ $x = 3\frac{1}{2}$

23. $16b - (6b - 13) = 3b - (b + 3) + (7b + 2)$ $b = -14$

24. $(0.5x - 3) + (0.5x + 1) = 0.7x - (4 - 0.3x) + 2x$ $1 = x$

25. $(2x - 1) - (x - 1) + (5x - 6) = 7x - 5$ $x = -1$

C

26. $[4a - (5a + 6)] + 2a = 12$ $a = 18$

27. $-(n^2 + 2n - 5) + (n^2 - 6n) = -4n + 21$ $n = -4$

28. $-[-6t + (9 - 5t)] + (4t - 7) = 8t - 30$ $t = -2$

29. $6b - [(3b + 1) - 2b] = 19 + b$ $b = 5$

30. $(t^2 + 1) + [(1 - t^2) + (4 + 2t)] = (5t + 3)$ $1 = t$

SELF-TEST 3

Write as a polynomial in standard form.

Section 12-8, p. 337 1. A triangle has sides of length $2ab + 1$, $3ab - 4$, and $4ab + 2$. Find the perimeter. $9ab - 1$

2. A square has sides of length $4x + 2$. Find the perimeter. $16x + 8$

3. One box of soap has volume $3x^2 - 5$. A smaller box has volume $x^2 + 4$. Find the difference. $2x^2 - 9$

Section 12-9, p. 339 Solve.

4. $6 + (3x + 1) = 19$ $x = 4$ 5. $-(x + 1) + (2x + 4) = 9$ $x = 6$

6. $(y - 3) + (3y + 1) = 10$ $y = 3$ 7. $(4n + 3) + (n + 5) = 23$ $n = 3$

Check your answers with those printed at the back of the book.

chapter summary

1. Polynomial expressions are sometimes classified by the number of terms. **Monomials** contain one term. **Binomials** contain two terms. **Trinomials** contain three terms.

2. A polynomial in one variable is in **standard form** when the terms are ordered so that the variable in the first term has the greatest exponent, the variable in the second term has the next greatest exponent, and so on.

3. The sum of polynomials is found by combining similar terms.

4. The commutative and associative properties of addition apply to polynomials.

5. Zero is the **identity element** for addition of polynomials. Every polynomial has an opposite. The sum of a polynomial and its opposite is 0.

6. The **opposite** (additive inverse) of a polynomial expression is the sum of the opposites of the terms of the polynomial.

7. Subtracting a polynomial is the same as adding its opposite.

chapter test

Tell whether the polynomial is a monomial, a binomial, or a trinomial.

1. -3 monomial

2. $\frac{1}{2}t$ monomial

3. $\frac{2}{3}x^2 + x + 1$ trinomial

4. $6 - c^2$ binomial

5. $7abc$ monomial

6. $2z^2 + z$ binomial

Write in standard form.

7. $a^3 - 8 + 2a^5$
$2a^5 + a^3 - 8$

8. $12 - 3a^2b$
$-3a^2b + 12$

9. $5x^2 - 4x^3 + x$
$-4x^3 + 5x^2 + x$

Add.

10. $7y^2 - 4y + 15$
$\underline{13y^2 - 9y + 18}$
$20y^2 - 13y + 33$

11. $11t^2 + 5$
$\underline{-2t^2 - 1}$
$9t^2 + 4$

Add.

12. $(2b + 7) + (4b - 5)$ $6b + 2$ **13.** $(-9m - 4) + (9m + 8)$ 4

Subtract the second polynomial from the first.

14. $5a^2 + a - 1$
$\underline{2a^2 - 3a + 4}$
$3a^2 + 4a - 5$

15. $4x^2 - 2x + 3$
$\underline{2x - 7}$
$4x^2 - 4x + 10$

Simplify.

16. $(2z^2 - 5) - (z^2 - 1)$ $z^2 - 4$ **17.** $(5b + 5) - (-3b - 2)$ $8b + 7$

Simplify. Remove grouping symbols and combine similar terms.

18. $(4t - 6) + (3t - 1)$ $7t - 7$ **19.** $(5s - 7r) - (7r - 5s)$ $10s - 14r$

20. $(4x + 5) - (2x - 6)$ $2x + 11$ **21.** $(a^2 - 2a) + (2a - 3)$ $a^2 - 3$

Write as a polynomial in standard form.

22. The perimeter of a square with sides of length $4t - 3$. $16t - 12$

23. The perimeter of a rectangle with length $\frac{1}{2}z + 4$ and width $z - 3$. $3z + 2$

Solve and check.

24. $(5a - 7) + 4a = 11$ $a = 2$ **25.** $k + (k - 4) = 0$ $k = 2$

26. $p - (4 - 2p) = 5$ $p = 3$ **27.** $(-3n + 1) + (2n - 1) = 5$ $n = -5$

challenge topics *Slope*

The slope of roads, sidewalks, or roof tops may be described by a ratio that compares **rise** and **run** in the form of a fraction, whole number, or percent.

EXAMPLE

rise = 40 m

run = 1000 m

$$\text{Slope} = \frac{\text{rise}}{\text{run}} = \frac{40}{1000} = 4\%$$

The idea of slope is important in mathematics. It is used to describe a line graphed on coordinate axes. A line that slopes upward from left to right is said to have **positive slope**. A line that slopes downward from left to right is said to have **negative slope**.

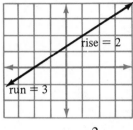

$$\text{Slope} = \frac{2}{3}$$

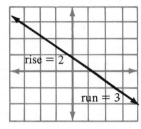

$$\text{Slope} = -\frac{2}{3}$$

If the coordinates of two points on a line are known, the slope may be found in this way:

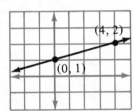

$$\text{Slope} = \frac{\text{difference in second coordinates}}{\text{difference in first coordinates}}$$

$$\text{Slope} = \frac{2 - 1}{4 - 0}$$

$$\text{Slope} = \frac{1}{4}$$

Tell whether the slope of the line is positive or negative. Then find the slope.

1.

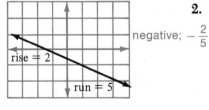

negative; $-\frac{2}{5}$

2.

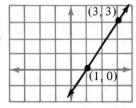

positive; $\frac{3}{2}$

3.

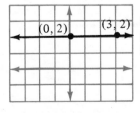

neither positive or negative; 0

4.

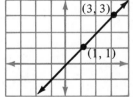

positive; 1

Match each expression in Column 1 with a corresponding expression in Column 2.

COLUMN 1

1. $4 + 4 + 4$ F
2. $3 \cdot 3 \cdot 3 \cdot 3$ A
3. $k \cdot h \cdot h$ C
4. $6 \cdot 6 \cdot 5 \cdot 5$ E
5. $5 + 5 + 5 + 5 + 4$ B
6. $b + b + c + c + c$ D

COLUMN 2

A. 3^4
B. $(4 \cdot 5) + 4$
C. kh^2
D. $2b + 3c$
E. $6^2 \cdot 5^2$
F. $3 \cdot 4$

Simplify.

7. 4^2 16
8. $4 \cdot 5$ 20
9. 6^3 216
10. $(a)(-5)$ $-5a$
11. $5 \cdot 5^2$ 125
12. $-7 \cdot 2$ -14

Simplify. Use the distributive property.

13. $(4 - t)3$ $12 - 3t$
14. $(c + d)b$ $cb + db$
15. $r(s - t)$ $rs - rt$
16. $-4(2 - x)$ $-8 + 4x$
17. $2m(n + p)$ $2mn + 2mp$
18. $2a(b - c)$ $2ab - 2ac$

Tell which of the following properties is illustrated: Commutative Property, Associative Property, Multiplicative Property of One, Property of Reciprocals.

19. $3 = 3 \cdot 1$ mult. prop. of one
20. $4 \cdot 3 = 3 \cdot 4$ comm. prop.
21. $a \cdot b = b \cdot a$ comm. prop.
22. $k \cdot 1 = k$ mult. prop. of one
23. $(2 \cdot 3)5 = 2(3 \cdot 5)$ assoc. prop.
24. $6 \cdot \dfrac{1}{6} = 1$ prop. of reciprocals

Solve.

25. $t \cdot 7 = 42$ $t = 6$
26. $t \cdot 2 = 2^3$ $t = 4$
27. $2^2 \cdot t = 2^3$ $t = 2$
28. $t \cdot 6^2 = 6^3$ $t = 6$
29. $7 \cdot t = 7^2$ $t = 7$
30. $7^2 \cdot t = 7^3$ $t = 7$

Simplify.

31. $\dfrac{22 \cdot 8 - 13 \cdot 8}{8}$ 9
32. $\dfrac{20 \cdot 5 - 12 \cdot 7}{4}$ 4

Left: Road builders, near Eighty Eight, Kentucky.

Right: Construction crew on modern highway.

13
Multiplication and Division of Polynomials

Multiplication by a Monomial

13-1 *Repeating Factors and Exponents*

> **OBJECTIVE**
> Apply a rule of exponents to simplify expressions: $m \cdot m \cdot m = m^3$;
> $\dfrac{a}{b} \cdot \dfrac{a}{b} = \left(\dfrac{a}{b}\right)^2$

Recall that a multiplication expression involving a repeating factor may be written with an exponent. The repeating factor is called the base.

$$7 \cdot 7 \cdot 7 \cdot 7 = 7^4 \blacktriangleleft \text{exponent}$$

base

$$m \cdot m \cdot m = m^3 \blacktriangleleft \text{exponent}$$

base

We read 7^4 as "seven to the fourth power." We read m^3 as either "m to the third power," or "m cubed."

EXAMPLE 1 $t \cdot t^3 = t(t \cdot t \cdot t)$ ◀ The factor t four times
$$= t^4$$

EXAMPLE 2 $rs \cdot rs \cdot rs \cdot rs \cdot rs = (rs)^5$ ◀ The factor rs five times

EXAMPLE 3 $rs \cdot s \cdot s \cdot s \cdot s = r(s \cdot s \cdot s \cdot s \cdot s)$ ◀ The factor r once
$$= rs^5 \qquad \qquad \text{The factor } s \text{ five times}$$

Compare Examples 2 and 3. In Example 2 we need the parentheses to show that the exponent applies to rs, not just to s.

Here is a rule for simplifying when a factor is repeated in a multiplication expression.

▶ For any directed number a, and all positive integers p and q:
$$a^p \cdot a^q = a^{p+q}$$

EXAMPLE 4 $n^4 \cdot n^3 = n^{4+3} = n^7$

EXAMPLE 5 $(ab)^2(ab) = (ab)^{2+1} = (ab)^3$

Simplify. Use exponents.

1. $m \cdot m \cdot m \cdot m \cdot m$ m^5
2. $6 \cdot x \cdot x^5 \cdot x$ $6x^7$
3. $n \cdot n \cdot n^4$ n^6
4. $2^3 \cdot 2$ 2^4
5. $(ab)^5 \cdot (ab)^4$ a^9b^9
6. $w \cdot w \cdot z \cdot z \cdot z$ w^2z^3
7. $5 \cdot r \cdot r \cdot s \cdot s \cdot s$ $5r^2s^3$
8. $(ef)^3 \cdot (ef)^4$ e^7f^7
9. $(x \cdot x)(y \cdot y \cdot y)$ x^2y^3
10. $p \cdot p^3 \cdot p$ p^5

Give the value. Use $x = 3$ and $z = 2$.

1. z^2 4
2. $10z^2$ 40
3. $(x \cdot x)(z \cdot z \cdot z)$ 72
4. x^3 27
5. $x \cdot x \cdot x \cdot z$ 54
6. $3 \cdot z \cdot z$ 12
7. $4 \cdot x \cdot x$ 36
8. $(x \cdot z)^2$ 36
9. $(3z)(xz)$ 36
10. $5z^4$ 80
11. $8x^3$ 216
12. x^3x^2 243

Simplify.   15. $(x + y + z)^2$

13. $(3 \cdot 4)(x \cdot x)$ $12x^2$
14. $(6 \cdot 12)(a \cdot a)$ $72a^2$
15. $(x + y + z)(x + y + z)$
16. $(z + 1)(1 + z)$ $(z + 1)^2$
17. $-5 \cdot 3(h \cdot h)$ $-15h^2$
18. $(-4)(x \cdot x \cdot x)$ $-4x^3$
19. $-7(z \cdot z \cdot z \cdot z)(2)$ $-14z^4$
20. $5 \cdot 3 \cdot 2(r \cdot r)$ $30r^2$
21. $7(-12)(y \cdot y \cdot y \cdot y)$ $-84y^4$
22. $12 \cdot 5(z \cdot z^2)$ $60z^3$
23. $3(x)(-2)(x)$ $-6x^2$
24. $9 \cdot 4 \cdot x^3 \cdot x^2$ $36x^5$
25. $(8 \cdot 3)(m^2 \cdot m^3 \cdot m)$ $24m^6$
26. $-3 \cdot 4(n \cdot n^2)$ $-12n^3$
27. $6(-7)t \cdot t$ $-42t^2$
28. $m \cdot m(-m) \cdot m$ $-m^4$

Sample: $8^3 \cdot 8^4$ Solution: $8^3 \cdot 8^4 = 8^{3+4} = 8^7$

29. $2^2 \cdot 2^1$ 2^3
30. $6^3 \cdot 6^4$ 6^7
31. $x^5 \cdot x^2$ x^7
32. $m \cdot m \cdot m^2$ m^4
33. $p \cdot p^4$ p^5
34. $-3 \cdot t \cdot t^3$ $-3t^4$
35. $(a^2)(a^3)(-1)$ $-a^5$
36. $m^4 \cdot m^5$ m^9
37. $y \cdot y^2 \cdot y^3$ y^6

Find the correct replacement for x.

Sample: $3^6 = 3^x \cdot 3^2$ Solution: $3^6 = 3^4 \cdot 3^2$; $x = 4$

38. $2^2 \cdot 2^4 = 2^x$ $x = 6$
39. $(y \cdot y)(y \cdot y) = y^x$ $x = 4$
40. $b^7 = b^2 \cdot b^x$ $x = 5$
41. $r^2 \cdot r^x \cdot r^4 = r^{12}$ $x = 6$
42. $(s + t)^x = (s + t)(s + t)^3$ $x = 4$
43. $(m - n)^6 = (m - n)^4(m - n)^x$ $x = 2$

13-2 *Products of Monomials*

Our rule of exponents allows us to simplify expressions such as $(3xy)(8x^2y)$.

EXAMPLE 1
$$\begin{aligned} (3xy)(8x^2y) &= (3 \cdot 8)(x \cdot x^2)(y \cdot y) \\ &= (24)(x^{1+2})(y^{1+1}) \\ &= 24x^3y^2 \end{aligned}$$

EXAMPLE 2
$$\begin{aligned} (a^3)(-3ab)(2b) &= (-3 \cdot 2)(a^3 \cdot a)(b \cdot b) \\ &= (-6)(a^{3+1})(b^{1+1}) \\ &= -6a^4b^2 \end{aligned}$$

EXAMPLE 3
$$\begin{aligned} \left(\tfrac{1}{2}mn\right)\left(\tfrac{1}{3}m^3\right) &= \left(\tfrac{1}{2} \cdot \tfrac{1}{3}\right)(m^{1+3})(n) \\ &= \tfrac{1}{6}m^4n \text{ or } \frac{m^4n}{6} \end{aligned}$$

EXERCISES

Simplify.

Sample 1: $2(-6y)$ *What you say:* $-12y$

Sample 2: $(-7)(-2)(a^2 \cdot a)(b \cdot b^2)$ *What you say:* $14a^3b^3$

1. $-7(5a)$ $-35a$
2. $(-t^2)(t)$ $-t^3$
3. $(k)(k^2)$ k^3
4. $(m^3)(m^4)$ m^7
5. $(m)^3(m)^2$ m^5
6. $(4t^2)(t)^2$ $4t^4$
7. $5(m^3 \cdot m^4)(n \cdot n^2)$ $5m^7n^3$
8. $7(x^4 \cdot x^5)(y^3 \cdot y^4)$ $7x^9y^7$
9. $a(x \cdot x^3)(d^3 \cdot d^3)$ ax^4d^6
10. $-4(a \cdot a^2)(b^3 \cdot b^5)$ $-4a^3b^8$
11. $-3(b^3 \cdot b^4)$ $-3b^7$
12. $2(r \cdot r^2)(n \cdot n \cdot n)$ $2r^3n^3$

For Extra Practice, see page 428.

Simplify.

Sample: $(-5xy)(3xy^2)$ Solution: $(-5xy)(3xy^2)$
$$= (-5 \cdot 3)(x \cdot x)(y \cdot y^2)$$
$$= -15x^2y^3$$

1. $(4xy)(7xy^2)$ $28x^2y^3$ 2. $(6ab)(8a^2b)$ $48a^3b^2$ **A**

3. $(-4b)(-4b)(3b)$ $48b^3$ 4. $(9xy^2)(y^3)(-3xy)$ $-27x^2y^6$

5. $(5mn)(3m)(-4m)$ $-60m^3n$ 6. $(11s^2)(-7s^3)$ $-77s^5$

7. $(x)(-6x)(5y)$ $-30x^2y$ 8. $(5a)(2ab)(-4b)$ $40a^2b^2$

9. $(-a)(-3ab)(-b)$ $-3a^2b^2$ 10. $m^2(-5mn)(2n)$ $-10m^3n^2$

11. $6s(rs)(-s^2)$ $-6rs^4$ 12. $(-4ab)(2bc)(-2b)$ $16ab^3c$

13. $(-12)(c)(-b)(-3c)$ $-36bc^2$ 14. $\left(\frac{1}{3}mn\right)\left(\frac{1}{5}mn\right)$ $\frac{1}{15}m^2n^2$

15. $(3xy)(-x)(-xy)$ $3x^3y^2$ 16. $-x^2y(xy)(-xy)$ $1x^4y^3 = x^4y^3$

17. $(-7y)(3x^3y^2)(-x^3y^5)$ $21x^6y^8$ 18. $-0.8x(20x^2y^5)(-x^2y)$ $16x^5y^6$ **B**

19. $-5a(0.2a^3b^5)(0.4ab^2)$ $-0.4a^5b^7$ 20. $(3^3ab^3)(-a^2b^5)$ $-27a^3b^8$

21. $(-ab)(-7a^2b^2)(-5c^8)$ 22. $(x^4)(x^2y)(y^3)$ x^6y^4

23. $(3n)^2(mn)(3m)^2$ $81m^3n^3$ 24. $(a^2d)(-2a^2)(d^2)$ $-2a^4d^3$

21. $-35a^3b^3c^8$

Sample: $(7x^2)(3y)(-2y) + (5xy)(-3xy)$

Solution: $(7x^2)(3y)(-2y) + (5xy)(-3xy)$
$$= (-42x^2y^2) + (-15x^2y^2) = -57x^2y^2$$

25. $(6x^2)(3y) + (7xy)(x)$ $25x^2y$ **C**

26. $(-3pq)(4p^2q) - (6q^2)(-p^3)$ $-6p^3q^2$

27. $(5a^2b^3)(-3ab) + (ab^2)(a^2b^2)$ $-14a^3b^4$

28. $(-2xy^2)(5x^2)(-4y) - (8xy^3)(5xy)$ $40x^3y^3 - 40x^3y^4$

29. $(-2m^2)(-6mn)(mn) - (2m^2n^2)(-4m)(m)$ $20m^4n^2$

30. $(-2a^2bc^3)(-ab^2c) + (5c^2)(-6a^3b^3)(ac^2)$ $2a^3b^3c^4 - 30a^4b^3c^4$

31. $\left(\frac{1}{2}xy^2\right)\left(\frac{2}{3}x\right) + \left(\frac{1}{3}y^2\right)\left(\frac{1}{2}x^2\right)$ $\frac{1}{2}x^2y^2$

32. $(0.3t)(-0.2t^2) + (0.2t^2)(0.2t)$ $-0.02t^3$

33. $\left(\frac{1}{4}mn\right)(2m^2) - (4m^3)\left(\frac{1}{4}n\right)$ $-\frac{1}{2}m^3n$

34. $\left(\frac{3}{5}xyz\right)\left(\frac{1}{2}x\right) - \left(\frac{2}{3}x^2\right)(3yz)$ $-1\frac{7}{10}x^2yz$

35. $(1.2a^2b)(2.7bc^2) - (6ab^2)(0.2ac^2)$ $2.04a^2b^2c^2$

13-3 *A Power of a Product*

OBJECTIVES

Simplify expressions such as $(3ab)^3$ and $(-2xy)^2$.

Find the value of such expressions as $(4^3)^2$ and $[(-2)^2]^3$.

When a product is raised to a power, each factor of the product is raised to that power.

EXAMPLE 1 $(3ab)^3 = 3ab \cdot 3ab \cdot 3ab = 3 \cdot 3 \cdot 3 \cdot a \cdot a \cdot a \cdot b \cdot b \cdot b$
$$= 3^3 a^3 b^3 = 27 a^3 b^3$$

EXAMPLE 2 $(-2xy)^2 = (-2)(-2)x \cdot x \cdot y \cdot y = 4x^2 y^2$

► For all directed numbers a and b and any positive integer p:

$$(ab)^p = a^p b^p$$

An expression such as $(4^2)^3$ is described as a **power of a power**.

EXAMPLE 3 $(4^2)^3 = 4^2 \cdot 4^2 \cdot 4^2 = 4^{2+2+2} = 4^{3 \cdot 2} = 4^6$

EXAMPLE 4 $(m^5)^2 = m^5 \cdot m^5 = m^{5+5} \blacktriangleleft m^{5+5} = m^{2 \cdot 5} = m^{5 \cdot 2}$
$$= m^{10}$$

► For every directed number a, and all positive integers p and q:

$$(a^p)^q = a^{p \cdot q}$$

EXERCISES

Answer *Yes* or *No*.

1. Is $(7a)^3$ always the same as $7a^3$? no
2. Is $4a^2$ always the same as $4^2 a^2$? no
3. Is $(2x^2)^2$ always the same as $4x^4$? yes
4. Is $9y^2$ always the same as $(3y)^2$? yes
5. Is $27t^3$ always the same as $(27t)^3$? no
6. Is $(x^2)^3$ always the same as x^5? no

For Extra Practice, see page 428.

Simplify.

1. $(3ab)^4$ $81a^4b^4$
2. $(4xy)^3$ $64x^3y^3$
3. $(3xyz)^4$ $81x^4y^4z^4$

4. $(-3xyz)^4$ $81x^4y^4z^4$
5. $(-8mn)^2$ $64m^2n^2$
6. $(-10xy)^3$

7. $(10x^2y^2)^2$ $100x^4y^4$
8. $(-5mn^2)^3$ $-125m^3n^6$
9. $(-ab)^3$ $-a^3b^3$

$-1000x^3y^3$

A

10. $(-2y^5)^3$ $-8y^{15}$
11. $(-r^2s^2)^4$ r^8s^8
12. $(3mnp)^2$

13. $(n^4)^4$ n^{16}
14. $(-a^5)^3$ $-a^{15}$
15. $(0 \cdot m^2n)^2$ 0

$9m^2n^2p^2$

Find the value. For a variable, use the given value.

Sample: $(5x^3)^2$; $x = 2$ *Solution:* $5^2x^{3\cdot2} = 5^2x^6 = 5^2 \cdot 2^6$
$= 25 \cdot 64 = 1600$

16. $(4^2)^2$ 256
17. $(10^2)^3$ $1{,}000{,}000$
18. $(-2^3)^2$ 64

19. $(2y^3)^2$; $y = 2$ 256
20. $(n^3)^5$; $n = 1$ 1
21. $(-3a^2)^2$; $a = 3$

729

22. $(5x)^4$; $x = 2$ $10{,}000$
23. $(-5y^3)^2$; $y = 10$ $25{,}000{,}000$

24. $(5n^3)^2$; $n = -2$ 1600
25. $(-5a^4)^3$; $a = 1$ -125

26. $(75y^3)^4$; $y = 0$ 0
27. $(0.5a^2)^2$; $a = 10$ 2500

28. $(8x^4)^2$; $x = \dfrac{1}{2}$ $\dfrac{1}{4}$
29. $(-8a^2b)^2$; $a = \dfrac{1}{2}$, $b = 10$

400

Square the monomial.

30. $5x$ $25x^2$
31. $-5xy^2$ $25x^2y^4$
32. $-0.4xy^3$

$0.16x^2y^6$

33. $0.2ab^2c^3$ $0.04a^2b^4c^6$
34. $-3x^3y^2z$ $9x^6y^4z^2$
35. $-12abc$ $144a^2b^2c^2$

Simplify.

Sample: $(3ab)^3 + (4a)(-5ab)(ab^2)$
Solution: $(3ab)^3 + (4a)(-5ab)(ab^2) = 27a^3b^3 - 20a^3b^3 = 7a^3b^3$

36. $(3xy)^4 + (3x)^3(5xy^2)(y)^2$ $216x^4y^4$

B

37. $(5a)^2(-2ab^2)^2 + (3a)^2(ab)^2(b)^2$ $109a^4b^4$

38. $(2x^2)(-5x^3) + (-4x)(2x^2)^2 + 10x^5$ $-16x^5$

39. $(m^2)m + (2m^2)(-2m) + (2m)^3$ $5m^3$

40. $(-2c)(-3cd)^3 + (3c)^2(cd)^2(-6d)$ 0

41. $(rs)^6(-6rs^3) + r(r^2s^3)^3 + (2r^2)^2(r^3s^9)$ $-r^7s^9$

42. $(3x)^2(-16x^4) + (-4x^4)(-x)^2 + (-10x^3)^2$ $-48x^6$

C

43. $(-at)^3(0.2a^2t) + (t^2a)^2(0.3a^3) + (t^2a)(ta^2)^2$ $1.1a^5t^4$

44. $(-0.1xy)^5(0.2xy^2) + (x^2y)^3(0.2y^2)^2 + (-x^3y)^2(-y)^4y$
$-0.959998x^6y^7$

13-4 A Monomial Times a Polynomial

> **OBJECTIVE**
> Complete multiplications such as
> $x(x^2 - 2xy + y)$ and
> $-3w^3(w^2 - 4w + 10)$.

We use the distributive property to complete ordinary multiplications such as 3×45. We use the same property to complete multiplications involving monomials and polynomials.

Compare Examples 1 and 2.

EXAMPLE 1

$$
\begin{aligned}
3 \times 45 &= 3(40 + 5) \\
&= 3(40) + 3(5) \\
&= 120 + 15 \\
&= 135
\end{aligned}
\quad \text{or} \quad
\begin{array}{r}
45 \\
3 \\
\hline
15 \\
120 \\
\hline
135
\end{array}
$$

EXAMPLE 2

$$
\begin{aligned}
3s(7s + 5) &= 3s(7s) + 3s(5) \\
&= 21s^2 + 15s
\end{aligned}
\quad \text{or} \quad
\begin{array}{r}
7s + 5 \\
3s \\
\hline
21s^2 + 15s
\end{array}
$$

EXAMPLE 3

$$
\begin{aligned}
x(x^2 - 2xy + y) & \\
= x \cdot x^2 + x(-2xy) + x \cdot y & \\
= x^3 - 2x^2y + xy &
\end{aligned}
\quad \text{or} \quad
\begin{array}{r}
x^2 - 2xy + y \\
x \\
\hline
x^3 - 2x^2y + xy
\end{array}
$$

EXAMPLE 4

$$
\begin{aligned}
&-3w^3(w^2 - 4w + 10) \\
&= -3w^3 \cdot w^2 + (-3w^3)(-4w) + (-3w^3)(10) \\
&= -3w^5 + 12w^4 - 30w^3
\end{aligned}
$$

Oral EXERCISES

Simplify. **5.** $-24x^5 - 8x^6$ **6.** $-k + 4b$ **9.** $21xy + 21y^2$

Sample: $5(x - 3)$ *What you say:* $5x - 15$

1. $3(2 + x)$ $6 + 3x$ **2.** $10(2x - 3y)$ $20x - 30y$ **3.** $-5(5 - x^2)$ $-25 + 5x^2$

4. $(8 - y)(-y)$ $-8y + y^2$ **5.** $-8x^5(3 + x)$ **6.** $-1(k - 4b)$

7. $(x - y)16$ $16x - 16y$ **8.** $(3x - 2)8$ $24x - 16$ **9.** $21y(x + y)$

10. $-2y(5y - 5)$ $-10y^2 + 10y$ **11.** $xy(4x + 7)$ $4x^2y + 7xy$ **12.** $12a(3a + 5)$ $36a^2 + 60a$

6. $6t^2 - 2t^4 - 4t^3$ **8.** $-5p^5 + p^4 - 2p^3 + 3p^2$

Multiply.   **3.** $5d^2 - 10ad + 5c^2$

1. $-x(x^2 + 6x - 7)$ **2.** $-6(2k^2 - 2k + 4)$

$-x^3 - 6x^2 + 7x$ $-12k^2 + 12k - 24$

3. $-5(-d^2 + 2ad - c^2)$ **4.** $-y(y^2 - 12xy + x^2)$ $-y^3 + 12xy^2 - x^2y$

A

5. $d^2(a - b - d)$ $ad^2 - bd^2 - d^3$ **6.** $2t^2(3 - t^2 - 2t)$

7. $(7 - 3x - x^3)(-x^2)$ **8.** $-p^2(5p^3 - p^2 + 2p - 3)$

$-7x^2 + 3x^3 + x^5$

9. $-y^3(y^2 - y^3 + 1)$ **10.** $ab(3ab - 4a + 2b)$ $3a^2b^2 - 4a^2b + 2ab^2$

$-y^5 + y^6 - y^3$

11. $5 + 2x - x^2$ **12.** $9y - 1 - 3y^3$

$\qquad\quad 5x$ $\qquad\quad -4y$

$\overline{25x + 10x^2 - 5x^3}$ $\overline{-36y^2 + 4y + 12y^4}$

13. $3k^2 - 4k - 10$ **14.** $6a + 4b - 2c^2$

$\qquad\qquad -0.2k$ $\qquad\qquad a^2$

$\overline{0.6k^3 + 0.8k^2 + 2k}$ $\overline{6a^3 + 4a^2b - 2a^2c^2}$

15. $-4m^2 + mn - 2n^2$ **16.** $4a^2 + 5b^2 + 6c^2$

$\qquad\qquad -m^2$ $\qquad\qquad 1.5a$

$\overline{4m^4 - m^3n + 2m^2n^2}$ $\overline{6a^3 + 7.5ab^2 + 9ac^2}$

Simplify.   **19.** $-6m^2 - 8mn$ **21.** $8a^2 + 8ab$

17. $x(2x + 3y) + 6(-xy)$ $2x^2 - 3xy$ **18.** $a(b - a) + 3a(b - a)$ $-4a^2 + 4ab$

19. $7m(-m - n) + m(m - n)$ **20.** $x(2y - 3x) + 3x(y + 2x)$ $3x^2 + 5xy$

21. $3a(a + b) + 5a(a + b)$ **22.** $6a(3a - 2b) + a(a + b)$ $19a^2 - 11ab$

23. $-2y(5x + y) + 4y(5y + x)$ $18y^2 - 6xy$

24. $4ab(a - 3b) - 8a(2ab + 8b^2)$ $-12a^2b - 76ab^2$

25. $2x^2(3w - 4x) + (wx + x^2)(2x)$ $-6x^3 + 8x^2w$

26. $(a + 3b)(ab) - a(-3b^2 - 2ab)$ $3a^2b + 6ab^2$

27. $xy(x^2 - xy^2) + (3y)(2x^3 + x^2y^2)$ $7x^3y + 2x^2y^3$

28. $(3m^2n)(m^2 - 2n) + m^2n(m^2 - n)$ $4m^4n - 7m^2n^2$

30. $160ab^2 + 120a^{11}b^4 - 40a^7b^{10} - 140a^3b^{12}$

Multiply.   **32.** $-0.75a^3b^3 - 6a^4b^3 + 7.5a^4b^2 + 1.5a^2b^4$

29. $6a^2b(10 - 4ab^4 + 6a^6b^4 - b^{10})$ $60a^2b - 24a^3b^5 + 36a^8b^5 - 6a^2b^{11}$ **B**

30. $10ab^2(16 + 12a^{10}b^2 - 4a^6b^8 - 14a^2b^{10})$

31. $-a^3b\left(\dfrac{1}{2}a^2b - \dfrac{1}{2}a^3b^2 + ab^2 - b^3\right) -\dfrac{1}{2}a^5b^2 + \dfrac{1}{2}a^6b$

32. $-1.5a^2b(0.5ab^2 + 4a^2b^2 - 5a^2b - b^3)$

33. $21x^4y^4(-3x^2 + 4xy - 5y^2)$ $-63x^6y^4 + 84x^5y^5 - 105x^4y^6$

Solve.

34. $-3x + 4(x - 2) = 11$ $x = 19$

35. $4y + 3(-y - 1) = 8$ $y = 11$

36. $2 = (10n + 6) - 6(n - 2)$ $-4 = n$

37. $22 = (4k + 3.5) - 2.5(k - 8)$ $-1 = k$

38. $0.8(7 - x) + 7x = 18$ $x = 2$

39. $34 = (-1)(10 - x) - 12x$ $-4 = x$

40. $7 - 3(a + 1) = 3a - 2$ $1 = a$

41. $-10(2 + 2x) + 312 = 3.5(8 - x)$ $x = 16$

C

42. $y - (y + 21) = 1.5y - 1.5(7 - y)$ $-3.5 = y$

43. $10a - 2(a - 8) = 28a - 14(2a + 8)$ $a = -16$

44. $0.1(3x - 12) - (3x - 2.5) = 6.5 + 0.25(3x - 7)$ $-1 = x$

45. $6(4k - 5) = -5(2 - 5k) - 7$ $-13 = k$

Problems

Express the answer as a polynomial in simplest form.

A

1. Find the area of the triangle.

$\left(Hint: A = \dfrac{1}{2}bh\right)$ $\dfrac{1}{2}m^2 + m$

2. Find the area of a square with sides of length $5t$. Find the perimeter. *area*, $25t^2$; *perimeter*, $20t$

3. Find the area of a rectangle with length $2x^2 + 5$ and width $3x$. $6x^3 + 15x$

4. Find the perimeter of a pentagon if each of the five sides has length $2x^2 + 3x + 1$. $10x^2 + 15x + 5$

B

5. A cylinder has radius 5 centimeters and height $2t + 9$. The volume is given by the formula $V = \pi r^2 h$. Find the volume. (You don't have to substitute a value for π.) $50\pi t + 225\pi$

6. Find the volume of the box. (*Hint:* $V = lwh$) $8x^4 + 3x^3$

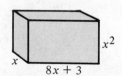

7. Find the area of the bottom of the box in Problem 6. $8x^2 + 3x^3$

8. A rectangular piece of sheet metal has four circular holes punched in it. The area of each circle is $c^2 + 1$. Find the area of the remaining piece of metal. $-3c^2 + 7c + 6$

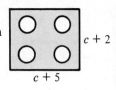

C

9. A cube has edges of length y^2. (a) Find its volume. (b) Find its surface area. volume, y^6; surface area, $6y^4$

10. (a) Find the area of the figure. (b) Find its perimeter.

area, $3a^2 + 6a$; perimeter, $8a + 12$

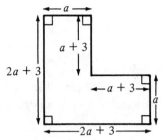

SELF-TEST 1

Be sure that you understand these terms.

exponent (p. 346)   base (p. 346)

Simplify.   Section 13-1, p. 346

1. $(-2 \cdot 5)(x \cdot x \cdot x)$ $-10x^3$ 2. $3 \cdot n(-n)n$ $-3n^3$

3. $4^2 \cdot 4^3$ 1024   4. $z^5 \cdot z^6$ z^{11}

5. $(3ts)(-2t^2s^2)$ $-6t^3s^3$ 6. $(-4m)(2m^2n)$ $-8m^3n$ Section 13-2, p. 348

7. $(2xy)^3$ $8x^3y^3$ 8. $(-a^2)^2$ a^4 Section 13-3, p. 350

Find the value.

9. $(2^4)^2$ 256   10. $(3^3)^2$ 729

Simplify. $-3x^3 - x^2 + 5x$ $4a^2 + 12ab + b^2$ Section 13-4, p. 352

11. $-x(3x^2 + x - 5)$ 12. $4a(a + 2b) + b(4a + b)$

Check your answers with those printed at the back of the book.

Multiplication of Polynomials

13-5 *A Polynomial Times a Polynomial*

OBJECTIVE

Multiply polynomials, such as
$(3x - 4)(x + 2)$ and
$(n^2 + 2n - 3)(n + 4)$.

To multiply two polynomials we may need to use the distributive property more than once and then combine like terms. This is the same method we use to multiply two numbers such as 45 and 23.

EXAMPLE 1

$$
\begin{array}{r}
45 \\
\times 23 \\
\hline
\end{array}
\qquad
\begin{array}{r}
40 + 5 \\
20 + 3 \\
\hline
120 + 15 \\
800 + 100 \\
\hline
800 + 220 + 15 = 1035
\end{array}
$$

◄ $3(40 + 5)$
◄ $20(40 + 5)$

If we arrange the work horizontally the steps are in different order.

$$
\begin{aligned}
(23)(45) &= (20 + 3)(40 + 5) \\
&= 20(40 + 5) + 3(40 + 5) \\
&= [(20 \cdot 40) + (20 \cdot 5)] + [(3 \cdot 40) + (3 \cdot 5)] \\
&= 800 + 100 + 120 + 15 \\
&= 1035
\end{aligned}
$$

EXAMPLE 2 $(3x - 4)(x + 2)$ ►

$$
\begin{array}{r}
3x - 4 \\
x + 2 \\
\hline
6x - 8 \\
3x^2 - 4x \\
\hline
3x^2 + 2x - 8
\end{array}
$$

◄ $2(3x - 4)$
◄ $x(3x - 4)$

Horizontally:

$$
\begin{aligned}
(3x - 4)(x + 2) &= 3x(x + 2) - 4(x + 2) \\
&= 3x^2 + 6x - 4x - 8 \\
&= 3x^2 + 2x - 8
\end{aligned}
$$

When one of the polynomials is a trinomial, it is usually easier to arrange the work in vertical form.

EXAMPLE 3 $(n^2 + 2n - 3)(n + 4)$ ►

$$
\begin{array}{r}
n^2 + 2n - 3 \\
n + 4 \\
\hline
4n^2 + 8n - 12 \\
n^3 + 2n^2 - 3n \\
\hline
n^3 + 6n^2 + 5n - 12
\end{array}
$$

Use the distributive property to name the expression as the sum or difference of two products. Do not multiply.

Oral
EXERCISES

Sample: $(n - 2)(3n + 7)$

What you say: $n(3n + 7) - 2(3n + 7)$

1. $3a(5a + 2) - 9(5a + 2)$

$3a(2a + 3) + 2(2a + 3)$

1. $(3a - 9)(5a + 2)$ 2. $(3a + 2)(2a + 3)$

3. $(3x + 5)(3x + 5)$ 4. $(7b - 3)(b + 4)$

5. $(4c + 14)(c - 3)$ 6. $(6t + 5)(t - 4)$

7. $(a + 7b)(2ab + 3a - b)$ 8. $(3n + 5)(3n - 5)$

  $a(2ab + 3a - b) + 7b(2ab + 3a - b)$ $3n(3n - 5) + 5(3n - 5)$

  3. $3x(3x + 5) + 5(3x + 5)$ 5. $4c(c - 3) + 14(c - 3)$

  4. $7b(b + 4) - 3(b + 4)$ 6. $6t(t - 4) + 5(t - 4)$

For Extra Practice, see page 429.

Multiply in two ways. Use both the vertical and the horizontal form.

Written
EXERCISES

Sample: $(5x + 3)(2x + 4)$

Solution:

$$
\begin{array}{r}
5x + 3 \\
2x + 4 \\
\hline
20x + 12 \\
10x^2 + 6x \\
\hline
10x^2 + 26x + 12
\end{array}
$$

$(5x + 3)(2x + 4)$
$= [5x(2x + 4)] + [3(2x + 4)]$
$= (10x^2 + 20x) + (6x + 12)$
$= 10x^2 + 26x + 12$

1. $(a + 5)(a + 4)$ $a^2 + 9a + 20$ 2. $(m + 2)(m + 7)$ $m^2 + 9m + 14$ **A**

3. $(6m + 4)(2m + 13)$ $12m^2 + 86m + 52$ 4. $(4a - 8)(a + 8)$ $4a^2 + 24a - 64$

5. $(x + 1.5)(x + 1.5)$ $x^2 + 3x + 2.25$ 6. $(2b - 9)(b + 10)$ $2b^2 + 11b - 90$

Multiply. Use either form.

10. $20m^2 - 14m + 2$

7. $(x - 9)(x + 9)$ $x^2 - 81$

8. $(m - 4)(m - 6)$ $m^2 - 10m + 24$

9. $(x + 2)(x - 2)$ $x^2 - 4$

10. $(5m - 1)(4m - 2)$

$24 + 59n + 36n^2$ **11.** $(3 + 4n)(8 + 9n)$

12. $(5x + 2)(7x - 3)$ $35x^2 - x - 6$

$18 - 36y + 16y^2$ **13.** $(6 - 4y)(3 - 4y)$

14. $(2x + 7)(2x + 7)$ $4x^2 + 28x + 49$

$9 - 3n - 20n^2$ **15.** $(4n + 3)(3 - 5n)$

16. $(2 - 3n)(3 - 2n)$ $6n^2 - 13n + 6$

17. $(5t + 1)(5t - 1)$ $25t^2 - 1$

18. $(z - 12)(2z + 1)$ $2z^2 - 23z - 12$

Multiply. Do it mentally if you can. Then substitute to check your answer. Use $x = 1$ and $y = 1$.

Sample: $(3x + y)(x + 2y)$
Solution: $3x^2 + 7xy + 2y^2$
Check: $\quad [3(1) + 1][1 + 2(1)] = 4 \cdot 3$
$$= 12$$
$$3(1^2) + 7(1 \cdot 1) + 2(1^2) = 3 + 7 + 2$$
$$= 12 \ \checkmark$$

$16x^2 + 44x + 10$

19. $(8x + 2)(2x + 5)$

20. $(y + 3)(y + 1)$ $y^2 + 4y + 3$

21. $(2x + 3)(x + 1)$ $2x^2 + 5x + 3$

22. $(2a - 5)(2a + 5)$ $4a^2 - 25$

23. $(1 - 2b)(2 + b)$ $2 - 3b - 2b^2$

24. $(p + 8)(p - 20)$ $p^2 - 12p - 160$

25. $(2s + 3)(s + 4)$ $2s^2 + 11s + 12$

26. $(m + n)(m + n)$ $m^2 + 2mn + n^2$

$8q^2 - 82q + 20$ **27.** $(8q - 2)(q - 10)$

28. $(2x - 2y)(x + y)$ $2x^2 - 2y^2$

$1.5w^2 - w - 2.5$ **29.** $(1.5w - 2.5)(w + 1)$

30. $(s + t)(2s + 2t)$ $2s^2 + 4ts + 2t^2$

34. $2a^2 + 15.5ab - 4b^2$ **35.** $6m^2 + 3.5mn + 0.5n^2$

Multiply. **36.** $4m^3 - 17m^2 - 11m - 20$ **37.** $x^3 - 9x^2 + 25x - 21$

$8 - 32m - 40m^2$

B **31.** $(8 + 8m)(1 - 5m)$

32. $(y + x)(8y + x)$ $8y^2 + 9xy + x^2$

33. $(2n - m)(2n + m)$ $4n^2 - m^2$

34. $(2a - 0.5b)(a + 8b)$

35. $(3m + n)(0.5n + 2m)$

36. $(4m^2 + 3m + 4)(m - 5)$

37. $(x - 3)(x^2 - 6x + 7)$

38. $(x + y)(x^2 - xy - y^2)$

39. $(s^2 + st + t^2)(s - t)$ $s^3 - t^3$

40. $(a^2 - 6a - 7)(a - 1)$

$36 + 13x^2 + x^4$ **41.** $(9 + x^2)(4 + x^2)$

42. $(x^2 - 10)(x^2 - 7)$ $x^4 - 17x^2 + 70$

43. $(a^2b^2 + 1)(a^2b^2 + 1)$

44. $(k^4 - 2)(k^4 - 3)$ $k^8 - 5k^4 + 6$

45. $(6 + r)(7 - 3r - 2r^2)$ $42 - 11r - 15r^2 - 2r^3$

46. $(m^2 + 3m + 4)(m^2 - 4m + 5)$ $m^4 - m^3 - 3m^2 - m + 20$

47. $(x - 2y + 1.5z)(2x + 2y - 2.5z)$

48. $(y - 1)(y^4 - y^3 - y^2 - y - 1)$ $y^5 - 2y^4 + 1$

38. $x^3 - 2xy^2 - y^3$ **40.** $a^3 - 7a^2 - a + 7$ **43.** $a^4b^4 + 2a^2b^2 + 1$

47. $2x^2 - 4y^2 - 3.75z^2 - 2xy + 0.5xz + 8yz$

Express the answer as a polynomial in simplest form.

1. A rectangle has length $2x + 1$ and width $x + 4$. (a) Find the area. (b) Find the perimeter. **a.** $2x^2 + 9x + 4$ **b.** $6x + 10$

2. The height of a triangle is 3 centimeters more than the length of the base. Find the area. $\left(Hint:\ A = \dfrac{1}{2}bh.\ \text{Express the}\right.$ answer in terms of $b.\Big)$ $\left(\dfrac{1}{2}b^2 + \dfrac{3}{2}b\right)cm^2$

3. A square has sides of length t. A rectangle has length 6 meters more than t and width 5 meters less than t. Find the area of the rectangle. $(t^2 - t - 30)m^2$

4. Find the area of each small rectangle. Find the area of the entire region by finding the product of two binomials. Check by comparing the result with the sum of the areas of the four small rectangles.
    $A = 2s^2;\ B = 4s;\ C = 3s;\ D = 6;$
    Area $= 2s^2 + 7s + 6$

    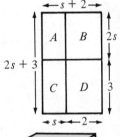

5. Find the volume of the box. Find the area of the bottom of the box. $c^2 - 4$

    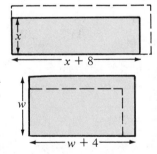

Write an equation representing the given information. Express each member of the equation as a polynomial in simplest form.

6. The length of a rectangle is 8 meters more than its width. If each dimension were 1 meter more, the area would be 35 square meters greater.
    $x^2 + 8x + 35 = x^2 + 10x + 9$

    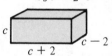

7. The length of a rectangle is 4 centimeters more than its width. If the length and width were each 2 centimeters shorter, the area would be 40 square centimeters less. $w^2 - 4 = w^2 + 4w + 40$

8. Jill has some pennies, nickels, and dimes. In all, she has \$3.92. The number of nickels is two less than the number of pennies. She has 13 more dimes than pennies. (*Hint:* the value of the nickels is $5(p - 2)$ cents.) $16p + 120 = 392$

13-6 *Special Polynomial Products*

OBJECTIVE

Complete multiplications such as $(3x + 2)^2$ and $(m + 5)(m - 5)$.

The exponent in an expression like $(3x + 2)^2$ tells us how many times the polynomial is to be used as a factor. We expand the expression by finding the product of the factors.

EXAMPLE 1
$$
\begin{aligned}
(3x + 2)^2 &= (3x + 2)(3x + 2) \\
&= 3x(3x + 2) + 2(3x + 2) \\
&= 9x^2 + 6x + 6x + 4 \\
&= 9x^2 + 12x + 4
\end{aligned}
$$

EXAMPLE 2
$$
\begin{aligned}
(2n - 3)^2 &= (2n - 3)(2n - 3) \\
&= 2n(2n - 3) - 3(2n - 3) \\
&= 4n^2 - 6n - 6n + 9 \\
&= 4n^2 - 12n + 9
\end{aligned}
$$

Examples 1 and 2 demonstrate a pattern for expanding the square of *any* binomial:

$$(\text{first term})^2 + (\text{twice product of terms}) + (\text{second term})^2$$

▶ For all directed numbers a and b, $(a + b)^2 = a^2 + 2ab + b^2$.

EXAMPLE 3
$$
\begin{aligned}
(5t + 3)^2 &= (5t)^2 + 2(5t)(3) + 3^2 \\
&= 25t^2 + 30t + 9
\end{aligned}
$$

We can also find a pattern for products of pairs of binomials like $(x + 7)(x - 7)$ and $(2n - 3)(2n + 3)$.

EXAMPLE 4
$$
\begin{aligned}
(x + 7)(x - 7) &= x(x - 7) + 7(x - 7) \\
&= x^2 - 7x + 7x - 49 \\
&= x^2 - 49 \blacktriangleleft x^2 - 7^2
\end{aligned}
$$

EXAMPLE 5
$$
\begin{aligned}
(2n - 3)(2n + 3) &= 2n(2n + 3) - 3(2n + 3) \\
&= 4n^2 + 6n - 6n - 9 \\
&= 4n^2 - 9 \blacktriangleleft (2n)^2 - 3^2
\end{aligned}
$$

In Examples 4 and 5 notice that the **terms are the same** in each pair of expressions but the **signs are different.** The product in each case is the difference of two squares.

▶ For all directed numbers a and b, $(a - b)(a + b) = a^2 - b^2$.

EXAMPLE 6 $(3m - 5)(3m + 5) = (3m)^2 - 5^2$
$$= 9m^2 - 25$$

Name the missing terms.

Oral EXERCISES

1. $(a + 3)^2 = (a + 3)(a + 3) = a^2 + \underline{\ ?\ } + 9$ $6a$
2. $(3x + 2y)^2 = (3x + 2y)(3x + 2y) = 9x^2 + 12xy + \underline{\ ?\ }$ $4y^2$
3. $(3 - 2x)^2 = (3 - 2x)(3 - 2x) = \underline{\ ?\ } - \underline{\ ?\ } + 4x^2$ $9; 12x$
4. $(2k + 3)^2 = (2k + 3)(2k + 3) = 4k^2 + \underline{\ ?\ } + \underline{\ ?\ }$ $12k; 9$
5. $(c + 5d)^2 = (c + 5d)(c + 5d) = c^2 + \underline{\ ?\ } + \underline{\ ?\ }$ $10cd; 25d^2$
6. $(1 - 5z)^2 = (1 - 5z)(1 - 5z) = \underline{\ ?\ } - \underline{\ ?\ } + 25z^2$ $1; 10z$
7. $(3x + 1)(3x - 1) = \underline{\ ?\ } - \underline{\ ?\ }$ $9x^2; 1$
8. $(a - 5b)(a + 5b) = \underline{\ ?\ } - \underline{\ ?\ }$ $a^2; 25b^2$

For Extra Practice, see page 429.

Multiply mentally. Then check your work on paper.

Written EXERCISES

Sample: $(5a + 2b)^2$

Solution: $25a^2 + 20ab + 4b^2$ Check:

$$\begin{array}{r} 5a + 2b \\ 5a + 2b \\ \hline 10ab + 4b^2 \\ 25a^2 + 10ab \\ \hline 25a^2 + 20ab + 4b^2 \end{array}$$

1. $(b + 4)^2$ $b^2 + 8b + 16$
2. $(x + y)^2$ $x^2 + 2xy + y^2$
3. $(x - a)^2$ $x^2 - 2ax + a^2$
4. $(y + 3b)^2$ $y^2 + 6by + 9b^2$
5. $(3x - 2c)^2$ $9x^2 - 12xc + 4c^2$
6. $(2a - 5)^2$ $4a^2 - 20a + 25$
7. $(x + y)(x - y)$ $x^2 - y^2$
8. $(4m - 3)(4m + 3)$ $16m^2 - 9$
9. $(3k - q)(3k + q)$ $9k^2 - q^2$
10. $(ax + 2by)(ax - 2by)$ $a^2x^3 - 4b^2y^2$
11. $(ac + bc)^2$ $a^2c^2 + 2abc^2 + b^2c^2$
12. $(ac + bc)(ac - bc)$ $a^2c^2 - b^2c^2$

A

See page A2 at the back of the book for Ex. 19–39.

Expand.

B

13. $(k^2 + 6)^2$ $k^4 + 12k^2 + 36$

14. $(16 - 3ab)^2$ $256 - 96ab + 9a^2b^2$

15. $(3x - 5wy)^2$ $9x^2 - 30wxy + 25a^2y^2$

16. $(2m^2 - 11)^2$ $4m^4 - 44m^2 + 121$

17. $(3 + xy^2)^2$ $9 + 6xy^2 + x^2y^4$

18. $\left(5p - \dfrac{1}{5}\right)^2$

19. $(4m + 10)^2$

20. $(x^2 + 8y^2)^2$

21. $(3bc + 10ab)^2$

22. $(mn^2 - 5)^2$

23. $(4m + 7b)^2$

24. $(0.8 - a)^2$

25. $(18 + 2ab)^2$

26. $(2x^2 + y^2)^2$

27. $(-st^2 + 2)^2$

28. $(3x^2 + 4z^2)^2$

29. $(0.4m^2 + 0.9n)^2$

30. $(4c - 0.1d)^2$

C

31. $(2y^3 - 1)^2$

32. $(z^2 - y^3)^2$

33. $(z^3 - x^3)^2$

34. $(p + 0.3q^2)^2$

35. $(0.2 + 0.5m^2)^2$

36. $(0.7 + 3x)^2$

37. $\left(4k - \dfrac{1}{8}\right)^2$

38. $\left(a + \dfrac{1}{2}b\right)^2$

39. $\left(2\dfrac{1}{2}p - \dfrac{1}{5}\right)^2$

18. $25p^2 - 2p + \dfrac{1}{25}$

Problems

Write the answer as a polynomial in simplest form.

1. What is the area of a square if the length of one side is $3x + 8$? $9x^2 + 48x + 64$

2. The length of a box is four meters longer than its width. The height is 3 meters shorter than the width. If the width is $2v$ meters, what is the volume? $(8v^3 + 4v^2 - 24v)\,m^3$

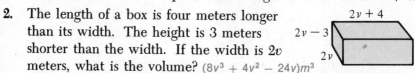

3. A square was cut from a rectangular piece of wood. The L-shaped scrap which was left has an area of $176\ \text{cm}^2$. The width of the scrap is 8 cm. How wide was the original piece of wood? (*Hint:* $x^2 = (x - 8)^2 + 176$) 15 cm

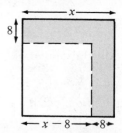

4. The difference between the square of a whole number and the square of the next greater whole number is 15. What are the numbers? 7; 8

5. The square of a whole number is 41 less than the square of the next greater whole number. What are the numbers? 20; 21

6. The product of an integer and the next greater integer exceeds the square of the lesser integer by 13. Find the integers. 13; 14

7. The product of an integer and the next greater integer is 36 less than the square of the greater integer. What are the integers? 35; 36

career capsule   *Sheet-Metal Worker*

Sheet-metal workers put together, install, and repair products made of sheet-metal such as drainpipes and furnace casings. They begin their jobs by reading blueprints to learn the dimensions of the product part and the type of metal needed. By putting measurement marks on the metal parts, they set up reference lines for cutting, straightening, or bending parts by machine. Workers then hand-hammer parts into shape, weld the parts together and finish the job by smoothing the joints with files or grinders.

A sheet-metal worker needs a background in mathematics, drafting, and sheet-metal working. Apprenticeship programs are available or skills can be acquired from on-the-job training. Applicants should have good health, balance, and manual dexterity.

13-7 *Multiplication Properties for Polynomials*

Changing the order in which two polynomials are multiplied does not change the result. Multiplication of polynomials is **commutative.**

EXAMPLE 1 $(4a^2b)(3ac) = 12a^3bc$
$(3ac)(4a^2b) = 12a^3bc$

EXAMPLE 2 $(2x + 1)(x - 8) = 2x^2 - 15x - 8$
$(x - 8)(2x + 1) = 2x^2 - 15x - 8$

When more than two polynomials are multiplied, changing the way in which they are grouped does not change the result. Multiplication of polynomials is **associative.**

EXAMPLE 3 $[(2xy)(4x^2)](3y) = (8x^3y)(3y) = 24x^3y^2$
$(2xy)[(4x^2)(3y)] = (2xy)(12x^2y) = 24x^3y^2$

We have already used the **distributive** property to multiply polynomials.

EXAMPLE 4 $(x + 3)(x - 7) = x(x - 7) + 3(x - 7)$
$= x^2 - 4x - 21$

1 is the **identity element** for multiplication of polynomials.

EXAMPLE 5 $(3m^2 + 5) \cdot 1 = 1 \cdot (3m^2 + 5) = 3m^2 + 5$

Recall that if the product of two numbers is 1, the numbers are **reciprocals** or **multiplicative inverses** of each other. Polynomials have multiplicative inverses.

EXAMPLE 6 $2y \cdot \dfrac{1}{2y} = 1$ ◄ Assume $y \neq 0$.

$2y$ and $\dfrac{1}{2y}$ are **reciprocals** if $y \neq 0$.

EXAMPLE 7 $-\dfrac{3xy^2}{2}\left(-\dfrac{2}{3xy^2}\right) = 1$ ◄ Assume $x \neq 0$ and $y \neq 0$.

$-\dfrac{3xy^2}{2}$ and $-\dfrac{2}{3xy^2}$ are **reciprocals** if $x \neq 0$ and $y \neq 0$.

Match each expression in Column 1 with its reciprocal in Column 2. Assume that no denominator is zero.

COLUMN 1

1. $\dfrac{1}{2a - b}$ **D**

2. $\dfrac{1}{2}a^2b^2$ **F**

3. $\dfrac{3a^2b}{3a + 2}$ **A**

4. $\dfrac{1}{a^3}$ **B**

5. $\dfrac{a - b}{a + b}$ **C**

6. $\dfrac{1}{a + b}$ **E**

COLUMN 2

A. $\dfrac{3a + 2}{3a^2b}$

B. a^3

C. $\dfrac{a + b}{a - b}$

D. $2a - b$

E. $a + b$

F. $\dfrac{2}{a^2b^2}$

Tell which property of multiplication justifies the sentence.

Sample: $(ax + b)(2xy) = (2xy)(ax + b)$

What you say: The commutative property

7. $(2x + y)(3x - 4) = 2x(3x - 4) + y(3x - 4)$ distributive

8. $\left(\dfrac{5x^2y^3}{z}\right)\left[(4az)\left(\dfrac{5x}{2y}\right)\right] = \left[\left(\dfrac{5x^2y^3}{z}\right)(4az)\right]\left(\dfrac{5x}{2y}\right)$ associative

9. $[(3at^2)(2s^2t^3)](2s^2t) = [(2s^2t^3)(3at^2)](2s^2t)$ commutative

10. $[(3x^2)(5 + a)]\left(\dfrac{1}{y^4}\right) = \left(\dfrac{1}{y^4}\right)[(3x^2)(5 + a)]$ commutative

11. $(d^2 - 1)(d^2 + 1) = (d^2 + 1)(d^2 - 1)$ commutative

12. $(x^3 + 1)(x^3 - 1) = (x^3 + 1)x^3 - (x^3 + 1) \cdot 1$ distributive

Name the reciprocal. Assume that no numerator or denominator is zero.

1. $\dfrac{bm}{a-b}$ $\dfrac{a-b}{bm}$

2. $\dfrac{\dfrac{ab}{b-ac}}{ab}$ $\dfrac{b-ac}{ab}$

3. $\dfrac{a-b}{m-b}$ $\dfrac{m-b}{a-b}$

4. $\dfrac{\dfrac{1+y}{1-y}}{1+y}$ $\dfrac{1-y}{1+y}$

5. $\dfrac{x^2+2x+1}{(x+1)^2}$ $\dfrac{(x+1)^2}{x^2+2x+1}$

6. $\dfrac{x^3-y^3}{x-y}$ $\dfrac{x-y}{x^3-y^3}$

7. $\dfrac{3}{4}(a+b^2)$ $\dfrac{4}{3(a+b^2)}$

8. $\dfrac{-5x}{7(y^2+1)}$ $\dfrac{7(y^2+1)}{-5x}$

Show that the sentence is correct.

Sample: $3x[(x+y)y^2] = [(3x)(x+y)]y^2$

Solution: $3x[(x+y)y^2] \stackrel{?}{=} [3x(x+y)]y^2$

$$
\begin{array}{c|c}
3x[xy^2+y^3] & [3x^2+3xy]y^2 \\
3x^2y^2+3xy^3 & 3x^2y^2+3xy^3 \quad \checkmark
\end{array}
$$

9. $ab[(a+b)(a-b)] = (a^2b+ab^2)(a-b)$ a^3b-ab^3

10. $4x^2(14z+7y) = (14z+7y)\cdot 4x^2$ $56x^2z+28x^2y$

11. $8mn(5x+4y) = 40mnx+32mny$ $40mnx+32mny$

12. $[(x+y)(x-y)](x+y) = (x-y)(x+y)^2$ $x^3+x^2y-xy^2-y^3$

B

13. $(0.5a+b)(b-0.5a)x = b^2x-(0.5a)^2x$ $b^2x-0.25a^2x$

14. $[(a^2b^2)(ab^2)](ab) = (ab^2)[(a^2b^2)(ab)]$ a^4b^5

15. $(3x^2+5x+2)(x-1) = (x^2-1)(3x+2)$ $3x^3+2x^2-3x-2$

16. $[(3m+2n)(3m-2n)](6m+2n) =$
$(3m+2n)[(3m-2n)(6m+2n)]$ $54m^3+18m^2n-24mn^2-8n^3$

Multiply. Tell which property of multiplication justifies each step.

Sample: $(4x^4)(9x^2)$

Solution:
$$
\begin{aligned}
(4x^4)(9x^2) &= 4(x^4\cdot 9)x^2 \qquad \blacktriangleleft\text{The Associative Property} \\
&= 4(9\cdot x^4)x^2 \qquad \blacktriangleleft\text{The Commutative Property} \\
&= (4\cdot 9)(x^4\cdot x^2) \blacktriangleleft\text{The Associative Property} \\
&= 36x^6
\end{aligned}
$$

See page A2 at the back of the book for Ex. 17–30.

C

17. $4y^3(y^4-1)$

18. $(mn)(mn+1)$

19. $(x^2+xy+y^2)(xy)$

20. $(12a^2b)(4ab)$

21. $\left(\dfrac{1}{2}mnp\right)\left(\dfrac{1}{4}m^2np^2\right)$

22. $(0.5a^2-b)(b+0.5a)$

23. $2.3x^2(2x^3+0.5x)$

24. $\dfrac{2}{3}m\left(\dfrac{1}{2}mn-\dfrac{n^2}{4}\right)$

See pages A3 and A4 at the back of the book for Ex. 25–30.

25. $\left(\dfrac{5n}{2} - \dfrac{1}{3}\right)\left(\dfrac{5n}{2} + \dfrac{1}{3}\right)$ **26.** $3.7y(2y + 0.3y - 1.2)$

27. $(0.1t + t^2)(3t - 0.5)$ **28.** $\dfrac{1}{4}x\left(x^3 - \dfrac{1}{3}x^2 + \dfrac{1}{2}\right)$

29. $\left(\dfrac{3}{5}rs - r\right)\left(\dfrac{5s}{r}\right)$ **30.** $\left(\dfrac{xy}{3}\right)\left(\dfrac{3x}{y}\right)$

SELF-TEST 2

Be sure you understand these terms.

expand (p. 360)   multiplicative inverse (p. 364)

Multiply.   Section 13-5, p. 356

1. $(b - 3)(b + 2)$ $b^2 - b - 6$ **2.** $(n + 1)(n - 5)$ $n^2 - 4n - 5$

3. $(2t + 7)(3t - 1)$ $6t^2 + 19t - 7$ **4.** $(4x^2 + x + 2)(x + 1)$ $4x^3 + 5x^2 + 3x + 2$

Expand.   Section 13-6, p. 360

5. $(n - 4)^2$ $n^2 - 8n + 16$ **6.** $(2b + 6)^2$ $4b^2 + 24b + 36$

Multiply.

7. $(n - 4)(n + 4)$ $n^2 - 16$ **8.** $(2b + 6)(2b - 6)$ $4b^2 - 36$

Tell which property justifies each step. Assume $y \neq 0$.   Section 13-7, p. 364

9. $3y^2\left(8y + \dfrac{1}{3y^2}\right) = (3y^2)(8y) + 3y^2\left(\dfrac{1}{3y^2}\right)$ distributive prop.

$= 3(y^28)y + 3y^2\left(\dfrac{1}{3y^2}\right)$ associative prop.

$= 3(8y^2)y + 3y^2\left(\dfrac{1}{3y^2}\right)$ commutative prop.

$= 24y^3 + 1$ substitution prop.

Check your answers with those printed at the back of the book.

13-8 *Dividing Monomials*

OBJECTIVE

Simplify expressions such as $\dfrac{14a^3b}{-2a}$ by division.

To divide one monomial by another we need to understand and use factors, exponents, and division.

Example 1 shows two methods of dividing. We can use either to divide monomials.

EXAMPLE 1 $4^5 \div 4^2 = \underline{\ ?\ }$

(1) $\dfrac{4^5}{4^2} = \dfrac{4 \cdot 4 \cdot 4 \cdot 4 \cdot 4}{4 \cdot 4} = \left(\dfrac{4}{4} \cdot \dfrac{4}{4}\right) 4 \cdot 4 \cdot 4$

$= (1 \cdot 1) 4 \cdot 4 \cdot 4 = \mathbf{4^3}$

(2) $\dfrac{4^5}{4^2} = \dfrac{4^2 \cdot 4^3}{4^2} = \dfrac{4^2}{4^2} \cdot 4^3 = 1 \cdot 4^3 = \mathbf{4^3}$

EXAMPLE 2 $t^6 \div t^2 = \underline{\ ?\ }$

(1) $\dfrac{t^6}{t^2} = \dfrac{t \cdot t \cdot t \cdot t \cdot t \cdot t}{t \cdot t} = \left(\dfrac{t}{t} \cdot \dfrac{t}{t}\right) t \cdot t \cdot t \cdot t$

$= (1 \cdot 1) t \cdot t \cdot t \cdot t = \mathbf{t^4}$

(2) $\dfrac{t^6}{t^2} = \dfrac{t^2 \cdot t^4}{t^2} = \left(\dfrac{t^2}{t^2}\right) t^4 = 1 \cdot t^4 = \mathbf{t^4}$

Examples 1 and 2 demonstrate a general rule of exponents that applies when the greater exponent appears in the **numerator.**

▶ For any directed number *a*, except 0, and all positive integers *p* and *q* such that $p > q$:

$$\frac{a^p}{a^q} = a^{p-q}$$

Now let's consider situations where the greater exponent appears in the **denominator.**

EXAMPLE 3 $\dfrac{8^2}{8^5} = \dfrac{8^2 \cdot 1}{8^2 \cdot 8^3} = \left(\dfrac{8^2}{8^2}\right) \cdot \dfrac{1}{8^3} = 1 \cdot \dfrac{1}{8^3} = \dfrac{1}{8^3}$

EXAMPLE 4 $\dfrac{x^3}{x^5} = \dfrac{x^3 \cdot 1}{x^3 \cdot x^2} = \left(\dfrac{x^3}{x^3}\right) \cdot \dfrac{1}{x^2} = 1 \cdot \dfrac{1}{x^2} = \dfrac{1}{x^2}$

For any directed number a, except 0, and for all positive integers p and q such that $p < q$:

$$\frac{a^p}{a^q} = \frac{1}{a^{q-p}}$$

You already know that any number divided by itself equals 1. This leads us to the following property.

▶ For any directed number a, except 0, and for all positive integers p:

$$\frac{a^p}{a^p} = 1.$$

Now let's use these rules of exponents to divide.

EXAMPLE 5 $\quad \dfrac{15a^3b^4}{-3ab^2} = \dfrac{15}{-3} \cdot \dfrac{a^3}{a} \cdot \dfrac{b^4}{b^2} = -5a^{3-1}b^{4-2} = \mathbf{-5a^2b^2}$

EXAMPLE 6 $\quad \dfrac{-9abc}{3b^3c} = \dfrac{-9}{3} \cdot a \cdot \dfrac{b}{b^3} \cdot \dfrac{c}{c} = -3a \cdot \dfrac{1}{b^{3-1}} \cdot 1 = \dfrac{\mathbf{-3a}}{\mathbf{b^2}}$

Name the missing factor.

Sample 1: $p^{10} = p^7 \cdot \underline{\ ?\ }$ $\qquad$ *What you say:* p^3

Sample 2: $\dfrac{m^5}{m^4} = \dfrac{m^2}{m} \cdot \underline{\ ?\ }$ $\qquad$ *What you say:* $\dfrac{m^3}{m^3}$ or 1

1. $5^8 = 5^5 \cdot \underline{\ ?\ }$ $\ 5^3$ $\qquad$ **2.** $x^5 = x \cdot \underline{\ ?\ }$ $\ x^4$ $\qquad$ **3.** $a^3 \cdot \underline{\ ?\ } = a^7$ $\ a^4$

4. $\dfrac{n^4}{n^3} = \dfrac{n^3}{n^2} \cdot \underline{\ ?\ }$ $\ 1$ $\qquad$ **5.** $\dfrac{7^5}{7^2} = \dfrac{7}{7} \cdot \underline{\ ?\ }$ $\ 7^3$ $\qquad$ **6.** $\dfrac{b^3}{b^2} = \dfrac{b}{b} \cdot \underline{\ ?\ }$ $\ b$

Simplify.

7. $\dfrac{3x \cdot 3x \cdot 3y}{3 \cdot 3 \cdot y}$ $\ 3x^2$ $\qquad\qquad$ **8.** $\dfrac{m \cdot n \cdot m}{m \cdot m \cdot m \cdot m \cdot n \cdot n}$ $\ \dfrac{1}{m^2 n}$

9. $\dfrac{x \cdot y \cdot y \cdot y \cdot y}{x \cdot y \cdot y}$ $\ y^2$ $\qquad\qquad$ **10.** $\dfrac{3 \cdot 3 \cdot a \cdot a \cdot a}{3 \cdot a \cdot a}$ $\ 3a$

11. $\dfrac{2 \cdot 2 \cdot 3}{2 \cdot 2 \cdot 2 \cdot 2 \cdot 2 \cdot 3}$ $\ \dfrac{1}{8}$ $\qquad\qquad$ **12.** $\dfrac{x \cdot x \cdot y}{x \cdot x \cdot y \cdot y \cdot y}$ $\ \dfrac{1}{y^2}$

For Extra Practice, see page 429.

Written EXERCISES

Simplify. Assume no denominator is 0.

Sample 1: $\dfrac{15a^4b^3}{-3ab}$

Solution: $\dfrac{15a^4b^3}{-3ab} = \dfrac{15}{-3} \cdot \dfrac{a^4}{a} \cdot \dfrac{b^3}{b} = -5a^3b^2$

Sample 2: $\dfrac{-10xy}{-5x^5}$ Solution: $\dfrac{-10xy}{-5x^5} = \dfrac{-10}{-5} \cdot \dfrac{x}{x^5} \cdot y = \dfrac{2y}{x^4}$

A

1. $\dfrac{x^9}{x^5}$ x^4

2. $\dfrac{m^7}{m^2}$ m^5

3. $\dfrac{(rs)^6}{(rs)^{11}}$ $\dfrac{1}{(rs)^5}$

4. $\dfrac{8x^9}{4x^5}$ $2x^4$

5. $\dfrac{6m^7}{-12m^{12}}$ $-\dfrac{1}{2m^5}$

6. $\dfrac{-x^{12}}{x^4}$ $-x^8$

7. $\dfrac{3q^6}{15q^4}$ $\dfrac{q^2}{5}$

8. $\dfrac{-6b^6}{-2b^2}$ $3b^4$

9. $\dfrac{12z^2}{9x^2z^3}$ $\dfrac{4}{3zx^2}$

10. $\dfrac{42a^2b}{49ab^2}$ $\dfrac{6a}{7b}$

11. $\dfrac{30mn^3}{18m^2n^2}$ $\dfrac{5n}{3m}$

12. $\dfrac{34xy^3}{51x^2y}$ $\dfrac{2y^2}{3x}$

13. $\dfrac{6xy^2}{9x^2y}$ $\dfrac{2y}{3x}$

14. $\dfrac{3mn^2}{15m^2n^2}$ $\dfrac{1}{5m}$

15. $\dfrac{26a^2b^3}{39ab^5}$ $\dfrac{2a}{3b^2}$

B

16. $\dfrac{10x^4y^2}{-2xy}$ $-5x^3y$

17. $\dfrac{64x^2y^2z^2}{(-4xyz)^3}$ $-\dfrac{1}{xyz}$

18. $\dfrac{-25ab^2}{-10a^2b}$ $\dfrac{5b}{2a}$

19. $\dfrac{21xyz}{7x^2y^2z^2}$ $\dfrac{3}{xyz}$

20. $\dfrac{-x^7}{-2xy}$ $\dfrac{x^6}{2y}$

21. $\dfrac{-27x^2y^3}{-9xyz}$ $\dfrac{3xy^2}{z}$

22. $\dfrac{2ab^4}{(2b^2)^2}$ $\dfrac{a}{2}$

23. $\dfrac{30bd^3}{(3d^2)^2}$ $\dfrac{10b}{3d}$

24. $\dfrac{-24a^2b^5}{-8ab^3}$ $3ab^2$

25. $\dfrac{21x^7y^9}{(-x^2y^4)^2}$ $21x^3y$

26. $\dfrac{(-0.6m^4n^5)^2}{6m^3n}$

27. $\dfrac{14r^4s^2}{-7s^2}$ $-2r^4$

28. $\dfrac{35x^6y^4z^2}{5x^3y^3z}$ $7x^3yz$

29. $\dfrac{58mn^2p^3}{87m^4n^3p^2}$ $\dfrac{2p}{3m^3n}$

30. $\dfrac{38a^2b^3c^4}{57a^3b^2c^2}$ $\dfrac{2bc^2}{3a}$

C

31. $\dfrac{3(ab)^3}{2.7ab^2}$ $\dfrac{10a^2b}{9}$

32. $\dfrac{(-3)^4x^9y^2}{27x^8y^4}$ $\dfrac{3x}{y^2}$

33. $\dfrac{-0.8x^2y^7}{-0.56x^{12}y^3}$ $\dfrac{10y^4}{7x^{10}}$

34. $\dfrac{5m^p}{5m}, p > 1$ m^{p-1}

35. $\dfrac{(-2)^3x^py^q}{x^qy^p}, p > q$ $\dfrac{-8x^{p-q}}{y^{p-q}}$

36. $\dfrac{2^6x^py^{p-1}}{16x^2y^{p-2}}, p > 2$ $4x^{p-2}y$

37. $\dfrac{(-69ab)^r}{(23ab)^r}$ $(-3)^r$

26. $0.06m^5n^9$

13-9 *Dividing a Polynomial by a Monomial*

OBJECTIVE

Simplify expressions such as $\dfrac{18r^2 - 6s}{3r}$ by division.

Recall that division is distributive over addition. We use this fact to divide a polynomial by a monomial. We divide each term of the polynomial by the monomial, then add the quotients.

Example 1 shows two ways to divide.

EXAMPLE 1 $(6x^2 + 18x) \div 3$

$$(1) \quad 3\overline{)\begin{array}{l} 2x^2 + 6x \\ 6x^2 + 18x \end{array}}$$

$$(2) \quad \frac{6x^2 + 18x}{3} = \frac{6x^2}{3} + \frac{18x}{3} = 2x^2 + 6x$$

EXAMPLE 2 $\dfrac{21x^2 - 14y^2}{7} = \dfrac{21x^2}{7} - \dfrac{14y^2}{7} = 3x^2 - 2y^2$

EXAMPLE 3 $\dfrac{4c^3 + 20bc}{4c} = \dfrac{4c^3}{4c} + \dfrac{20bc}{4c}$ ◀ $c \neq 0$

$\qquad\qquad\qquad = c^2 + 5b$

EXAMPLE 4 $\dfrac{18r^2 - 6s}{3r} = \dfrac{18r^2}{3r} - \dfrac{6s}{3r}$ ◀ $r \neq 0$

$\qquad\qquad\qquad = 6r - \dfrac{2s}{r}$

EXAMPLE 5 $\dfrac{12ac^2 + 9a^2c}{3ac} = \dfrac{12ac^2}{3ac} + \dfrac{9a^2c}{3ac}$ ◀ $a \neq 0$
$\qquad\qquad\qquad\qquad\qquad\qquad\qquad\qquad$ ◀ $c \neq 0$

$\qquad\qquad\qquad = 4c + 3a$

EXERCISES

Match each expression in Column 1 with an equivalent expression in Column 2.

COLUMN 1

1. $\dfrac{ab^3}{b^2}$ **F**

2. $\dfrac{x^3a^2b}{x^3a^3b^2}$ **E**

3. $\dfrac{6a + 3b}{3}$ **D**

4. $\dfrac{14a}{3} + \dfrac{15}{3}b$ **A**

5. $\dfrac{b^2 + ab}{b}$ **B**

6. $\dfrac{63a + 5b}{9}$ **C**

COLUMN 2

A. $\dfrac{14a + 15b}{3}$

B. $\dfrac{b^2}{b} + \dfrac{ab}{b}$

C. $\dfrac{63a}{9} + \dfrac{5b}{9}$

D. $\dfrac{6a}{3} + \dfrac{3b}{3}$

E. $\dfrac{1}{ab}$

F. ab

Written EXERCISES

Simplify. Assume no denominator is 0.

Sample: $\dfrac{1}{7}(14x + 21y)$ *Solution:* $\dfrac{1}{7}(14x + 21y) = \dfrac{14x}{7} + \dfrac{21y}{7}$

$$= 2x + 3y$$

A

1. $\dfrac{1}{6}(24x - 18)$ $4x - 3$

2. $\dfrac{-2}{3}(18a + 9)$ $-12a - 6$

3. $\dfrac{1}{2}(-14b - 8)$ $-7b - 4$

4. $\dfrac{1}{r}(rs^2 - r^2)$ $s^2 - r$

5. $(-54x^2 - 108)\dfrac{1}{9}$ $6x^2 - 12$

6. $(ab^2 + b^3)\dfrac{1}{b^2}$ $a + b$

7. $\left(\dfrac{-1}{3}\right)(6a^2 + 6a + 9)$ $-2a^2 - 2a - 3$

8. $\left(\dfrac{-1}{6}\right)(36x^2 + 12x - 42)$ $-6x^2 - 2x + 7$

Divide. Assume no denominator is 0.

9. $\dfrac{6y - 12}{6}$ $y - 2$

10. $\dfrac{7m + 28}{14}$ $\dfrac{m}{2} + 2$

11. $\dfrac{7k + 63a}{7}$ $k + 9a$

12. $\dfrac{9p^2 + 6p}{3p}$ $3p + 2$

13. $\dfrac{16q + 72q^2}{8}$ $2q + 9q^2$

14. $\dfrac{3a^5 + 7a^7}{a^4}$ $3a + 7a^3$

15. $\dfrac{48b^2 + 64b^4}{16b^2}$ $3 + 4b^2$

16. $\dfrac{30x^6 + 7x^4}{3x^3}$ $10x^3 + \dfrac{7x}{3}$

17. $\dfrac{9m^3 + 21}{-3}$ $-3m^3 - 7$

18. $\dfrac{4y^2 + 6y}{2y}$ $2y + 3$

19. $\dfrac{4b^2 - 8b}{2b}$ $2b - 4$

20. $\dfrac{64k^3 - 24k}{-8k}$ $-8k^2 + 3$

21. $\dfrac{t^3 + s^2t^2 + t}{st}$ $\dfrac{t^2}{s} + st + \dfrac{1}{s}$

22. $\dfrac{9a^2b + 12ab - 15ab^2}{3ab}$ $3a + 4 - 5b$

B

23. $\dfrac{xy^2 - x^2y + x^3}{x}$ $y^2 - xy + x^2$

24. $\dfrac{18t^2 + 15t - 33}{3}$ $6t^2 + 5t - 11$

25. $\dfrac{14t^3 - 56t^2 - 35t}{7t}$ $2t^2 - 8t - 5$

26. $\dfrac{ax^5 - bx^4 + cx^3 - x^2}{-x}$ $-ax^4 + bx^3 - cx^2 + x$

27. $\dfrac{-s^3 + s^2 - 27s}{-s}$ $s^2 - s + 27$

28. $\dfrac{14a^3b^3 - 12a^2b^2 + 10ab}{2ab}$ $7a^2b^2 - 6ab + 5$

29. $\dfrac{mx^3 + nx^2 + ax^4}{x^2}$ $mx + n + ax^2$

30. $\dfrac{xyp + xyq}{xy}$ $p + q$

31. $\dfrac{4x^2y^4 + 8xy^3}{x^3y}$ $\dfrac{4y^3}{x} + \dfrac{8y^2}{x^2}$

32. $\dfrac{st^4 + 2s^3t^2 + s^2t^2}{s^2t^2}$ $\dfrac{t^2}{s} + 2s + 1$

ᴄᴏɴsᴜᴍᴇʀ ɴᴏᴛᴇs *Recycling*

Recycling refuse is one way consumers can fight pollution. Recycling is sometimes done with the aid of a conveyor belt. Large chunks of waste are placed on a belt where a machine shreds them. Some metal items are sorted out by a magnet. Paper and cloth are separated by a column of air. Glass is sorted into colors by an optical scanner. Aluminum is separated by an electrostatic field. The combustibles left, such as food, rubber, and rags can be burned as fuel. The noncombustibles can be reused by companies. Efforts are being made to encourage consumers to deposit waste in containers, lessening the cost of refuse collection and the cost of environmental pollution. Can you think of other ways to recycle waste?

13-10 *Dividing by a Binomial*

OBJECTIVE

Simplify expressions like $\dfrac{x^2 - 8x - 9}{x + 1}$ by division.

Division by a binomial is a process similar to "long division." Read each step of Example 1 carefully and be sure you follow it.

EXAMPLE 1 $\dfrac{x^2 + 7x + 12}{x + 3}$ ▶ $x + 3 \overline{)x^2 + 7x + 12}$

$$x + 3 \overline{)\overset{\textstyle x + 4}{x^2 + 7x + 12}}$$

Think: $x \cdot \underline{\ ?\ } = x^2$. ▶

Multiply $x(x + 3)$. ▶ $\quad x^2 + 3x$

Subtract $x^2 + 3$ from $x^2 + 7x + 12$. ▶ $\quad 4x + 12$

Think: $x \cdot \underline{\ ?\ } = 4x$. ▶ $\quad 4x + 12$

Multiply $4(x + 3)$ and subtract. ▶ $\qquad\quad 0$

Check: $(x + 3)(x + 4) = x^2 + 7x + 12$ ✓

EXAMPLE 2 $\dfrac{10m^2 - 19m - 15}{2m - 5} = \underline{\ ?\ }$ ▶ $2m - 5 \overline{)\overset{\textstyle 5m + 3}{10m^2 - 19m - 15}}$

$$\begin{array}{r} 10m^2 - 25m \\ \hline 6m - 15 \\ 6m - 15 \\ \hline 0 \end{array}$$

Check: $(2m - 5)(5m + 3) = 10m^2 - 19m - 15$ ✓

EXAMPLE 3 $\dfrac{4x^2 - 9}{2x - 3} = \underline{\ ?\ }$

$$2x - 3 \overline{)\overset{\textstyle 2x \qquad\quad +\, 3}{4x^2 \qquad\quad -\, 9}}$$ ◀ Leave space for $0 \cdot x$.

$$\begin{array}{r} 4x^2 - 6x \\ \hline 6x - 9 \\ 6x - 9 \\ \hline 0 \end{array}$$

Check: $(2x - 3)(2x + 3) = 4x^2 - 9$

Divide and check.

1. $\dfrac{x^2 + 7x + 12}{x + 4}$ $x + 3$

2. $\dfrac{a^2 + 5a + 6}{a + 2}$ $a + 3$

3. $\dfrac{b^2 - 5b + 6}{b - 3}$ $b - 2$

4. $\dfrac{t^2 + 3t - 10}{t - 2}$ $t + 5$

5. $\dfrac{y^2 + 3y - 10}{y + 5}$ $y - 2$

6. $\dfrac{m^2 + 12m + 27}{m + 9}$ $m + 3$

7. $\dfrac{m^2 + 8m + 15}{m + 3}$ $m + 5$

8. $\dfrac{n^2 - 10n + 24}{n - 4}$ $n - 6$

9. $\dfrac{p^2 + 7p - 18}{p - 2}$ $p + q$

10. $\dfrac{4a^2 + 12a + 9}{2a + 3}$ $2a + 3$

11. $\dfrac{3x^2 - 4x - 4}{x - 2}$ $3x + 2$

12. $\dfrac{x^2 - 8x + 15}{x - 3}$ $x - 5$

13. $\dfrac{6c^2 - 12c + 6}{2c - 2}$ $3c - 3$

14. $\dfrac{9z^2 + 24z + 16}{3z + 4}$ $3z + 4$

15. $\dfrac{a^2 - 2ab + b^2}{a - b}$ $a - b$

16. $\dfrac{s^2 + 2st + t^2}{s + t}$ $s + t$

17. $\dfrac{4b^2 + 4bc - 3c^2}{2b + 3c}$ $2b - c$

18. $\dfrac{8m^2 - 10am - 3a^2}{4m + a}$ $2m - 3a$

19. $\dfrac{s^2 - 4}{s - 2}$ $s + 2$

20. $\dfrac{p^4 - 16}{p^2 + 4}$ $p^2 - 4$

21. $\dfrac{6t^2 - 6}{2t - 2}$ $3t + 3$

22. $\dfrac{y^4 + y^2 - 2}{y^2 - 1}$ $y^2 + 2$

Divide. First rewrite the polynomial in standard form if necessary.

23. $\dfrac{6a^2 + 2 + 7a}{3a + 2}$ $2a + 1$

24. $\dfrac{12k^2 - 32k + 5}{6k - 1}$ $2k - 5$

25. $\dfrac{12b^2 - 1 - b}{4b + 1}$ $3b - 1$

26. $\dfrac{10a^2 + 48a + 54}{5a + 9}$ $2a + 6$

27. $\dfrac{x^2 - 4ax - 5a^2}{x - 5a}$ $x + a$

28. $\dfrac{y^2 + 14by + 45b^2}{9b + y}$ $5b + y$

29. $\dfrac{15a^2 - 8b^2 - 2ab}{3a + 2b}$ $5a - 4b$

30. $\dfrac{20p^2 - 43pq + 21q^2}{-3q + 4p}$ $5p - 7q$

Divide.

C

31. $\dfrac{y^3 - y^2 + 2y - 2}{y - 1}$ $y^2 + 2$

32. $\dfrac{x^3 + 5x^2 + 2x + 10}{x + 5}$ $x^2 + 2$

33. $\dfrac{3t^3 + 6t^2 - t - 2}{3t^2 - 1}$ $t + 2$

34. $\dfrac{t^3 - 6t^2 + 12t - 8}{t - 2}$ $t^2 - 4t + 4$

35. $\dfrac{3b^2 + b - 2}{b + 1}$ $3b - 2$

36. $\dfrac{m^3 + 2m^2 - 2m - 4}{m + 2}$ $m^2 - 2$

37. $\dfrac{x^3 - x^2 + x - 1}{x - 1}$ $x^2 + 1$

38. $\dfrac{2r^2t^2 + r^3t + 2t^2 + rt}{2t + r}$ $t + tr^2$

SELF-TEST 3

Section 13-8, p. 368

Simplify. Assume no denominator is 0.

1. $\dfrac{x^6}{x^2}$ x^4

2. $\dfrac{2mn^2}{m}$ $2n^2$

3. $\dfrac{-3zx}{z^2x^3}$ $-\dfrac{3}{zx^2}$

4. $\dfrac{15xy}{3x^2y}$ $\dfrac{5}{x}$

Section 13-9, p. 371

Divide. Assume no denominator is 0.

5. $\dfrac{4z + 8}{2}$ $2z + 4$

6. $\dfrac{9b^2 - 18}{3b}$ $3b - \dfrac{6}{b}$

7. $\dfrac{-8n^2 + 4n}{2n}$ $-4n + 2$

8. $\dfrac{c^2d + cd^2}{cd}$ $c + d$

Section 13-10, p. 374

9. $\dfrac{m^2 + 6m + 9}{m + 3}$ $m + 3$

10. $\dfrac{4n^2 - 4n - 15}{2n + 3}$ $2n - 5$

Check your answers with those printed at the back of the book.

chapter summary

1. **Exponents** are used to simplify multiplication expressions that contain repeating factors.

2. For all directed numbers a and b, and positive integers p and q:

$$a^p \cdot a^q = a^{p+q}$$
$$(ab)^p = a^p \cdot b^p$$
$$(a^p)^q = a^{pq}$$

3. For all directed numbers a and b:

$$(a + b)^2 = a^2 + 2ab + b^2$$
$$(a - b)(a + b) = a^2 - b^2$$

4. Multiplication of polynomials is **commutative** and **associative.** Also, multiplication of polynomials is distributive over addition.

5. The product of a polynomial and 1 is that polynomial. 1 is the **identity element** for multiplication of polynomials.

6. An expression is the **multiplicative inverse (reciprocal)** of a polynomial if its product with the polynomial is 1.

7. For all directed numbers a and b, and positive integers p and q.

$$\text{If } p > q: \frac{a^p}{a^q} = a^{p-q}$$

$$\text{If } p < q: \frac{a^p}{a^q} = \frac{1}{a^{q-p}}$$

$$\text{If } p = q: \frac{a^p}{a^q} = 1$$

8. To divide a polynomial by a monomial we divide each term of the polynomial by the monomial, then add the quotients.

9. To divide a polynomial by a binomial, we use a process similar to long division.

challenge topics

Know Your Angles

Use heavy paper or cardboard to make a 30–60–90 triangle and a 45–90–45 triangle.

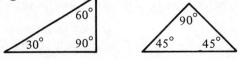

By using both triangles at the same time you can draw many figures quickly and accurately.

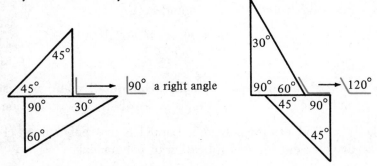

Use the triangles to duplicate each figure.

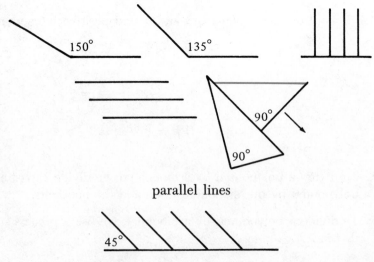

parallel lines

parallel lines intersecting a line at 45°

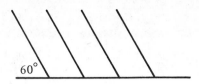

parallel lines intersecting a line at 60°

chapter test

Simplify. Use exponents.

1. $x^4 \cdot x^3$ x^7

2. $(2x \cdot x \cdot x)(3 \cdot y \cdot y)$ $6x^3y^2$

3. $a^3(-2a^2b^2)(5b)$ $-10a^5b^3$

4. $(2xz)^3$ $8x^3z^3$

5. $(2mn)^2$ $4m^2n^2$

Multiply.

6. $-n(n^2 - 2n + 3)$ $-n^3 + 2n^2 - 3n$

7. $pq(p^2 - 2pq + 3)$ $p^3q - 2p^2q^2 + 3pq$

8. $5x(xy^2 + x^2y)$ $5x^2y^2 + 5x^3y$

9. $2a(3a + 4) + 3a(2a + 3)$ $12a^2 + 17a$

10. $10x(x + 6)$ $10x^2 + 60x$

11. $(t - 5)^2$ $t^2 - 10t + 25$

12. $(5x + 2)(5x - 2)$ $25x^2 - 4$

13. $-3a(a^2 + 5a - 3)$ $-3a^3 - 15a^2 + 9a$

14. $(s - 4)(2s + 3)$ $2s - 5s - 12$

15. $(x - 4)(2x^2 - 9x + 1)$
    $2x^3 - 17x^2 + 37x - 4$

Express the answer as a polynomial in simplest form.

16. A rectangle has length $2y - 1$ and width $y + 4$. Find the area. $2y^2 + 7y - 4$

17. A square has sides of length $4z + 2$. Find the area. $16z^2 + 16z + 4$

Divide. Assume no denominator is 0.

18. $\dfrac{z^5}{z^2}$ z^3

19. $\dfrac{-4x^2z}{xz}$ $-4x$

20. $\dfrac{15a^2 + 10a}{5a}$ $3a + 2$

21. $\dfrac{21y^2 - 12y + 9}{3y}$ $7y - 4 + \dfrac{3}{y}$

22. $\dfrac{m^2 - 8m + 16}{m - 4}$ $m - 4$

23. $\dfrac{a^2 - 7a + 10}{a - 5}$ $a - 2$

Review of skills

Name the factors and the greatest common factor.

Sample: 12 and 14   *Solution:* 12: 1, 2, 3, 4, 6, and 12
14: 1, 2, 7, and 14
GCF of 12 and 14 is 2.

1. 6 and 36 GCF: 6   **2.** 4 and 8 GCF: 4   **3.** 4 and 19 GCF: 1

4. 3 and 8 GCF: 1   **5.** 10 and 15 GCF: 5   **6.** 12 and 16 GCF: 4

7. 14 and 35 GCF: 7   **8.** 5 and 11 GCF: 1   **9.** 7 and 28 GCF: 7

Tell whether or not the number is prime.

10. 8 not prime   **11.** 13 prime   **12.** 19 prime

13. 35 not prime   **14.** 43 prime   **15.** 27 not prime

Complete.

Sample: $28 = 2 \cdot 2 \cdot \underline{\ ?\ }$ *Solution:* $28 = 2 \cdot 2 \cdot 7$

16. $35 = 5 \cdot \underline{\ ?\ }$ 7   **17.** $80 = 2 \cdot 2 \cdot 2 \cdot 2 \cdot \underline{\ ?\ }$ 5

18. $106 = \underline{\ ?\ } \cdot 53$ 2   **19.** $50 = \underline{\ ?\ } \cdot 5 \cdot 5$ 2

20. $141 = 3 \cdot \underline{\ ?\ }$ 47   **21.** $27 = 3 \cdot 3 \cdot \underline{\ ?\ }$ 3

Write as the product of prime numbers.

Sample: 45   *Solution:* $45 = 3 \cdot 3 \cdot 5$

22. 42 $2 \cdot 3 \cdot 7$ **23.** 55 $5 \cdot 11$ **24.** 58 $2 \cdot 29$

25. 121 $11 \cdot 11$ **26.** 144 $2 \cdot 2 \cdot 2 \cdot 2 \cdot 3 \cdot 3$ **27.** 100 $2 \cdot 2 \cdot 5 \cdot 5$

28. 142 $2 \cdot 71$ **29.** 120 $2 \cdot 2 \cdot 2 \cdot 3 \cdot 5$ **30.** 270 $2 \cdot 3 \cdot 3 \cdot 3 \cdot 5$

Tell whether the statement is *always* true, *never* true, or *sometimes* true.

31. The GCF of two different prime numbers is 1. always

32. The GCF of two numbers which are not prime is 1. sometimes

33. The product of two prime numbers is also prime. never

34. The GCF of two consecutive counting numbers is 1. always

Left: Uniformed security guards, 1900.

Right: Central security office at convention center in Chicago.

14 Products and Factoring

Factoring

14-1 *Factoring Whole Numbers*

OBJECTIVE
Write the prime factorization of a composite number.

A positive number which has exactly two different positive factors, itself and 1, is called a **prime** number. A number which can be expressed as the product of two or more prime numbers is called a **composite** number. 1 is neither a prime number nor a composite number.

EXAMPLE 1 15 is a composite number. $15 = 3 \cdot 5$
primes

A composite number written as the product of primes is in **prime factorization** form.

EXAMPLE 2 The prime factorization of 15 is $3 \cdot 5$.

EXAMPLE 3 $42 = 2 \cdot 3 \cdot 7$
prime factorization of 42

EXAMPLE 4 $36 = 2 \cdot 2 \cdot 3 \cdot 3 = 2^2 \cdot 3^2$
prime factorization of 36

 **Oral EXERCISES** | Tell whether the number is prime or composite.

1. 12 composite
2. 17 prime
3. 16 composite
4. 7 prime
5. 18 composite
6. 5 prime
7. 10 composite
8. 19 prime
9. 11 prime

Name all the factors. Then express the number as the product of primes. ce = cannot express as product of primes

Sample: 6   *Solution:*   The factors are 1, 2, 3, and 6.
$$6 = 2 \cdot 3$$

1. 5 1; 5; ce   **2.** 9 1, 3, 9; 3 · 3   **3.** 4 1, 2, 4; 2 · 2   **A**

4. 17 1, 17; ce   **5.** 21 1, 3, 7, 21; 3 · 7   **6.** 35 1, 5, 7, 35; 5 · 7

7. 45 1, 3, 5, 9, 15, 45; 3 · 3 · 5   **8.** 29 1, 29; ce   **9.** 16 1, 2, 4, 8, 16; 2 · 2 · 2 · 2

10. 12 1, 2, 3, 4, 6, 12; 2 · 2 · 3   **11.** 30 1, 2, 3, 5, 6, 10, 15, 30; 2 · 3 · 5   **12.** 49 1, 7, 49; 7 · 7

Complete the prime factorization. **13.** 5; $2^2 \cdot 3 \cdot 5$

Sample: $28 = 2 \cdot 2 \cdot \underline{\ ?\ } = \underline{\ ?\ }$ *Solution:* $28 = 2 \cdot 2 \cdot 7 = 2^2 \cdot 7$

13. $60 = 2 \cdot 2 \cdot 3 \cdot \underline{\ ?\ } = \underline{\ ?\ }$ **14.** $20 = 2 \cdot \underline{\ ?\ } \cdot \underline{\ ?\ } = \underline{\ ?\ }$ 2, 5; $2^2 \cdot 5$

15. $100 = 2 \cdot 2 \cdot \underline{\ ?\ } \cdot \underline{\ ?\ } = \underline{\ ?\ }$ **16.** $52 = 2 \cdot \underline{\ ?\ } \cdot \underline{\ ?\ } = \underline{\ ?\ }$ 2, 13; $2^2 \cdot 13$

17. $45 = 3 \cdot 3 \cdot \underline{\ ?\ } = \underline{\ ?\ }$ 5; $3^2 \cdot 5$ **18.** $68 = 2 \cdot \underline{\ ?\ } \cdot \underline{\ ?\ } = \underline{\ ?\ }$ 2 · 17; $2^2 \cdot 17$

19. $141 = 3 \cdot \underline{\ ?\ }$ 47   **20.** $88 = \underline{\ ?\ } \cdot 11$ 2^3

15. 5, 5; $2^2 \cdot 5^2$

Write the prime factorization.

21. 125 5^3 **22.** 235 5 · 47   **23.** 144 $2^4 \cdot 3^2$ **B**

24. 42 2 · 3 · 7   **25.** 270 2 · 3^3 · 5   **26.** 625 5^4

27. 1000 $2^3 \cdot 5^3$ **28.** 243 3^5 **29.** 120 $2^3 \cdot 3 \cdot 5$

30. 111 3 · 37   **31.** 194 2 · 97   **32.** 57 3 · 19

calculator corner

Here is a calculator race that you can have with one of your friends. The first person to add the whole numbers from 1 to 50 with the calculator wins. One of you can simply add while the other can use a trick from algebra. Add the first and last numbers (1 and 50) and multiply the answer by half of the last number (25). Who won?

You may want to use this method to add the whole numbers from 1 to 1000. Can you see why this method works?

14-2 *Factoring Monomials*

OBJECTIVES

Factor a monomial.

Find the GCF of two monomials.

The monomial $8a^2$ can be written as a product in several ways. A few are shown below. Each is called a **factorization**.

$$2a \cdot 2a \cdot 2 \qquad 8 \cdot a \cdot a \qquad (-2)(-4)a \cdot a \qquad 2 \cdot 2 \cdot 2 \cdot a \cdot a$$

We call $2 \cdot 2 \cdot 2 \cdot a \cdot a$ the **complete factorization** of $8a^2$. None of the factors can be factored any further. We **factor** a monomial by finding its complete factorization.

EXAMPLE 1 $10mn^2 = 2 \cdot 5 \cdot m \cdot n \cdot n$

complete factorization

To find the greatest common factor (GCF) of two monomials, compare their complete factorizations.

EXAMPLE 2 $10mn^2$ and $15mn$

$$10mn^2 = 2 \cdot \boxed{5 \cdot m \cdot n} \cdot n$$
$$15mn = 3 \cdot \boxed{5 \cdot m \cdot n}$$

The GCF of $10mn^2$ and $15mn$ is $5 \cdot m \cdot n = 5mn$.

EXAMPLE 3 $4r^2s$ and $-10s$

$$4r^2s = \quad 2 \cdot \boxed{2} \cdot r \cdot r \cdot \boxed{s}$$
$$-10s = -1 \cdot \boxed{2} \cdot 5 \cdot \quad \boxed{s} \quad \blacktriangleleft \text{Use } -1 \text{ as the first factor.}$$

The GCF of $4r^2s$ and $-10s$ is $2s$.

EXAMPLE 4 $8x^3y$ and $6x^2y$

$$8x^3y = \boxed{2} \cdot 2 \cdot 2 \cdot \boxed{x \cdot x} \cdot x \cdot \boxed{y}$$
$$6x^2y = \boxed{2} \cdot 3 \cdot \quad \boxed{x \cdot x} \cdot \quad \boxed{y}$$

The GCF of $8x^3y$ and $6x^2y$ is $2x^2y$.

True or false?

Sample: $4m^2$ is a factor of $12m^2n$.   *What you say:*   True

1. $3w$ is a factor of $21w^2$. true
2. $5t$ is a factor of $5t^2$. true
3. $6y$ is a factor of $12x$. false
4. bc is a factor of $3bc$. true
5. x^2y^2 is a divisor of $-7x^2y^3$. true
6. $-2m$ is a divisor of $-10mn$. true
7. $-3a$ is a divisor of $9b^2$. false
8. $-8y$ is a divisor of $16x^2y$. true
9. $3b$ is a common factor of $15ab$ and $7a^2b$. false
10. $2n$ is a common factor of $8m$ and $6mn$. false
11. x^2 is a common factor of $3x^2y$ and $2x^2$. true
12. $2k^3$ is a common factor of $24k^3r$ and $-4k^3t$. true

Complete.

Sample: $21x^3y = 3x^3(\underline{\ ?\ })$ *Solution:* $21x^3y = 3x^3(7y)$

1. $26y^2 = 2y(\underline{\ ?\ })$ 13y
2. $24x^4 = 3x(\underline{\ ?\ })$ 8x³
3. $12mn = 6(\underline{\ ?\ })$ 2mn
4. $25s^2t = 5t(\underline{\ ?\ })$ 5s²
5. $15b^5c^5 = 5b^5(\underline{\ ?\ })$ 3c⁵
6. $30x^2y^2 = -6xy(\underline{\ ?\ })$ −5xy
7. $-15r^3s^5 = -3rs(\underline{\ ?\ })$ 5r²s⁴
8. $-20xz^2 = 2(2)(\underline{\ ?\ })(xz^2)$ −5
9. $36a^2b^3c^4 = -4abc(\underline{\ ?\ })$ −9ab²c³
10. $-51rst^2 = -3rt(\underline{\ ?\ })$ 17st

Write two different factorizations.

Sample: $-18ax^2$ *Solution:* $-9 \cdot 2ax^2$; $3a(-6x^2)$
(Other answers are possible.)

11. $16t$ 4 · 4t; 2 · 8t
12. $30x$ 3 · 10x; (5)(6)(x)
13. $12mn$ 2m(6n); (3m)(4n)
14. $-20rs$ −2 · 10rs; 4r(−5s)
15. $-18r^2s^2$ −9rs · 2rs; −6r · 3rs²
16. $50ab^2$ 2a · 5²b²; 2ab · 25b
17. $5cd^2$ (−5c)(−d²); 5d · cd
18. $7m^2n^2$ 7m · mn²; −7mn(−mn)
19. $-3xyz$ (−x)(3yz); −3z · xy

Write the complete factorization of each monomial. Then name the GCF.

Sample: $4a^2$ and $6ab^2$ *Solution:* $4a^2 = 2 \cdot 2 \cdot a \cdot a$
$$6ab^2 = 2 \cdot 3 \cdot a \cdot b \cdot b$$
The GCF is $2a$.

20. $2r^3$ and $8r^2$ GCF: $2r^2$ **21.** $5ab$ and $10ab$ GCF: $5ab$

22. $3xy^2$ and $12x^2y$ GCF: $3xy$ **23.** $14m^2n$ and $7mn$ GCF: $7mn$

24. $12s^2$ and $60st$ GCF: $12s$ **25.** $18w^2$ and $-3w$ GCF: $3w$

26. $2x^2$ and $6xy^4z$ GCF: $2x$ **27.** 15 and $3mn$ GCF: 3

Name the GCF.

Sample 1: $12r^3$ and $36rs^2t$ *Solution:* $12r$

Sample 2: $4c$, $2ac$, and $3c^2$ *Solution:* c

B

28. $8r^2$ and $16r$ $8r$ **29.** $6x^2y$ and $-24y$ $6y$

30. $3k^2$ and 10 1 **31.** $15x^3$ and $5y^3$ 5

32. $3x^2$ and $-6x$ $3x$ **33.** $2b$, $3b^2$, $2ab$ b

34. $14m^2$, $6n$, and 8 2 **35.** $4x^2$, $-8x$, and x^3 x

Write each monomial as a product whose first factor is the GCF of the monomials listed.

Sample: $8m^2n^2$ and $14mn^2$ *Solution:* $8m^2n^2 = (2mn^2)(4m)$
$$14mn^2 = (2mn^2)(7)$$

C

36. $6axy$ and $2xy^2$ $2xy(3a)$; $2xy(y)$ **37.** $7r^2s$ and $-28r^4s^2$

38. $13x^5y^3$ and x^2y^2 **39.** $4rst$ and $-2x$

40. $15x^2$, $-10xy$, and $5y$ **41.** $5s^3$, rs^2, and $5r$
$5(3x^2)$; $5(-2xy)$; $5(y)$ $1 \cdot 5s^3$; $1 \cdot rs^2$; $1 \cdot 5r$

37. $7r^2s(1)$; $7r^2s(-4r^2s)$

38. $x^2y^2(13x^3y)$; $x^2y^2(1)$

39. $2(2rst)$; $2(-x)$

Time out

Andrea and Brian are sister and brother. Andrea is twice as old as Brian was six years ago. Brian's age is one-twelfth Andrea's height in centimeters. Their street number is twice their combined ages and ten less than Brian's weight. The sum of their parents' ages is five times Brian's age. Brian is 14. How old is Andrea?
(*Hint:* Read the problem very carefully. Decide how much of the information you really need.) 16

14-3 *Factoring Polynomials*

OBJECTIVE

Factor polynomials like $6n^2 + 4$ and $16m^3 + 40m^2 - 24m$.

To factor a polynomial, we first identify the greatest common factor of its terms. Then we use the distributive property to "factor out" the GCF.

EXAMPLE 1 $6n^2 + 4$

$$6n^2 = 2 \cdot 3 \cdot n \cdot n \qquad 4 = 2 \cdot 2$$

GCF of $6n^2$ and 4 is 2.

$$6n^2 + 4 = 2(3n^2) + 2 \cdot 2 = 2(3n^2 + 2)$$

EXAMPLE 2 $12x^2 - 15xy$ ◀ GCF is $3x$.

$$12x^2 - 15xy = 3x(4x) - 3x(5y)$$
$$= 3x(4x - 5y)$$

EXAMPLE 3 $16m^3 + 40m^2 - 24m$ ◀ GCF is $8m$.

$$16m^3 + 40m^2 - 24m = 8m(2m^2) + 8m(5m) + 8m(-3)$$
$$= 8m(2m^2 + 5m - 3)$$

It is always a good idea to check the answer by multiplying the factors.

EXAMPLE 4 $7y^2 - 21y = 7y(y) - 7y(3)$
$$= 7y(y - 3)$$
Check: $7y(y - 3) = 7y^2 - 21y$ ✓

Name the GCF of the terms.

Sample 1: $2ab + 8bc$ *What you say:* GCF is $2b$.

Sample 2: $16n^2 - 5$ *What you say:* GCF is 1.

1. $12x + 3$ 3

2. $5a^2b^2 + 2a^2$ a^2

3. $6n^2 + 12$ 6

4. $7mn + 10m^2$ m

5. $24t^2 - 8t$ $8t$

6. $12y^2 - 3z^2$ 3

7. $3s^2 + 15$ 3

8. $3a^2 + 18c^2$ 3

9. $3w^2 + 10$ 1

10. $b^4c^3 + b^2c^2$ b^2c^2

11. $5t^2 - 35r$ 5

12. $x^2 + 7$ 1

Oral EXERCISES

Complete.

Sample: $21x^2 - 3y^2 = 3(\underline{\ ?\ })$

Solution: $21x^2 - 3y^2 = 3(7x^2 - y^2)$

4. $2m + 3n$ **6.** $1 + 2xy$

A

1. $5k^2 + 15 = 5(\underline{\ ?\ })$ $k^2 + 3$
2. $3b^2 + 21b = 3b(\underline{\ ?\ })$ $b + 7$
3. $28r^2 + 16r = 4r(\underline{\ ?\ })$ $7r + 4$
4. $24m^3 + 36m^2n = 12m^2(\underline{\ ?\ })$
5. $45b - 9b^2 = 9b(\underline{\ ?\ })$ $5 - b$
6. $15xy + 30x^2y^2 = 15xy(\underline{\ ?\ })$
7. $10y^2 + 15y = 5y(\underline{\ ?\ })$ $2y + 3$
8. $3a^3b^3 + 12b^2 = 3b^2(\underline{\ ?\ })$
   $a^3b + 4$

Factor and check.

Sample: $4x^2 - 10xy$ Solution: $2x(2x - 5y)$

Check: $2x(2x - 5y) = 4x^2 - 10xy$ √

9. $5m - 15$ $5(m - 3)$
10. $8t^2 + 12$ $4(2t^2 + 3)$
11. $6a^2 + 7a$ $a(6a + 7)$
12. $10ab + 30a^2$ $10a(b + 3a)$
13. $ab^2 - a^2b$ $ab(b - a)$
14. $3x^3 - 15x$ $3x(x^2 - 5)$
15. $20a^2b + 3ab^2$ $ab(20a + 3b)$
16. $8x^2 + 4x$ $4x(2x + 1)$
17. $9xyz - xy$ $xy(9z - 1)$
18. $3rs^2 + 12r^2s^2$ $3rs^2(1 + 4r)$
19. $az^2 - 12awz$ $az(z - 12w)$
20. $14bxy + 49by^2$ $7by(2x + 7y)$
21. $3y^2 - 3y$ $3y(y - 1)$
22. $by^2 + bx^2$ $b(y^2 + x^2)$
23. $50mn - 33n^2$ $n(50m - 33n)$

Complete.

24. $aw^2 - 12awz + 36aw = aw(\underline{\ ?\ })$ $aw(w - 12z + 36)$
25. $6m^3 + 3m^2 - 3m = 3m(\underline{\ ?\ })$ $3m(2m^2 + m - 1)$
26. $14 - 7z - 21z^2 = 7(\underline{\ ?\ })$ $7(2 - z - 3z^2)$
27. $t^3 + t - t^2 = t(\underline{\ ?\ })$ $t(t^2 + 1 - t)$
28. $4b^3 - 4b^2 - 4b = 4b(\underline{\ ?\ })$ $4b(b^2 - b - 1)$
29. $21y^2 + 42x^2y^2 - 14xy^2 = 7y^2(\underline{\ ?\ })$ $7y^2(3 + 6x^2 - 2x)$

Factor.

B

30. $15y + 25y^2 - 20$ $5(3y + 5y^2 - 4)$
31. $3x^3 - x^2 + 5x$ $x(3x^2 - x + 5)$
32. $2x^2 - 6x - 4$ $2(x^2 - 3x - 2)$
33. $4t^2 - 28t + 28$ $4(t^2 - 7t + 7)$
34. $20y^2 + 43xy - 14y$ $y(20y + 43x - 14)$
35. $x^3 + x^2 - x$ $x(x^2 + x - 1)$
36. $y^3 - y^2 + y$ $y(y^2 - y + 1)$
37. $6v^3 + 26v^2 + 8v$ $2v(3v^2 + 13v + 4)$
38. $x^2y + 2bx - x^2$ $x(xy + 2b - x)$
39. $3m^4 - 57m^3 + 111m^2$ $3m^2(m^2 - 19m + 37)$

14-4 *Factoring Polynomials by Grouping Terms*

OBJECTIVE
Factor polynomials like $ac + ad + 4bc + 4bd$.

Suppose all the terms of a polynomial do not have a common factor. We may be able to group pairs of terms to express the polynomial as one which can be factored by using the distributive property.

EXAMPLE 1 $3rs + 15st + 2r^2 + 10rt$
$= (3rs + 15st) + (2r^2 + 10rt)$ ◀ Group terms.
$= [3s(r + 5t)] + [2r(r + 5t)]$ ◀ Factor in pairs.
$= (3s + 2r)(r + 5t)$ ◀ $(r + 5t)$ is a common factor.

EXAMPLE 2 $xy - 2x + 3y - 6$
$= (xy - 2x) + (3y - 6)$ ◀ Group terms.
$= [x(y - 2)] + [3(y - 2)]$ ◀ Factor in pairs.
$= (x + 3)(y - 2)$ ◀ $(y - 2)$ is a common factor.

EXAMPLE 3 $ac + ad + 4bc + 4bd$
$= (ac + ad) + (4bc + 4bd)$ ◀ Group terms.
$= [a(c + d)] + [4b(c + d)]$ ◀ Factor in pairs.
$= (a + 4b)(c + d)$ ◀ $(c + d)$ is a common factor.

EXAMPLE 4 $mn + 6mt + 6t + n$
$= mn + n + 6mt + 6t$ ◀ Rearrange terms.
$= (mn + n) + (6mt + 6t)$ ◀ Group terms.
$= [n(m + 1)] + [6t(m + 1)]$ ◀ Factor in pairs.
$= (n + 6t)(m + 1)$ ◀ $(m + 1)$ is a common factor.

There is often more than one way to rearrange and group terms. Compare Example 4 and Example 5.

EXAMPLE 5 $mn + 6mt + 6t + n$
$= 6mt + mn + 6t + n$ ◀ Rearrange terms.
$= (6mt + mn) + (6t + n)$ ◀ Group terms.
$= m(6t + n) + 1(6t + n)$ ◀ Factor in pairs.
$= (m + 1)(6t + n)$ ◀ $6t + n$ is a common factor.

Oral EXERCISES

Simplify.

Sample: $m(m + 3) + 6n(m + 3)$

What you say: $m^2 + 3m + 6mn + 18n$

$k^2 + 2k + tk + 2t$

$2rs + 4rt - 4s - 8t$

$3xy + 6x + 5y + 10$

1. $k(k + 2) + t(k + 2)$
2. $3x(y + 2) + 5(y + 2)$
3. $2r(s + 2t) - 4(s + 2t)$
4. $2t(1 + 2s) + s(1 + 2s)$
5. $2a(b + 2) + 5(b + 2)$
   $2ab + 4a + 5b + 10$
6. $n^2(n - 2) + 3(n - 2)$
   $n^3 - 2n^2 + 3n - b$

4. $2t + 4st + s + 2s^2$

Written EXERCISES

A

Write the factored form.

Sample: $5y(x + 2) - 4(x + 2)$ Solution: $(5y - 4)(x + 2)$

$(x - w)(y + 3)$

$(10a + c)(2 + b)$

$(5 - a)(b + c)$

$(x^2 - 5)(y + 7)$

1. $3s(r + 4) + t(r + 4)$ $(3s + t)(r + 4)$
2. $5b(c - 1) + 2a(c - 1)$
3. $x(y + 3) - w(y + 3)$
4. $2p(r + 7) - 4q(r + 7)$
5. $10a(2 + b) + c(2 + b)$
6. $4m(n + 3) - r(n + 3)$
7. $5(b + c) - a(b + c)$
8. $w(y - 5) - 3(y - 5)$
9. $x^2(y + 7) - 5(y + 7)$
10. $(4 - x)3t + (4 - x)r$
    $(4 - x)(3t + r)$

2. $(5b + 2a)(c - 1)$
4. $(2p - 4q)(r + 7)$
6. $(4m - r)(n + 3)$
8. $(w - 3)(y - 5)$

Factor.

Sample: $(4rt + 5st) + (4rx + 5sx)$

Solution: $(4rt + 5st) + (4rx + 5sx)$
$= t(4r + 5s) + x(4r + 5s)$
$= (t + x)(4r + 5s)$

$(2t + r)(s + 4)$

$(m + r)(n - 3)$

$(n - t)(m + 3)$

11. $(2st + 8t) + (rs + 4r)$
12. $(2ab + 2a) + (bc + c)$
13. $(mn - 3m) + (nr - 3r)$
14. $(5xy + 10x) + (3y + 6)$
15. $(mn + 3n) - (mt + 3t)$
16. $(2x + 2y) - (mx + my)$
17. $(r^2s - 2r^2) + (st^2 - 2t^2)$
    $(r^2 + t^2)(s - 2)$
18. $(b^2c - 2b^2) - (cd - 2d)$
    $(b^2 - d)(c - 2)$

12. $(2a + c)(b + 1)$
14. $(5x + 3)(y + 2)$
16. $(2 - m)(x + y)$

Factor.

Sample: $2b^2 + 2c - 4bc - b$

Two possible solutions are given.

Solution 1: $2b^2 + 2c - 4bc - b$
$= (2b^2 - b) + (-4bc + 2c)$
$= b(2b - 1) - 2c(2b - 1)$
$= (b - 2c)(2b - 1)$

Solution 2: $2b^2 + 2c - 4bc - b$
$$= (2b^2 - 4bc) - 1(b - 2c)$$
$$= 2b(b - 2c) - 1(b - 2c)$$
$$= (2b - 1)(b - 2c)$$

19. $(q + 5s)(2p - 3r)$
20. $(5y + k)(x - 4)$

19. $2pq - 3qr + 10ps - 15rs$

20. $5xy + xk - 4k - 20y$ **B**

21. $ac - 2d + cd - 2a\ (c - 2)(a + d)$

22. $2r - 3st - rt + 6s$ $(2 - t)(r + 3s)$

23. $mn^2 + 3n^2 - 3m - 9$
    $(n^2 - 3)(m + 3)$

24. $ms - mt - 3s + 3t$ $(m - 3)(s - t)$

SELF-TEST 1

Be sure you understand these terms.

composite (p. 382)   prime factorization (p. 382)
complete factorization (p. 384)   factor (p. 384)

Write the prime factorization.   **Section 14-1, p. 382**

1. 8 $2 \cdot 2 \cdot 2$ **2.** 18 $2 \cdot 3 \cdot 3$ **3.** 42 $2 \cdot 3 \cdot 7$

Write the complete factorization.   **Section 14-2, p. 384**

4. $35mn$ $5 \cdot 7 \cdot m \cdot n$ **5.** $-12x^2y$ **6.** $10a^2b^3$ $2 \cdot 5 \cdot a \cdot a \cdot b \cdot b \cdot b$
                        $2 \cdot 2 \cdot 3 \cdot x \cdot x \cdot y$

Name the GCF.

7. $8a^2b$ and $4ab$ $4ab$ **8.** $9xy^3$ and $-3x^2y$ $3xy$

Complete.   **Section 14-3, p. 387**

9. $7w^2 + 14 = 7(\underline{\ ?\ })$ $w^2 + 2$ **10.** $20x - 5x^2 = 5x(\underline{\ ?\ })$ $4 - x$

Factor.   **Section 14-4, p. 389**

11. $x(2 + x) + y(2 + x)$ **12.** $4n(m - 2) + m(m - 2)$ $(4n + m)(m - 2)$
       $(x + y)(2 + x)$

Check your answers with those printed at the back of the book.

Factoring Special Polynomials

14-5 *Difference of Two Squares*

OBJECTIVE
Factor polynomials like $m^2 - 64$ and $9x^2 - 1$.

Recall the pattern you learned for multiplying an expression like $(x - 4)(x + 4)$. The product of the sum and difference of two numbers is the difference of their squares.

EXAMPLE 1 $(x - 4)(x + 4) = x^2 - 4^2 = x^2 - 16$

To factor a polynomial that is the difference of two squares, we simply reverse the procedure.

EXAMPLE 2 $x^2 - 16 = (x - 4)(x + 4)$

EXAMPLE 3 $m^2 - 64$ ◄ $64 = 8^2$
$m^2 - 64 = (m + 8)(m - 8)$

EXAMPLE 4 $9x^2 - 1$ ◄ $9x^2 = (3x)^2; 1 = 1^2$
$9x^2 - 1 = (3x - 1)(3x + 1)$

EXAMPLE 5 $3y^2 - 12 = 3(y^2 - 4)$ ◄ 3 is a common factor.
$= 3(y - 2)(y + 2)$ $y^2 - 4$ is a difference of squares.

 Oral
EXERCISES

Tell whether or not the expression is a square. If it is, name the factors.

Sample: 25 *What you say:* Yes, $25 = 5 \cdot 5$

1. 36 yes, $6 \cdot 6$
2. m^2 yes, $m \cdot m$
3. t^4 yes, $t^2 \cdot t^2$
4. 81 yes, $9 \cdot 9$
5. 20 no
6. x^3 no

Multiply.

7. $(y + 7)(y - 7)$ $y^2 - 49$
8. $(r - 1)(r + 1)$ $r^2 - 1$
9. $(2x + 2)(2x - 2)$ $4x^2 - 4$
10. $(b - 3)(b + 3)$ $b^2 - 9$
11. $(2t + 1)(2t - 1)$ $4t^2 - 1$
12. $(z - 2)(z + 2)$ $z^2 - 4$

Complete.

Sample: $r^2 - 64 = (r + \underline{\ ?\ })(r - \underline{\ ?\ })$

Solution: $r^2 - 64 = (r + 8)(r - 8)$

1. $h^2 - 9 = (h + \underline{\ ?\ })(h - \underline{\ ?\ })$ 3; 3
2. $n^2 - 49 = (n + \underline{\ ?\ })(n - \underline{\ ?\ })$ 7; 7
3. $4k^2 - 1 = (2k + \underline{\ ?\ })(2k - \underline{\ ?\ })$ 1; 1
4. $x^2 - 25 = (\underline{\ ?\ } + 5)(\underline{\ ?\ } - 5)$ x; x
5. $4 - t^2 = (\underline{\ ?\ } - t)(\underline{\ ?\ } + t)$ 2; 2
6. $81 - y^2 = (\underline{\ ?\ } - y)(\underline{\ ?\ } + y)$ 9; 9

Factor.   13. $(3p + 12)(3p - 12)$ 14. $(mn + 3)(mn - 3)$

Sample: $a^2b^2 - 4$ *Solution:* $a^2b^2 - 4 = (ab + 2)(ab - 2)$

7. $z^2 - 4$ $(z - 2)(z + 2)$ 8. $y^2 - 1$ $(y + 1)(y - 1)$ 9. $a^2 - 36$ $(a - 6)(a + 6)$
10. $9 - t^2$ $(3 - t)(3 + t)$ 11. $r^2 - 900$ $(r + 30)(r - 30)$ 12. $16n^2 - 121$ $(4n - 11)(4n + 11)$
13. $9p^2 - 144$ 14. $m^2n^2 - 9$ 15. $t^4 - 100$ $(t^2 - 10)(t^2 + 10)$
16. $169 - r^4$ 17. $k^4 - 1$ 18. $49 - x^2y^4$
$(13 - r^2)(13 + r^2)$ $(k^2 + 1)(k^2 - 1)$ $(7 - xy^2)(7 + xy^2)$

Write as the difference of two squares. Then factor.

Sample: $-4 + m^2$ *Solution:* $-4 + m^2 = m^2 - 4$
$$= (m + 2)(m - 2)$$

$(y - 4)(y + 4)$ $(m^2 - 7)(m^2 + 7)$
19. $-16 + y^2$ 20. $-49 + m^4$ 21. $-36 + m^2n^2$ $(mn - 6)(mn + 6)$
22. $-1 + x^2$ $(x - 1)(x + 1)$ 23. $-r^2 + 64$ 24. $-4n^2 + 25$ $(5 + 2n)(5 - 2n)$
25. $-m^2 + n^2$ $(n + m)(n - m)$ 26. $-x^2y^2 + z^2$ $(z - xy)(z + xy)$ 27. $-x^4 + y^4$ $(y^2 - x^2)(y^2 + x^2)$
                                          23. $(8 - r)(8 + r)$

Factor completely.

Sample: $3y^2 - 27$ *Solution:* $3y^2 - 27 = 3(y^2 - 9)$
$$= 3(y + 3)(y - 3)$$

      30. $2(6 - z)(6 + z)$
$6(x + 2)(x - 2)$ $5(n + 2)(n - 2)$
28. $6x^2 - 24$ 29. $5n^2 - 20$ 30. $72 - 2z^2$ **B**
31. $3r^2 - 3s^2$ $3(r - s)(r + s)$ 32. $48 - 3r^2$ $3(4 - r)(4 + r)$ 33. $4 - 4r^2$ $4(1 - r)(1 + r)$
34. $5m^2 - 500$ 35. $2m^2n^2 - 162$ 36. $-36 + 9x^2$ $9(x - 2)(x + 2)$
37. $-75 + 3x^2$ 38. $xy^2 - xz^2$ 39. $\pi x^2 - \pi y^2$ $\pi(x + y)(x - y)$

40. $x^2 - \dfrac{1}{4}$ 41. $\dfrac{x^2}{9} - 1$ 42. $\dfrac{n^2}{4} - 25$ **C**
$\left(x - \dfrac{1}{2}\right)\left(x + \dfrac{1}{2}\right)$ $\left(\dfrac{x}{3} + 1\right)\left(\dfrac{x}{3} - 1\right)$ $\left(\dfrac{n}{2} - 5\right)\left(\dfrac{n}{2} + 5\right)$

34. $5(m + 10)(m - 10)$ 35. $2(mn - 9)(mn + 9)$ **PRODUCTS AND FACTORING / 393**
37. $3(x + 5)(x - 5)$ 38. $x(y - z)(y + z)$

14-6 *Factoring Trinomial Squares*

OBJECTIVE

Factor trinomial squares like
$m^2 + 4m + 4$ and $9x^2 + 12x + 4$.

Recall the pattern for expanding an expression like $(x + 5)^2$.

$$(x + 5)^2 = \underbrace{x^2}_{\substack{\text{square of the}\\\text{first term}}} + \underbrace{10x}_{\substack{\text{twice the product}\\\text{of the terms}}} + \underbrace{25}_{\substack{\text{square of the}\\\text{second term}}}$$

A polynomial such as $x^2 + 10x + 25$ is called a **trinomial square**. To factor a trinomial square, we reverse the process.

EXAMPLE 1 $m^2 + 4m + 4$ ◄ Think: $m^2 = m \cdot m$, $4 = 2 \cdot 2$, and $4m = 2(m \cdot 2)$.
$$m^2 + 4m + 4 \text{ is a trinomial square.}$$
$$m^2 + 4m + 4 = (m + 2)(m + 2)$$

EXAMPLE 2 $9x^2 + 12x + 4 = (3x + 2)(3x + 2)$ ◄ $9x^2 = 3x \cdot 3x$
$$4 = 2 \cdot 2, \ 12x = 2(3x \cdot 2)$$
$$= (3x + 2)^2$$

EXAMPLE 3 $4s^2 + 28s + 49 = (2s + 7)(2s + 7)$
$$= (2s + 7)^2$$

EXAMPLE 4 $t^2 + 2st + s^2 = (t + s)(t + s)$
$$= (t + s)^2$$

EXERCISES

Tell how to complete.

Sample: $4y^2 + 20y + 25 = (2y + \underline{\ ?\ })^2$
What you say: $5^2 = 25$ and $2(2y \cdot 5) = 20y$.
$$4y^2 + 20y + 25 = (2y + 5)^2$$

1. $4k^2 + 8k + 4 = (2k + \underline{\ ?\ })^2$ 2
2. $25t^2 + 10t + 1 = (\underline{\ ?\ } + 1)^2$ 5t
3. $m^2 + 6m + 9 = (\underline{\ ?\ } + 3)^2$ m
4. $9a^2 + 6a + 1 = (3a + \underline{\ ?\ })^2$ 1
5. $x^2 + 2xy + y^2 = (x + \underline{\ ?\ })^2$ y

Complete.

1. $4x^2 = (\underline{\;?\;})^2$ 2x   2. $m^2 = (\underline{\;?\;})^2$ m   3. $m^2n^2 = (\underline{\;?\;})^2$ mn

4. $r^2t^2 = (\underline{\;?\;})^2$ rt   5. $100 = (\underline{\;?\;})^2$ 10   6. $225 = (\underline{\;?\;})^2$ 15

Name the missing term.

Sample: $n^2 + \underline{\;?\;} + 9 = (n + 3)^2$ *Solution:* $2(n \cdot 3)$ or $6n$

7. $x^2 + \underline{\;?\;} + 9 = (x + 3)^2$ 6x   8. $r^2 + \underline{\;?\;} + 16 = (r + 4)^2$ 8r

9. $z^2 + \underline{\;?\;} + 1 = (z + 1)^2$ 2z   10. $m^2 + \underline{\;?\;} + n^2 = (m + n)^2$ 2mn

11. $9w^2 + \underline{\;?\;} + 1 = (3w + 1)^2$ 6w

12. $16y^2 + \underline{\;?\;} + 1 = (4y + 1)^2$ 8y

13. $4s^2 + \underline{\;?\;} + t^2 = (2s + t)^2$ 4st

14. $4a^2 + \underline{\;?\;} + 9b^2 = (2a + 3b)^2$ 12ab

Expand.

15. $(c + 2)^2$ $c^2 + 4c + 4$ 16. $(2d + 1)^2$ $4d^2 + 4d + 1$ 17. $(k + 5)^2$ $k^2 + 10k + 25$

18. $(a + b)^2$ 19. $(3n + 4m)^2$ 20. $(2b + 5c)^2$
$a^2 + 2ab + b^2$ $9n^2 + 24n + 16m^2$ $4b^2 + 20bc + 25c^2$

Factor. Check by multiplying.

Sample: $h^2 + 12h + 36$ *Solution:* $(h + 6)^2$

Check: $(h + 6)(h + 6) = h^2 + 12h + 36$

21. $p^2 + 2p + 1$ $(p + 1)^2$ 22. $b^2 + 4b + 4$ $(b + 2)^2$

23. $t^2 + 10t + 25$ $(t + 5)^2$ 24. $4r^2 + 4r + 1$ $(2r + 1)^2$

25. $m^2 + 4m + 4$ $(m + 2)^2$ 26. $a^2 + 24a + 144$ $(a + 12)^2$

27. $100 + 20x + x^2$ $(10 + x)^2$ 28. $q^2 + 18q + 81$ $(q + 9)^2$

29. $y^2 + 16y + 64$ $(y + 8)^2$ 30. $9w^2 + 6w + 1$ $(3w + 1)^2$

31. $z^2 + 30z + 225$ $(z + 15)^2$ 32. $25b^2 + 30b + 9$ $(5b + 3)^2$

33. $16c^2 + 24c + 9$ $(4c + 3)^2$ 34. $9m^2 + 42m + 49$ $(3m + 7)^2$

35. $m^2n^2 + 2mn + 1$ $(mn + 1)^2$ 36. $p^2 + 2pr + r^2$ $(p + r)^2$ B

37. $k^2 + 16kt + 64t^2$ $(k + 8t)^2$ 38. $4b^2 + 4bd + d^2$ $(2b + d)^2$

39. $100x^2 + 20xy + y^2$ $(10x + y)^2$ 40. $144r^2 + 120rs + 25s^2$ $(12r + 5s)^2$

41. $9r^2s^2 + 30rst + 25t^2$ $(3rs + 5t)^2$ 42. $25a^2b^2c^2 + 70abc + 49$ $(5abc + 7)^2$

43. $9x^2 + 60xy + 100y^2$ $(3x + 10y)^2$ 44. $16a^2b^2 + 8abc + c^2$ $(4ab + c)^2$

45. $x^2 + x + \dfrac{1}{4}$ $\left(x + \dfrac{1}{2}\right)^2$ 46. $m^2 + \dfrac{2m}{5} + \dfrac{1}{25}$ $\left(m + \dfrac{1}{5}\right)^2$ C

14-7 *Square of a Binomial Difference*

OBJECTIVE

Factor trinomial squares like $x^2 - 10x + 25$ and $4x^2 - 12x + 9$.

When a binomial expression such as $(x + 7)^2$ is expanded, all terms in the result are positive. When $(x - 7)^2$ is expanded, the middle term is always **negative**.

$$(x - 7)^2 = \underbrace{x^2}_{\substack{\text{square of} \\ \text{the first term}}} - \underbrace{14x}_{\substack{\text{twice the product} \\ \text{of the terms}}} + \underbrace{49}_{\substack{\text{square of} \\ \text{the last term}}}$$

To factor a trinomial square like $x^2 - 14x + 49$, we reverse this process.

EXAMPLE 1 $x^2 - 10x + 25$ ◀ Think: $x^2 = x \cdot x$, $25 = 5 \cdot 5$, and $10x = 2(x \cdot 5)$.
$x^2 - 10x + 25$ is a trinomial square.

$$x^2 - 10x + 25 = (x - 5)(x - 5)$$
$$= (x - 5)^2$$

EXAMPLE 2 $9t^2 - 12t + 4 = (3t - 2)(3t - 2)$ ◀ $9t^2 = 3t \cdot 3t$
$$= (3t - 2)^2 \qquad\qquad 4 = 2 \cdot 2, \ 12t = 2(3t \cdot 2)$$

EXAMPLE 3 $x^2 - 2xy + y^2 = (x - y)(x - y)$
$$= (x - y)^2$$

EXERCISES

Add or multiply as indicated.

Sample: $(-4) + (-4)$ *What you say:* $(-4) + (-4) = -8$
$(-4)(-4)$ $(-4)(-4) = 16$

1. $(-7) + (-7)$ -14 **2.** $(-3) + (-3)$ -6 **3.** $(-10) + (-10)$ -20
$(-7)(-7)$ 49 $(-3)(-3)$ 9 $(-10)(-10)$ 100

4. $(-1) + (-1)$ -2 **5.** $(-x) + (-x)$ $-2x$ **6.** $(-m) + (-m)$ $-2m$
$(-1)(-1)$ 1 $(-x)(-x)$ x^2 $(-m)(-m)$ m^2

$-2xy$

7. $(-xy) + (-xy)$ **8.** $(-2ab) + (-2ab)$ **9.** $(-3n) + (-3n)$ $-6n$

x^2y^2 $(-xy)(-xy)$ $\quad(-2ab)(-2ab)$ $4a^2b^2$ $(-3n)(-3n)$ $9n^2$

$-4ab$

$-10m$

10. $(-5m) + (-5m)$ **11.** $(-8rs) + (-8rs)$ **12.** $(-6t) + (-6t)$ $-12t$

$\quad(-5m)(-5m)$ $25m^2$ $\quad(-8rs)(-8rs)$ $64r^2s^2$ $(-6t)(-6t)$ $36t^2$

$-16rs$

13. $(-2t) + (-2t)$ $-4t$ **14.** $(-3ab) + (-3ab)$ **15.** $(-4z) + (-4z)$ $-8z$

$\quad(-2t)(-2t)$ $4t^2$ $\quad(-3ab)(-3ab)$ $-6ab$ $(-4z)(-4z)$ $16z^2$

$9a^2b^2$

Simplify.

Sample 1: $(-m)(-m)$ Solution: m^2

Sample 2: $(-2t)(-2t)$ Solution: $4t^2$

1. $(-s)(-s)$ s^2 **2.** $(-b)(-b)$ b^2 **3.** $(-5y)(-5y)$ $25y^2$

A

4. $(-7x)(-7x)$ $49x^2$ **5.** $(-8)(-8)$ 64 **6.** $(-10)(-10)$ 100

7. $(3x)(3x)$ $9x^2$ **8.** $(4t)(4t)$ $16t^2$ **9.** $(-10k)(-10k)$ $100k^2$

10. $(-ab)(-ab)$ a^2b^2 **11.** $(-xy)(-xy)$ x^2y^2 **12.** $(-2xy)(-2xy)$ $4x^2y^2$

Complete with positive factors and then with negative factors.

Sample: $x^2 = \underline{\ ?\ } \cdot \underline{\ ?\ } = \underline{\ ?\ } \cdot \underline{\ ?\ }$

Solution: $x^2 = x \cdot x = -x(-x)$

13. $z^2 = \underline{\ ?\ } \cdot \underline{\ ?\ } = \underline{\ ?\ } \cdot \underline{\ ?\ }$ $z \cdot z = (-z)(-z)$

14. $d^2 = \underline{\ ?\ } \cdot \underline{\ ?\ } = \underline{\ ?\ } \cdot \underline{\ ?\ }$ $d \cdot d = (-d)(-d)$

15. $w^2 = \underline{\ ?\ } \cdot \underline{\ ?\ } = \underline{\ ?\ } \cdot \underline{\ ?\ }$ $w \cdot w = (-w)(-w)$

16. $4t^2 = \underline{\ ?\ } \cdot \underline{\ ?\ } = \underline{\ ?\ } \cdot \underline{\ ?\ }$ $2t \cdot 2t = (-2t)(-2t)$

17. $25y^2 = \underline{\ ?\ } \cdot \underline{\ ?\ } = \underline{\ ?\ } \cdot \underline{\ ?\ }$ $5y \cdot 5y = (-5y)(-5y)$

18. $b^2c^2 = \underline{\ ?\ } \cdot \underline{\ ?\ } = \underline{\ ?\ } \cdot \underline{\ ?\ }$ $bc \cdot bc = (-bc)(-bc)$

19. $x^2y^2 = \underline{\ ?\ } \cdot \underline{\ ?\ } = \underline{\ ?\ } \cdot \underline{\ ?\ }$ $xy \cdot xy = (-xy)(-xy)$

20. $9a^2b^2 = \underline{\ ?\ } \cdot \underline{\ ?\ } = \underline{\ ?\ } \cdot \underline{\ ?\ }$ $3ab \cdot 3ab = (-3ab)(-3ab)$

Expand.

$r^2 - 4r + 4$

21. $(r - 2)^2$ **22.** $(m - 1)^2$ **23.** $(k - 5)^2$

$m^2 - 2m + 1$ $k^2 - 10k + 25$

24. $(4 - x)^2$ **25.** $(1 - y)^2$ **26.** $(2c - 1)^2$ $4c^2 - 4c + 1$

27. $(a - b)^2$ **28.** $(x - y)^2$ **29.** $(2x - 3)^2$ $4x^2 - 12x + 9$

30. $(t - 4s)^2$ **31.** $(2s - y)^2$ **32.** $(2x - 5y)^2$

$t^2 - 8st + 16s^2$ $4s^2 - 4sy + y^2$ $4x^2 - 20xy + 25y^2$

24. $16 - 8x + x^2$ **25.** $1 - 2y + y^2$

27. $a^2 - 2ab + b^2$ **28.** $x^2 - 2xy + y^2$

Factor. Check by multiplication.

Sample: $m^2 - 6m + 9$

Solution: $(m - 3)^2$

Check: $(m - 3)(m - 3) = m^2 - 6m + 9$

33. $t^2 - 8t + 16 \ (t - 4)^2$ 34. $n^2 - 4n + 4 \ (n - 2)^2$

35. $s^2 - 10s + 25 \ (s - 5)^2$ 36. $m^2 - 2m + 1 \ (m - 1)^2$

37. $c^2 - 6c + 9 \ (c - 3)^2$ 38. $25b^2 - 60b + 36 \ (5b - 6)^2$

39. $r^2 - 6r + 9 \ (r - 3)^2$ 40. $9 - 12s + 4s^2 \ (3 - 2s)^2$

41. $9k^2 - 6k + 1 \ (3k - 1)^2$ 42. $a^2 - 12a + 36 \ (a - 6)^2$

43. $x^2 - 14x + 49 \ (x - 7)^2$ 44. $4y^2 - 12y + 9 \ (2y - 3)^2$

B

45. $16a^2 - 8ab + b^2 \ (4a - b)^2$ 46. $a^2 - 2ab + b^2 \ (a - b)^2$

47. $9m^2 - 6mn + n^2 \ (3m - n)^2$ 48. $4x^2 - 4xy + y^2 \ (2x - y)^2$

49. $36y^2 - 36yz + 9z^2 \ (6y - 3z)^2$ 50. $16r^2 - 40rs + 25s^2 \ (4r - 5s)^2$

Factor completely.

C

51. $3n^2 - 6n + 3 \ 3(n - 1)(n - 1)$ 52. $20r^2 - 20r + 5$
$5(2r - 1)(2r - 1)$

53. $2m^2 - 20m + 50 \ 2(m - 5)^2$ 54. $5a^3 - 10a^2 + 5a$
$5a(a - 1)(a - 1)$

55. $k^2 - k + \dfrac{1}{4} \ \left(k - \dfrac{1}{2}\right)\left(k - \dfrac{1}{2}\right)$ 56. $x^2 - 0.2x + 0.01$
$(x - 0.1)(x - 0.1)$

Andrija Mohorovičić   1857–1936

Andrija Mohorovičić discovered one of the most important principles of geology while studying wave patterns of a Balkan earthquake in 1909. He found that waves which penetrated deeper into the earth arrived sooner than waves traveling along the surface. He deduced the fact that the earth possessed a layered structure having sharp separations. Attempts to drill through the first layer of the earth's surface were considered in the 1960's. The project, called *Mohole*, was named after Mohorovičić.

Factoring Other Polynomials

14-8 *Product of Binomial Sums or Differences*

> **OBJECTIVE**
> Factor trinomials like
> $x^2 + 7x + 12$ and $x^2 - 10x + 9$.

We are now ready to factor products of binomial sums and binomial differences. First let's review multiplication of binomials. Notice the signs in the result.

$(x + 5)(x + 7) = x^2 + 12x + 35$ ◀ all terms positive
$(x - 2)(x - 3) = x^2 - 5x + 6$ ◀ middle term negative, last term positive

These processes are reversed to factor trinomials that are not squares.

EXAMPLE 1 $x^2 + 6x + 8$ ◀ Product of binomial *sums*
  Step 1 (   )(   )   ◀ Set up parentheses.
  Step 2 $(x$ $)(x$)   ◀ Factor x^2 as $x \cdot x$.
  Step 3 $(x +$ $)(x +$)   ◀ Signs are $+$.
  Step 4 $(x + 4)(x + 2)$ ◀ Factor 8 as $4 \cdot 2$. Note that $4 + 2 = 6$.

EXAMPLE 2 $x^2 - 8x + 12$ ◀ Product of binomial *differences*
  Step 1 (   )(   )   ◀ Set up parentheses.
  Step 2 $(x$ $)(x$)   ◀ Factor x^2 as $x \cdot x$.
  Step 3 $(x -$ $)(x -$)   ◀ Signs are $-$.
  Step 4 $(x - 6)(x - 2)$ ◀ Factor 12 as $(-6)(-2)$. Note that
                   $(-6) + (-2) = -8$.

Complete both statements with the same positive numbers.

Sample: $15 = (\underline{\ ?\ })(\underline{\ ?\ })$ *What you say:* $15 = (5)(3)$
        $8 = (\underline{\ ?\ }) + (\underline{\ ?\ })$ $8 = 5 + 3$

Oral
EXERCISES

1. $10 = (\underline{\ ?\ })(\underline{\ ?\ })$ 2, 5
    $7 = (\underline{\ ?\ }) + (\underline{\ ?\ })$ 2, 5

2. $6 = (\underline{\ ?\ })(\underline{\ ?\ })$ 2, 3
    $5 = (\underline{\ ?\ }) + (\underline{\ ?\ })$ 2, 3

3. $9 = (\underline{\ ?\ })(\underline{\ ?\ })$ 3, 3
    $6 = (\underline{\ ?\ }) + (\underline{\ ?\ })$ 3, 3

4. $9 = (\underline{\ ?\ })(\underline{\ ?\ })$ 1, 9
    $10 = (\underline{\ ?\ }) + (\underline{\ ?\ })$ 1, 9

5. $12 = (\underline{\ ?\ })(\underline{\ ?\ })$ 2, 6
    $8 = (\underline{\ ?\ }) + (\underline{\ ?\ })$ 2, 6

6. $20 = (\underline{\ ?\ })(\underline{\ ?\ })$ 4, 5
    $9 = (\underline{\ ?\ }) + (\underline{\ ?\ })$ 4, 5

Complete both statements with the same negative numbers.

7. $12 = (\underline{\ ?\ })(\underline{\ ?\ }) -3, -4$
   $-7 = (\underline{\ ?\ }) + (\underline{\ ?\ }) -3, -4$

8. $7 = (\underline{\ ?\ })(\underline{\ ?\ }) -1, -7$
   $-8 = (\underline{\ ?\ }) + (\underline{\ ?\ }) -1, -7$

9. $24 = (\underline{\ ?\ })(\underline{\ ?\ }) -3, -8$
   $-11 = (\underline{\ ?\ }) + (\underline{\ ?\ }) -3, -8$

10. $15 = (\underline{\ ?\ })(\underline{\ ?\ }) -3, -5$
    $-8 = (\underline{\ ?\ }) + (\underline{\ ?\ }) -3, -5$

11. $9 = (\underline{\ ?\ })(\underline{\ ?\ }) -1, -9$
    $-10 = (\underline{\ ?\ }) + (\underline{\ ?\ }) -1, -9$

12. $12 = (\underline{\ ?\ })(\underline{\ ?\ }) -1, -12$
    $-13 = (\underline{\ ?\ }) + (\underline{\ ?\ }) -1, -12$

Written EXERCISES

Complete.

Sample: $x^2 + 5x + 6 = (x + \underline{\ ?\ })(x + \underline{\ ?\ })$

Solution: $x^2 + 5x + 6 = (x + 2)(x + 3)$

A

1. $x^2 + 6x + 8 = (x + \underline{\ ?\ })(x + \underline{\ ?\ })\ 4, 2$

2. $n^2 - 3n + 2 = (n - \underline{\ ?\ })(n - \underline{\ ?\ })\ 1, 2$

3. $t^2 - 10t + 16 = (t - \underline{\ ?\ })(t - \underline{\ ?\ })\ 2, 8$

4. $a^2 + 9a + 20 = (a + \underline{\ ?\ })(a + \underline{\ ?\ })\ 5, 4$

5. $w^2 - 14w + 24 = (w - \underline{\ ?\ })(w - \underline{\ ?\ })\ 12, 2$

6. $x^2 + 12x + 20 = (x + \underline{\ ?\ })(x + \underline{\ ?\ })\ 2, 10$

7. $b^2 - 14b + 40 = (b - \underline{\ ?\ })(b - \underline{\ ?\ })\ 4, 10$

8. $c^2 + 12c + 32 = (c + \underline{\ ?\ })(c + \underline{\ ?\ })\ 8, 4$

9. $s^2 - 11s + 24 = (s - \underline{\ ?\ })(s - \underline{\ ?\ })\ 8, 3$

10. $x^2 - 14x + 33 = (x - \underline{\ ?\ })(x - \underline{\ ?\ })\ 11, 3$

14. $(a - 5)(a - 3)$
15. $(x - 4)(x - 5)$
16. $(r + 16)(r + 2)$
18. $(d + 1)(d + 8)$
19. $(z - 4)(z - 3)$
21. $(c + 8)(c + 3)$
22. $(x - 2)(x - 9)$

Factor.

11. $n^2 + 3n + 2$ — $(n + 1)(n + 2)$
12. $y^2 + 5y + 4$ — $(y + 4)(y + 1)$
13. $k^2 - 5k + 6$ — $(k - 3)(k - 2)$

14. $a^2 - 8a + 15$
15. $x^2 - 9x + 20$
16. $r^2 + 18r + 32$

17. $t^2 - 14t + 24$ — $(t - 12)(t - 2)$
18. $d^2 + 9d + 8$
19. $z^2 - 7z + 12$

20. $w^2 + 10w + 21$ — $(w + 7)(w + 3)$
21. $c^2 + 11c + 24$
22. $x^2 - 11x + 18$

B

23. $r^2 + 9rs + 20s^2$ $(r + 5s)(r + 4s)$
24. $b^2 - 3bc + 2c^2$ $(b - 2c)(b - c)$

25. $m^2 + 8mn + 15n^2$ — $(m + 5n)(m + 3n)$
26. $x^2 + 14xy + 13y^2$ $(x + 13y)(x + y)$

27. $16 - 10x + x^2$ $(8 - x)(2 - x)$
28. $48 + 14s + s^2$ $(8 + s)(6 + s)$

29. $4 + 5q + q^2$ $(4 + q)(1 + q)$
30. $30 - 17w + w^2$ $(15 - w)(2 - w)$

C

31. $b^2 + 20 - 12b$ $(b - 2)(b - 10)$
32. $30 + x^2 - 11x$ $(x - 6)(x - 5)$

33. $-6m + m^2 + 8$ $(m - 2)(m - 4)$
34. $-13y + 30 + y^2$ $(y - 3)(y - 10)$

14-9 *More About Factoring Trinomials*

OBJECTIVE

Factor trinomials such as
$x^2 + 5x - 14$ and $x^2 - 3x - 18$.

The method for factoring $x^2 + 5x - 14$ is very similar to the factoring of other trinomials. Note that the last term is negative. When you factor it, remember that one factor must be positive and one negative.

EXAMPLE 1 $x^2 + 5x - 14$

Step 1   (   )(   ) ◀ Set up parentheses.
Step 2 $(x$ $)(x$ $)$ ◀ Factor x^2 as $x \cdot x$.
Step 3 $(x +$ $)(x -$ $)$ ◀ One factor of -14 must be positive
                                  and the other negative.
Step 4 $(x + 7)(x - 2)$ ◀ Factor -14 as $(7)(-2)$ since $7 + (-2) = 5$.

EXAMPLE 2 $x^2 - 3x - 18$

Step 1   (   )(   ) ◀ Set up parentheses.
Step 2 $(x$ $)(x$ $)$ ◀ Factor x^2 as $x \cdot x$.
Step 3 $(x +$ $)(x -$ $)$ ◀ One factor of -18 must be positive
                                  and the other negative.
Step 4 $(x + 3)(x - 6)$ ◀ Factor -18 as $(3)(-6)$ since $3 + (-6) = -3$.

Complete. Use the same numbers in both statements.

EXERCISES

Sample: $-10 = (\underline{\ ?\ })(\underline{\ ?\ })$
         $-3 = (\underline{\ ?\ }) + (\underline{\ ?\ })$

What you say: $-10 = 2(-5)$
              $-3 = 2 + (-5)$

1. $-18 = (\underline{\ ?\ })(\underline{\ ?\ })$
    $7 = (\underline{\ ?\ }) + (\underline{\ ?\ })$ $-2, 9$

2. $-24 = (\underline{\ ?\ })(\underline{\ ?\ })$
    $5 = (\underline{\ ?\ }) + (\underline{\ ?\ })$ $-3, 8$

3. $-12 = (\underline{\ ?\ })(\underline{\ ?\ })$
    $-1 = (\underline{\ ?\ }) + (\underline{\ ?\ })$ $3, -4$

4. $-5 = (\underline{\ ?\ })(\underline{\ ?\ })$
    $-4 = (\underline{\ ?\ }) + (\underline{\ ?\ })$ $-5, 1$

5. $-12 = (\underline{\ ?\ })(\underline{\ ?\ })$
    $1 = (\underline{\ ?\ }) + (\underline{\ ?\ })$ $-3, 4$

6. $-14 = (\underline{\ ?\ })(\underline{\ ?\ })$
    $5 = (\underline{\ ?\ }) + (\underline{\ ?\ })$ $-2, 7$

7. $-14 = (\underline{\ ?\ })(\underline{\ ?\ })$
    $-5 = (\underline{\ ?\ }) + (\underline{\ ?\ })$ $2, -7$

8. $-30 = (\underline{\ ?\ })(\underline{\ ?\ })$
    $1 = (\underline{\ ?\ }) + (\underline{\ ?\ })$ $-5, 6$

Written EXERCISES

Complete and check.

Sample: $x^2 + 2x - 35 = (x + \underline{})(x - \underline{})$

Solution: $x^2 + 2x - 35 = (x + 7)(x - 5)$

Check: $(x + 7)(x - 5) = x^2 + 2x - 35$

A

1. $n^2 + 7n - 18 = (n + \underline{})(n - \underline{})$ $(n + 9)(n - 2)$

2. $y^2 + 3y - 18 = (y + \underline{})(y - \underline{})$ $(y + 6)(y - 3)$

3. $x^2 - x - 42 = (x + \underline{})(x - \underline{})$ $(x + 6)(x - 7)$

4. $x^2 + x - 42 = (x - \underline{})(x + \underline{})$ $(x - 6)(x + 7)$

5. $a^2 - 2a - 63 = (a - \underline{})(a + \underline{})$ $(a - 9)(a + 7)$

6. $m^2 + 3m - 40 = (m + \underline{})(m - \underline{})$ $(m + 8)(m - 5)$

7. $m^2 - 3m - 40 = (m + \underline{})(m - \underline{})$ $(m + 5)(m - 8)$

8. $w^2 + 7w - 44 = (w + \underline{})(w - \underline{})$ $(w + 11)(w - 4)$

9. $z^2 - 7z - 44 = (z - \underline{})(z + \underline{})$ $(z - 11)(z + 4)$

10. $s^2 + 11s - 26 = (s + \underline{})(s - \underline{})$ $(s + 13)(s - 2)$

11. $(t + 9)(t - 7)$
12. $(x - 6)(x + 3)$
13. $(r + 11)(r - 2)$
14. $(a - 7)(a + 2)$
15. $(s - 3)(s + 2)$
16. $(y - 7)(y + 6)$
17. $(c - 8)(c + 7)$
18. $(m + 9)(m - 5)$
19. $(n - 15)(n + 3)$
20. $(q + 8)(q - 4)$

21. $(w + 8)(w - 6)$ 22. $(r - 20)(r + 3)$ 23. $(b + 9)(b - 4)$
24. $(s - 12)(s + 3)$

25. $(x + 8)(x - 5)$

Factor.   26. $(t + 14)(t - 3)$ 27. $(b + 17)(b - 1)$ 28. $(n - 15)(n + 4)$

11. $t^2 + 2t - 63$ 12. $x^2 - 3x - 18$ 13. $r^2 + 9r - 22$

14. $a^2 - 5a - 14$ 15. $s^2 - s - 6$ 16. $y^2 - y - 42$

17. $c^2 - c - 56$ 18. $m^2 + 4m - 45$ 19. $n^2 - 12n - 45$

20. $q^2 + 4q - 32$ 21. $w^2 + 2w - 48$ 22. $r^2 - 17r - 60$

23. $b^2 + 5b - 36$ 24. $s^2 - 9s - 36$ 25. $x^2 + 3x - 40$

26. $t^2 + 11t - 42$ 27. $b^2 + 16b - 17$ 28. $n^2 - 11n - 60$
$(mn - 25)(mn + 4)$

B

29. $y^2 + 22xy - 23x^2$ 30. $m^2n^2 - 21mn - 100$

31. $r^2 + 48ar - 100a^2$ 32. $b^2 - 10bc - 75c^2$

33. $y^2z^2 - 16yz - 57$ 34. $m^2 + 24mn - 81n^2$
$(m + 27n)(m - 3n)$

35. $a^2 - 14ac - 72c^2$ 36. $p^2 - 13pq - 68q^2$

37. $x^2 + 8xy - 65y^2$ 38. $r^2 - 15rs - 54s^2$ $(r - 18s)(r + 3s)$

39. $b^2 + 17bc - 38c^2$ 40. $m^2 - 10ms - 39s^2$
$(m - 13s)(m + 3s)$

29. $(y + 23x)(y - x)$
31. $(r + 50a)(r - 2a)$
32. $(b - 15c)(b + 5c)$
33. $(yz - 19)(yz + 3)$
35. $(a - 18c)(a + 4c)$

36. $(p - 17q)(p + 4q)$ 37. $(x + 13y)(x - 5y)$ 39. $(b + 19c)(b - 2c)$

Write in standard form and factor.
$(r - 27s)(r + 3s)$

C

41. $16xy - 57y^2 + x^2$ 42. $-24rs + r^2 - 81s^2$

43. $5z + z^2 - 50$ 44. $-8x^2 - 2xy + y^2$

45. $b^2 - 24a^2 + 10ab$ 46. $-3bc + b^2 - 10c^2$

41. $(x + 19y)(x - 3y)$ 43. $(z + 10)(z - 5)$ 44. $(y - 4x)(y + 2x)$

45. $(b + 12a)(b - 2a)$ 46. $(b - 5c)(b + 2c)$

SELF-TEST 2

Factor.

1. $b^2 - 9$ $(b + 3)(b - 3)$
2. $x^2 - 25$ $(x + 5)(x - 5)$ **Section 14-5, p. 392**
3. $z^2 + 10z + 25$ $(z + 5)(z + 5)$
4. $y^2 + 4y + 4$ $(y + 2)(y + 2)$ **Section 14-6, p. 394**
5. $x^2 - 12x + 36$ $(x - 6)(x - 6)$
6. $m^2 - 18m + 81$ $(m - 9)^2$ **Section 14-7, p. 396**
7. $x^2 + 5x + 6$ $(x + 3)(x + 2)$
8. $n^2 - 9n + 14$ $(n - 7)(n - 2)$ **Section 14-8, p. 399**
9. $c^2 + 2c - 8$ $(c + 4)(c - 2)$
10. $c^2 - 2c - 8$ $(c - 4)(c + 2)$ **Section 14-9, p. 401**

Check your answers with those printed at the back of the book.

---

chapter summary

1. A **prime number** has exactly two different positive factors, itself and one.

2. A number which can be expressed as the product of two or more prime numbers is called a **composite number.**

3. The prime factorization of a number is its expression as a product of prime factors.

4. It may be possible to factor a polynomial in one of the following ways:
   a. Factor out the GCF of the terms.
   b. Group terms. Then factor out the GCF.
   c. Identify and factor as the difference of two squares.
   d. Identify and factor as the square of a binomial sum or a binomial difference.
   e. Identify and factor as the product of binomial sums or binomial differences.
   f. Identify and factor as the product of a binomial sum and a binomial difference.

chapter test

Write the prime factorization.

1. 28 $2 \cdot 2 \cdot 7$ **2.** 30 $2 \cdot 3 \cdot 5$ **3.** 27 $3 \cdot 3 \cdot 3$

Name the GCF.

4. $14x^2, 6xy$ $2x$ **5.** 10 and $8n$ 2 **6.** $16x^2y, 12xy^2$ $4xy$

Factor.

                                                  $2(3x^2 + 6x - 5)$

7. $6 + 8n^2$ $2(3 + 4n^2)$ **8.** $8x^2 - 8y$ $8(x^2 - y)$ **9.** $6x^2 + 12x - 10$

10. $10y^2 + 50x$ **11.** $m^2n + n$ $n(m^2 + 1)$ **12.** $4st^2 - 12st + 4s$

       $10(y^2 + 5x)$ $4s(t^2 - 3t + 1)$

Group terms. Then factor.

13. $xy + 2x + 6 + 3y$ $(y + 2)(x + 3)$ **14.** $mn + 5n - 10 - 2m$ $(m + 5)(n - 2)$

Factor.   **15.** $(w + 8)(w - 8)$ **16.** $(x + 3)^2$ **18.** $(t - 6)(t + 6)$ **19.** $(r + 5)^2$

15. $w^2 - 64$ **16.** $x^2 + 6x + 9$ **17.** $y^2 - 2y + 1$ $(y - 1)^2$

18. $t^2 - 36$ **19.** $r^2 + 10r + 25$ **20.** $a^2 + 7a + 10$ $(a + 5)(a + 2)$

21. $4m^2 - n^2$ **22.** $b^2 - 8b + 16$ **23.** $t^2 - 8t + 15$ $(t - 5)(t - 3)$

21. $(2m - n)(2m + n)$ **22.** $(b - 4)^2$

Factor. Check by multiplication.

24. $n^2 + 6n - 16$ $(n + 8)(n - 2)$ **25.** $r^2 - 3r - 18$ $(r - 6)(r + 3)$

26. $c^2 + 6c - 27$ $(c + 9)(c - 3)$ **27.** $t^2 - 2t - 35$ $(t - 7)(t + 5)$

 # challenge topics

Similar Triangles

When two triangles have the same shape, we say they are **similar.** In this illustration $\triangle ABC$ is similar to $\triangle PQR$. We use the symbol $\sim$ to mean "is similar to" and write $\triangle\mathbf{ABC} \sim \triangle\mathbf{PQR}$.

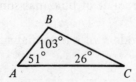

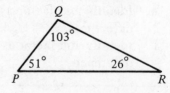

In $\triangle ABC$ and $\triangle PQR$ shown on page 404.

$\overline{AB}$ corresponds to $\overline{PQ}$ $\angle A$ corresponds to $\angle P$

$\overline{AC}$ corresponds to $\overline{PR}$ $\angle B$ corresponds to $\angle Q$

$\overline{BC}$ corresponds to $\overline{QR}$ $\angle C$ corresponds to $\angle R$

When two triangles are similar, the measures of their **corresponding angles** are **equal.**

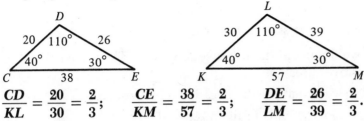

$$\frac{CD}{KL} = \frac{20}{30} = \frac{2}{3}; \qquad \frac{CE}{KM} = \frac{38}{57} = \frac{2}{3}; \qquad \frac{DE}{LM} = \frac{26}{39} = \frac{2}{3}.$$

For similar triangles, the ratios between the measures of pairs of corresponding sides are equal.

1. If $\triangle KLR \sim \triangle BCD$, what is the length of KL?

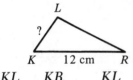 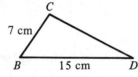

$$\frac{KL}{BC} = \frac{KR}{BD}; \qquad \frac{KL}{7} = \frac{12}{15}; \qquad KL = \underline{\ ?\ }\ \text{5.6 cm}$$

2. $\triangle XYZ$ and $\triangle BTR$ are similar. What are the measures of $\overline{YZ}$ and $\overline{BR}$? 10 m; 21.6 m

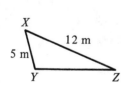 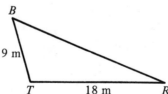

3. $\triangle BGH$ and $\triangle BKT$ are similar. What is the measure of $\angle KTH$? $\angle BHG$? $\angle BKT$? $\angle GKT$? $\angle BGH$?

145; 35; 75; 105; 75

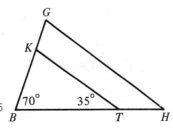

cumulative review

Name the directed number.

1. The positive number 7 units from 0. ₇

2. The negative number 5 units from 0. ₋₅

3. The negative number 13 units from 0. ₋₁₃

Make a number line sketch to show the moves. Tell where you finish.

4. Start at 0. Move 6 units in the negative direction. Then move 4 units in the positive direction. ₋₂

5. Start at 2 and move 4 units in the negative direction. ₋₂

6. Start at −1 and move 3 units in the positive direction. ₂

Complete to make a true statement. Use $<$ or $>$.

7. $-6 \underline{\ ?\ } -10$ $>$ **8.** $0.5 \underline{\ ?\ } -0.56$ $>$

9. $\dfrac{1}{2} \underline{\ ?\ } -\dfrac{3}{2}$ $>$ **10.** $0.4 \underline{\ ?\ } 0.47$ $<$

Write and graph the solution set. The replacement set is {the directed numbers}.

11. $q < 3$ **12.** $t \geq -3$

13. $s > 0.5$ **14.** $r \leq 3.7$

Solve.

15. $t = 4 + (-4)$ $t = 0$ **16.** $15 + n = 4$ $n = -11$

17. $14 + q = 5$ $q = -9$ **18.** $m + 5 = 15$ $m = 10$

Simplify.

19. $-(-7)$ ₇   **20.** $-12 + (-3)$ ₋₁₅

21. $-(5 + 7)$ ₋₁₂   **22.** $-(-4 + 3)$ ₁

Complete to make a true statement. Name the property illustrated.

23. $4 = (-2 + 2) + \underline{\ ?\ }$ ₄   **24.** $13 + \underline{\ ?\ } = 13$ 0; Add. Prop. of 0

25. $\underline{\ ?\ } + 7 = 7 + 5$ ₅   **26.** $14 + (-14) = \underline{\ ?\ }$ 0; Add. Prop. of Inv.

23. Add. Prop of Inv., Add. Prop. of 0   **25.** Comm. Prop. of Add.

Solve. Begin by writing the equivalent addition equation.

27. $15 - (-3) = n$ 18

28. $7 - 20 = d$ -13

29. $3.2 - (-0.2) = g$ 3.4

30. $-11 - 3 = h$ -14

Complete according to the given function equation.

31. $f(q) = q - (-1)$: $\{(-1, 0), (0, 1), (1, \underline{\ ?\ }), (2, \underline{\ ?\ }), (3, \underline{\ ?\ })\}$ (1, 2), (2, 3), (3, 4)

32. $f(t) = 13 - t$: $\{-2, 15), (-1, \underline{\ ?\ }), (0, \underline{\ ?\ }), (1, \underline{\ ?\ }), (2, \underline{\ ?\ })\}$

(−1, 14), (0, 13), (1, 12), (2, 11)

Simplify.

33. $4(-11)$ -44

34. $-\dfrac{52}{4}$ -13

35. $-\dfrac{21}{7}$ -3

36. $-4(-4)(-4)$ -64

37. $-4(2 + 7)$ -36

38. $a^2 + 2 + 3a^2 + 1$ $4a^2 + 3$

39. $-48 \div -6$ 8

40. $4 \div \dfrac{1}{-3}$ -12

41. $\dfrac{3}{-8} \div \dfrac{3}{4}$ $-\dfrac{1}{2}$

Solve.

42. $-6x = -18$ 3

43. $4b = -32$ -8

44. $\dfrac{4}{5}y = -8$ -10

45. $-\dfrac{1}{y} = 10$ $-\dfrac{1}{10}$

Complete the set of number pairs by using the given function equation.

46. $f(t) = 2t - 6$: $\{(-1, \underline{\ ?\ }), (0, \underline{\ ?\ }), (1, \underline{\ ?\ }), (2, \underline{\ ?\ })\}$

(−1, −8), (0, −6), (1, −4), (2, −2)

Solve.

47. $-6 + b = -18$ -12

48. $4x + 3x = -28$ -4

Solve. Use the formula $A = lw$.

49. A table top has area $8m$ and length $4m$. Find the width. 2

Solve. Graph the solution set.

50. $k - 4 \leq 7$ $k \leq 11$

51. $-2w + 3 > 9$ $w < -3$

52. $6r < 30$ $r < 5$

53. $\dfrac{3}{5}s \geq -6$ $s \geq -10$

Tell whether the polynomial is a monomial, a binomial, or a trinomial.

54. $17t + w$ **55.** $q + r - t$ **56.** $2lw$
     binomial   trinomial   monomial

Write in standard form.

57. $2 + 2x + x^2$ $x^2 + 2x + 2$ **58.** $x^3 - x^4 + x + 1$ $-x^4 + x^3 + x + 1$

59. $4x^4 + 2 - x^6$ $-x^6 + 4x^4 + 2$ **60.** $-10x^3 + 2 + x$ $-10x^3 + x + 2$

Add.

61. $t - 6$ **62.** $t^3 + \ \ t^2 + 1$
      $2t + 2$ $7t^2$
    $\overline{3t - 4}$ $\overline{t^3 + 8t^2 + 1}$

63. $(7k + 3) + (-5k - 2)$ $2k + 1$ **64.** $(b + 5) + (4b - 7)$ $5b - 2$

Subtract the second polynomial from the first.

65. $6k + 2c$ **66.** $7h^2 - 2h$
     $4k - 3c$ $2h^2 - 3h$
    $\overline{2k + 5c}$ $\overline{5h^2 + h}$

67. $2b^2 + b - 1$ **68.** $4k + 7y$
      $b^2 - b$ $3k + 2y + 5$
   $\overline{b^2 + 2b - 1}$ $\overline{k + 5y - 5}$

Simplify.

69. $(2t - 3) - (8t - 4)$ $-6t + 1$ **70.** $(t^2 + 5) - (2t^2 - 8)$ $-t^2 + 13$

71. $(11t^2 - 3t) - (t + 2)$ **72.** $(4 + t) - (10 + 3t)$ $-2t - 6$
                  $11t^2 - 4t - 2$

Write as a polynomial in standard form.

73. The perimeter of a rectangle with length $x^2 + 4$ and width $x^2 - 16$.
                                               $4x^2 - 24$

Solve and check.

74. $7x + (x - 3) = 13$ 2 **75.** $10 = 2t - (10 + 3t)$ -20

76. $(4n + 11) + (5 + n) = 21$ 1 **77.** $0 = (3m + 3) - 2m$ -3

Simplify. Use exponents.

78. $(2 + t)(t + 2)$ $(2 + t)^2$ **79.** $-4(q \cdot q \cdot q)(-3)$ $12q^3$

80. $-2y(-4)(y)$ $8y^2$ **81.** $(2t^2x)(3t^2x)$ $6t^4x^2$

82. $[(-4)^2]^2$ 16^2 **83.** $(-3y^2)^2$ $9y^4$

Multiply.

84. $-2x(x^2 - 3x + 2)$

85. $t(-t^3 + 2t^2 + 1)$ $-t^4 + 2t^3 + t$

86. $(x + 2)(x - 3)$

87. $(2r - 4)(r + 3)$ $2r^2 + 2r - 12$

88. $(y + 1)(y^2 + y - 1)$

89. $(2y + 2)(2y + 2)$ $4y^2 + 8y + 4$

90. $(3a - b)(3a - b)$

91. $(t - 3)(t + 3)$ $t^2 - 9$

92. $(2t - 4)(2t + 4)$

93. $(7a + 1)(7a - 1)$ $49a^2 - 1$

84. $-2x^3 + 6x^2 - 4x$ 86. $x^2 - x - 6$ 88. $y^3 - 2y^2 - 1$
90. $9a^2 - 6ab + b^2$ 92. $4t^2 - 16$

Divide. Assume no denominator is 0.

94. $\dfrac{2ab^2}{ab}$ $2b$

95. $\dfrac{6a^2 + 10a}{2a}$ $3a + 5$

96. $\dfrac{x^2 - 4x - 12}{x}$ $x - 4 - \dfrac{12}{x}$

97. $\dfrac{ab^2 - 2b + 2}{2b}$ $\dfrac{ab}{2} - 1 + \dfrac{1}{b}$

98. $\dfrac{s^2 - 4s + 4}{s - 2}$ $s - 2$

99. $\dfrac{x^2 - 4x - 12}{x - 6}$ $x + 2$

Write the prime factorization.

100. 12 $2^2 \cdot 3$

101. 72 $2^3 \cdot 3^2$

102. 121 11^2

103. 110 $2 \cdot 5 \cdot 11$

104. 98 $2 \cdot 7^2$

105. 126 $2 \cdot 3^2 \cdot 7$

Name the GCF.

106. $3s^3$, $4s^2t$ s^2

107. $2ab^2$, $2a^2b$ $2ab$

108. $2rq$, $22r^3q^2$ $2rq$

109. $49x^2$, $7xy$ $7x$

Factor.

110. $7k^2 - 14k$ $7k(k - 2)$

111. $39k^4 + 13k - 13$ $13(3k^4 + k - 1)$

112. $2q(t + 1) - 3q(t + 1)$

113. $2t + 2s + rt + rs$ $(2 + r)(t + s)$

114. $(4b^2 + 4b) - (3b + 3)$

115. $25 - b^2$ $(5 + b)(5 - b)$

116. $h^2 + 14h + 49$ $(h + 7)^2$

117. $4t^2 - 12t + 9$ $(2t - 3)^2$

118. $x^2 + 7x + 10$ $(x + 2)(x + 5)$

119. $x^2 + 11x + 30$ $(x + 5)(x + 6)$

120. $x^2 - 10x + 21$ $(x - 3)(x - 7)$

121. $x^2 - 13x + 22$ $(x - 11)(x - 2)$

122. $x^2 + 6x - 16$ $(x + 8)(x - 2)$

123. $x^2 - 2x - 35$ $(x - 7)(x + 5)$

124. $x^2 + 4x - 77$ $(x + 11)(x - 7)$

125. $x^2 - 5x - 24$ $(x - 8)(x + 3)$

112. $(2q - 3q)(t + 1)$ 114. $(4b - 3)(b + 1)$

Extra Practice Exercises

For use with Section 1-1.

Simplify.

1. $(5 \times 6) - 8$ 22
2. $18 + 9 + 6$ 33
3. $144 \div 6$ 24
4. $\dfrac{20}{8 + 12}$ 1
5. $\dfrac{54 - 34}{2}$ 10
6. $\dfrac{56}{8}$ 7
7. $64 \div 8$ 8
8. $2^2 + 5$ 9
9. $(3 + 7) + 4$ 14
10. 23×1 23
11. $(55 - 0) \times 1$ 55
12. $24 - (3 \times 6)$ 6
13. $10 - \dfrac{15}{3}$ 5
14. $\dfrac{90}{15}$ 6
15. $4 + \dfrac{16}{4}$ 8
16. $3^2 + 4^2$ 25
17. $2 \times 3 \times 4$ 24
18. 21×3 63
19. $6 \times (36 - 24)$ 72
20. $7^2 - (6^2 + 1)$ 12
21. $65 + 0$ 65
22. $\dfrac{28 + 34}{5^2 + 6}$ 2
23. $(6 \times 7) - 42$ 0
24. $\dfrac{56 - 24}{8 \times 4}$ 1
25. $(18 - 16) \times 24$ 48
26. $(12 + 9) - 11$ 10
27. $(2^2 + 5^2) - 9$ 20
28. $\dfrac{52}{20 \cdot 7}$ $\dfrac{13}{35}$
29. $\dfrac{3 \times 6}{13 - 4}$ 2
30. $\dfrac{27}{3} + 18$ 27
31. $200 \div 25$ 8
32. $(34 - 12) + 20$ 42
33. $5 \times 3 \times 4$ 60

For use with Section 1-6.

Solve.

34. $s + 2 = 8$ 6
35. $16 - r = 10$ 6
36. $m + 15 = 30$ 15
37. $n - 12 = 7$ 19
38. $20 + t = 34$ 14
39. $36 = 18 + x$ 18
40. $y = 8 + 9 + 10$ 27
41. $28 - z = 16$ 12
42. $8 + 4 = m$ 12
43. $x - 23 = 7$ 30
44. $46 = t + 19$ 27
45. $h - 13 = 2$ 15
46. $72 + t = 100$ 28
47. $k - 4 = 28$ 32
48. $11 + 22 = r$ 33
49. $33 - y = 17$ 16
50. $q + 15 = 45$ 30
51. $49 - 18 = n$ 31
52. $10 + 8 = 9 + t$ 9
53. $12 - s = 8 + 4$ 0
54. $56 + 49 = 32 + t$ 73
55. $24 + 14 = 46 - m$ 8
56. $3 + h = 4 + 12$ 13
57. $k - 10 = 72 - 47$ 35
58. $21 - 6 = t + 7$ 8
59. $8 + m = 26 - 10$ 8

For use with Section 1-7.

Solve.

1. $r \times 6 = 24$ 4
2. $50 = 25 \times t$ 2
3. $90 = 15 \times m$ 6

4. $\dfrac{48}{y} = 16$ 3
5. $\dfrac{56}{r} = 8$ 7
6. $12 \times h = 144$ 12

7. $12 = 4 \times h$ 3
8. $125 = 25 + q$ 5
9. $72 = s \times 12$ 6

10. $\dfrac{156}{x} = 3$ 52
11. $13 = \dfrac{169}{r}$ 13
12. $\dfrac{63}{t} = 9$ 7

13. $k \times 13 = 39$ 3
14. $8 \times h = 88$ 11
15. $12 \times b = 48$ 4

16. $\dfrac{360}{m} = 60$ 6
17. $24 = \dfrac{240}{r}$ 10
18. $\dfrac{425}{k} = 17$ 25

19. $t \times 2 = 666$ 333
20. $8 \times h = 16$ 2
21. $\dfrac{49}{m} = 7$ 7

22. $4 \times m = 36$ 9
23. $7 \times r = 98$ 14
24. $21 \times s = 84$ 4

25. $\dfrac{s}{8} = 15$ 120
26. $\dfrac{55}{r} = 5$ 11
27. $9 = \dfrac{z}{5}$ 45

For use with Section 3-4.

Write in lowest terms.

28. $\dfrac{4}{6}$ $\dfrac{2}{3}$
29. $\dfrac{10}{50}$ $\dfrac{1}{5}$
30. $\dfrac{12}{24}$ $\dfrac{1}{2}$
31. $\dfrac{4}{8}$ $\dfrac{1}{2}$

32. $\dfrac{16}{24}$ $\dfrac{2}{3}$
33. $\dfrac{25}{100}$ $\dfrac{1}{4}$
34. $\dfrac{10}{10,000}$ $\dfrac{1}{1000}$
35. $\dfrac{25}{20}$ $\dfrac{5}{4}$

36. $\dfrac{8}{12}$ $\dfrac{2}{3}$
37. $\dfrac{14}{32}$ $\dfrac{7}{16}$
38. $\dfrac{21}{56}$ $\dfrac{3}{8}$
39. $\dfrac{12}{16}$ $\dfrac{3}{4}$

40. $\dfrac{100}{40}$ $\dfrac{5}{2}$
41. $\dfrac{6 \times 3}{42}$ $\dfrac{3}{7}$
42. $\dfrac{4 + 3}{35}$ $\dfrac{1}{5}$
43. $\dfrac{36}{9}$ 4

Replace the variable to name the equivalent fraction.

44. $\dfrac{3}{4} = \dfrac{x}{8}$ $\dfrac{6}{8}$
45. $\dfrac{18}{36} = \dfrac{s}{2}$ $\dfrac{1}{2}$
46. $\dfrac{20}{60} = \dfrac{k}{6}$ $\dfrac{2}{6}$

47. $\dfrac{5}{75} = \dfrac{m}{15}$ $\dfrac{1}{15}$
48. $\dfrac{y}{8} = \dfrac{16}{32}$ $\dfrac{4}{8}$
49. $\dfrac{4}{9} = \dfrac{m}{63}$ $\dfrac{28}{63}$

50. $\dfrac{8}{64} = \dfrac{s}{16}$ $\dfrac{2}{16}$
51. $\dfrac{2}{46} = \dfrac{x}{23}$ $\dfrac{1}{23}$
52. $\dfrac{15}{60} = \dfrac{y}{12}$ $\dfrac{3}{12}$

For use with Section 3-6.

Write as a decimal.

1. $\frac{1}{2}$ 0.5
2. $\frac{7}{8}$ 0.875
3. $\frac{57}{100}$ 0.57
4. $\frac{1}{5}$ 0.2

5. $\frac{3}{5}$ 0.6
6. $\frac{3}{3}$ 1
7. $\frac{25}{50}$ 0.5
8. $\frac{3}{10}$ 0.3

9. $\frac{9}{10}$ 0.9
10. $\frac{8}{12}$ 0.6
11. $\frac{5}{8}$ 0.625
12. $\frac{6}{16}$ 0.375

13. $\frac{4}{25}$ 0.16
14. $\frac{2}{40}$ 0.05
15. $\frac{18}{100}$ 0.18
16. $\frac{16}{60}$ $0.2\overline{6}$

Write as a fraction in lowest terms.

17. 0.2 $\frac{1}{5}$
18. 2.6 $\frac{13}{5}$
19. 1.8 $\frac{9}{5}$
20. 0.75 $\frac{3}{4}$
21. 0.375 $\frac{3}{8}$
22. 0.051 $\frac{51}{1000}$
23. 2.16 $\frac{54}{25}$
24. 0.85 $\frac{17}{20}$
25. 3.125 $\frac{25}{8}$
26. 0.15 $\frac{3}{20}$
27. 1.01 $\frac{101}{100}$
28. 0.90 $\frac{9}{10}$

For use with Section 3-7.

Write as a decimal and as a percent.

29. $\frac{45}{100}$ 0.45 = 45%
30. $\frac{6}{8}$ 0.75 = 75%
31. $\frac{3}{10}$ 0.3 = 30%
32. $\frac{2}{5}$ 0.4 = 40%

33. $\frac{12}{12}$ 1 = 100%
34. $\frac{600}{1000}$ 0.6 = 60%
35. $\frac{2}{16}$ 0.125 = 12.5%
36. $\frac{260}{1000}$ 0.26 = 26%

37. $\frac{25}{50}$ 0.5 = 50%
38. $\frac{7}{8}$ 0.875 = 87.5%
39. $\frac{6}{10}$ 0.6 = 60%
40. $\frac{1}{40}$ 0.025 = 2.5%

41. $\frac{8}{25}$ 0.32 = 32%
42. $\frac{37}{100}$ 0.37 = 37%
43. $\frac{520}{1000}$ 0.52 = 52%
44. $\frac{8}{20}$ 0.4 = 40%

Write as a fraction in lowest terms.

45. 16% $\frac{4}{25}$
46. 60% $\frac{3}{5}$
47. 24% $\frac{6}{25}$
48. 80% $\frac{4}{5}$
49. 36% $\frac{9}{25}$
50. 64% $\frac{16}{25}$
51. 45% $\frac{9}{20}$
52. 72% $\frac{18}{25}$
53. 12% $\frac{3}{25}$
54. 3% $\frac{3}{100}$
55. 31.5% $\frac{63}{200}$
56. 18% $\frac{9}{50}$
57. 120% $\frac{6}{5}$
58. 8% $\frac{2}{25}$
59. 64.8% $\frac{81}{125}$
60. 72.4% $\frac{181}{250}$
61. 420% $\frac{21}{5}$
62. 28.5% $\frac{57}{200}$

For use with Section 4-1.

Add or subtract. Simplify the result.

1. $\dfrac{2}{7} + \dfrac{3}{7}$ $\frac{5}{7}$

2. $\dfrac{13}{14} - \dfrac{6}{14}$ $\frac{1}{2}$

3. $\dfrac{1}{3} + \dfrac{1}{3}$ $\frac{2}{3}$

4. $\dfrac{8}{9} - \dfrac{5}{9}$ $\frac{1}{3}$

5. $\dfrac{3}{8} + \dfrac{1}{8} + \dfrac{2}{8}$ $\frac{3}{4}$

6. $5\frac{3}{4} - \frac{1}{4}$ $5\frac{1}{2}$

7. $\dfrac{7}{12} + \dfrac{1}{12}$ $\frac{2}{3}$

8. $3\frac{1}{2} - \frac{1}{2}$ 3

9. $\dfrac{15}{16} - \dfrac{8}{16}$ $\frac{7}{16}$

10. $4\frac{1}{6} + 1\frac{3}{6}$ $5\frac{2}{3}$

11. $9\frac{2}{3} - \frac{1}{3}$ $9\frac{1}{3}$

12. $\frac{2}{7} + 5\frac{3}{7}$ $5\frac{5}{7}$

13. $5\frac{6}{8} - 4\frac{4}{8}$ $1\frac{1}{4}$

14. $6\frac{1}{4} - 6$ $\frac{1}{4}$

15. $3\frac{5}{8} - \frac{5}{8}$ 3

Solve. Simplify the result.

16. $x = \dfrac{1}{4} + \dfrac{1}{8}$ $\frac{3}{8}$

17. $t = 2\frac{1}{3} + 3\frac{1}{2}$ $5\frac{5}{6}$

18. $\dfrac{1}{3} + \dfrac{1}{6} + \dfrac{1}{6} = h$ $\frac{2}{3}$

19. $4\frac{5}{6} - \frac{1}{3} = h$ $4\frac{1}{2}$

20. $\dfrac{15}{16} - \dfrac{3}{8} = m$ $\frac{9}{16}$

21. $\dfrac{5}{7} - \dfrac{1}{2} = r$ $\frac{3}{14}$

22. $t = \dfrac{1}{4} + \dfrac{1}{3}$ $\frac{7}{12}$

23. $h = 6\frac{5}{9} - 4\frac{1}{6}$ $2\frac{7}{18}$

24. $s = \dfrac{5}{6} + \dfrac{1}{9}$ $\frac{17}{18}$

25. $3\frac{5}{12} + 4\frac{1}{4} = y$ $7\frac{2}{3}$

26. $\frac{1}{3} + 1\frac{2}{9} = q$ $1\frac{5}{9}$

27. $k = 8\frac{1}{10} - 7\frac{3}{10}$ $\frac{4}{5}$

28. $v = \dfrac{1}{2} + \dfrac{4}{9}$ $\frac{17}{18}$

29. $w = 2\frac{3}{7} + 4\frac{1}{3}$ $6\frac{16}{21}$

30. $d = \dfrac{9}{10} - \dfrac{1}{2}$ $\frac{2}{5}$

31. $1\frac{7}{12} - \frac{1}{4} = m$ $1\frac{1}{3}$

32. $\dfrac{15}{16} - \dfrac{3}{4} = n$ $\frac{3}{16}$

33. $9\frac{1}{4} - 3\frac{3}{4} = b$ $5\frac{1}{2}$

Find the value when $x = 3$, $y = 4$, $z = 2$. Simplify the result.

34. $\dfrac{1}{x} + \dfrac{1}{x} + \dfrac{1}{x}$ 1

35. $\dfrac{1}{x} - \dfrac{1}{y}$ $\frac{1}{12}$

36. $\dfrac{y - x}{z}$ $\frac{1}{2}$

37. $\dfrac{x}{10} + \dfrac{z}{10}$ $\frac{1}{2}$

38. $\dfrac{x + x}{z}$ 3

39. $\dfrac{2}{x} - \dfrac{1}{y}$ $\frac{5}{12}$

40. $\dfrac{y}{7} - \dfrac{z}{7}$ $\frac{2}{7}$

41. $\dfrac{z}{x} - \dfrac{1}{y}$ $\frac{5}{12}$

42. $\dfrac{y}{5} - \dfrac{y}{10}$ $\frac{2}{5}$

43. $\dfrac{x + z}{x + y + z}$ $\frac{5}{9}$

44. $\dfrac{x}{7} + \dfrac{1}{z}$ $\frac{13}{14}$

45. $\dfrac{x}{y} - \dfrac{z}{x}$ $\frac{1}{12}$

For use with Section 4-2.

Multiply. Simplify the result.

1. $\frac{1}{4} \times \frac{1}{3}$ $\frac{1}{12}$

2. $\frac{2}{3} \times \frac{1}{4}$ $\frac{1}{6}$

3. $\frac{1}{5} \times \frac{5}{6} \times \frac{1}{2}$ $\frac{1}{12}$

4. $2\frac{1}{2} \times 1\frac{1}{3}$ $3\frac{1}{3}$

5. $\frac{2}{5} \times \frac{3}{6}$ $\frac{1}{5}$

6. $4\frac{1}{2} \times 2$ 9

7. $\frac{2}{9} \times \frac{3}{9}$ $\frac{2}{27}$

8. $\frac{1}{7} \times \frac{1}{3} \times \frac{2}{3}$ $\frac{2}{63}$

9. $\frac{5}{7} \times \frac{7}{5}$ 1

10. $\frac{3}{4} \times \frac{2}{8}$ $\frac{3}{16}$

11. $1\frac{1}{4} \times 3\frac{1}{5}$ 4

12. $4 \times 2\frac{1}{2}$ 10

13. $\frac{2}{7} \times \frac{1}{2} \times \frac{7}{2}$ $\frac{1}{4}$

14. $\frac{3}{3} \times \frac{4}{9}$ $\frac{4}{9}$

15. $1\frac{3}{8} \times \frac{4}{2}$ $\frac{11}{4}$

Solve. Simplify the result.

16. $x = \frac{1}{2} \div \frac{1}{2}$ 1

17. $y = \frac{3}{4} \div \frac{2}{3}$ $1\frac{1}{8}$

18. $a = 1\frac{1}{3} \div 2\frac{1}{6}$ $\frac{8}{13}$

19. $b = \frac{3}{4} \times 4$ 3

20. $d = \frac{4}{5} \div \frac{1}{3}$ $2\frac{2}{5}$

21. $2\frac{4}{9} \times 3 = c$ $7\frac{1}{3}$

22. $m = 8 \div \frac{1}{2}$ 16

23. $n = \frac{2}{9} \div \frac{2}{9}$ 1

24. $p = 1\frac{3}{8} \times 8$ 11

25. $\frac{3}{8} \div \frac{6}{4} = t$ $\frac{1}{4}$

26. $2\frac{5}{6} \times \frac{2}{4} = s$ $\frac{17}{12}$

27. $k = \frac{4}{3} \div \frac{8}{9}$ $1\frac{1}{2}$

28. $h = 3\frac{1}{9} \times \frac{2}{3}$ $\frac{56}{27}$

29. $\frac{5}{8} \div \frac{3}{4} = z$ $\frac{5}{6}$

30. $q = 1\frac{7}{8} \div \frac{3}{16}$ 10

31. $3\frac{5}{6} \times \frac{1}{7} = x$ $\frac{23}{42}$

32. $f = 4\frac{2}{3} \times 1\frac{7}{8}$ $8\frac{3}{4}$

33. $\frac{4}{9} \div \frac{8}{3} = m$ $\frac{1}{6}$

Find the value when $s = 1$, $t = 2$, $u = 3$, $v = 5$.

34. $\frac{s}{t} \div \frac{s}{t}$ 1

35. $\frac{t}{u} \times \frac{s}{v}$ $\frac{2}{15}$

36. $\frac{s}{t} \times \frac{t}{u} \times \frac{u}{v}$ $\frac{1}{5}$

37. $\frac{v}{u} \div \frac{u}{t}$ $1\frac{1}{9}$

38. $\frac{t}{v} \div v$ $\frac{2}{25}$

39. $\frac{s}{v} \div \frac{t}{v}$ $\frac{1}{2}$

40. $\frac{v}{t} \times \frac{t}{u} \times u$ 5

41. $\frac{v}{s} \times \frac{t}{u}$ $3\frac{1}{3}$

42. $\frac{s}{v} \div \frac{u}{t}$ $\frac{2}{15}$

43. $\frac{u}{t} \div t$ $\frac{3}{4}$

44. $\frac{t}{s} \times \frac{u}{v} \times u$ 6

45. $\frac{v}{u} \div \frac{v}{t}$ $\frac{2}{3}$

For use with Section 4-3.

Add or subtract. Estimate to check your work.

1. $1.715 + 2.657$ 4.372
2. $6.9 - 0.7$ 6.2
3. $0.01 + 3.328$ 3.338
4. $23.1 + 0.067 + 9.16$ 32.327
5. $8.5 - 5.004$ 3.496
6. $54.08 - 1.76$ 52.32
7. $3.1 - 0.09$ 3.01
8. $12.073 + 7.6$ 19.673
9. $14.96 - 3.002$ 11.958
10. $200 - 56.8$ 143.2
11. $18.9 + 6.83$ 25.73
12. $0.001 + 6.92$ 6.921
13. $161.02 - 9.88$ 151.14
14. $27.9 + 30.01 + 0.002$ 57.912
15. $13.57 - 2.34$ 11.23
16. $8.997 - 6.899$ 2.098

Solve.

17. $s = 64.021 - 0.9$ 63.121
18. $21.7 + 0.02 + 6.89 = t$ 28.61
19. $115.75 = m + 110.33$ 5.42
20. $42.006 - k = 21.8$ 20.206
21. $0.042 - 0.007 = v$ 0.035
22. $39.62 - 0.07 = h$ 39.55

For use with Section 4-4.

Solve. Check by estimating the solution.

23. $n = 6.32 \times 0.07$ 0.4424
24. $14.04 \div 12 = s$ 1.17
25. $63.57 \div 3 = t$ 21.19
26. $0.506 \times 27.2 = b$ 13.7632
27. $54.02 \times 0.008 = v$ 0.43216
28. $0.168 \div 4 = m$ 0.042
29. $\dfrac{1.56}{6} = x$ 0.26
30. $\dfrac{14.078}{2} = y$ 7.039
31. $14.375 \div 5 = k$ 2.875
32. $74.058 \times 0.03 = p$ 2.22174

Solve.

33. $\$23.96 \times 7 = m$ $167.72
34. $s = 0.032 \times 14.82$ 0.47424
35. $\$19.05 \div 3 = k$ $6.35
36. $22\% \times 89 = q$ 19.58
37. $w = 0.8 \times 17.1 \times 0.02$ 0.2736
38. $48.72 \times 100 = t$ 4872
39. $\dfrac{22.5}{15} = b$ 1.5
40. $\dfrac{\$10.24}{8} = r$ $1.28

For use with Section 4-5.

Simplify.

1. $(5 + 7)(2 + 6)$ 96

2. $[(10 + 4)(2)] \div 7$ 4

3. $(0.01 \times 15.6) \div 3$ 0.052

4. $15 \div (5 \times 3 \times 10)$ 0.1

5. $23 + (10.7 - 0.09)$ 33.61

6. $(144 \div 12) + 13$ 25

7. $(\frac{1}{3} + \frac{1}{6})12$ 6

8. $(1\frac{3}{4} - \frac{1}{2}) + \frac{2}{3}$ $1\frac{11}{12}$

9. $[(6)(9)] \div 18$ 3

10. $(12.05 - 8.167) + 0.05$ 3.933

Solve.

11. $m = 5^2(7 + 3)$ 250

12. $[(15 + 20) \div 5] + 6 \cdot 2^2 = x$ 31

13. $\left(\frac{1}{8} + \frac{3}{4}\right) \div \frac{3}{16} = s$ $4\frac{2}{3}$

14. $\dfrac{(5 + 6 + 10.7)10}{7} = b$ 31

15. $(8 \times 9) + 14.382 = c$ 86.382

16. $r = 8[9 - (5 + 1 + 2)]$ 8

17. $(12.032 - 0.08) + 1.7 = p$ 13.652

18. $t = (4^2 + 4.48) - 9.03$ 11.45

19. $b = (4 \times 3 \times 4) \div 6$ 8

20. $(121 \div 11) + (6.2)(7) = n$ 54.4

For use with Section 5-2.

Make a table to find the solution set. Use the given replacement set.

21. $4m + (56 \div 7) = 4^2$ $\{1, 2, 3\}$ {2}

22. $6 - k = 24 \div 8$ $\{3, 5, 7\}$ {3}

23. $5.8 + s = 14.23 - 7.15$ $\{1.28, 6.0, 10\}$ {1.28}

24. $3(t + 9) = 6 \times 7$ $\{0, 4, 5\}$ {5}

25. $\dfrac{(9 \times 4) - 4}{h} = 2^3$ $\{2, 4, 6\}$ {4}

26. $(18 - 12)5 = b \div 2$ $\{1, 15, 60\}$ {60}

27. $4x = 2$ $\left\{0, \frac{1}{2}, 1, 2\right\}$ $\left\{\frac{1}{2}\right\}$

28. $w(4 + 3) = 5w + (100 \div 10)$ $\left\{\frac{1}{4}, 5, 7, 10\right\}$ {5}

29. $v^3 - 1 = 13 - 6$ $\{2, 3, 4\}$ {2}

30. $\left(\frac{3}{8} - \frac{1}{4}\right)(8) = \frac{k}{k}$ $\{5, 8\}$ {5, 8}

31. $3d + d + 4d = 64 \div 2$ $\{1.5, 2, 4\}$ {4}

32. $(0 \cdot p) + 5 = p - 12$ $\{3, 6, 7.4, 17\}$ {17}

For use with Section 5-3.

Add or subtract the given number to or from both members.

Simplify the result.

1. $16 - 8 = 2 + 6$; add 3 $\quad 11 = 11$
2. $10.5 = 4.6 + 5.9$; subtract 0.2 $\quad 10.3 = 10.3$
3. $\dfrac{14.82}{2} = 3 \times 2.47$; subtract 2.39 $\quad 5.02 = 5.02$
4. $5 + 8 = 6 + 7$; subtract 0 $\quad 13 = 13$
5. $6 + 8 = 2 \times 7$; subtract 12 $\quad 2 = 2$
6. $14 - 5 = 3 \times 3$; add 3 $\quad 12 = 12$
7. $8 - 7 = 36 \div 36$; add 0.5 $\quad 1.5 = 1.5$

Add or subtract. Simplify the result.

8. $k + 8$; subtract 8 $\quad k + 8 - 8 = k$ 9. $15 + m$; subtract 15 $\quad 15 - 15 + m = m$
10. $t - 5.4$; add 5.4 $\quad t - 5.4 + 5.4 = t$ 11. $s - 8.62$; add 8.62 $\quad s - 8.62 + 8.62 = s$
12. $r + 20$; subtract 20   13. $h - 9.2$ add 9.2 $\quad h - 9.2 + 9.2 = h$
$\qquad\qquad\qquad\qquad r + 20 - 20 = r$

For use with Section 5-4.

Multiply or divide both members. Simplify the result.

14. $5 + 3 = 8$; divide by 4 $\quad 2 = 2$ 15. $28 = 4 \cdot 7$; divide by 14 $\quad 2 = 2$
16. $24 = 4 \cdot 6$; divide by 3 $\quad 8 = 8$ 17. $25 \cdot 2 = 5 \cdot 10$; multiply by 2 $\quad 100 = 100$
18. $10 = 7 + 3$; multiply by 4 $\quad 40 = 40$ 19. $3^2 = 9$; multiply by 5 $\quad 45 = 45$
20. $\dfrac{15}{3} = \dfrac{25}{5}$; multiply by 6 $\quad 30 = 30$ 21. $7 \cdot 8 = 28 \cdot 2$; multiply by $\dfrac{1}{2}$ $\quad 28 = 28$
22. $4^2 = 32 \div 2$; multiply by 3 $\quad 48 = 48$ 23. $11 + 9 = 4 \cdot 5$; divide by 10 $\quad 2 = 2$

To solve, multiply or divide as indicated.

24. $9t = 72$; divide by 9 $\quad 8$ 25. $s \cdot 7 = 63$; divide by 7 $\quad 9$
26. $\dfrac{1}{4} \cdot x = 48.6$; multiply by 4 $\quad 194.4$ 27. $6 = \dfrac{m}{6}$; multiply by 6 $\quad 36$
28. $144 = 12k$; divide by 12 $\quad 12$ 29. $n \cdot 8 = 48$; divide by 8 $\quad 6$
30. $\dfrac{n}{10} = 12$; multiply by 10 $\quad 120$ 31. $\dfrac{a}{2} = 64$; multiply by 2 $\quad 128$
32. $169 = p \cdot 13$; divide by 13 $\quad 13$ 33. $150 = 25c$; divide by 25 $\quad 6$

For use with Section 5-6.

Solve.

1. A rectangle has width 6.2 cm and length 3.8 cm. Find the area. 23.56 cm²

2. The diameter of a circle is 14 m. Find the circumference. 44 m

3. Find the area of a square whose side measures 14 cm. 196 cm²

4. The lengths of the sides of a triangle are 2.2 m, 0.3 m, and 5.25 m. Find the perimeter. 7.75 m

5. A circle has radius 7 mm. Find the area. 154 mm²

6. A triangle has a base of length 24 cm and height 10 cm. Find the area. 120 cm²

7. A square has a side of length 6.04 m. Find the perimeter. 24.16 m

8. A rectangle has length 9.72 m and width 4.6 m. Find the perimeter. 28.64 m

9. A circle has radius 38.5 cm. Find the circumference. 242 cm

10. The perimeter of a square is 128 mm. Find the length of each side. 32 mm

11. A rectangle has length 0.5 m and width 0.4 m. Find the area. 0.2 m²

12. A triangle has a base of 0.52 m and a height of 1.5 m. Find the area. 0.39 m²

13. The circumference of a circle is 198 cm. Find the diameter. 63 cm

14. The length of a rectangle is 3.2 m and the width is 1.15 m. Find the perimeter. 8.7 m

15. If the area of a square is 169 cm², what is the length of its side? 13 cm

16. The diameter of a circle is 14 cm. Find the area. 154 cm²

17. The perimeter of a triangle is 90 mm. All three sides are the same length. Find the length of each side. 30 mm

18. A triangle has base 8 m and height 8 m. What is the area of the triangle? 32 m²

For use with Section 5-7. In Ex. 1–20, numbers which produce true statements are circled.

For each member of the replacement set, show whether a true or a false statement results when the variable in the sentence is replaced by the number.

1. $4m < 18$; {⓪,①,③,5}
2. $72 \div t > 9$, {①,8, 12}
3. $6b - 10 > 3$; {2,④,⑦}
4. $s^2 > 2s$; {1, 2,③,④}
5. $16 < (x + 8)(0.5)$; {4, 8,㉚}
6. $24.65 - r < 10$; {4.65, 14,⓴.㉓}
7. $(5 + 3) - n > 2$; {⓪,④,7, 8}
8. $9 < 6 + b$; {0, 2,⑥,㉚}
9. $10k > 5$; $\left\{0, \dfrac{1}{2}, \dfrac{1}{5}, ②\right\}$
10. $3 < 18c$; $\left\{\dfrac{1}{3}, \dfrac{1}{2}, ⑨, ⑫\right\}$

For use with Section 5-8.

For each member of the replacement set, show whether a true or a false statement results when the variable is replaced by the number.

11. $k + 8 \geq 12$; {2,④,⑥}
12. $3p \geq 22$; {5, 7,⑧,⑩}
13. $s \leq 16.5$; {⑦,⑯, 17}
14. $^-7 \leq c$; {$^-$9,⊖,⓪,①}
15. $6r \geq 42$; {2,⑦,⑩,⑫}
16. $4(8 - n) \neq 32$; {0,②,⑤}
17. $(27 - k) + 1 \leq 20$; {0, 7,⑧,⑩}
18. $y + 14 \leq 36$; {⑩,㉒, 30}
19. $50h \geq 25$; $\left\{0, \dfrac{1}{5}, \dfrac{1}{2}, \dfrac{4}{5}\right\}$
20. $12 \leq 10 + d$; {½, 1½, ②, ②½}

For use with Section 7-1.

Combine similar terms to simplify.

21. $15r + 9r$ 24r
22. $22t - 10t + 8t$ 20t
23. $2(7 - 6m)$ 14 − 12m
24. $a + b + 2a + 3b$ 3a + 4b
25. $10vw + 4(vw + 7)$ 14rw + 28
26. $(9 + 3)(5k - 2)$ 60k − 24
27. $(s + 5s) - 2s$ 4s
28. $16xy + 7x - 2y - 8xy + 2y$ 8xy + 7x
29. $3(x + 2) + 2(2 + x)$ 5x + 10
30. $120 + 80 + 10m - 200 - 10m$ 0
31. $7(a + 3 + 2c) + 6(2c + 1)$
32. $4(4b + 6) - b$ 15b + 24
33. $14p + 2p - 8p$ 8p
34. $28r - 16r - 8r$ 4r
35. $6z + 6y + 8y - 3z$ 3z + 14y
36. $10k + m - 6k + 2 + 3m$ 4m + 4k + 2
37. $48qr - 3st - 12qr + 5st$
38. $5(6q + 7f) + 2f(6 + 1)$ 30q + 49f
39. $(7w + 2)8 + (6 + 4w)3$ 68w + 34
40. $12a + 3b - 8a + 7b - 6$ 4a + 10b − 6
41. $f + 2f + g + 5f - g$ 8f
42. $4(36c + 10d) - 11d - 46c$ 98c + 29d

31. 7a + 26c + 27   37. 36qr + 2st

For use with Section 7-2.

Solve. Begin by stating what number should be added to or subtracted from both members.

1. $m + 7 = 15$ 8

2. $0.5 + t = 1.5$ 1

3. $s - 4 = 16$ 20

4. $p - 13 = 23$ 36

5. $\dfrac{7}{8} = x - 1$ $1\frac{7}{8}$

6. $a - \dfrac{2}{3} = 2$ $2\frac{2}{3}$

7. $\dfrac{4}{7} = k + \dfrac{2}{7}$ $\frac{2}{7}$

8. $\dfrac{5}{9} = b - \dfrac{1}{9}$ $\frac{2}{3}$

Solve and check.

9. $q - 10.5 = 6.4$ 16.9

10. $17 - m = 9$ 8

11. $14 + k = 43$ 29

12. $23.65 - p = 18.31$ 5.34

13. $s + 115 = 694$ 579

14. $r + 67 = 81$ 14

15. $96 - d = 42$ 54

16. $w - 18.6 = 2.5$ 21.1

17. $39 = n + 13$ 26

18. $6.05 = x - 2.38$ 8.43

For use with Sections 7-3 and 7-4.

Solve. Check your answer.

19. $6r = 30$ 5

20. $0.4t = 1.2$ 3

21. $3q = 42$ 14

22. $\dfrac{1}{3} = 2m$ $\frac{1}{6}$

23. $\dfrac{n}{2} = 16$ 32

24. $8t = \dfrac{8}{9}$ $\frac{1}{9}$

25. $\dfrac{15y}{3} = 0.25$ 0.05

26. $7r = 49$ 7

27. $\dfrac{5b}{6} = 10$ 12

28. $\dfrac{c}{3} = 5 + 1$ 18

29. $\dfrac{3}{8}g = \dfrac{6}{4}$ 4

30. $7 + 2x = 55$ 24

31. $9c - 8 = 46$ 6

32. $\dfrac{t}{8} = 7$ 56

33. $\dfrac{k}{100} = 0.05$ 5

34. $\dfrac{w}{18} = 2$ 36

35. $6v - 13.2 = 11.4$ 4.1

36. $\dfrac{4a}{5} = 16$ 20

37. $\dfrac{f}{81} = 1$ 81

38. $\dfrac{3}{4}t = \dfrac{7}{16}$ $\frac{7}{12}$

39. $2.1 + 4x = 3.7$ 0.4

40. $11z = 121$ 11

41. $8e \div 4 = 4.8$ 2.4

42. $5j + 21 = 86$ 13

43. $\dfrac{g}{12} = 168$ 2016

44. $\dfrac{2}{5}h = \dfrac{16}{25}$ $1\frac{3}{5}$

45. $\dfrac{k}{1.05} = 4.8$ 5.04

For use with Sections 7-5, 7-6, 7-7.　　**6.** $\frac{1}{2}z + 17 = 26$; 18

Write and solve an equation to answer the question.　　$5n + 8 = 38$; 6

1. The sum of five times a number and 8 is 38. What is the number?

2. If twelve times a number is decreased by 16, the result is equal to 20. Find the number. $12x - 16 = 20$; $x = 3$

3. If 1.56 is added to four times a number, the result is equal to 3.64. Find the number. $4y + 1.56 = 3.64$; $y = 0.52$

4. The sum of a whole number and a number 3 greater than the first is 19. Find the two numbers. $w + w + 3 = 19$; 8 and 11

5. Brandon is 7 cm shorter than Teresa. The sum of their heights is 337 cm. How tall is each person? $s + s - 7 = 337$; Teresa: 172 cm, Brandon: 165 cm

6. If half a number is increased by 17, the result is 26. Find the number.

7. Jon worked 1 more hour on his project than Mia. Together they worked 8 hours. How long did each person work? $h + h + 1 = 8$; Jon: 4.5 hr, Mia: 3.5 hr

8. Soap A costs 23¢ more than Soap B. Soap A costs 89¢. How much does Soap B cost? $89 - 23 = b$; Soap B: 66¢

9. The sum of three consecutive even numbers is 30. What are the three numbers? $x + 1 + x + 3 + x + 5 = 30$; 8, 10, 12

10. Vincent has scored 5 more goals than Orlando. Together they have scored 13 goals. How many goals has Vincent scored? $g + g + 5 = 13$; 9 goals

11. April worked 15 more hours this week than last. She worked 27 hours last week. How many hours did she work this week? $27 + 15 = j$; 42 hr

12. Two jugs hold a total of 4.8 liters. The large jug holds twice as much as the small jug. How much does each jug hold? $x + 2x = 4.8$; 1.6 ℓ and 3.2 ℓ

13. A scientist had 9 clean test tubes at the end of an experiment. This was three-fourths of the original number of test tubes. How many test tubes were used? $\frac{3}{4}t = 9$; 3 test tubes

14. If a number is increased by 27, the result is the same as four times the number. Find the number. $k + 27 = 4k$; 9

15. The length of a rectangle is 1 cm more than four times the width. The perimeter is 42 cm. Find the length and width. $P = 2l + 2w$; $l = 17$ cm, $w = 4$ cm

16. Eight times a number, decreased by four, is the same as 2.5 times the same number increased by 12.5. Find the number. $8n - 4 = 2.5n + 12.5$; 3

17. Angela has three more dimes than quarters. She has 27 coins in all. How many dimes does she have? $y + y + 3 = 27$; 15 dimes

18. The sum of an even number and the next two consecutive odd numbers is 64. Find the numbers. $p + 1 + p + 2 + p + 4 = 64$; 20, 21, 23

For use with Section 9-3.

Solve.

1. $-10 = 2 + t$ -12
2. $m = -(-8 + 37)$ -29
3. $q + 8 = -14$ -22
4. $-7 + (-9) = r$ -16
5. $-[12 + (15 - 2)] = a$ -25
6. $4 + (-3) = s$ 1
7. $11 = 7 + w$ 4
8. $-(15 + 20) = b$ -35
9. $y = -\dfrac{1}{3} + \dfrac{1}{4} - \dfrac{1}{12}$
10. $1\frac{1}{2} + (-3\frac{1}{2}) = f$ -2
11. $-3 = k + 10$ -13
12. $-[24 + (-12)] = h$ -12
13. $14 = x + 8.75$ 5.25
14. $-27 = -5 + u$ -22

For use with Section 9-4.
Simplify.

15. $-10 + 6 + (-17) + 2$ -19
16. $7 + (-3) + (-4) + 7$ 7
17. $3\frac{3}{8} + 2\frac{1}{4} + (-2\frac{1}{4})$ $3\frac{3}{8}$
18. $-4\frac{2}{5} + 9\frac{1}{5} + (-9\frac{1}{5})$ $-4\frac{2}{5}$
19. $27 + (-27) + 3$ 3
20. $-36 + 14 + 7 + (-14) + (-7)$ -36
21. $56 + (-10) + 2 + (-56)$ -8
22. $-12 + 6 + 8 + (-6)$ -4
23. $-11 + (-6) + 24$ 7
24. $5 + 2\frac{7}{9} + (-5) + (-2\frac{7}{9})$ 0

For use with Section 9-5.
Solve.

25. $17 - (-2) = m$ 19
26. $p = 4 - (-14)$ 18
27. $1 - 13 = x$ -12
28. $8 - 9 = k$ -1
29. $6 - (-6) = s$ 12
30. $36 - 27 = r$ 9
31. $38 - 40 = w$ -2
32. $1\frac{3}{4} - (-1\frac{1}{4}) = a$ 3
33. $15 - (-18) = v$ 33
34. $-0.45 - (-0.15) = b$ -0.3
35. $21 - (-7) = c$ 28
36. $f = 9.4 - (-1.5)$ 10.9
37. $96 - 13 = k$ 83
38. $c = 25 + (-18)$ 7
39. $7 - 16 = q$ -9
40. $x = 21 - 30$ -9
41. $1\frac{1}{3} - (-\frac{1}{3}) = a$ $1\frac{2}{3}$
42. $b = 100 + (-50)$ 50
43. $g = 16 - (-32)$ 48
44. $48 + (-12) = h$ 36
45. $14 - 6 = j$ 8
46. $38 - 56 = s$ -18
47. $61.7 - (-2.3) = w$ 64.0
48. $x = 17 + (-30)$ -13
49. $-1.65 - (-3.86) = h$ 2.21
50. $26 - 49 = r$ -23

For use with Section 10-1.

Solve.

1. $(-6 \cdot 3) + (-3 \cdot 6) = x$ -36 **2.** $8(-6) = p$ -48

3. $-9 \cdot 6 = z$ -54 **4.** $s = (11)(12)$ 132

5. $a = -0.165(3)$ -0.495 **6.** $-155 \cdot 0 = b$ 0

7. $-16 \cdot \dfrac{7}{8} = f$ -14 **8.** $-\dfrac{1}{2} \cdot 50 = g$ -25

9. $h + 3(-4) = 12$ 24 **10.** $(-5 \cdot 7) + y = 5$ 40

11. $(-5.2)(4) + 1.2 = w$ -19.6 **12.** $(-9 \cdot 3) + (-3) = k$ -30

13. $\left(-\dfrac{4}{5}\right)\left(\dfrac{15}{4}\right) = d$ -3 **14.** $\left(\dfrac{1}{7}\right)\left(-\dfrac{5}{3}\right) = e$ $-\dfrac{5}{21}$

15. $-2(0.985) = j$ -1.97 **16.** $r = (14.83)(1)$ 14.83

17. $(6 \cdot 0) + 1.05 = n$ 1.05 **18.** $27 + (-9 \cdot 2) = m$ 9

19. $\left(\dfrac{2}{3}\right)(-4) = b$ $-\dfrac{8}{3}$ **20.** $\left(-\dfrac{3}{4}\right)\left(\dfrac{2}{7}\right) = q$ $-\dfrac{3}{14}$

21. $r = (-1)(27)$ -27 **22.** $-7 \cdot 8 = c$ -56

23. $5(-3) + 3(-2) = a$ -21 **24.** $f = (-0.54)(3)$ -1.62

For use with Section 10-3.

Tell whether the expression represented is positive, negative, or zero. Then simplify. **26.** P; 60   **31.** P; $\dfrac{1}{16}$ **32.** N; $-\dfrac{1}{3}$ **35.** N; -75 **38.** P; 168

25. $-3 \cdot 2 \cdot 5$ N; -30 **26.** $-10 \cdot 1(-6)$ **27.** $(-4)^2$ P; 16

28. $-8(-2)(-1)$ N; -16 **29.** $0 \cdot 7(-9)$ Z; 0   **30.** $(-3)^3$ N; -27

    N; $-\dfrac{4}{105}$

31. $\left(-\dfrac{1}{2}\right)\left(-\dfrac{1}{2}\right)\left(\dfrac{1}{4}\right)$ **32.** $\dfrac{2}{3} \cdot \dfrac{1}{2}(-1)$ **33.** $\left(-\dfrac{2}{5}\right)\left(-\dfrac{1}{7}\right)\left(-\dfrac{2}{3}\right)$

34. $(-1.5)(-2)(3)$ P; 9   **35.** $(-3)(-5)^2$ **36.** $(2.6)(0.5)(4)$ P; 5.2

37. $(4)(-3)(9)$ N; -108 **38.** $-7 \cdot 3(-8)$ **39.** $(-2)6 \cdot 0$ Z; 0

Evaluate. Let $k = -2$, $h = -3$, $m = -1$, $n = 4$, $p = 2$.

40. $k \cdot h \cdot m$ -6 **41.** $(-n)(-4)$ 16 **42.** $(-3)h - p$ 7

43. $kn + mp$ -10 **44.** $(-p) + np$ 6 **45.** $(k)(-n)(-1)$ -8

46. $\dfrac{n}{p} + (-n)$ -2 **47.** $\dfrac{1}{n}(h)$ $-\dfrac{3}{4}$ **48.** $\dfrac{k}{n}(p)$ -1

49. $4 \cdot k \cdot n \cdot p$ -64 **50.** $h \cdot k + m$ -7 **51.** $mn \cdot 0$ 0

For use with Section 10-4.

Simplify by combining similar terms.

1. $7x + 3(-2x - 4y)$ $x - 12y$
2. $3m^2 + m + (-m^2)$ $2m^2 + m$
3. $12ab - 4c + (-10ab)$ $2ab - 4c$
4. $(5f - 10g)2 + 20g + (-10f)$ 0
5. $\left(\dfrac{2}{3}m + \dfrac{2}{5}p\right)15 - (-8m)$ $18m + 6p$
6. $\dfrac{1}{2}(k + 4h) + (-h)$ $\dfrac{1}{2}k + h$
7. $6w + (-7u) - 3 + 4u - 2w$
8. $10(pq + 6) - 7pq - 30$ $3pq + 30$
9. $18d + [-12(d - 2)]$ $6d + 24$
10. $-2s + 7t - 4t + 8s$ $6s + 3t$
11. $3r^2 + 11 + (r \cdot r) - 7$ $4r^2 + 4$
12. $9mn - n + 3m - 2mn + n$ $7mn + 3m$
13. $\dfrac{7}{8}h - \dfrac{5}{7}j + \left(-\dfrac{3}{8}h\right) + \dfrac{4}{7}j$
14. $\dfrac{3}{4}(8a + c) + (-3a) + \dfrac{1}{4}c$ $3a + c$
15. $2.5d - 6f + (-0.5d) + 10f$
16. $16x - 10y + 20x + 5y$ $36x - 5y$
17. $-7w + 4(-3w + 6)$ $-19w + 24$
18. $9e^2 + (-e^2) + 15$ $8e^2 + 15$
19. $\dfrac{3}{2}(a + b) - \dfrac{a}{2} + 2\dfrac{1}{2}b$ $a + 4b$
20. $\dfrac{3}{4}s + \dfrac{5}{6}p - \dfrac{1}{4}s + \left(-\dfrac{4}{6}p\right)$ $\dfrac{1}{6}p + \dfrac{1}{2}s$

For use with Section 10-5.

7. $-3u + 4w - 3$ 13. $\dfrac{1}{2}h - \dfrac{1}{7}j$ 15. $2d + 4f$

Simplify.

21. $-64 \div 8$ -8
22. $-8\overline{)72}$ -9
23. $-32 \div 2$ -16
24. $-12\overline{)132}$ -11
25. $-100 \div 5$ -20
26. $-9\overline{)36}$ -4
27. $\dfrac{-24}{-8 + 2}$ 4
28. $\dfrac{35}{-7}$ -5
29. $\dfrac{-8 + (-7)}{3}$ -5
30. $\dfrac{-56}{-8}$ 7
31. $\dfrac{-70}{-1 + (-9)}$ 7
32. $\dfrac{-40 + (-4)}{16 + (-5)}$ -4

For use with Section 10-6.

Simplify. Use reciprocals.

33. $\dfrac{-3}{4} \div \dfrac{1}{8}$ -6
34. $6 \div \dfrac{1}{-5}$ -30
35. $\dfrac{2}{3} \div \dfrac{-8}{9}$ $-\dfrac{3}{4}$
36. $\dfrac{4}{-7} \div \dfrac{8}{14}$ -1
37. $\dfrac{5}{6} \div \dfrac{-1}{2}$ $-1\dfrac{2}{3}$
38. $-28 \div \dfrac{7}{4}$ -16
39. $\dfrac{3}{-8} \div \dfrac{1}{4}$ $-1\dfrac{1}{2}$
40. $-36 \div \dfrac{12}{3}$ -9
41. $\dfrac{-1}{2} \div \dfrac{1}{4}$ -2
42. $\dfrac{-5}{3} \div \dfrac{1}{6}$ -10
43. $\dfrac{2}{5} \div \dfrac{-1}{10}$ -4
44. $-81 \div \dfrac{9}{2}$ -18

For use with Section 11-1.

Solve and check.

1. $m + 7 = 12$ 5
2. $-56 = u + 7$ −63
3. $-2 - b = 15$ −17
4. $z + (-15) = 24$ 39
5. $-\dfrac{5}{8} + t = \dfrac{3}{8}$ 1
6. $c + \dfrac{3}{5} = -4$ $-4\frac{3}{5}$
7. $-18 = a + (-14)$ −4
8. $-d - 3.5 = 4.2$ −7.7
9. $y - \dfrac{3}{2} = -1$ $\frac{1}{2}$
10. $-f + (-2\frac{3}{8}) = \frac{5}{8}$ −3
11. $g + 3.76 = 5.02$ 1.26
12. $-15.7 + k = 32.9$ 48.6
13. $-45 + (-t) = 10$ −55
14. $-87 - e = 6$ −93

For use with Section 11-2.

Solve and check.

15. $9a = 81$ 9
16. $38 = 2b$ 19
17. $-\dfrac{2}{3}t = -12$ 18
18. $\dfrac{1}{4}m = -16$ −64
19. $7c = 42.63$ 6.09
20. $s(-8) = -96$ 12
21. $4k = -11$ $-2\frac{3}{4}$
22. $-14h = 15.4$ −1.1
23. $\dfrac{-5}{8}u = -10$ 16
24. $-16 = \dfrac{2}{5}w$ −40
25. $36 = 18p$ 2
26. $-48 = 6q$ −8
27. $54.6 = 6f$ 9.1
28. $g(-7) = 21$ −3

For use with Section 11-3.

Solve and check.

29. $5x + 6x = 121$ 11
30. $16c = 14c + 8$ 4
31. $-4 - (-2m) = -10m$ $\frac{1}{3}$
32. $26a - 11a = -3a + 54$ 3
33. $\dfrac{2}{3}f + \dfrac{4}{3}f = -18$ −9
34. $-\dfrac{4}{5}r + 6 = -\dfrac{1}{5}r + 4$ $3\frac{1}{3}$
35. $6.42s + 3.58s = 160$ 16
36. $7w = -81 - 2w$ −9
37. $\dfrac{3}{7}(-h) + \left(-\dfrac{2}{7}\right)h = 15$ −21
38. $-\dfrac{8}{3} = \dfrac{1}{3}c + \left(-\dfrac{5}{3}\right)c$ $\frac{2}{3}$
39. $2d + 4d = 48$ 8
40. $-225 = -42v - 33v$ 3

For use with Section 11-4.

Solve. Use the formula $V = lwh$ (Volume = length × width × height).

1. The volume of a box is 10.5 m³. The length is 3.5 m and the width is 2 m. Find the height. 1.5 m

2. The dimensions of a carton are as follows: length, 0.75 m; width, 0.5 m; height, 1.5 m. Find the volume. 0.5625 m³

Solve. Use the formula $I = prt$ (Interest = principal × rate × time).

3. A bank made a two year loan of $1400 at 7% interest to a small business firm. How much interest did the company pay? $196

4. A couple borrowed money at the rate of 8% for a period of 3 years. The interest charged was $480. What was the amount (principal) of the loan? $2000

5. Randy borrowed $4000 from a local bank. The interest rate was 7.5%. He paid $450 interest. How long did he keep the loan? $1\frac{1}{2}$ years

6. Gloria borrowed $1560 from the bank for a period of 10 months ($\frac{5}{6}$ year). She paid $117 in interest. What was the rate of interest? 9%

Solve. Use the formula $V = \frac{1}{3}Bh$ (Volume = $\frac{1}{3}$ × area of base × height).

7. The area of the base of a triangular pyramid is 72 cm². The height is 14 cm. Find the volume. 336 cm³

8. If the volume of a triangular pyramid is 400 cm³ and the height is 20 cm, find the area of the base. 60 cm²

Solve. Use the formula $A = lw$ (Area = length × width).

9. The dimensions of a rectangle are as follows: length, 24.5 m and width, 20.6 m. What is the area? 504.7 m²

10. A florist wants a garden to cover an area of 192 m². The length of the plot is 16 m. Find the width. 12 m

11. The length of a frame on a filmstrip is 1.8 cm and the width is 1.5 cm. Find the area of the frame. 2.7 cm²

Solve. Use the formula Area = $\frac{1}{2}$ × altitude × sum of bases, $A = \frac{1}{2}as$.

12. The measure of the two bases of a trapezoid are 15.3 m and 16.7 m. If the altitude equals 14 m, what is the area? 224 m²

13. The area of a trapezoid is 86 cm². The sum of the two bases is 43 cm. Find the altitude. 4 cm

For use with Section 12-4.

Add.

1. $\begin{array}{r} 5s + 11 \\ 6s + 5 \\ \hline 11s + 16 \end{array}$

2. $\begin{array}{r} 4r^2 + 2m^2 + p^2 \\ 3r^2 - m^2 + 5p^2 \\ \hline 7r^2 + m^2 + 6p^2 \end{array}$

3. $\begin{array}{r} 9x^2 + 14 \\ 23z - 10 \\ \hline 9x^2 + 23z + 4 \end{array}$

4. $\begin{array}{r} -7f + 4b - 24 \\ -8f - 3b + 36 \\ \hline -15f + b + 12 \end{array}$

5. $\begin{array}{r} h^3 + 14k^2 - 13 \\ -3h^3 - 2k^2 + 5 \\ \hline -2h^3 + 12k^2 - 8 \end{array}$

6. $\begin{array}{r} y^3 - z^2 + w - 17 \\ -y^3 + z^2 - w + 17 \\ \hline 0 \end{array}$

7. $\begin{array}{r} 8a + 9c \\ -6a - 7c \\ \hline 2a + 2c \end{array}$

8. $\begin{array}{r} -10d - 4e + 6 \\ 3d + 7e - 10 \\ \hline -7d + 3e - 4 \end{array}$

9. $\begin{array}{r} 8.9y + 1.5x + 2.3 \\ -0.4y - 7.2x + 4.8 \\ \hline 8.5y - 5.7x + 7.1 \end{array}$

For use with Section 12-7.

Subtract. Check by addition.

10. $\begin{array}{r} 20m + 15n \\ 6m + 9n \\ \hline 14m + 6n \end{array}$

11. $\begin{array}{r} -2s^2 + 7s - 19 \\ -5s^2 + 3s - 17 \\ \hline 3s^2 + 4s - 2 \end{array}$

12. $\begin{array}{r} 16r + 5t - 9 \\ 8r - t + 11 \\ \hline 8r + 6t - 20 \end{array}$

13. $\begin{array}{r} 3f + 2q^2 \\ -5f + 8q^2 \\ \hline 8f - 6q^2 \end{array}$

14. $\begin{array}{r} 4 + e \\ 9 - 2e \\ \hline -5 + 3e \end{array}$

15. $\begin{array}{r} -25p + 15q^2 \\ + 8q^2 \\ \hline -25p + 7q^2 \end{array}$

16. $\begin{array}{r} -7k + 12j - 6 \\ -4k - 3j + 18 \\ \hline -3k + 15j - 24 \end{array}$

17. $\begin{array}{r} + w^2 - 18 \\ -40t + 9 \\ \hline 40t + w^2 - 27 \end{array}$

18. $\begin{array}{r} 6x^3 + z^2 \\ - x^3 + 14 \\ \hline 7x^3 + z^2 - 14 \end{array}$

For use with Section 12-8.

Write the answer as a polynomial in standard form.

19. The length of a rectangular pool is $8 - 6p$ and the width is $3 + 2p$. Find the perimeter of the pool. $22 - 8p$

20. A truck driver drove a distance of $3x^2 + x - 9$. Then he realized he has passed his stop. He retraced a distance of $x^2 - 2x + 8$. What was the actual distance from the beginning point to his destination? $2x^2 + 3x - 17$

21. A plumber has a length of tubing measuring $7a - 2b + c$. She only needs a piece $-4a + 3b + c$ long. How much should she cut off? $11a - 5b$

22. A real estate agent purchased two pieces of property. The first had an area of $10w^3 - 14$ and the second an area of $w^3 + 4w^2 + 6w$. Find the combined area. $11w^3 + 4w^2 + 6w - 14$

23. A sports car was using gasoline at the rate of $-2f + 14$ before a tune-up and $12f - 5$ after the tune-up. By how much did the gasoline usage improve? $14f - 19$

For use with Section 12-9.

Solve. Check your solution.

1. $48 = (2t + 8) + (4t - 2)$ 7
2. $r + (r - 7) + 5r + 3 = 31$ 5
3. $-(10m + 8) + 14m = 32$ 10
4. $-72 = -q + 3(-1 + 2q) + 1$ -14
5. $x^2 + 6 - (9x + x^2) = -12$ 2
6. $0.4y + 2(10 - 1.3y) = 64$ -20
7. $(3f + 7) - (19 + f) = 0$ 6
8. $50 = (4g - 27) + 2 + 21g$ 3
9. $a - (9 + a) + (3a - 6) = 9$ 8
10. $-(8 + 20b) + 4 = 12(b - 1)$ $\frac{1}{4}$
11. $3(k - 2) - (8 - k) = 14$ 7
12. $-7n + 13 + 2(n - 2) = -16$ 5
13. $-4c - (7 + 2c) = 35 + c$ -6
14. $(8d + 52) - (6d + 8) = 68 - 4d$ 4
15. $2.3 - 9(0.9 + f) = 0.6 - f$ -0.8
16. $25 = 3p - (8p + 5)$ -6
17. $(x - 5) - (8x + 6) = x + 21$ -4
18. $3r + 4(5 - 2r) = 2r - 1$ 3
19. $17w + 6(8 - 3w) = 2w$ 16
20. $-1.1 + 2(j + 6) = 8.9$ -1

For use with Sections 13-1, 13-2, 13-3.

Simplify.

21. $(x + 3)(x + 3)(x + 3)$ $(x + 3)^3$
22. $(12 \cdot 6)(b \cdot b)$ $72b^2$
23. $-7(14)(z \cdot z \cdot z \cdot z \cdot z)$ $-98z^5$
24. $s \cdot s(-s) \cdot s(-s)$ s^5
25. $3(f \cdot f \cdot f)$ $3f^3$
26. $4(-12)(w^2)(w^4)$ $-48w^6$
27. $b^4 \cdot b \cdot b^3$ b^8
28. $(12 \cdot 11)(c - 2)(c - 2)$ $132(c - 2)^2$
29. $(8 \cdot 9)(p \cdot p \cdot 2p)$ $144p^3$
30. $2^3 \cdot 2^2 \cdot 2$ 64
31. $(14xyz)(xy)$ $14x^2y^2z$
32. $(m \cdot m)(m \cdot m \cdot m)$ m^5
33. $2(j \cdot j \cdot j)(-3)$ $-6j^3$
34. $(h^3)(h^2)(-5)$ $-5h^5$
35. $7(-8y)$ $-56y$
36. $c(d \cdot d^2)(e^3 \cdot e^3)$ cd^3e^6
37. $(4f \cdot f)(g^2)(g \cdot g^3)$ $4f^2g^6$
38. $(0.7qr)(-0.7q^2r^3)$ $-0.49q^3r^4$
39. $(-st)(8s^2t^2)(s^3)$ $-8s^6t^3$
40. $(u^3v)(-4uv^3)(8c)$ $-32cr^4u^4$
41. $(-36i)(2h^2i)(h^3i^4)$ $-72h^5i^6$
42. $(2.5k^2m)(-0.5km^3)$ $-1.25k^3m^4$
43. $\left(\frac{2}{3}pr\right)\left(\frac{1}{7}pr\right)$ $\frac{2}{21}p^2r^2$
44. $\left(\frac{3}{4}x^2yz^3\right)\left(\frac{1}{6}xy^2z\right)$ $\frac{1}{8}x^3y^3z^4$
45. $(4st)^3$ $256s^3t^3$
46. $(3f^2g)^4$ $81f^8g^4$
47. $(-ab)(6a^2b^5)(-3ac)$ $18a^4b^6c$
48. $(-5j^3k^2)^3$ $-125j^9k^6$
49. $(-cd)\left(\frac{1}{3}d^2\right)\left(\frac{5}{2}c^3d^3\right)$ $-\frac{5}{6}c^4d^6$
50. $\left(\frac{1}{8}mn^2p^3\right)\left(-\frac{3}{5}m^3np^2\right)$ $-\frac{3}{40}m^4n^3p^5$
51. $(e^4)^5$ e^{20}
52. $(7ab)^2 + (2a^3b)(9ab^2)$ $49a^2b^2 + 18a^4b^3$
53. $c(c^3d^2)^4 + 5(cd)^3$ $c^{13}d^8 + 5c^3d^3$
54. $n^2(m^3n^4) + (mn^2)^3$ $2m^3n^6$

For use with Sections 13-4, 13-5.

Simplify. **3.** $40mn^2 + 100m^3n^5 - 120m^4n^7$ **4.** $4s^2t - 1.6st^3 + s^2t^3$

1. $f(e - f) + 6f(e + f)$ $7ef + 5f^2$ **2.** $7ab(a + 2b) - 3a(b^2 + 2ab)$ $a^2b + 11ab^2$
3. $20mn^2(2 + 5m^2n^3 - 6m^3n^5)$ **4.** $1.6(s^2t - st^3) + s^2(2.4t + t^3)$
5. $-uv\left(\dfrac{3}{4}u^2 + \dfrac{1}{5}v\right) + \dfrac{1}{5}u\left(v^2 + \dfrac{5}{4}u^2v\right)$ **6.** $\dfrac{3}{8}j(2j + 3k) + \dfrac{1}{4}(-j^2 + k)$
7. $6cd(c^2d + 10) - cd(4 + 2c^2d)$ **8.** $(xy + 5x)8 - 9x(2xy - 7)$

5. $-\dfrac{1}{2}u^3r$ **6.** $\dfrac{9}{8}jk + \dfrac{1}{4}k$ **7.** $4c^3d^2 + 56cd$ **8.** $8xy + 103x - 18x^2y$

Multiply. Use either the vertical or horizontal form.

9. $(a + b)(a + b)$ $a^2 + 2ab + b^2$ **10.** $(x + 4)(x - 4)$ $x^2 - 16$
11. $(4 + 7n)(3 + 6n)$ $12 + 45n + 42n^2$ **12.** $(q - r)(2q^2 + qr + r^2)$ $2q^3 - q^2r - r^3$
13. $(1.2t + 2)(7.4 - 3t)$ **14.** $(6c + 5)(3c + 8)$ $18c^2 + 63c + 40$
15. $(5f + 9)(5f - 9)$ $25f^2 - 81$ **16.** $(d - 1)(2d^2 + d + 4)$ $2d^3 - d^2 + 3d - 4$

13. $14.8 + 2.88t - 3.6t^2$

For use with Sections 13-6, 13-8.

Multiply.

17. $(f + 6)^2$ $f^2 + 12f + 36$ **18.** $(4q + 7)(4q - 7)$ $16q^2 - 49$
19. $(eg + eh)^2$ $e^2g^2 + 2e^2gh + e^2h^2$ **20.** $(2uv + 16)(2uv - 16)$ $4u^2r^2 - 256$
21. $\left(9s + \dfrac{2}{5}t\right)\left(9s - \dfrac{2}{5}t\right)$ $81s^2 - \dfrac{4}{25}t^2$ **22.** $\left(\dfrac{1}{2}w + \dfrac{2}{3}\right)^2$ $\dfrac{1}{4}w^2 + \dfrac{2}{3}w + \dfrac{4}{9}$
23. $(z + 3y)^2$ $z^2 + 6yz + 9y^2$ **24.** $(8b - 6c)^2$ $64b^2 - 96bc + 36c^2$
25. $(6 - 2j)(6 + 2j)$ $36 - 4j^2$ **26.** $(11ab - cd)(11ab + cd)$ $121a^2b^2 - c^2d^2$

Simplify. Assume no denominator is 0.

27. $\dfrac{t^5}{t^3}$ t^2 **28.** $\dfrac{24a^5b^3}{12a^2b}$ $2a^3b^2$ **29.** $\dfrac{-25m^6}{75m^7}$ $-\dfrac{1}{3m}$
30. $\dfrac{-q^3r^7}{q^5r^4}$ $-\dfrac{r^3}{q^2}$ **31.** $\dfrac{64x^2y^7z^3}{(-2xyz)^3}$ $-\dfrac{8y^4}{x}$ **32.** $\dfrac{0.3kh^4}{1.2k^3h^2j}$ $\dfrac{h^2}{4k^2j}$
33. $\dfrac{(-4e^3f^2g)^3}{8e^{10}f^4g}$ $-\dfrac{8f^2g^2}{e}$ **34.** $\dfrac{(uv)^5}{u^3v^2}$ u^2r^3 **35.** $\dfrac{-8a^7b^9c}{56a^2b^5}$ $-\dfrac{a^5b^4c}{7}$
36. $\dfrac{(6z)^2w^5y^3}{(4w)^3zy}$ $\dfrac{9w^2y^2z}{16}$ **37.** $\dfrac{1.44s^4r^3t^5}{1.2sr^5t^3}$ $\dfrac{1.2s^3t^2}{r^2}$ **38.** $\dfrac{(5d^2e^3f)^2}{30d^2f^4}$ $\dfrac{5d^2e^6}{6f^2}$
39. $\dfrac{(2ij^2)^3}{(-8i^3j)^2}$ $\dfrac{j^4}{8i^3}$ **40.** $\dfrac{19p^3r^5t^7}{-38p^5r^2t^4}$ $-\dfrac{r^3t^3}{2p^2}$ **41.** $\dfrac{21x^{10}}{14x^6}$ $\dfrac{3x^4}{2}$

Extra Practice: Word Problems

1. There are 9000 people in Port City. 4200 are registered to vote. In the last town election 2800 people voted. What is the ratio of the number of registered voters to the total number of people in town? What is the ratio of the number of people who voted to the number registered? $\frac{7}{15}$; $\frac{2}{3}$

2. The graph shows how one homemaker spends her work time. What fraction of her time is spent buying and preparing food? What percent of her time is spent housekeeping? $\frac{1}{5}$; 30%

3. Carla bought two medicines at the drugstore. One cost $6.95. The other cost $11.25. Estimate her change from a $20 bill to the nearest dollar. Then find the exact amount. $2; $1.80

4. A plumber spent 4.5 hours on a job. The charge for labor was $112.50. What is the plumber's hourly rate? $25/hr

5. The area of a rectangle is 35rs. The length is 7s. Find the width. 5r

6. The area of a rectangle is at least 96 square centimeters. The length is 8 centimeters. What numbers may represent the width in centimeters? $w \geqslant$ 12 cm

7. Renee has d dimes and twice as many nickels in her pocket. Write an algebraic expression that shows how much money she has in cents. 20 d

8. Craig uses at least 8 pieces of scrap metal to make a mobile. He is making 5 mobiles for a craft fair. What numbers represent the possible number of pieces he will need? $n \geqslant$ 40

9. The sum of eight times a number and 7 is 55. What is the number? 6

10. The sum of an odd number and the next greater odd number is 72. Find the numbers. (Hint: If n is an odd number, the next greater odd number is $n + 2$.) 35, 37

11. The width of a rectangle is twice its length. The area of the rectangle is 128 m². Find the width. Use $A = lw$. 16 m

12. The sum of three consecutive even numbers is 150. What are the three numbers? 48, 50, 52

13. One day the Burger Shop offered free soft drinks during lunch hour. Of the 185 people who came in, 157 bought lunch. How many people only took the free soft drink? 28

14. If 35 is decreased by seven times a number, the result is the same as ten times the number increased by 1. Find the number. 2

15. Dwan has 4 more classical albums than country-western, and 9 more rock albums than classical. He has 35 albums altogether. How many of each kind does he have? 10 classical, 6 country-western, 19 rock

16. A tank containing 1040 liters of heating oil springs a leak. It loses 56 liters of oil before the leak is stopped. Express the amount of oil left in the tank as a sum of directed numbers. Then find the sum. $1040 + {}^-56 = 984$

17. Window washers finished the windows on the thirtieth floor of a building. They lowered their scaffold to check some windows 9 floors down. Express the number of the floor they were on then as a sum of directed numbers. Then find the sum. $30 + {}^-9 = 21$

18. The product of $^-8$ and some number is 24. What is the number? $^-3$

19. The area of a trapezoid is given by the formula Area $= \frac{1}{2} \times$ altitude $\times$ sum of bases. The area of one trapezoid is 432 cm^2. The sum of the bases is 48 cm. Find the altitude. 18 cm

20. A bolt of fabric has length $3a^2 + 2a - 1$. A tailor cuts off a piece of length $a^2 - 5$. How long is the remaining piece? $2a^2 + 2a + 4$

21. The area of a rectangular field is $2x^2 - 5x + 3$. One square section has area $x^2 - 2x + 1$. What is the area of the remaining section? $x^2 - 3x + 2$

22. Find the area of a triangle with height y and base $y + 6$. Use $A = \frac{1}{2} bh$. $\frac{1}{2}y^2 + 3y$

23. Find the volume of a box with length $2x$, width $x^2 - 1$, and height $3x^2$. Use $V = lwh$. $6x^5 - 6x^3$

24. Find the volume of a cube with sides of length $2x^2$. Use $V = lwh$. Then find the area of each face and add to find the surface area. Use $A = lw$. $8x^6$; $24x^4$

25. Find the area of a rectangle with length $3y - 1$ and width $2y + 4$. $6y^2 + 10y - 4$

26. Find the volume of a box that has edges of length x, $x + 3$, and $x - 2$. $x^3 + x^2 - 6x$

27. The square of a whole number is 31 less than the square of the next greater whole number. What are the numbers? 15, 16

28. The product of an integer and the next greater integer is 20 less than the square of the greater integer. What are the integers? 19, 20

Extra Practice: Fractions and Decimals

Write the fraction in lowest terms.

1. $\dfrac{12}{36}$ $\dfrac{1}{3}$
2. $\dfrac{15}{24}$ $\dfrac{5}{8}$
3. $\dfrac{2}{18}$ $\dfrac{1}{9}$
4. $\dfrac{16}{12}$ $\dfrac{4}{3}$

5. $\dfrac{15}{36}$ $\dfrac{5}{12}$
6. $\dfrac{25}{50}$ $\dfrac{1}{2}$
7. $\dfrac{14}{49}$ $\dfrac{2}{7}$
8. $\dfrac{24}{72}$ $\dfrac{1}{3}$

9. $\dfrac{27}{33}$ $\dfrac{9}{11}$
10. $\dfrac{24}{60}$ $\dfrac{2}{5}$
11. $\dfrac{30}{100}$ $\dfrac{3}{10}$
12. $\dfrac{11}{66}$ $\dfrac{1}{6}$

Replace the variable to name the equivalent fraction.

13. $\dfrac{36}{5} = \dfrac{x}{10}$ $\dfrac{72}{10}$
14. $\dfrac{42}{7} = \dfrac{y}{2}$ $\dfrac{12}{2}$
15. $\dfrac{3}{27} = \dfrac{x}{18}$ $\dfrac{2}{18}$

16. $\dfrac{5}{7} = \dfrac{a}{35}$ $\dfrac{25}{35}$
17. $\dfrac{3}{4} = \dfrac{t}{24}$ $\dfrac{18}{24}$
18. $\dfrac{12}{42} = \dfrac{b}{21}$ $\dfrac{6}{21}$

Write as a decimal.

19. $\dfrac{3}{5}$ 0.6
20. $\dfrac{1}{2}$ 0.5
21. $\dfrac{13}{13}$ 1.0
22. $\dfrac{9}{10}$ 0.9

23. $\dfrac{10}{25}$ 0.4
24. $\dfrac{7}{8}$ 0.875
25. $\dfrac{7}{9}$ $0.\overline{7}$
26. $2\dfrac{1}{4}$ 2.25

Write as a fraction in lowest terms.

27. 0.8 $\dfrac{4}{5}$
28. 0.45 $\dfrac{9}{20}$
29. 0.65 $\dfrac{13}{20}$

30. 0.50 $\dfrac{1}{2}$
31. 0.12 $\dfrac{3}{25}$
32. 0.72 $\dfrac{18}{25}$

33. 1.5 $\dfrac{3}{2}$
34. 0.005 $\dfrac{1}{200}$
35. 1.25 $\dfrac{5}{4}$

Write as a decimal and as a percent.

36. $\dfrac{28}{100}$ 0.28 = 28%
37. $\dfrac{3}{10}$ 0.3 = 30%
38. $\dfrac{4}{5}$ 0.8 = 80%
39. $\dfrac{18}{20}$ 0.9 = 90%

40. $\dfrac{3}{8}$ 0.375 = 37.5%
41. $\dfrac{250}{1000}$ 0.25 = 25%
42. $\dfrac{20}{25}$ 0.8 = 80%
43. $\dfrac{13}{25}$ 0.52 = 52%

Write as a fraction in lowest terms.

44. 35% $\dfrac{7}{20}$
45. 70% $\dfrac{7}{10}$
46. 24% $\dfrac{6}{25}$
47. 16% $\dfrac{4}{25}$

48. 44% $\dfrac{11}{25}$
49. 62% $\dfrac{31}{50}$
50. 28% $\dfrac{7}{25}$
51. 55% $\dfrac{11}{20}$

52. 52% $\dfrac{13}{25}$
53. 68% $\dfrac{17}{25}$
54. 45% $\dfrac{9}{20}$
55. 85% $\dfrac{17}{20}$

Add or subtract. Simplify the result.

56. $\dfrac{3}{8} + \dfrac{2}{8}$ $\dfrac{5}{8}$

57. $\dfrac{5}{9} - \dfrac{2}{9}$ $\dfrac{1}{3}$

58. $\dfrac{3}{10} + \dfrac{5}{10}$ $\dfrac{4}{5}$

59. $\dfrac{7}{12} + \dfrac{5}{12}$ 1

60. $\dfrac{6}{7} - \dfrac{1}{7}$ $\dfrac{5}{7}$

61. $\dfrac{9}{10} + \dfrac{1}{10}$ 1

62. $2\dfrac{3}{4} - 1\dfrac{1}{4}$ $1\dfrac{1}{2}$

63. $5\dfrac{1}{5} + 6\dfrac{3}{5}$ $11\dfrac{4}{5}$

64. $4\dfrac{7}{8} - 2\dfrac{3}{8}$ $2\dfrac{1}{2}$

65. $\dfrac{1}{3} + \dfrac{1}{4}$ $\dfrac{7}{12}$

66. $\dfrac{1}{2} - \dfrac{1}{5}$ $\dfrac{3}{10}$

67. $\dfrac{1}{4} + \dfrac{5}{8}$ $\dfrac{7}{8}$

68. $12.25 + 3.7$ 15.95

69. $11.34 - 4.6$ 6.74

70. $0.08 + 9.02$ 9.1

71. $7.22 + 6.14$ 13.36

72. $4.725 - 3.5$ 1.225

73. $5.24 - 2.36$ 2.88

74. $0.05 + 6.66$ 6.71

75. $35 - 6.18$ 28.82

76. $4.065 + 3.82$ 7.885

Multiply.

77. $\dfrac{1}{2} \times \dfrac{5}{6}$ $\dfrac{5}{12}$

78. $\dfrac{2}{5} \times \dfrac{3}{7}$ $\dfrac{6}{35}$

79. $\dfrac{1}{2} \times \dfrac{4}{5}$ $\dfrac{2}{5}$

80. $\dfrac{5}{9} \times \dfrac{1}{4}$ $\dfrac{5}{36}$

81. $\dfrac{3}{8} \times \dfrac{1}{3}$ $\dfrac{1}{8}$

82. $1\dfrac{1}{2} \times 2$ 3

83. $\dfrac{5}{6} \times \dfrac{2}{3}$ $\dfrac{5}{9}$

84. $\dfrac{3}{5} \times \dfrac{2}{3}$ $\dfrac{2}{5}$

85. $\dfrac{3}{4} \times \dfrac{5}{8}$ $\dfrac{15}{32}$

86. 1.7×3.5 5.95

87. 4.5×6.8 30.6

88. 2.9×3.2 9.28

89. 1.23×0.005 0.00615

90. 8.25×5.42 44.715

91. 5.23×6.02 31.4846

92. 3.17×0.108 0.34236

93. 44.73×0.65 29.0745

94. 0.98×0.072 0.07056

Solve.

95. $\dfrac{2}{3} \div \dfrac{4}{5} = a$ $\dfrac{5}{6}$

96. $\dfrac{3}{8} \div \dfrac{4}{5} = s$ $\dfrac{15}{32}$

97. $x = \dfrac{1}{2} \div \dfrac{1}{2}$ 1

98. $\dfrac{1}{2} \div \dfrac{5}{8} = y$ $\dfrac{4}{5}$

99. $\dfrac{2}{3} \div \dfrac{1}{4} = n$ $\dfrac{8}{3}$

100. $\dfrac{3}{4} \div \dfrac{1}{3} = r$ $\dfrac{9}{4}$

101. $z = 6 \div \dfrac{2}{3}$ 9

102. $\dfrac{1}{2} \div 1\dfrac{1}{2} = t$ $\dfrac{1}{3}$

103. $x = 4 \div \dfrac{3}{4}$ $\dfrac{16}{3}$

104. $18.6 \div 2 = m$ 9.3

105. $a = 15.45 \div 5$ 3.09

106. $38.07 \div 9 = k$ 4.23

107. $w = 15.48 \div 0.6$ 25.8

108. $n = 8.82 \div 0.7$ 12.6

109. $16.4 \div 0.8 = a$ 20.5

110. $x = 45.92 \div 14$ 3.28

111. $0.047 \div 0.2 = z$ 0.235

112. $3.249 \div 0.9 = y$ 3.61

113. $33.32 \div 0.136 = l$ 245

114. $b = 1.925 \div 0.05$ 38.5

115. $d = 0.624 \div 240$ 0.0026

Extra Practice: Formulas

Use the given perimeter formula to complete.

1.

$w = 10$ m
$l = 11$ m

$P = 2l + 2w$
$P = ?$ 42 m

2.

$s = 8.2$ m

$P = 4s$
$P = ?$ 32.8 m

3.

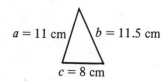

$a = 11$ cm $b = 11.5$ cm
$c = 8$ cm

$P = a + b + c$
$P = ?$ 30.5 cm

4.

s

$P = 4s$
$P = 64$ km
$s = ?$ 16 km

5.

$a = 5.6$ m $b = 7$ m
c

$P = a + b + c$
$P = 20.6$ m
$c = ?$ 8 m

6.

w
$l = 14$ mm

$P = 2l + 2w$
$P = 42$ mm
$w = ?$ 7 mm

Use the given circumference formula to complete.

7.

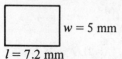

$r = 21$ cm

$C = 2\pi r$
$\pi = \dfrac{22}{7}$
$C = ?$ 132 cm

8.

$d = 4$ m

$C = \pi d$
$\pi = 3.14$
$C = ?$ 12.56 m

9.

r

$C = 2\pi r$
$C = 22$ cm, $\pi = \dfrac{22}{7}$
$r = ?$ 3.5 cm

Use the given area formula to complete.

10.

$w = 5$ mm
$l = 7.2$ mm

$A = lw$
$A = ?$ 36 mm²

11.

$s = 6.1$ km

$A = s^2$
$A = ?$ 37.21 km²

12.

$r = 14$ m

$A = \pi r^2$
$\pi = \dfrac{22}{7}$; $A = ?$ 616 m²

13.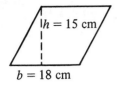

$A = bh$

$A = ?$ 270 cm²

14.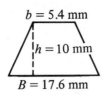

$A = \frac{1}{2}h(b + B)$

$A = ?$ 115 mm²

15.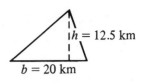

$A = \frac{1}{2}bh$

$A = ?$ 125 km²

16.

$A = s^2$
$A = 64$ cm²
$s = ?$ 8 cm

17.

$A = \pi r^2$
$A = 12.56$ m, $\pi = 3.14$
$r = ?$ 2 m

18.

$A = lw$
$A = 17.5$ cm
$w = ?$ 3.5 cm

Use the given volume formula to complete.

19.

$V = \frac{1}{3}Bh = \frac{1}{3}\pi r^2 h$
$V = ?$ 18.84 cm³

20.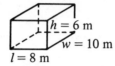

$V = lwh$
$V = ?$ 480 m³

21.

$V = Bh = \pi r^2 h$
$V = ?$ 706.5 mm³

22.

$V = Bh$

$V = 50.4$ m³
$B = 12.6$ m²
$h = ?$ 4 m

23.

$V = lwh$

$V = 64$ cm³
$l = w = h$
$l = ?$ 4 cm

24.

$V = \frac{1}{3}Bh$

$V = 2156$ m³
$h = 14$ m
$B = ?$ 462 m²

Extra Practice: Factoring

Factor by rearranging, if necessary, and then grouping.

1. $xy + 2x + 3y + 6$ $(x + 3)(y + 2)$
2. $st + 5s + 2t + 10$ $(s + 2)(t + 5)$
3. $7ab + 14b - 2 - a$ $(7b - 1)(a + 2)$
4. $2xy - 6y + 4x - 12$ $2(y + 2)(x - 3)$
5. $2mn + 6m + 6 + 2n$ $2(m + 1)(n + 3)$
6. $12 - 3z - 4z + z^2$ $(z - 3)(z - 4)$
7. $d^2 - 35f - 7d + 5df$ $(d + 5f)(d - 7)$
8. $4b^2 + 5c + 10bc + 2b$ $(2b + 5c)(2b + 1)$
9. $xy - 12 + 6x - 2y$ $(x - 2)(y + 6)$
10. $m^2 - 6p - 2mp + 3m$ $(m - 2p)(m + 3)$
11. $6kn + 8n + 3k + 4$ $(3k + 4)(2n + 1)$
12. $6st + 9s - 2t - 3$ $(2t + 3)(3s - 1)$
13. $15nx - 10n - 9x + 6$ $(5n - 3)(3x - 2)$
14. $8yz + 6 + 24y + 2z$ $2(4y + 1)(z + 3)$

Factor completely.

15. $x^2 - 25$ $(x + 5)(x - 5)$
16. $y^2 - 49$ $(y + 7)(y - 7)$
17. $m^2 - 36$ $(m + 6)(m - 6)$
18. $d^2 - 100$ $(d + 10)(d - 10)$
19. $x^2y^2 - 1$ $(xy + 1)(xy - 1)$
20. $x^2 - 9z^2$ $(x + 3z)(x - 3z)$
21. $m^2n^2 - 16$ $(mn + 4)(mn - 4)$
22. $w^2z^4 - 64$ $(wz^2 + 8)(wz^2 - 8)$
23. $5c^2 - 45$ $5(c + 3)(c - 3)$
24. $400s^2 - 81$ $(20s + 9)(20s - 9)$
25. $16y^2 - 25$ $(4y + 5)(4y - 5)$
26. $100m^2 - 9$ $(10m + 3)(10m - 3)$
27. $48z^2 - 27$ $3(4z + 3)(4z - 3)$
28. $72t^2 - 8$ $8(3t + 1)(3t - 1)$
29. $x^2 + 10x + 25$ $(x + 5)^2$
30. $x^2 + 22x + 121$ $(x + 11)^2$
31. $y^2 + 18y + 81$ $(y + 9)^2$
32. $m^2 + 6m + 9$ $(m + 3)^2$
33. $b^2 + 26b + 169$ $(b + 13)^2$
34. $w^2 + 14w + 49$ $(w + 7)^2$
35. $x^2y^2 + 4xy + 4$ $(xy + 2)^2$
36. $x^2 + 2xy + y^2$ $(x + y)^2$
37. $4a^2b^2 + 4ab + 1$ $(2ab + 1)^2$
38. $4x^2 + 12x + 9$ $(2x + 3)^2$
39. $25x^2 + 10x + 1$ $(5x + 1)^2$
40. $9t^2 + 6tv + v^2$ $(3t + v)^2$
41. $4c^2 + 12cd + 9d^2$ $(2c + 3d)^2$
42. $36g^2 + 36gh + 9h^2$ $(6g + 3h)^2$
43. $n^2 - 6n + 9$ $(n - 3)^2$
44. $m^2 - 20m + 100$ $(m - 10)^2$

45. $t^2 - 18t + 81$ $(t - 9)^2$

46. $c^2 - 16c + 64$ $(c - 8)^2$

47. $c^2d^2 - 2cd + 1$ $(cd - 1)^2$

48. $x^2y^2 - 10xy + 25$ $(xy - 5)^2$

49. $s^2t^2 - 4st + 4$ $(st - 2)^2$

50. $4x^2 - 4x + 1$ $(2x - 1)^2$

51. $25x^2 - 20x + 4$ $(5x - 2)^2$

52. $16z^2 - 24z + 9$ $(4z - 3)^2$

53. $9n^2 - 24n + 16$ $(3n - 4)^2$

54. $4s^2 - 28st + 49t^2$ $(2s - 7t)^2$

55. $16x^2 - 24xy + 9y^2$ $(4x - 3y)^2$

56. $100w^2 - 60wz + 9z^2$ $(10w - 3z)^2$

57. $y^2 + 9y + 8$ $(y + 1)(y + 8)$

58. $z^2 + 5z + 6$ $(z + 2)(z + 3)$

59. $x^2 + 12x + 35$ $(x + 5)(x + 7)$

60. $p^2 + 10p + 21$ $(p + 3)(p + 7)$

61. $y^2 + 11y + 18$ $(y + 2)(y + 9)$

62. $p^2 + 9p + 20$ $(p + 4)(p + 5)$

63. $k^2 + 10k + 16$ $(k + 2)(k + 8)$

64. $t^2 + 13t + 42$ $(t + 6)(t + 7)$

65. $z^2 + 8z + 15$ $(z + 3)(z + 5)$

66. $n^2 + 12n + 32$ $(n + 4)(n + 8)$

67. $b^2 + 14b + 13$ $(b + 1)(b + 13)$

68. $n^2 + 12n + 20$ $(n + 2)(n + 10)$

69. $d^2 + 10d + 24$ $(d + 4)(d + 6)$

70. $z^2 + 9z + 14$ $(z + 2)(z + 7)$

71. $x^2 - 3x + 2$ $(x - 1)(x - 2)$

72. $y^2 - 9y + 8$ $(y - 1)(y - 8)$

73. $z^2 - 5z + 6$ $(z - 2)(z - 3)$

74. $t^2 - 8t + 7$ $(t - 1)(t - 7)$

75. $w^2 - 7w + 10$ $(w - 2)(w - 5)$

76. $k^2 - 11k + 10$ $(k - 1)(k - 10)$

77. $x^2 - 7x + 12$ $(x - 3)(x - 4)$

78. $y^2 - 10y + 21$ $(y - 3)(y - 7)$

79. $p^2 - 10p + 9$ $(p - 1)(p - 9)$

80. $m^2 - 10m + 16$ $(m - 2)(m - 8)$

81. $d^2 - 12d + 27$ $(d - 3)(d - 9)$

82. $z^2 - 10z + 24$ $(z - 4)(z - 6)$

83. $b^2 - 8b + 15$ $(b - 3)(b - 5)$

84. $n^2 - 7n + 6$ $(n - 1)(n - 6)$

85. $x^2 + 2x - 15$ $(x - 3)(x + 5)$

86. $y^2 + 5y - 14$ $(y - 2)(y + 7)$

87. $z^2 + z - 6$ $(z - 2)(z + 3)$

88. $t^2 + 2t - 35$ $(t - 5)(t + 7)$

89. $a^2 + 7a - 8$ $(a - 1)(a + 8)$

90. $m^2 + m - 2$ $(m - 1)(m + 2)$

91. $n^2 - 9n - 10$ $(n + 1)(n - 10)$

92. $z^2 + 11z - 12$ $(z - 1)(z + 12)$

93. $c^2 + 4c - 21$ $(c - 3)(c + 7)$

94. $w^2 + 3w - 18$ $(w - 3)(w + 6)$

95. $d^2 - 7d - 18$ $(d + 2)(d - 9)$

96. $k^2 + 4k - 32$ $(k - 4)(k + 8)$

97. $g^2 + g - 20$ $(g - 4)(g + 5)$

98. $h^2 + 6h - 16$ $(h - 2)(h + 8)$

Metric System of Measurement

Metrics: Past and Present

The metric system of weights and measures was developed in France during the late 18th century. It was created to overcome the confusion caused by the widespread use of non-standardized units of length, liquid capacity, and mass. In the year 1791, the French Academy of Sciences recommended the new system based on the unit of length called the *meter*. The meter was defined at the time as $\dfrac{1}{10,000,000}$ of the distance from the North Pole to the equator, measured along the meridian of the earth running near Dunkirk in France and Barcelona in Spain.

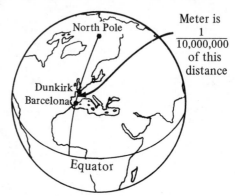

Original Definition of Meter

France adopted the metric system of measurement in 1795 and made its use compulsory in 1840. The optional use of the system was made legal in Great Britain in 1864. Other European countries adopted the system later in the 19th century. The metric system of measurement is now used by an overwhelming majority of the world's population.

In 1875, seventeen countries from around the world met and signed a treaty which established the International Bureau of Weights and Measures. The Bureau meets at regular six-year intervals to refine and improve the metric system. The Bureau in recent times has redefined the meter in terms of wave lengths of light.

Multiples of the metric units of measurement are designated by Greek and Latin prefixes.

Prefixes of Greek Origin	
kilo-	1000
hecto-	100
deka-	10

Prefixes of Latin Origin	
deci-	$\dfrac{1}{10}$, or 0.1
centi-	$\dfrac{1}{100}$, or 0.01
milli-	$\dfrac{1}{1000}$, or 0.001

Thus: 1 kilometer (km) is equivalent to 1000 meters (m).
   1 hectometer (hm) is equivalent to 100 meters.
   1 dekameter (dam) is equivalent to 10 meters.

   1 decimeter (dm) is equivalent to $\frac{1}{10}$, or 0.1, meter.

   1 centimeter (cm) is equivalent to $\frac{1}{100}$, or 0.01, meter.

   1 millimeter (mm) is equivalent to $\frac{1}{1000}$, or 0.001, meter.

   The metric unit of mass was derived from the mass of a cubic centimeter of water. This unit was named the *gram* (g), and the standard of mass became the *kilogram* (1000 g). The metric unit of capacity was chosen as

$$1L = (10 \text{ cm})^3 = 1000 \text{ cm}^3$$

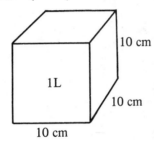

the volume of a cube with sides of 10 cm and bears the name *liter* (L). The prefixes listed above are also used with the gram and the liter.

Try these exercises which apply the metric system.

Problems

Sample: A pair of corduroy pants was 80 cm long. After being washed and dried, the pants shrank 2 cm. How long are the pants now?

Solution: Let x = length of the pants now.
   Then $x = 80 - 2$
       $x = 78$
   The pants are now 78 cm long.

1. A boat travels up a river 23 km. Then it travels down the river 9 km. How far is it from its original position? 14 km

2. The length of a rectangle is 1 m less than three times the width. The perimeter is 22 m. Find the length and the width. 8 m; 3 m

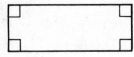

3. Joshua is 10 cm shorter than Brendan. The sum of their heights is 3 m. How tall is each boy? 145 cm; 155 cm

4. The perimeter of a square is 20 cm. What is the area of the square? 25 cm²

5. One container holds three times as much as another container. The two containers together hold 10.8 L. How much does each container hold? 8.1 L; 2.7 L

Introduction to BASIC Programming

1. Multiples; PRINT, INPUT, FOR-NEXT loop; RUN

People control computers by giving them lists of statements (instructions), called *programs,* which are written in one of several *programming languages.* Since BASIC is a programming language that is generally available, it is the language that we have chosen for this book.

BASIC uses these signs for the arithmetical operations:

$$+ \text{ for addition} \qquad * \text{ for multiplication}$$
$$- \text{ for subtraction} \qquad / \text{ for division}$$

You will ordinarily use a calculator for simple arithmetical computation. Computers are most effectively used when a great many repetitive operations are to be done, since they can perform these with enormous speed.

A BASIC program is made up of a list of *numbered statements.* These statements are usually numbered 10, 20, . . . to allow intermediate statements to be inserted later if necessary. These statements may be organized into blocks. For instance, the following program for finding multiples (see page 56) may be said to have two blocks:

The first block (lines 10–70) describes the program and provides for the INPUT.

The second block (lines 80–100) is a *loop.* In this program, it provides the computation and the output. Some versions of BASIC require an END statement (line 110).

```
 10    PRINT "THIS PROGRAM WILL"
 20    PRINT "PRINT THE FIRST TWELVE"
 30    PRINT "MULTIPLES OF A NUMBER N."
 40    PRINT
 50    PRINT "WHAT IS YOUR VALUE OF N";
 60    INPUT N
 70    PRINT
 80    FOR I=1 TO 12
 90    PRINT I,I*N
100    NEXT I
110    END
```

The PRINT statements will print exactly any words or symbols that are enclosed in quotation marks (see lines 10-30 and 50). Lines 40 and 70 will "print" blank lines.

BASIC handles *variables* much as you do in algebra. There are several ways of giving a *value* to a variable. Lines 50 and 60 ask for a value of N. The INPUT statement in line 60 will print a question mark and wait for you to type in the value asked for in line 50.

In line 80, I is also a variable, but it is given the values 1, 2, . . . , 12 in succession by the program. Line 90 computes the multiples, I*N, and causes the values of I and I*N to be printed for each value of I. Line 100 ends the loop. Such a loop is called a FOR-NEXT loop.

Punctuation is important in PRINT statements. The *semicolon* at the end of line 50 will cause the question mark from line 60 to be printed right after the words in quotation marks. On the other hand, the *comma* between I and I*N in line 90 will cause their values to be spaced apart.

You type the *command* RUN to run, or *execute*, a program.

EXERCISES

Type the program into your computer, and RUN it for these values of N. Copy down each list.

1. N = 2	**2.** N = 3	**3.** N = 4
4. N = 5	**5.** N = 6	**6.** N = 8

From the lists that you made for Exercises 1–6, find the first two common multiples of:

7. 2 and 3	**8.** 3 and 4	**9.** 4 and 5
10. 5 and 6	**11.** 4 and 6	**12.** 6 and 8

By locating each of the following numbers in the lists that you made for Exercises 1–6, tell which of the numbers in those exercises are factors of these numbers.

13. 12	**14.** 18	**15.** 20	**16.** 24

2. Function machine, ordered pairs

We can write a program that will make our computer work like a "function machine" (see page 110). The following program, for example, finds ordered pairs when the input number, X, has the whole number values 0, 1, . . . , 10. (There will be 11 ordered pairs.) A FOR-NEXT loop is used to find the output numbers. The rule for finding the output numbers is given in line 40.

```
10   PRINT "TO FIND ORDERED PAIRS:"
20   FOR X=0 TO 10
30   PRINT "(";X;",";
40   PRINT X+2;")"
50   NEXT X
60   END
```

EXERCISES

1. RUN the program above as written and compare the results with those given on page 110.

Change line 40 each time and find the ordered pairs when the output number is given by:

2. $X - 3$

3. $4 + X$

4. $3 * X$

5. $X/2$

6. $2 * (X + 1)$

7. $(X + 2)/3$

3. Open sentences in one variable, graphs; IF-THEN-ELSE, GOTO, REM; LIST

There is a construction in BASIC that will test a sentence and provide one branch (sequence of statements) if it is *true* and another branch if it is *false*. This is often called the IF-THEN-ELSE construction. It will be used in our next program.

In Chapter 5, you used true and false statements to find solution sets for equations and inequalities. The following program uses the replacement set

$$\{0, 1, 2, 3, 4, 5, 6, 7, 8, 9, 10\} \quad \text{(line 110)}$$

and the

IF-THEN-ELSE construction (lines 120–170).

The sentence to be tested is in line 120. If it is true for the tested value of X, then the computer moves to line 170. If it is false, the computer moves on down the list of statements to lines 130 and 140. Then the GOTO statement in line 150 tells the computer to "skip over" the "then" branch to line 180.

The REM (remark) statements in lines 130 and 160 emphasize the ELSE and THEN branches. REM statements are often used when extra comments will clarify the program. They do not affect the running of the program.

```
10    PRINT "TO TEST MEMBERS OF"
20    PRINT "A REPLACEMENT SET"
30    PRINT "IN A SENTENCE TO SEE"
40    PRINT "WHICH WILL MAKE IT"
50    PRINT "TRUE AND WHICH FALSE:"
60    PRINT
70    PRINT "CONSIDER THE REPLACEMENT"
80    PRINT "SET OF WHOLE NUMBERS"
90    PRINT "FROM 0 TO 10:"
100   PRINT
110   FOR X=0 TO 10
120   IF X+9=14 THEN 170
130   REM: ELSE (FALSE)
140   PRINT "X = ";X,"FALSE"
150   GOTO 180
160   REM: THEN (TRUE)
170   PRINT "X = ";X,"TRUE"
180   NEXT X
190   END
```

EXERCISES

1. Copy and RUN the program above. Which member of the replacement set is a solution of the equation in line 120?

2. Change line 120 to test $X + 9 = 20$. You will find that no member of the chosen replacement set is a solution. What is the solution?

Change line 120 to find solutions of the following inequalities.

3. $X > 2$ 4. $X < 7$ 5. $2*X > 11$ 6. $3*X < 18$

Change the program above as follows:

```
10 PRINT "TO PRINT A GRAPH"
20 PRINT "OF THE SOLUTION SET"
30 PRINT "OF THE GIVEN SENTENCE:"
```

Delete lines 40 and 50.

140 PRINT "−";	184 PRINT "0 "; (4 spaces)
170 PRINT "*";	186 PRINT "5 "; (4 spaces)
182 PRINT ">"	188 PRINT "10"

7. Type the command LIST to see the revised program, and then RUN it.

8–11. RUN the revised program for the sentences in Exercises 3–6.

4. Average; LET

Another way of giving a value to a variable is by using a LET statement; for example:

$$\text{LET A} = 5$$

This is read from right to left and means, "Give the value 5 to A."
LET statements may also be used with formulas; for example:

$$\text{LET A} = \text{L} * \text{W}$$

Another use of a LET statement is in setting up a *counter:*

$$\text{LET C} = \text{C} + 1$$

This means, "Take the value of C, add 1 to it, and then give C this new value."

The following program will find the average of test scores that may range from 0 to 100. Since these scores are not negative, a negative number, such as −1, can be used to end the INPUT (lines 50 and 100). The variable T is used for each test score.

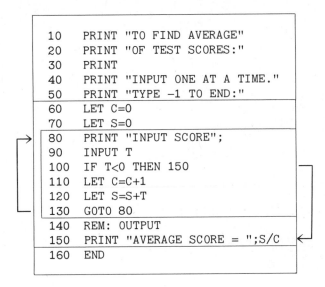

```
10    PRINT "TO FIND AVERAGE"
20    PRINT "OF TEST SCORES:"
30    PRINT
40    PRINT "INPUT ONE AT A TIME."
50    PRINT "TYPE -1 TO END:"
60    LET C=0
70    LET S=0
80    PRINT "INPUT SCORE";
90    INPUT T
100   IF T<0 THEN 150
110   LET C=C+1
120   LET S=S+T
130   GOTO 80
140   REM: OUTPUT
150   PRINT "AVERAGE SCORE = ";S/C
160   END
```

Variables C and S are given initial values in lines 60 and 70. C counts the number of scores entered. Line 120 accumulates the sum of the scores in S. Lines 80–130 form a *loop*. Line 100 ends the loop. Line 150 prints the result.

EXERCISES

1. RUN the program on page 445 for these scores:

$$78, 98, 88, 85, 96, 100, 70, 75$$

2. Explain the meaning of line 120.

You can use the program on page 445 to find the average of any number of nonnegative numbers.

5. Division, rounding; INT

BASIC has an interesting special function called the "greatest integer function." It is written INT. Thus:

$$INT(N)$$

will give the greatest integer less than or equal to N. For example:

$$INT(4) = 4, \quad INT(4.4) = 4, \quad INT(4.9) = 4.$$

It can be used in a program that will simulate division with a remainder. That is:

$$D1/D2: Q, \text{ remainder } R$$

```
10    PRINT "TO EXPRESS D1/D2 AS"
20    PRINT "  Q, REMAINDER R:"
30    PRINT "WHAT IS D1";
40    INPUT D1
50    PRINT "WHAT IS D2";
60    INPUT D2
70    LET Q1=D1/D2
80    LET Q=INT (Q1)
90    LET R=D1-D2*Q
100   PRINT D1;"/";D2;":   ";
110   PRINT Q;
120   PRINT ", REMAINDER";R
130   END
```

EXERCISES

RUN the program for these values of D1 and D2.

1. 768, 16	**2.** 768, 24	**3.** 768, 27	**4.** 768, 128
5. 768, 256	**6.** 768, 320	**7.** 768, 360	**8.** 768, 480
9. 768, 384	**10.** 768, 280	**11.** 768, 150	**12.** 768, 192

INT can also be used to round off numbers. Make the changes listed at the right in the preceding program.

```
20    PRINT "A DECIMAL ROUNDED"
25    PRINT "TO HUNDREDTHS:"
80    LET Q=INT(100*Q1+.5)/100
90
120   PRINT
```

13–24. RUN the revised program for the values given in Exercises 1–12.

25. Write out in words what line 80 does.

26. Rewrite line 80 to round to tenths. (Change line 25 also.)

Mixed Review

Chapters 1–4

Solve.

1. $15 - x = 10$ 5

2. $n \times 5 = 20$ 4

3. $34 = 24 + e$ 10

4. $\dfrac{24}{m} = 2$ 12

5. $12 = h - 10$ 22

6. $3 = \dfrac{z}{8}$ 24

7. Write the common factors and the GCF of 8 and 12. 1, 2, 4; 4

8. Write the ratio, 9 out of 24, as a fraction in lowest terms. $\frac{3}{8}$

9. Complete the whole number pattern: 56, 57, __?__, __?__, __?__, 61. 58, 59, 60

10. When $V = 6 \times e$ and $e = 3$, $V =$ __?__. 18

11. Find the least common multiple of 10 and 15. 30

Simplify.

12. $20 \div (2 + 8)$ 2

13. $\dfrac{6 + (9 \times 2)}{6}$ 4

14. $\dfrac{7 + 14}{35}$ $\frac{3}{5}$

15. List the integers between $^-1$ and $^+3$. 0, 1, 2

16. Write $64 = n + 9$ in words.
The sum of some number and 9 is 64.

17. Name the product: $8(m)(n)(6)$.
48mn

Add, subtract, multiply, or divide.

18. $(3)(5)(4 + 1)$ 75   **19.** $75.24 \div 3$ 25.08   **20.** 3.75×0.85 3.1875

21. $6.0 + 0.2 + 0.015$ 6.215   **22.** $3\frac{3}{4} - 1\frac{1}{2}$ $2\frac{1}{4}$ **23.** $1\frac{1}{2} \times 4$ 6

24. $(0.4)(5.0)(m)(n)$ 2mn   **25.** $\frac{3}{8} + \frac{1}{4}$ $\frac{5}{8}$ **26.** $6 \div \frac{2}{3}$ 9

27. Use $<$ or $>$ to complete $^+4$ __?__ $^-2$. $>$ **28.** Find a solution of $x^2 = 9$. 3

Find each value when $a = 4$, $b = 6$, and $c = 3$.

29. $a^2 + b$ 22   **30.** $ab + c$ 27   **31.** $a^2 + b^2 + c^2$ 61

32. Graph {the integers between $^-4$ and $^+1$, including $^+1$} Check students' graph.

33. Draw axes and graph $(^-3, 4)$. Check students' graph.

34. Use the rule, $a - 2$, to complete the set of ordered pairs,
{(6, 4), (7, 5), (8, __?__), (9, __?__), (10, __?__), (11, 9)}. 6, 7, 8

35. Write 250% as a fraction in lowest terms. $\frac{5}{2}$

36. Show whether $6 \times 4 = 12 + 10$ is true or false. $24 \neq 22$; false

37. Replace x to name an equivalent fraction: $\frac{4}{9} = \frac{x}{45}$. 20

38. List all factors of 18. 1, 2, 3, 6, 9, 18

39. List all whole numbers less than 7. 0, 1, 2, 3, 4, 5, 6

40. When $R = \frac{E}{I}$, $E = 72$, and $I = 12$, $R =$ __?__. 6

41. Complete 364 mm = __?__ cm and __?__ mm. 36; 4

42. Write $\frac{12}{30}$ in lowest terms. $\frac{2}{5}$ **43.** Write $\frac{32}{x}$ in words.

Thirty-two divided by some number

Chapters 5–7

Solve.

1. $x + 7 = 15$ 8   **2.** $\frac{3}{4}x = 12$ 16   **3.** $7d + d = 72$ 9

4. $2(6 + m) = 12$ 0   **5.** $x + x + x + x = 4$ 1   **6.** $7r = 56$ 8

7. Show that $\left(2 + \frac{1}{3}\right) + \frac{2}{3} = 2 + \left(\frac{1}{3} + \frac{2}{3}\right)$. Name the property shown. associative prop.
of addition

8. Use the transitive property to complete "If $x^2 - 1 = 8$ and $8 = p^3$,
then __?__." $x^2 - 1 = p^3$

9. Is {multiples of 4} closed under addition, subtraction, multiplication,
or division? yes, no, yes, no

10. Simplify $7x + 6y - x - y + 3$ by combining similar terms. $6x + 5y + 3$

11. Write whether each replacement from {3, 4, 5} makes $x + 2 \leq 6$ true or false. true, true, false

12. Show that $9(5 + 3) = 9(5) + 9(3)$. Name the property shown. distributive prop.

13. Let $f(x) = 3x - 1$ have replacement set $\left\{0, \frac{1}{3}, 1, 1\frac{1}{3}\right\}$. Write the set of ordered pairs for the function. $\left\{(0, -1), \left(\frac{1}{3}, 0\right), (1, 2), \left(1\frac{1}{3}, 3\right)\right\}$

14. For the problem "The sum of 9 and some number is greater than 32. What are the possible values of the number?": write an inequality and write the solution set if the replacement set is {numbers of arithmetic}. $9 + x > 32$; {numbers of arithmetic greater than 23}

15. Simplify each expression by using the substitution principle.
    a. $(3 \cdot 4) + (3 \cdot 6)$ b. $(30 - 5) + 12 - 20$
    12 $\underline{}$ + $\underline{}$ 18   25 $\underline{}$ + 12 - 20
         30 $\underline{}$ 17 $\underline{}$

16. Simplify $5k + 3t + 2k + 7t$ by combining similar terms. $7k + 10t$

Solve.

17. $\frac{2}{3}d = 1$ $\frac{3}{2}$ 18. $3x = 3$ 1   19. $6c = 35 + c$ 7

20. $3 + 5x = 38$ 7   21. $t - 3 = 7$ 10   22. $2x + 1 = 13$ 6

23. For the problem "During a storm, the temperature dropped 2.5°C to 28°C. What was the temperature before the storm?": write and solve an equation to answer the question. $x - 2.5 = 28$; $x = 30.5$; 30.5°C

24. Show that $(2.5)(4) = (4)(2.5)$. Name the property shown. comm. prop. of mult.

25. Tell whether each replacement makes the sentence true or false.
    a. $3a + 1 > 7$; {1, 2, 3} false, false, true   b. $k \div 2 < 9$; {0, 4, 6, 8}
                                                          true, true, true, true
26. For the problem "The lengths of the sides of a square are 10 cm. Find the area.": write and use a formula to solve the problem. $A = s^2$; $A = 100$ cm²

27. Write an algebraic expression for "3 less than x." $x - 3$

28. Make a table to find the solution of $d + 2d = 6$ for the replacement set {0, 1, 2, 3}. 2

Chapters 8–11

Add, subtract, multiply, or divide.

1. $-7 + (-8)$ ⁻15   2. $\left(-\frac{2}{3}\right)\left(\frac{9}{10}\right)$ $\frac{-3}{5}$ 3. $15 - (-7)$ 22

4. $-72 \div 8$ ⁻9   5. $\left(-\frac{3}{8}\right) \div 4$ $\frac{-3}{32}$ 6. $\frac{-18}{8 + (-5)}$ ⁻6

Solve.

7. $-5 + r = 12$ 17

8. $\dfrac{n}{8} = 3$ 24

9. $-0.8 = -2.4 + v$ 1.6

10. Graph {directed numbers greater than -1}. Check students' graph.

11. Simplify $-5a^2 + 3a + 2a^2 - 2$ by combining similar terms. $-3a^2 + 3a - 2$

12. Use $a = 2$, $b = 3$, and $c = 4$ to illustrate $a(b + c) = ab + ac$. $2(3 + 4) = 14 = 2 \cdot 3 + 2 \cdot 4$

13. For $f(x) = 3x$, complete $(2, \underline{\quad?\quad})$, $(-2, \underline{\quad?\quad})$, $\left(-\frac{1}{3}, \underline{\quad?\quad}\right)$, $\left(\frac{2}{3}, \underline{\quad?\quad}\right)$ 6, ⁻6, ⁻1, 2

Solve.

14. $5x + 6 = 3x + 16$ 5

15. $\frac{3}{5}a = -15$ ₋25

16. $-3 - r = -17$ 14

17. Write the solution set of $k \leq \frac{5}{2}$ with replacement set $\{-2, 0, 2\}$. {⁻2, 0, 2}

Simplify.

18. $6 + 9 + 2 + (-15)$ 2

19. $-(-12)$ 12

20. $-a + 6b + 3a - 4b$ 2a + 2b

21. A field has length 35 m and width 26 m. Use $P = 2L + 2W$ to find its perimeter. 122 m

22. Write the number that lies the same distance from 0 as -6 on the number line. 6

23. Evaluate $ab + ac$ for $a = -2$, $b = 4$, and $c = -1$. ⁻6

24. The function equation $f(n) = \frac{2}{3}n$ has replacement set $\{-6, -3, 0, 3, 6\}$. Write the ordered pairs of the function. (⁻6, ⁻4), (⁻3, ⁻2), (0, 0), (3, 2), (6, 4)

Use $<$ or $>$ to make each statement true.

25. $-3 \underline{\quad?\quad} -4$ >

26. $0 \underline{\quad?\quad} -6$ >

27. $-5 \underline{\quad?\quad} 2$ <

28. Show that $-3(7 + 5) = (-3)(7) + (-3)(5)$. ⁻3(7 + 5) = ⁻36; (⁻3)(7) + (⁻3)(5) = ⁻36

29. Simplify $x(2x - 3) + (x^2 - 6)$ by combining similar terms. 3x² − 3x − 6

30. For $x \leq 2$ with replacement set $\{-2, -1, 0, 1, 2\}$, write and graph the solution set. {⁻2, ⁻1, 0, 1, 2}; Check students' graph.

Solve.

31. $-9k = 2k + 55$ ₋5

32. $-4 = m + (-7)$ 3

33. $3m = -27$ ₋9

34. Complete $(-8 + 8) + 18 = \underline{\quad?\quad}$ to make the statement true. 18

Simplify.

35. $(-2)(-3)(-6)$ ₋36

36. $\dfrac{1}{-6} \div \dfrac{1}{-3}$ ½

37. $6 + \left(-2\frac{1}{2}\right)$ 3½

38. Complete $(3, \underline{\quad?\quad})$, $(1, \underline{\quad?\quad})$, $(^-1, \underline{\quad?\quad})$, $(^-3, \underline{\quad?\quad})$ for the function $f(n) = -5 + n$. ⁻2, ⁻4, ⁻6, ⁻8

Solve each inequality and write each solution set.

39. $12 + r \geq -18$
$r \geq {}^-30;$
{directed numbers greater than or equal to $^-30$}

40. $-2m - 3 < 15.$
$m > {}^-9;$
{directed numbers greater than $^-9$}

Chapters 12–14

Add.

1. $m^2 + n^2 - 5p^2$
$\underline{3m^2 - 2n^2 + p^2}$
$4m^2 - n^2 - 4p^2$

2. $-(4x + 3y)$ and $-(-7x - 2y)$
$3x - y$

Solve.

3. $x + (x - 2) + (3x - 4) = 9$ 3

4. $(5k + 7) - (3k - 11) = (9 - 5k) - (5k + 3)$ $_{-1}$

Simplify.

5. $(2 - 9d + 5d^2) - (3 + 2d^2)$
$3d^2 - 9d - 1$

6. $(5a)(3ab)(-6b)$
$-90a^2b^2$

7. $8 \cdot 3 \cdot x^2 \cdot x^5$
$24x^7$

Multiply.

8. $-x^2(4x^3 - 2x^2 + x - 5)$
$-4x^5 + 2x^4 - x^3 + 5x^2$

9. $(5 + 6m)(1 - 3m)$
$-18m^2 - 9m + 5$

Subtract.

10. $4x^2 + 7y$
$\underline{x^2 - 6y + 3}$
$3x^2 + 13y - 3$

11. $2x^3 + 11x^2$
$\underline{-x^3 - 8}$
$3x^3 + 11x^2 + 8$

Factor.

12. $x^2 + 8x + 15$
$(x + 3)(x + 5)$

13. $x(3 + x) + y(3 + x)$
$(x + y)(3 + x)$

14. $t^2 - 3t - 40$
$(t - 8)(t + 5)$

15. Solve: A board has length $2 - 3a - 4a^2$. A carpenter sawed off a piece of length $3 - 2a^2$. How long is the remaining piece? $-2a^2 - 3a - 1$

Divide. Assume that denominators are not zero.

16. $\dfrac{x^2 - 3x - 10}{x + 2}$ $x - 5$

17. $\dfrac{-50a^3b^4}{-5a^2b^2}$ $10ab^2$

18. $\dfrac{15t^2 + 20t - 35}{5}$ $3t^2 + 4t - 7$

Expand.

19. $(2x - 5)^2$ $4x^2 - 20x + 25$

20. $(x + 4y)^2$ $x^2 + 8xy + 16y^2$

21. Write $-(2x^2 + 10x - 9)$ in standard form without parentheses. $-2x^2 - 10x + 9$

Find the GCF.

22. 16 and $8mn$ 8

23. $6r^2$ and $15r$ $3r$

Find each prime factorization.

24. 50 $2 \cdot 5^2$

25. 135 $5 \cdot 3^3$

Simplify.

26. $(2^3)^2$ 64

27. $(-ab)(-8a^2b^3)(-3c^7)$ $-24a^3b^4c^7$

28. For the function equation $f(d) = 5 - 3d + d^2$ with replacement set $\{-2, -1, 0, 1, 2\}$, write the ordered pairs for the function.
$(-2, 15), (-1, 9), (0, 5), (1, 3), (2, 3)$

Glossary

absolute value (p. 261). The distance from 0 of a number on the number line. The absolute value of any directed number is always a positive number or zero.

addition property of equality (p. 186). For every number r, every number s, and every number t, if $r = s$, then $r + t = s + t$.

addition property of inequality (p. 305). For all directed numbers r, s, and t, if $r < s$, then $r + t < s + t$; if $r > s$, then $r + t > s + t$.

additive identity element (p. 173). Zero. When you add 0 to any number, the sum is the number that was added to 0.

additive inverse (p. 241). The opposite of a number. *See also* opposite.

additive property of 0 (p. 173). For every number r, $r + 0 = 0 + r = r$.

associative property of addition (p. 162). For every number r, every number s, and every number t, $r + (s + t) = (r + s) + t$.

associative property of multiplication (p. 163). For every number r, every number s, and every number t, $r(st) = (rs)t$.

base (p. 346). A repeating factor in a multiplication expression. For example, in 7^4, 4 is the exponent and 7 is the base.

binomial (p. 316). A polynomial expression having two terms.

centimeter (p. 78). A metric unit of length, 0.01 of a meter.

closure property (p. 157). A set of numbers is closed under an operation if performing the operation on any two numbers of the set results in a member of the set.

coefficient (p. 106). Any factor of an expression can be called the coefficient of the remaining factors.

common factor (p. 59). A number which is a factor of two or more numbers.

common multiple (p. 56). A number which is a multiple of two or more numbers.

commutative property of addition (p. 162). For every number r and every number s, $r + s = s + r$

commutative property of multiplication (p. 162). For every number r and every number s, $r \cdot s = s \cdot r$

complete factorization (p. 384). An expression in which none of the factors can be factored any further.

composite (p. 382). A number which can be expressed as the product of two or more prime numbers.

congruent figures (p. 27). Figures having exactly the same size and shape.

coordinate (p. 31). The number matched with a point on the number line.

coordinate axes (p. 47). Two perpendicular number lines used to graph ordered pairs.

coordinates (p. 47). The numbers in an ordered pair.

counting numbers (p. 56). The numbers 1, 2, 3, 4, 5, 6, 7, and so on.

degree of a polynomial in one variable (p. 318). The greatest exponent of the polynomial.

directed numbers (p. 212). The positive numbers, the negative numbers, and 0.

distributive property (p. 166, 169). For every number r, every number s, and every number t:
1. $r(s + t) = rs + rt$ and $(s + t)r = sr + tr$
2. $r(s - t) = rs - rt$ and $(s - t)r = sr - tr$

division property of equality (p. 189). For every number r, every number s, and every number t except 0, if $r = s$ then $\dfrac{r}{t} = \dfrac{s}{t}$.

empty set (p. 30). The set with no members. Ø

equation (p. 5). A number sentence that consists of two expressions joined by the "is equal to" symbol, =.

equivalent equations (p. 250). Equations having the same solution set.

equivalent fractions (p. 65). Fractions that name the same number.

even numbers (p. 62). The numbers in $\{0, 2, 4, 6, 8, 10, 12, \ldots\}$. Numbers that can be expressed in the form $2 \times n$, where n is a whole number.

expand (p. 360). To find the product of the factors of an expression.

exponent (p. 346). The number of times a base is used as a factor. For example, in m^3, m is the base and 3 is the exponent.

exponential notation (p. 312). A simplified way of writing a number with repeating factors. For example, $3 \times 3 \times 3 \times 3 = 3^4$ in exponential notation. 3 is the base, 4 is the exponent.

factorization (p. 106). An indicated multiplication. For example, $2 \cdot 5$ is a factorization of 10.

factors (p. 106). Two or more numbers that are multiplied to name a product.

first coordinate (p. 47). The first number in an ordered pair.

formulas (p. 137). Equations used to solve problems that occur frequently.

fraction (p. 86). A number in form $\frac{a}{b}$, where $b \neq 0$. For example: $\frac{1}{8}, \frac{8}{10}$.

function (p. 110). A set of ordered pairs, in which no two different pairs have the same first element.

graph of an equation (p. 124–125). The graph of the equation's solution set.

graph of an inequality (p. 228). The graph of the inequality's solution set.

graph of a number (p. 37). The point on the number line that corresponds to the number.

graph of an ordered pair (p. 47). The point on a graph that corresponds to the ordered pair. Coordinate axes are used to graph ordered pairs.

greatest common factor (GCF) (p. 59). The greatest member of the set of common factors of two or more numbers.

honest (p. 288). In probability, coins (or dice) are honest if all possible outcomes are equally likely when the coins (or dice) are tossed.

identity element or addition (p. 240). *See* additive identity element.

identity element for multiplication (p. 364). *See* multiplicative identity element.

inductive reasoning (p. 233). A method of reasoning by drawing a general conclusion from a particular example.

inequality (p. 120). A number sentence which contains one of the symbols $<, >, \leq, \geq,$ or $\neq$.

integers (p. 34). The numbers in $\{\ldots -4, -3, -2, -1, 0, +1, +2, +3, +4 \ldots\}$. The whole numbers (including zero) and their opposites.

inverse operations (p. 276). Operations that "undo" each other, such as addition and subtraction, or multiplication and division.

least common multiple (LCM) (p. 56). The least member of the set of common multiples of two or more numbers.

like fractions (p. 86). Fractions having a common denominator.

like terms (p. 184). *See* similar terms.

lowest terms (p. 65). A fraction is in lowest terms when the greatest common factor of the numerator and denominator is 1.

magnitude (p. 212). The distance between 0 and any number on the number line.

members of an equation (p. 17). The expressions joined by the "is equal to" symbol, $=$.

meter (p. 78). The basic unit of length in the metric system.

millimeter (p. 78). A metric unit of length, 0.001 of a meter.

monomial (p. 316). A polynomial expression having one term.

multiple of a whole number (p. 56). The product of the whole number and any counting number.

multiplication property of equality (p. 191). For every number r, every number s, and every number t, if $r = s$, then $rt = st$.

multiplication property of inequality (p. 308). For all directed numbers r, s, and t:
(1) If t is positive and $r < s$, $rt < st$.
 If t is positive and $r > s$, $rt > st$.
(2) If t is negative and $r < s$, $rt > st$.
 If t is negative and $r > s$, $rt < st$.
(3) If $t = 0$ and $r < s$ or $r > s$, then $rt = st = 0$

multiplicative identity element (p. 173). One. When you multiply any number by 1, the product is the number that was multiplied by 1.

multiplicative inverses (p. 364). *See* reciprocals.

multiplicative property of 1. (p. 173). For every number r, $r \cdot 1 = 1 \cdot r = r$.

multiplicative property of 0 (p. 173). For every number r, $r \cdot 0 = 0 \cdot r = r$.

negative numbers (p. 34, 212). The numbers naming points to the left of 0 on the number line. On a vertical number line, the numbers below 0.

negative slope (p. 343). A slope that is downward from left to right.

numbers of arithmetic (p. 127). The numbers in {0 and all numbers to the right of 0 on the number line}.

odd numbers (p. 62). The numbers in $\{1, 3, 5, 7, 9, 11, 13, \ldots\}$. Numbers that can be expressed in the form $(2 \times n) + 1$, where n is a whole number.

opposite of a sum (p. 243, 246). The opposite of a sum is the sum of the opposites. That is, for all directed numbers r and s, $-(r + s) = -r + (-s)$.

opposites (p. 212). Two different numbers that are the same distance from 0 on the number line. Also called *additive inverses*.

ordered pair (p. 47). A pair of numbers in which the order is important.

origin (p. 30, 47). The point marked "0" on the number line. The point $(0, 0)$ where coordinate axes intersect.

outcome (p. 288). The result of an event.

percent (p. 74). An expression which names the ratio of a number to 100.

polynomial (p. 316). A polynomial in one variable takes the form $ax^m + bx^n + cx^p + \cdots + d$ where a, b, c, and d are directed numbers and m, n, and p are positive integers.

positive numbers (p. 34, 212). The numbers naming points to the right of 0 on the number line. On a vertical number line, the numbers above 0.

positive slope (p. 343). A slope that is upward from left to right.

power of a power (p. 350). An expression of the form $(a^p)^q$ where a is a directed number and p and q are positive integers.

prime factorization (p. 382). A form of a composite number in which the number is written as the product of primes.

prime number (p. 52). A whole number that has exactly two different whole number factors, itself and 1.

ratio (p. 68). A comparison of two quantities or numbers.

reciprocal property (p. 278). For every directed number r except 0, $r \cdot \dfrac{1}{r} = \dfrac{1}{r} \cdot r = 1$.

reciprocals (p. 90). Two numbers whose product is 1.

reflexive property of equality (p. 154). For any number r, $r = r$.

replacement set for a variable (p. 104). The set of numbers that the variable may represent.

scientific notation (p. 312). A way of writing a number as the product of a number between 1 and 10 and a power of 10.

second coordinate (p. 47). The second number in an ordered pair.

sign of a product (p. 271). The product of two or more directed numbers is positive if there is an even number of negative factors. The product is negative if there is an odd number of negative factors.

similar figures (p. 404). Figures having the same shape.

similar terms (p. 184). Terms which contain the same variables, or terms which contain no variables.

slope (p. 342–343). The steepness of a non-vertical line defined by the quotient $\dfrac{\text{difference in second coordinates}}{\text{difference in first coordinates}}$.

solution (p. 124). A member of the replacement set for an open sentence that makes the sentence true.

solution set (p. 124). The set of members of the replacement set for an open sentence that make the sentence true.

standard form of a polynomial (p. 318). A polynomial in one variable is in standard form when the terms are ordered so that the variable in the first term has the greatest exponent, the variable in the second term has the next greatest exponent, and so on. A polynomial with more than one variable is in standard form when the terms are ordered according to the exponents of one of the variables.

substitution principle (p. 157). A numeral may be substituted for any other numeral that names the same number.

subtraction property of equality (p. 186). For every number r, every number s, and every number t, if $r = s$, then $r - t = s - t$.

symmetric property of equality (p. 154). For any numbers r and s, if $r = s$, then $s = r$.

terms (p. 106). The parts of an expression that are separated by a plus or minus sign.

transformations (p. 186). The successive changes made in an equation to produce an equivalent equation in which the variable stands alone as one member.

transitive property of equality (p. 154). For any numbers r, s, and t, if $r = s$ and $s = t$, then $r = t$.

triangular number (p. 151). The nth triangular number is the sum of the first n counting numbers.

trinomial (p. 316). A polynomial expression having three terms.

trinomial square (p. 394). A trinomial obtained by squaring a binomial.

unlike fractions (p. 86). Fractions having different denominators.

unlike terms (p. 184). Terms of an expression that contain different variables or different powers of the same variable.

values of an expression (p. 104). The values the expression takes when members of the replacement set are substituted for the variables.

variable (p. 8). A letter used in algebra to represent one or more numbers.

whole numbers (p. 30). The numbers 0, 1, 2, 3, 4, 5, and so on.

Index

Line of Symmetry, 83
Lobachevski, Nikolai Ivanovich, 317

Magnitude, 212, 240
Members of an equation, 17
Metric system, 78, 438–440
Mohorovičić, Andrija, 398
Monomial(s), 316
  division by, 371
  division, 368–369
  factoring, 384
  multiplication by polynomial, 352
  product of, 348
  raised to a power, 350
Multiples, 56
  common, 56
  least common (LCM), 56
Multiplication
  associative property of, 163, 267
  of binomials, 360–361, 399
  commutative property of, 162, 267
  of decimals, 96
  of directed numbers, 264, 267, 270–271, 283
  of fractions, 90
  monomial times polynomial, 352
  of monomials, 348, 350
  of polynomials, 356–357
  property of equality, 191
  property of inequality, 308
Multiplicative Identity element, 173
Multiplicative property
  of 0 and 1, 173, 264, 267

Negative numbers, 34, 212
Number(s)
  of arithmetic, 127
  counting, the, 56
  directed, 212–289
  even and odd, 62
  magnitude of, 212
  on the number line, 40
  positive and negative, 34, 212
  prime, 62
  triangular, 151
  whole, 30
Number line, 30–31, 34, 212
  addition, 236
  comparing directed numbers, 221

number on the, 40
  solving equations, 243
  subtraction, 250
Numbers of arithmetic, the, 127

Odd numbers, 62
Operations, order of, 100, 101
Opposite, 34, 212, 240–241
  of a polynomial, 330
  of a sum, 243, 246
Ordered pair, 47, 110
Origin
  coordinate axes, 47
  number line, 30
Outcome, 288

Percent(s), 74, 96
  written as decimal, 74
  written as fraction, 74
Pi(π), value of, 40
Polynomial(s) 316–342
  addition, 323
  degree, 318
  division, 368–369, 371, 374
  factoring, 387, 389, 392, 394, 396, 399, 401
  multiplication, 352, 356–357, 360–361
  properties of addition, 326
  properties of multiplication, 364–365
  simplifying to solve equations, 339
  standard form, 318
  subtraction, 333
  type of, 316
Positive and negative numbers, 34, 212
Power
  of a power, 350
  of a product, 350
Prime factorization, 382
Prime number, 62, 382
Probability, 288–289
Problem solving
  addition of directed numbers, 239
  area, 109, 137–139
  circle graph, 76–77
  decimal, 98
  directed numbers, 281
  equations, 196–200, 202–203, 239
  using estimation, 95
  factors, 109

Standard form of a polynomial, 318
  in addition, 323
Substitution principle, 157
Subtraction
  of decimals, 93, 117
  of directed numbers, 250
  of fractions, 86
  of polynomials, 333
  property of equality, 186
Summary, *see* Chapter Summary
Symbol(s)
  for expressions and equations, 14
  used in graphing, 44
  grouping, 100
Symmetric property of equality, 154

Terms, 106
  similar or like, 184
  unlike, 184
  combining, 184, 194, 273
Tests, *see* Chapter Tests *and* Self-Tests

Time Out, 43, 67, 95, 172, 179, 220, 386
Transformations, 186
  order of, 191
  used to solve equations, 194, 201
Transitive property of equality, 154
Triangular number, 151
Trinomial(s), 316
  factoring, 399, 401
Trinomial square(s), 394
  factoring, 394, 396

Unlike terms, 184

Value
  of an expression, 8
  of a function, 175
Variable(s), 8

Walker, Maggie Lena, 103
Whole numbers, 30
  multiples of, 56

Photo Credits

Page 1. (left) THE BETTMAN ARCHIVE; (right) Tyrone Hall. Page 4. CULVER PICTURES, INC. Page 16. (left) UPI; (right) Patricia Hollander Gross-STOCK, BOSTON. Page 29. (left) THE BETTMAN ARCHIVE; (right) Digital Corporation. Page 55. (left) CULVER PICTURES, INC.; (right) John Hamilton Burke. Page 85. (left) BROWN BROTHERS; (right) Tyrone Hall. Page 89. Tyrone Hall. Page 103. Independent Order of St. Luke. Page 119. (left) THE BETTMAN ARCHIVE; (right) New England Telephone. Page 123. (left) U.S. Department of Labor; (right) Tyrone Hall. Page 136. THE BETTMAN ARCHIVE. Page 153. (left) Paul Thomspon-FREE LANCE PHOTOGRAPHERS GUILD, INC.; (right) DeWYS, INC. Page 161. Tyrone Hall. Page 183. (left) THE BETTMAN ARCHIVE; (right) Michael Kennedy-Harvard College Observatory. Page 211. (left) THE BETTMAN ARCHIVE; (right) Tyrone Hall. Page 215. (both) American Telephone and Telegraph. Page 227. Bryn Mawr College. Page 235. (left) Smithsonian Institute; (right) American Textile Manufacturers Institute. Page 253. (left) Elliot Erwitt-MAGNUM PHOTOS, INC.; (right) Tyrone Hall. Page 263. (left) THE BETTMAN ARCHIVE; (right) UPI. Page 266. SEKAI BUNKA PHOTO. Page 282. Tyrone Hall. Page 291. (left) National Archives; (right) National Weather Service. Page 294. BROWN BROTHERS. Page 315. (right) San Francisco Bay Area Transit Authority. Page 317. THE BETTMAN ARCHIVE. Page 345. (left) University of Louisville Photographic Archives; (right) Ellis Herwig-STOCK, BOSTON. Page 363. (left) Tyrone Hall; (right) UPI. Page 381. (left) Pinkerton's, Inc.; (right) Paul Sequeira-PHOTO RESEARCHERS, INC. Page 398. Courtesy D.S. Halacy from THEY GAVE THEIR NAMES TO SCIENCE by D.S. Halacy, © 1967, G.P. Putnam's Sons.

Answers to Odd-Numbered Exercises

Chapter 1. Working with Integers

Pages 3–4 Written Exercises A **1.** 45 **3.** 17 **5.** 0 **7.** 20 **9.** 25 **11.** 23 **13.** 3 **15.** 64 B **17.** 34
19. 5 **21.** 49 **23.** 1 **25.** 8, 20, 5 **27.** 2, 4, 12 **29.** 15, 1, 1 C **31.** 4, 13 **33.** 45, 3

Pages 6–7 Written Exercises A **1.** True **3.** False **5.** False **7.** True **9.** False B **11.** False **13.** True
15. False **17.** 5 **19.** 2 **21.** 20 **23.** 2 **25.** 6 C **27.** 54 **29.** 48 **31.** 16 **33.** 2 **35.** 2 **37.** 3 **39.** 5
41. Any whole number **43.** 1

Page 9 Written Exercises A **1.** 36 **3.** 84 **5.** 7 **7.** 3 **9.** 7 **11.** 3 **13.** 13 B **15.** 2 **17.** 35 **19.** 52
21. 125 C **23.** 3 **25.** 10 **27.** 625 **29.** Answers may vary. Examples: $m + y + w$, $y^2 + z$
31. Answers may vary. Examples: $(w \times z) + (y \times z)$, $w^2 + m + y$
33. Answers may vary. Examples: $(y \times w) - m$, $z^2 - w - y$

Pages 11–12 Written Exercises A **1.** 92 **3.** 32 **5.** 10,000 **7.** 270 **9.** 8 **11.** 1110 B **13.** 400 **15.** 400
17. 256 **19.** 153 **21.** 125 C **23.** 72

Page 12 Problems **1.** 204 **3.** 60 **5.** 15

Page 15 Written Exercises A **1.** The sum of six and nine **3.** Sixteen divided by some number
5. The difference of some number and six **7.** The product of five and eight
9. The product of some number and ten **11.** The difference between twenty-five and some number
13. The sum of ten and some number is equal to forty-two **15.** The difference of thirty-five and some
number is equal to nineteen. **17.** The sum of thirty-seven and some number is equal to ninety-five.
19. Forty-five divided by nine is equal to some number. B **21.** The difference of some number and
sixty-eight is equal to fourteen. **23.** Thirty-nine is equal to the sum of some number and ten.
25. Seventy-five is equal to the difference of ninety-two and some number. **27.** $n + 16 = 17$
29. $36 = 6 \times n$ **31.** $\frac{n}{15} = 10$ **33.** $0 + n = 12$

Pages 18–19 Written Exercises A **1.** 12 **3.** 36 **5.** 6 **7.** 8 **9.** 60 **11.** 4 **13.** 12 **15.** 40 **17.** 1
B **19.** 8 **21.** 2 **23.** 5 **25.** 6 **27.** 50 **29.** 9 C **31.** 15 **33.** 29 **35.** 9

Pages 21–22 Written Exercises A **1.** $6 \times n = 30$ **3.** $8 = \frac{32}{n}$ **5.** $42 = 6 \times n$ **7.** 12; 5; 60
9. 7; 6; 42 **11.** 17; 153; 9 B **13.** 50 **15.** 5 **17.** 24 **19.** 10 C **21.** 16 **23.** 4 **25.** 20 **27.** 44 **29.** 1

Pages 24–25 Written Exercises A **1.** 16; 25; 36; 49; 64; 81; 100
3. 484; 529; 576; 625; 676; 729; 784; 841; 900 **5.** 343; 512; 729; 1000 **7.** $10 \times 10 \times 10$; 1000
9. $15 \times 15 \times 15$; 3375 **11.** 28×28; 784 **13.** 19×19; 361 **15.** $4 \times 4 \times 4$; 64 B **17.** 5 **19.** 20
21. 24 **23.** 12 **25.** 8 **27.** 9 **29.** 14 C **31.** 5 **33.** 4

Page 28 Review of Skills **1.** $<$ **3.** $<$ **5.** $>$ **7.** Forty-seven hundredths
9. Three and one hundred twenty-five thousandths **11.** One hundredth **13.** 10.65 **15.** 0.05 **17.** 1343
19. $11\frac{4}{5}$ **21.** 269 **23.** $7\frac{3}{4}$ **25.** 5110 **27.** $\frac{3}{10}$ **29.** 17.75 **31.** 8 **33.** $\frac{1}{3}$ **35.** 50

Chapter 2. Positive and Negative Numbers

Pages 32–33 Written Exercises A 1. 8 3. 7 5. 4 7. 1 9. 2 11. Q 13. H 15. M B 17. M 19. B
21. {1} 23. {6, 7} 25. {0, 1, 2, 3, 4, 5} 27. {0, 1, 2, 3, 4, 5, 6, 7, 8}
C 29. The whole numbers between 2 and 7 31. The whole numbers between 9 and 13
33. The whole numbers greater than 6 35. The whole numbers greater than 20 37. 3 39. 8 41. 13
43. The average of 3 consecutive whole numbers is the middle number. 45. 13, 14 47. 9, 10, 11

Pages 35–36 Written Exercises A 1. $^+4$ 3. $^+7$ 5. 0 7. $^-7$ 9. $^+1$ 11. X 13. R 15. G 17. $^-3, ^-2, ^+2, ^+3$
19. $^-1, 0, ^+1, ^+2$ 21. $^-2, ^-1, 0$ 23. 0, $^+1, ^+2$ B 25. {$^-5, ^-4, ^-3, ^-2, ^-1, 0$} 27. {$^+1, ^+2, ^+3, ^+4, ^+5, \ldots$}
29. {$\ldots, ^-5, ^-4, ^-3, ^-2, ^-1$} 31. {0} 33. {0} 35. The integers between $^-3$ and $^+1$
37. The integers between $^-2$ and $^+2$ C 39. Negative 41. Positive

Pages 38–39 Written Exercises A 1. {$^+1, ^+2, ^+3, ^+4, ^+5$} 3. {$^-3, ^-2, ^+2, ^+3$} 5. {$^-6, ^-4, ^-2$}

7. [number line] 9. [number line]
11. [number line] B 13. [number line] 15. [number line]
17. [number line]

C 19. {the integers greater than $^+99$} 21. {the integers between $^+9$ and $^+14$}
23. {the integers less than $^+1$} 25. {the integers between $^-5$ and 0}
27. {the integers between $^-10$ and $^-7$ and their opposites}
29. {the even integers between 0 and $^+6$ inclusive} 31. {the integers greater than $^+4$}

Pages 42–43 Written Exercises A 1. $^-1\frac{1}{2}$ 3. $^+2\frac{3}{4}$ 5. $^-1\frac{3}{4}$ 7. $^-\frac{3}{4}$ 9. $^+\frac{3}{4}$ 11. Left, $<$ 13. Left, $<$
15. Left, $<$ 17. $^-7, ^-5, 0, ^+2, ^+8$ 19. $^-6, ^-3, ^-2, 0, ^+1$ 21. $^-6, ^-1\frac{1}{2}, 0, ^+1, ^+5$ 23. $^-7, ^-5, ^-4, ^-3, ^-1$
B 25. $^-1\frac{1}{2}, ^-\frac{2}{3}, ^+\frac{1}{2}, ^+\frac{4}{5}$ 27. $^-75, ^-\frac{3}{4}, 0, ^+1, ^+5$

29. [number line] 31. [number line]
33. [number line] 35. [number line]
37. [number line] 39. $>$ 41. $<$

Pages 45–46 Written Exercises A 1. [number line] 3. [number line]
5. [number line] 7. [number line] 9. [number line]
11. [number line] 13. [number line]
15. [number line] 17. [number line]
19. [number line] B 21. [number line]
23. [number line] 25. [number line]
27. [number line] 29. [number line]

Pages 48–50 Written Exercises A **1.** B **3.** D **5.** A **7.** $(^+1, ^+5)$ **9.** $(^-4, 0)$ **11.** $(^+5, 0)$ **13.** $(^-2, ^-5)$
15. $(^+2, ^-2)$

17. 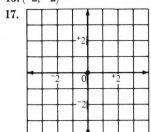 **19.** 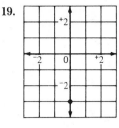 **21.**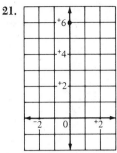

23. 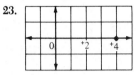 **25.**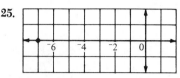

B **27.** $(^+2, 0)$ **29.** $(^-8, ^+5)$ **31.** $(0, ^+6)$ **33.** $(0, ^-6)$ **35.** $(^-3, ^-3), (^-2, ^-2), (^-1, ^-1), (0, 0)$
37. $(^-3, 0), (^-2, 0), (^-1, 0), (0, 0), (^+1, 0), (^+2, 0)$

Page 54 Review of Skills **1.** 9 **3.** 4 **5.** 8 **7.** 4 **9.** 4 **11.** hundreds **13.** hundredths **15.** tenths
17. thousandths **19.** 13 **21.** 19 **23.** 15 **25.** 6 **27.** 2 **29.** 2.5 **31.** 0.6 **33.** 0.5 **35.** 0.25

Chapter 3. Factors and Multiples

Pages 57–58 Written Exercises A **1.** $a = 9, h = 18, m = 27, w = 36; \{9, 18, 27, 36, \ldots\}$
3. $t = 4, r = 8, p = 12, n = 16; \{4, 8, 12, 16, \ldots\}$ **5.** 6
7. $\{3, 6, 9, 12, \ldots\}, \{4, 8, 12, 16, \ldots\}, \{12, 24, 36\}; 12$ **9.** 15 **11.** 12 **13.** 12 **15.** 10 **17.** 16 B **19.** 18
21. 60 **23.** 60 **25.** 55 **27.** 10 **29.** 40 **31.** 20 **33.** 40 **35.** 270 **37.** 96 **39.** 80 **41.** 144

Pages 60–61 Written Exercises A **1.** D **3.** A **5.** B **7.** $v = 35, w = 7; \{1, 5, 7, 35\}$
9. $k = 17; \{1, 17\}$ **11.** 1, 2, 13, 26 **13.** 1, 2, 7, 14 **15.** 1, 13 **17.** 1, 2, 5, 10, 25, 50 B **19.** $\{1, 2, 4\}; 4$
21. $\{1, 2, 4\}; 4$ **23.** $\{1, 3, 9\}; 9$ **25.** $\{1\}; 1$ **27.** 3 **29.** 2 **31.** 1 C **33.** True **35.** True

Pages 63–64 Written Exercises A **1.** 2×7 **3.** Not possible **5.** Not possible **7.** 2×36
9. Not possible **11.** Not possible **13.** $(2 \times 11) + 1$ **15.** $(2 \times 20) + 1$ **17.** $(2 \times 27) + 1$
19. $6 + 8 = 14$; even + even = even **21.** $4 + 0 = 4$; even + even = even
23. $7 + 9 = 16$; odd + odd = even **25.** $5 + 1 = 6$; odd + odd = even
27. $8 + 9 = 17$; even + odd = odd **29.** 5, 7, 9 B **31.** $\{1, 2, 7, 14\}; 2, 7$ **33.** $\{1, 11\}; 11$
35. $\{1, 2, 4\}; 2$ **37.** $\{1, 3, 19, 57\}; 3, 19$ **39.** Even **41.** Odd **43.** Even C **45–49.** Answers may vary.
Examples: **45.** $3 + 5$ **47.** $31 + 19$ **49.** $97 + 3$

Pages 66–67 Written Exercises A **1.** $\frac{5}{10}$ **3.** $\frac{8}{10}$ **5.** $\frac{1}{8}$ **7.** $\frac{3}{8}$ **9.** 1 **11.** $\frac{3}{4}$ **13.** $\frac{7}{10}$ **15.** $\frac{1}{2}$ **17.** $\frac{3}{5}$ **19.** $\frac{3}{4}$ **21.** $\frac{7}{10}$
23. $\frac{1}{5}$ **25.** 21 **27.** 90 **29.** 5 B **31.** $r = 20, t = 15$ **33.** $h = 5, a = 50$ **35.** $x = 25, y = 40$

Page 69 Written Exercises A **1.** $\frac{2}{3}$ **3.** $\frac{19}{100}$ **5.** $\frac{5}{6}$ **7.** $\frac{9}{5}$ **9.** $\frac{1}{4}$ **11.** $\frac{1}{5}$ B **13.** 2, 5 **15.** 3, 5 **17.** 3, 7

Pages 69–70 Problems **1.** $\frac{1}{5}, \frac{4}{5}$ **3.** $\frac{7}{3}, \frac{10}{7}$ **5.** $\frac{1}{75}$ **7.** $\frac{3}{1}$

Pages 72–73 Written Exercises A 1. five tenths 3. twenty-three hundredths
5. eight and forty hundredths 7. twenty-five thousandths 9. eight tenths
11. twenty and zero hundredths 13. 0.2 15. 0.4 17. 0.8 19. 0.7 21. 0.18 23. 0.125
25. $\frac{2}{10} = \frac{20}{100} = \frac{200}{1000}$ 27. $\frac{13}{100} = \frac{130}{1000} = \frac{1300}{10,000}$ 29. $0.7 = 0.70 = 0.700 = 0.7000$
31. $0.12 = 0.120 = 0.1200 = 0.12000$ 33. $\frac{2}{5}$ 35. $\frac{3}{4}$ B 37. $\frac{7}{8}$ 39. $\frac{1}{40}$ 41. $1\frac{1}{20}$ 43. $0.1\overline{6}$ 45. $0.\overline{5}$ 47. $0.8\overline{3}$
49. 2.75

Pages 75–76 Written Exercises A 1. $0.37 = 37\%$ 3. $0.70 = 70\%$ 5. $0.25 = 25\%$ 7. $0.625 = 62.5\%$
9. $1.00 = 100\%$ 11. $0.45 = 45\%$ 13. $\frac{3}{20}$ 15. $\frac{1}{10}$ 17. $\frac{1}{4}$ 19. $\frac{21}{50}$ 21. 1 B 23. $12.5\% = 0.125 = \frac{125}{1000} = \frac{1}{8}$
25. $87.5\% = 0.875 = \frac{875}{1000} = \frac{7}{8}$ 27. $5\% = 0.05 = \frac{5}{100} = \frac{1}{20}$ 29. $325\% = 3.25 = 3\frac{25}{100} = 3\frac{1}{4}$ C 31. 0.035
33. 0.0275 35. 0.0075

Pages 76–77 Problems 1. Boys: $\frac{2}{5}$, girls: $\frac{3}{5}$ 3. Nonfiction: $\frac{1}{2}$, fiction: $\frac{1}{4}$,
reference: $\frac{1}{8}$; nonfiction or fiction: $\frac{3}{4}$

Pages 79–80 Written Exercises A 1. 9.0 cm 3. 4.5 cm 5. 6.7 cm 7. 9.5 cm B 19. 0.95 m 21. 0.6 m
23. 4 m 9 cm 25. 5.0 cm 27. 10.0 cm 29. 12 cm 5 mm C 31. 2.6 cm 33. 94 mm 35. 6 mm
37. 0.15 m 39. 75 cm 41. 60 cm

Page 84 Review of Skills 1. $3\frac{1}{2}$ 3. $4\frac{1}{5}$ 5. $\frac{7}{3}$ 7. $\frac{8}{7}$ 9. 6.367 11. 34.5 13. 1 15. $\frac{1}{7}$ 17. 52.48 19. 17.23
21. 1.7 23. 230 25. 216 27. (3, 2) 29. (5, 4)

Chapter 4. Working with Fractions and Decimals

Pages 87–88 Written Exercises A 1. $\frac{7}{9}$ 3. $\frac{3}{4}$ 5. $\frac{3}{5}$ 7. $\frac{1}{2}$ 9. 1 11. 4 13. $7\frac{7}{10}$ 15. $\frac{1}{4}$ 17. $1\frac{1}{2}$ 19. 4; $t = \frac{5}{8}$
21. 2; $m = \frac{1}{2}$ B 23. $\frac{11}{40}$ 25. $\frac{1}{4}$ 27. $2\frac{1}{6}$ 29. $11\frac{3}{8}$ 31. 1 33. $2\frac{1}{3}$ 35. $\frac{1}{2}$ 37. 1 39. $\frac{1}{5}$

Pages 91–92 Written Exercises A 1. $\frac{2}{27}$ 3. $3\frac{3}{10}$ 5. $\frac{1}{2}$ 7. $\frac{3}{20}$ 9. $\frac{5}{6}$ 11. $\frac{3}{8}$ 13. $\frac{5}{6}$ 15. $\frac{1}{2}$; 3 B 17. $\frac{1}{8}$ 19. $\frac{2}{5}$
21. 9 23. $\frac{1}{7}$ 25. 1 27. $\frac{n}{m}$ 29. $\frac{5}{a}$ 31. $1\frac{1}{5}$ 33. $2\frac{1}{4}$ 35. 1 37. $\frac{6}{7}$ 39. 2 41. $1\frac{7}{10}$ 43. $\frac{1}{4}$ 45. $\frac{1}{24}$ 47. 4

Pages 94–95 Written Exercises A 1. 19.18 3. 57.7 5. 0.005 7. 11.607 9. 4.203 11. 48.6 13. 10.97
15. 6.25 17. 6.908 B 19. 436.28 21. 7.404 23. 11.396 25. 248.77 C 27. 2.5 29. 2.54 31. 11.996

Page 95 Problems 1. $13.16 3. $6.14 5. $159.80

Pages 97–98 Written Exercises A 1. 4.6265 3. 57.998 5. 0.00832 7. 11.25 9. 43.82 11. 3.1
13. 33.67 15. 1.91 17. 18 19. 112 B 21. 11.59 23. 2.55 25. $312 27. $968.40 29. 13.6 31. 768
33. 5.0275; 50.275; 502.75; 5027.5 35. 34.28; 342.8; 3428; 34,280

Page 98 Problems 1. $2700 3. 717.01 cm^2 5. $4.73 per kilogram

Pages 101–102 Written Exercises A 1. 72 3. 31 5. 32 7. 46 9. $4\frac{1}{2}$ 11. $4\frac{1}{3}$ 13. 5 15. 67 B 17. 210
19. 71 21. 175 23. $5\frac{3}{4}$ 25. $2\frac{7}{8}$ 27. 10 29. 36 31. 30 C 33. $5\frac{1}{3}$ 35. $(6 \times 4) - 4 = 20$
37. $17 = (3 \times 5) + 2$ 39. $4 + [(2 \times 3) \times 1] = 10$ 41. $(12 \div 4) - 3 = 0$ 43. $8 - (2 \times 3) = 2$

Page 105 Written Exercises A 1. 11, 13, 15, 17 3. 3, 5, 7, 9 5. $\frac{1}{2}, \frac{3}{2}, \frac{5}{2}, \frac{7}{2}$ 7. 3, 9, 15, 21
9. 1, 9, 25, 49 11. 10, 14, 18, 22 13. 16 15. 30 17. 5 B 19. 38 21. 17 23. 5 25. 10 C 27. 37
29. 18 31. 5

Pages 107–109 Written Exercises A **1.** s, $2s$, $5s$ **3.** $\frac{1}{8}$, $\frac{t}{4}$, $\frac{t}{2}$ **5.** yz, xz, xy **7.** $2m$, m, 2

9. $8k = 8(4) = 32$

$\quad 4(2k) = 4(8) = 32$

$\quad 2(4k) = 2(16) = 32$

11. $\dfrac{kt}{3} = \dfrac{4(6)}{3} = 8$

$\quad \frac{1}{3}(kt) = \frac{1}{3}(24) = 8$

$\quad \frac{t}{3}(k) = 2(4) = 8$

13. $\frac{1}{2}kt = \frac{1}{2}(4)(6) = 12$

$\quad \dfrac{k}{2}\cdot t = 2\cdot 6 = 12$

$\quad k\cdot \dfrac{t}{2} = 4\cdot 3 = 12$

15. $6x$ **17.** $9ab$ B **19.** $P = 2l + 2w$ **21.** $A = \frac{1}{2}bh$ **23.** $54m^2$ **25.** $\frac{3}{5}bc$ **27.** $\dfrac{5pq}{2}$

C **29.** $8(rs)$, $8r(s)$, $8s(r)$ **31.** $3rs\left(\dfrac{1}{t}\right)$, $3r\left(\dfrac{s}{t}\right)$, $3\left(\dfrac{rs}{t}\right)$ **33.** $2(d)$, $\frac{1}{2}(4d)$, $6\left(\dfrac{d}{3}\right)$ **35.** $3t(\frac{1}{10})$, $3\left(\dfrac{t}{10}\right)$, $\frac{3}{5}\left(\dfrac{t}{2}\right)$

37. $a\left(\dfrac{1}{cd}\right)$, $\dfrac{1}{c}\left(\dfrac{a}{d}\right)$, $\dfrac{1}{d}\left(\dfrac{a}{c}\right)$ **39.** $\frac{4}{5}(cd)$, $4c\left(\dfrac{d}{5}\right)$, $\dfrac{d}{5}(4c)$

Page 109 Problems **1.** $2d$ **3.** $6ab$

Pages 112–114 Written Exercises A **1.** $(7, 2)$, $(8, 3)$, $(9, 4)$ **3.** $(8, 1\frac{3}{5})$, $(12, 2\frac{2}{5})$, $(16, 3\frac{1}{5})$, $(20, 4)$, $(24, 4\frac{4}{5})$
5. $(2, 4)$, $(3, 9)$, $(6, 36)$, $(7, 49)$

7.

9.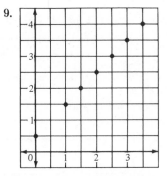

11. C **13.** E B **15.** $(0, 0.5)$ **17.** 2.0; $(1.5, 2.0)$ **19.** 3.0; $(2.5, 3.0)$ **21.** 2; $(4, 2)$ **23.** $3\frac{1}{3}$; $(8, 3\frac{1}{3})$
25. $4\frac{2}{3}$; $(12, 4\frac{2}{3})$ C **27.** $\frac{1}{2}x$ **29.** x^2 **31.** $\frac{1}{10}x$

Page 118 Review of Skills **1.** 10^2 **3.** 5^5 **5.** n^3 **7.** 45 **9.** 490 **11.** 0 **13.** 64 **15.** 1008 **17.** True
19. True **21.** 7 **23.** 9 **25.** 8 **27.** $\$412.38$ **29.** $\frac{3}{4}$ **31.** $\frac{1}{4}$ **33.** $\frac{3}{5}$ **35.** $\frac{3}{4}$

Chapter 5. Equations and Inequalities

Pages 121–122 Written Exercises A **1.** C **3.** F **5.** H **7.** B **9.** $n + 8 = 23$ **11.** $3x^3 > 7$
13. $3.4n > 100$ B **15.** $2n^2 > 5$ **17.** $4n + 1 < 20$ **19-25.** Answers may vary.
One example is given for each. **19.** 7 is less than some number minus 2.
21. The sum of 5 squared and s is not equal to 8. **23.** The product of 2 and r added to 3 is less than 11.
25. 3 subtracted from the square of a is greater than 10.

Pages 125–126 Written Exercises A **1.** $\{12\}$ **3.** $\{6\}$ **5.** $\{2\}$ **7.** $\{3\}$ **9.** $\{2\}$
B **11.** $\{5\}$ 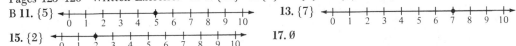 **13.** $\{7\}$
15. $\{2\}$ **17.** $\emptyset$

19. $\{0, 2\}$ 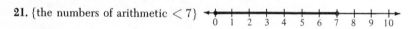   **21.** $\{2\}$

23. $\{0\}$ **25.** $\emptyset$

Pages 128–129   Written Exercises   A   **1.** $13 = 13$ **3.** $16 = 16$ **5.** $6 = 6$ **7.** $7 = 7$ **9.** $31 = 31$
11. $12 = 12$ **13.** $x + 10 - 10 = x$ **15.** $n - 17 + 17 = n$ **17.** $10 + z - 10 = z$
19. $3.9 + m - 3.9 = m$ **21.** 7 **23.** 30 B   **25.** $7\frac{1}{2}$ **27.** 5 **29.** $3\frac{1}{3}$ C   **31.** $25; 5$ **33.** $36; 6$ **35.** 21 **37.** 79
39. $6\frac{1}{2}$

Pages 131–133   Written Exercises   A   **1.** $2 = 2$ **3.** $30 = 30$ **5.** $4 = 4$ **7.** $24 = 24$ **9.** $6 = 6$ **11.** $8 = 8$
13. $\frac{8n}{8} = n$ **15.** $\frac{4x}{4} = x$ **17.** $7 \cdot \frac{h}{7} = h$ **19.** 21 **21.** 18 **23.** 294 B   **25.** 13 **27.** 400 **29.** 32.5 **31.** 5
33. $22\frac{1}{2}$ **35.** 6

Pages 135–136   Written Exercises   A   **1.** $\frac{2}{3}a$ **3.** $c^2 - 1$ **5.** $\frac{3}{4}m + 2$ **7.** $x - 2$ **9.** $y + 5$
11. $\frac{1}{3}z - 1$ or $\frac{z}{3} - 1$ **13.** $10a + 5$ **15.** $t + b$ **17.** $100 - 5r$ **19.** $5n - 2k$ B   **21.** $\frac{1}{4}mn$ or $\frac{mn}{4}$ **23.** $n - 2$
25. $\frac{a + b + c}{3}$ C   **27.** $10d + 5n + p$

Pages 138–139   Written Exercises   A   **1.** $A = lw$; 115 cm^2 **3.** $A = \frac{1}{2}bh$; 65 cm^2
5. $P = a + b + c$; 4.4 m **7.** $P = 4s$; 36 cm **9.** $A = lw$; 336.2 cm^2 B   **11.** $C = \pi d$; 21 cm
C   **13. a.** $C = \pi d$; 80 cm **b.** $C = \pi d$; 80 cm

Pages 141–142   Written Exercises   A   **1.** True   **3.** False   **5.** False   **7.** False   **9.** False; True; True
11. True; True; True; False   **13.** True; True; True; False   B   **15.** False; False; False; True
17. True; True; True   **19.** True; False; False; True

21. {the numbers of arithmetic < 7}

23. {the numbers of arithmetic > 4}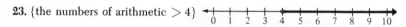

25. {the numbers of arithmetic > 2.5}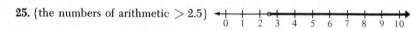

27. {the numbers of arithmetic > 11}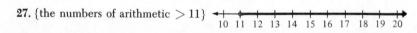

29. {the numbers of arithmetic}

C   **31.** {the numbers of arithmetic > 21}

33. {the numbers of arithmetic > 0}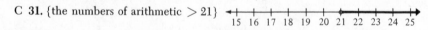

35. {the numbers of arithmetic whose squares are greater than 47}

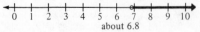
about 6.8

37. {the numbers of arithmetic > 10}

39. {the numbers of arithmetic $> 1\frac{1}{2}$}

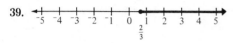

41. {the numbers of arithmetic > 2}

Pages 144–145 Written Exercises A **1.** True; True; False **3.** False; False; True **5.** True; True; True
7. True; False; False **9.** False; True; True **11.** True; False; False B **13.** True; True; False
15. False; False; True; True **17.** B **19.** A **21.** D **23.** {the numbers of arithmetic ≥ 1}
25. {the numbers of arithmetic $\leq 4\frac{1}{2}$} **27.** {the numbers of arithmetic ≤ 5}
C **29.** {the numbers on the number line ≤ 0} **31.** {the numbers on the number line $\leq {}^-3$}
33. {the numbers on the number line, except 3}
35.   **37.**
39.

Pages 147–148 Problems A **1.** $4s < 50$; {the numbers of arithmetic $< 12\frac{1}{2}$}
3. $n \geq 10$ and $n \leq 18$; {the numbers of arithmetic between 10 and 18, inclusive}
5. $2\frac{1}{3} + n < 14$; {the numbers of arithmetic $< 11\frac{2}{3}$}
B **7.** $2l + 2 \cdot 10 \leq 100$; {the numbers of arithmetic ≤ 40}
9. $\frac{22}{7}d \geq 44$; {the numbers of arithmetic ≥ 14}; $2 \cdot \frac{22}{7}r \geq 44$; {the numbers of arithmetic ≥ 7}

Page 152 Review of Skills **1.** 72 **3.** 210 **5.** $\frac{23}{20}$ or $1\frac{3}{20}$ **7.** 6 **9.** 11 **11.** 125 **13.** 25 **15.** 0 **17.** 33
19. True **21.** False **23.** True **25.** True **27.** True **29.** 8 **31.** 0 **33.** 2.5 **35.** 0 **37.** 12 **39.** {0} **41.** {4}

Chapter 6. Axioms and Properties

Pages 155–156 Written Exercises A **1.** $m = m$ **3.** $a + b^3 = a + b^3$ **5.** $10 = m + 4$
7. $p = b + 2c$ **9.** $k(k^2 + k + 1) = k^3 + k^2 + k$ **11.** $7 = 5 + 2$ **13.** $48 \div 12 = 2^2$ B **15.** $q = 13$
17. $x^2 + 1 = p^3$ **19.** $m = r^2 + t$ C **21.** Symm. Prop.; Trans. Prop. **23.** Trans. Prop.; Trans. Prop.
25. Symm. Prop.; Trans. Prop. **27.** Trans. Prop; Trans. Prop.

Pages 158–160 Written Exercises A **1.** Closed under add. and mult.; not closed under subtr. and div.
3. Closed under add. and mult.; not closed under subtr. and div.
5. Closed under mult.; not closed under add., subtr., and div.
7. Closed under mult. and div.; not closed under add. and subtr.
9. Not closed under add., subtr., mult., and div. **11.** 1; $\frac{1}{2}$; 1 **13.** 18, 14; 32 **15.** $\frac{3}{4}$; $1\frac{1}{2}$; $2\frac{1}{4}$
17. 6.1; 13.8; 19.4 **19.** $\frac{2}{9}$, $\frac{4}{9}$; $\frac{2}{3}$ **21.** 2, $\frac{1}{2}$; $2\frac{1}{2}$

Pages 164–165 Written Exercises A **1.** $27 = 27$; Assoc. Prop. of Add.
3. $21.2 = 21.2$; Comm. Prop. of Add. **5.** $6\frac{1}{6} = 6\frac{1}{6}$; Comm. Prop. of Mult.
7. $2\frac{1}{4} = 2\frac{1}{4}$; Assoc. Prop. of Add. **9.** $1.83 = 1.83$; Comm. Prop. of Mult.
B **11.** $1.23 = 1.23$; Comm. Prop. of Add. **13.** $5.23 = 5.23$; Assoc. Prop. of Add.
15. $0.156 = 0.156$; Comm. Prop. of Mult. **17.** $5.23 = 5.23$; Assoc. Prop. of Add.
19. $0.144 = 0.144$; Comm. Prop. of Mult.
21. Comm. Prop. of Add.; Assoc. Prop. of Add.; Subst. Principle; Subst. Principle
23. Comm. Prop. of Mult.; Assoc. Prop. of Mult.; Subst. Principle; Subst. Principle **25.** = **27.** =
29. $\neq$ **31.** = **33.** = **35.** = **37.** =

Pages 167–168 Written Exercises A **1.** $27 = 27$ **3.** $104 = 104$ **5.** $161 = 161$ **7.** $12 = 12$
9. $3.18 = 3.18$ B **11.** 1836 **13.** 922 **15.** 3708 **17.** 2933 **19.** $12{,}496$ **21.** $37{,}017$
23. Comm. Prop. of Mult.; Subst. Principle; Dist. Prop.; Subst. Principle; Subst. Principle
25. Subst. Principle; Dist. Prop.; Subst. Principle, Comm. Prop. of Add.; Subst. Principle C **27.** True
29. True

Pages 170–171 Written Exercises A **1.** $28 = 28$ **3.** $123 = 123$ **5.** $165 = 165$ **7.** $33.5 = 33.5$
9. $19\frac{49}{50} = 19\frac{49}{50}$ B **11.** 117 **13.** 56 **15.** 433 **17.** 71 **19.** 49 **21.** 107 **23.** 53.4 **25.** $79\frac{1}{2}$ **27.** $=$ **29.** $\neq$

Page 174 Written Exercises A **1.** 1 **3.** 0 **5.** 0 **7.** 0 **9.** 1 **11.** 2 **13.** 0 **15.** 0 B **17.** 0 **19.** 0
21. Every number is a solution. **23.** Every number is a solution **25.** Every number is a solution
27. Every number except 0 is a solution. C **29.** No solution **31.** 2 **33.** 1
35. Every number is a solution **37.** 0

Pages 176–178 Written Exercises A **1.** $(0, 0)$ **3.** $(2, 1)$ **5.** 2; $(4, 2)$ **7.** $(0, \frac{1}{2})$ **9.** $6\frac{1}{2}$; $(6, 6\frac{1}{2})$
11. $12\frac{1}{2}$; $(12, 12\frac{1}{2})$ **13.** $(0, 1)$ **15.** 13; $(4, 13)$ **17.** 31; $(10, 31)$ **19.** 0; $(0, 0)$ **21.** 6; $(3, 6)$ **23.** 30; $(6, 30)$
B **25.** $\{(0, 1), (1, 2), (2, 5)\}$ **27.** $\{(0, \frac{1}{2}), (3, 1\frac{1}{2}), (6, 2\frac{1}{2})\}$

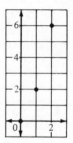

29. $\{(0, \frac{3}{4}), (1, 1), (2, \frac{5}{4}), (3, \frac{3}{2})\}$ **31.** $\{(0, 0), (1, 2), (2, 6)\}$

33. $(0, 1)$; $(3, 7)$; $(6, 13)$; $(9, 19)$ **35.** $(2, 0)$; $(5, 21)$; $(8, 60)$; $(11, 117)$ **37.** $(0, 29)$; $(1, 28)$; $(2, 21)$; $(3, 2)$
C **39.** 6; 12; 20 **41.** 4; 8; 14; 22 **43.** $9\frac{1}{3}$; $16\frac{1}{4}$; $25\frac{1}{5}$

Page 182 Review of Skills **1.** 40 **3.** $16x + 32$ **5.** $12a + 18$ **7.** $15n + 10$ **9.** $10m$ **11.** $5y$ **13.** $10\frac{5}{8}$
15. $\frac{3}{5}$ **17.** 0.82 **19.** 0 **21.** 30 **23.** 3 **25.** 1 **27.** 2 **29.** 1 **31.** 3 **33.** 8 **35.** 5
37. Every number is a solution. **39.** Every number is a solution. **41.** 2

Chapter 7. Equations and Problem Solving

Page 185 Written Exercises A **1.** $11r$ **3.** $7s$ **5.** $5p + 14$ **7.** $4cd$ **9.** $4t + 5$ **11.** $9k + 12t$ **13.** $5m$
15. $7mn + 8$ B **17.** $11k$ **19.** $4a + 5b$ **21.** $7s + 3$ **23.** $12q$ **25.** $12ab + 14a$ **27.** $11a + 11b$
29. $9k + 9m$ **31.** $46x + 3$ **33.** $42f + 14g$ **35.** $18x + 4y + 32$ C **37.** $32k + 11m$ **39.** $45 + 30c + 66d$
41. $8m^2 + 32m + 10mn + 8n$

Pages 187–188 Written Exercises A 1. 9 3. 6 5. $5\frac{1}{3}$ 7. 26 9. 8 11. 5 13. $3\frac{1}{2}$ 15. $\frac{4}{5}$ 17. 67
B 19. $1\frac{5}{6}$ 21. $1\frac{1}{2}$ 23. $\frac{3}{4}$ 25. 39 27. 42 29. 74 31. 52 33. 0.9 35. 0.01 C 37. $\frac{1}{2}$ 39. $\frac{3}{5}$

Page 190 Written Exercises A 1. 7 3. 4 5. $11\frac{1}{2}$ 7. $6\frac{4}{7}$ 9. $\frac{1}{20}$ 11. $\frac{1}{8}$ 13. 0 15. 6 17. 8 B 19. 4
21. 11 23. 0 25. 20 27. $5\frac{1}{2}$ C 29. $1\frac{1}{2}$ 31. 0.29 33. 3.06

Pages 192–193 Written Exercises A 1. 16 3. 2.5 5. 0.18 7. 4 9. $1\frac{3}{5}$ 11. 51 13. $4\frac{1}{2}$ 15. 6
B 17. 0.175 19. 0.08 21. 9 23. $4\frac{2}{3}$ 25. 0.6 27. $1\frac{2}{3}$ C 29. 0.32 31. 6 33. 0.1 35. 2 37. 44 39. 4

Pages 195–196 Written Exercises A 1. 7 3. 12 5. 5 7. 2 9. 5 11. 3 13. 6 15. 6 B 17. 6 19. 3
21. 2 23. $3\frac{1}{3}$ 25. $6\frac{1}{4}$ 27. 1 29. $23\frac{2}{3}$ 31. 40 C 33. 5.5 35. 1

Pages 196–197 Problems 1. 4 3. 50 5. 25 and 26 7. Rollie: 150, Maria: 155 9. 42 and 44

Pages 199–200 Written Exercises A 1. $15 3. 8 runs 5. 79¢ B 7. 44 pieces 9. 51, 52, 53
11. 36 and 38 C 13. width: 21, length: 27 15. Sam: 9, brother: 2 17. 42 km 19. 26 and 28

Page 202 Written Exercises A 1. 3 3. 7 5. 1.9 7. 9 9. 24 11. $\frac{1}{2}$ 13. 3 15. 7 17. 4.3 19. 2 21. 5
B 23. 2 25. $\frac{1}{3}$ 27. $\frac{1}{2}$ C 29. 2 31. 1 33. 11

Pages 202–203 Problems 1. 15 3. 7 5. width: 8, length: 15 7. 8 nickels

Pages 206–209 Cumulative Review 1. 5 3. 0 5. False 7. 18 9. 104 11. 17 13. 3 15. 7 17. 3
19. $^-1$ 21. $^+17$ 23. $^-4$ 25. 0, 1, 2, 3, 4, 5

27. 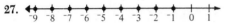 29. $>$ 31.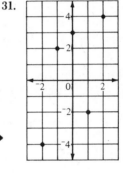

33. LCM = 12, GCF = 3
35. $(2 \times 10) + 1 = 21$ 37. 12 39. $\frac{1}{4}$ 41. 0.15, 15%
43. 0.99, 99% 45. 1000 47. $\frac{3}{22}$ 49. $\frac{6}{7}$ 51. 6 53. 1
55. 7.7 57. 2 59. 14 61. {14, 15, 18}
63. $14(lm)$, $7(2lm)$, $2l(7m)$ 65. $2.7x = 7.1$
67. {the numbers of arithmetic $<$ 4}
69. 14 71. 26 73. 12 cm²
75. {the numbers of arithmetic $\geq$ 3}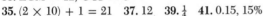

77. Assoc. Prop. of Addition 79. Comm. Prop. of Addition
81. Trans. Prop. of Equality
83. Closure Property 85. Comm. Prop. of Mult. 87. Distributive Property 89. Additive Identity
91. Mult. Identity 93. {(1, 2), (2, 4), (3, 6)} 95. 38 97. 1 99. 4 101. 9

Page 210 Review of Skills 1. H 3. K 5. M 7. 5 9. $^-2$ 11. 1 13. $^-5$

Chapter 8. Working with Directed Numbers

Page 214 Written Exercises A 1. 2 3. $^-1\frac{1}{4}$ 5. $1\frac{1}{4}$ 7. $^-1\frac{1}{2}$ 9. $\frac{1}{4}$ 11. $^-12$ 13. $^-100$ 15. 22.8 17. $^-0.1$
19. 0 21. 5 23. 20.9 B 25. $^-15$ 27. $^-45\frac{7}{9}$ 29. 11 and $^-11$ 31. $5\frac{1}{9}$ and $^-5\frac{1}{9}$ C 33. 1 and 9
35. $^-10$ and 16

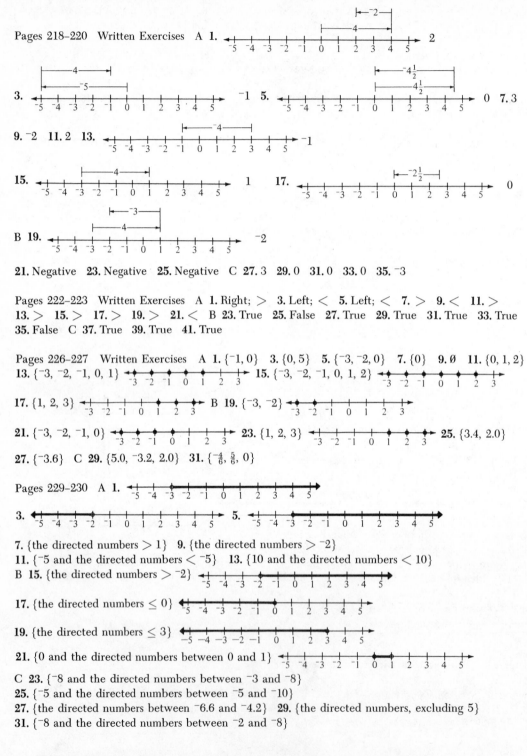

Pages 218–220 Written Exercises A 1. ... 2

3. ... ⁻1 5. ... 0 7. 3

9. ⁻2 11. 2 13. ... ⁻1

15. ... 1 17. ... 0

B 19. ... ⁻2

21. Negative 23. Negative 25. Negative C 27. 3 29. 0 31. 0 33. 0 35. ⁻3

Pages 222–223 Written Exercises A 1. Right; > 3. Left; < 5. Left; < 7. > 9. < 11. >
13. > 15. > 17. > 19. > 21. < B 23. True 25. False 27. True 29. True 31. True 33. True
35. False C 37. True 39. True 41. True

Pages 226–227 Written Exercises A 1. {⁻1, 0} 3. {0, 5} 5. {⁻3, ⁻2, 0} 7. {0} 9. ∅ 11. {0, 1, 2}
13. {⁻3, ⁻2, ⁻1, 0, 1} ... 15. {⁻3, ⁻2, ⁻1, 0, 1, 2} ...

17. {1, 2, 3} ... B 19. {⁻3, ⁻2} ...

21. {⁻3, ⁻2, ⁻1, 0} ... 23. {1, 2, 3} ... 25. {3.4, 2.0}

27. {⁻3.6} C 29. {5.0, ⁻3.2, 2.0} 31. {⁻$\frac{4}{6}$, $\frac{5}{6}$, 0}

Pages 229–230 A 1. ...

3. ... 5. ...

7. {the directed numbers > 1} 9. {the directed numbers > ⁻2}
11. {⁻5 and the directed numbers < ⁻5} 13. {10 and the directed numbers < 10}
B 15. {the directed numbers > ⁻2} ...

17. {the directed numbers ≤ 0} ...

19. {the directed numbers ≤ 3} ...

21. {0 and the directed numbers between 0 and 1} ...

C 23. {⁻8 and the directed numbers between ⁻3 and ⁻8}
25. {⁻5 and the directed numbers between ⁻5 and ⁻10}
27. {the directed numbers between ⁻6.6 and ⁻4.2} 29. {the directed numbers, excluding 5}
31. {⁻8 and the directed numbers between ⁻2 and ⁻8}

Page 234   Review of Skills   **1.** Fall   **3.** South   **5.** 0   **7.** 3   **9.** 1   **11.** 15   **13.** $x + 1$ **15.** $R + 8$ **17.** 9
19. 11   **21.** 13   **23.** (7, 2), (11, 6), (13, 8)

Chapter 9. Addition and Subtraction of Directed Numbers

Pages 237–238   Written Exercises   A **1.**

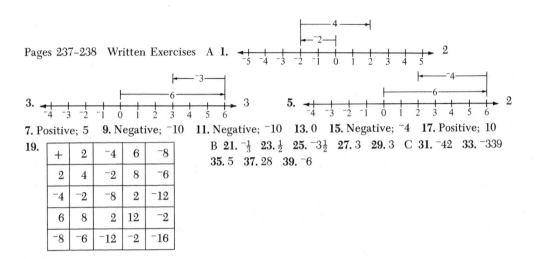

7. Positive; 5   **9.** Negative; ⁻10   **11.** Negative; ⁻10   **13.** 0   **15.** Negative; ⁻4   **17.** Positive; 10
19.

+	2	⁻4	6	⁻8
2	4	⁻2	8	⁻6
⁻4	⁻2	⁻8	2	⁻12
6	8	2	12	⁻2
⁻8	⁻6	⁻12	⁻2	⁻16

B **21.** $-\frac{1}{3}$ **23.** $\frac{1}{2}$ **25.** $-3\frac{1}{2}$ **27.** 3   **29.** 3   C **31.** ⁻42   **33.** ⁻339
35. 5   **37.** 28   **39.** ⁻6

Page 239   Problems   **1.** 26 bales   **3.** 915 m   **5.** Fifth floor   **7.** 88 cm   **9.** 3270 m

Pages 241–242   Written Exercises   A **1.** $-(^-4)$; 4   **3.** -3; ⁻3   **5.** $-(^-16)$; 16   **7.** $-(^-3.4)$; 3.4
9. $-(-6.08)$; 6.08   **11.** 5, ⁻5   **13.** 20, ⁻20   **15.** 0, ⁻10   **17.** Yes   **19.** No   **21.** No   **23.** Yes   **25.** Yes   **27.** 6
B **29.** 3   **31.** 0   **33.** $-2\frac{2}{3}$ **35.** $\frac{2}{3}$ **37.** 4   **39.** 0   **41.** The opposite of ⁻16; 16
43. The opposite of 1.42; ⁻1.42   **45.** The opposite of ⁻0.63; 0.63   C **47.** 3   **49.** ⁻3   **51.** 1.3

Pages 244–245   Written Exercises   A **1.** -7 **3.** -39 **5.** -16 **7.** -8 **9.** -1

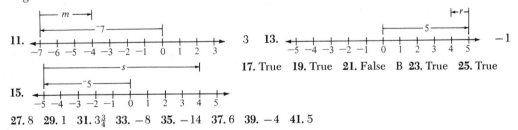

17. True   **19.** True   **21.** False   B **23.** True   **25.** True

27. 8   **29.** 1   **31.** $3\frac{3}{4}$ **33.** -8 **35.** -14 **37.** 6   **39.** -4 **41.** 5

Pages 247–249   Written Exercises   A **1.** Comm. Prop.   **3.** Add. Prop. of Inverses
5. Add. Prop. of Zero   **7.** 16   **9.** 0   **11.** 0   **13.** 7   **15.** 17   **17.** -15 **19.** 7   **21.** -27 **23.** 23   **25.** $3\frac{1}{4}$
27. -5 **29.** 33   B **31.** $10 = 10$ **33.** $-8\frac{1}{3} = -8\frac{1}{3}$ **35.** $21 = 21$ **37.** 2   **39.** 7   C **41.** 74   **43.** -71
45. 5.7

Pages 251–252   Written Exercises   A **1.** 12   **3.** -11 **5.** 17   **7.** 33   **9.** 14   **11.** 14   **13.** 8   **15.** 9   **17.** -5
19. -25 **21.** $-4\frac{1}{2}$ **23.** -10 **25.** -17 **27.** 7   **29.** 14   B **31.** -26 **33.** 11   **35.** -42 **37.** -1 C **39.** 20
41. -25 **43.** -22 **45.** -15 **47.** 6   **49.** 0

Pages 256–257 Written Exercises A 1. (2, 4) 3. (0, 0) 5. (−2, −4)
7. {(−3, 0), (0, 3), (3, 6), (6, 9), (9, 12)} 9. {(2, 1), (4, 3), (6, 5), (8, 7)}
11. {(6, −4), (4, −6), (2, −8)} 13. {(6, 6), (8, 8), (−4, −4), (−6, −6)} 15. No 17. Yes B 19. B
21. A 23. C 25. {(5, 7), (10, 12), (15, 17), (20, 22), (25, 27)}
27. {(−21, −28), (−14, −21), (−7, −14), (0, −7), (7, 0), (14, 7), (21, 14)}
C 29. {($\frac{1}{2}$, −$\frac{1}{2}$), (1, −1), (1$\frac{1}{2}$, −1$\frac{1}{2}$), (2$\frac{3}{4}$, −2$\frac{3}{4}$)} 31. {(−$\frac{1}{2}$, $\frac{1}{2}$), (−$\frac{1}{4}$, $\frac{3}{4}$), (0, 1), ($\frac{1}{4}$, 1$\frac{1}{4}$)}
33. {(−1, −3), (−$\frac{1}{2}$, −2$\frac{1}{2}$), ($\frac{1}{2}$, −1$\frac{1}{2}$), (1, −1)}

Page 262 Review of Skills 1. $\frac{5}{8}$ 3. $\frac{5}{2}$ 5. 41 7. −12 9. 0, −4, −8 11. True 13. True 15. False
17. 16y 19. 5t + 5n

Chapter 10. Multiplication and Division of Directed Numbers

Pages 265–266 Written Exercises A 1. −24 3. 143 5. −36 7. −5w 9. $\frac{-p}{5}$ 11. −0.30

13. −0.963 15. $\frac{5}{6}$
19. −24 21. −6 23. 1 25. −54 B 27. −18
29. −15 31. 1 33. −60 35. −17 37. −1
C 39. 3 41. 33 43. −3

17.

×	−1	−3	−5	−7
1	−1	−3	−5	−7
3	−3	−9	−15	−21
5	−5	−15	−25	−35
7	−7	−21	−35	−49
9	−9	−27	−45	−63

Pages 268–269 Written Exercises A 1. 15 = 15 3. −63 = −63 5. −56 = −56 7. −24 = −24
9. −1 = −1 11. −7 = −7 13. −40 = −40 15. −9 = −9 17. −30 = −30 19. −9 = −9
21. −15 = −15 23. −35 = −35 B 25. −9 = −9 27. −$\frac{5}{4}$ = −$\frac{5}{4}$ 29. −5 = −5 31. −6 = −6
33. 0 = 0 C 35. −648 37. −112 39. −745 41. −192 43. −1488

Pages 271–272 Written Exercises A 1. Negative; −18 3. Negative; −30 5. Positive; 30
7. Negative; −12 9. Positive; 21 11. Positive; 25 13. −6 = −6 15. −24 = −24 17. 6 = 6
19. −$\frac{5}{2}$ = −$\frac{5}{2}$ 21. 35 = 35 23. 3 = 3 B 25. 6 27. 15 29. −11 31. −9 33. 9 C 35. −21
37. −30 39. −1 41. −27 43. −512

Pages 274–275 Written Exercises A 1. 4b − 3a 3. −9xy − 3p 5. 3m − 6x − 2 7. 2a + 2c + 6
9. ab − 5 11. −2c² + 7c B 13. 4x² − 2x + 5 15. ab + 3c + 1 17. −13s + 6 19. $\frac{3}{5}$m − $\frac{1}{3}$m²
C 21. 7x² − 4x + 5 23. −110a² − 80a 25. 150 27. −6 29. −4 31. 14 33. 0 35. 194.5

Page 277 Written Exercises A 1. −5 3. −12 5. −6 7. −4 9. 7 11. 4 13. −3 15. 5 17. −7
19. −0.6 21. −0.6 23. 0.2 25. 0.12 27. 0.75 29. −0.75 31. False 33. True 35. True

Pages 280–281 Written Exercises A 1. 1 3. −3 5. −1 7. −$\frac{1}{2}$ 9. −$\frac{1}{2}$ 11. −28 13. 1 15. −49
17. $\frac{25}{36}$ 19. −2 21. −7 23. −5 25. 1 27. 10 29. $\frac{1}{2}$ B 31. −$\frac{1}{6}$ 33. −25 35. $\frac{1}{21}$ 37. −15a 39. $\frac{m}{15}$
41. $\frac{t}{2}$

Page 281 Problems 1. −$\frac{1}{2}$ 3. −$\frac{7}{5}$ 5. −$\frac{7}{5}$ 7. 8

Pages 284–286 Written Exercises A 1. 2; $(-1, 2)$ 3. -4; $(2, -4)$ 5. $-\frac{2}{3}$; $(1, -\frac{2}{3})$ 7. $\frac{2}{3}$; $(-1, \frac{2}{3})$
9. 2; $(-3, 2)$ 11. 0; $(0, 0)$ 13. $\frac{4}{3}$; $(-2, \frac{4}{3})$ 15. $(0, 0)$, $(1, \frac{1}{3})$ 17. $(0, 0)$, $(1, -\frac{1}{5})$
19. $(-2, \frac{4}{5})$, $(0, 0)$, $(1, -\frac{2}{5})$ B 21. $\{(-2, 8), (-1, 5), (0, 2), (2, -4), (3, -7), (4, -10)\}$
23. $\{(-2, 0), (-1, \frac{1}{2}), (0, 1), (6, 4)\}$ 25. $\{(-5, -\frac{6}{5}), (-4, -\frac{3}{2}), (-3, -2), (-2, -3), (-1, -6)\}$
27. $\{(-4, \frac{7}{4}), (-3, \frac{7}{3}), (-2, \frac{7}{2}), (3, -\frac{7}{3})\}$ 29. $\{(-2, 20), (-1, 5), (0, 0), (2, 20)\}$

Page 290 Review of Skills 1. -5 3. -12 5. -14 7. 0 9. -15 11. -10 13. -5 15. -9 17. 15
19. 8 21. 33 23. $m(8 - 3) = 5m$ 25. $y(5 + 4) = 9y$ 27. $<$ 29. $>$ 31. $<$ 33. $<$ 35. False
37. True

Chapter 11. Solving Equations and Inequalities

Pages 293–294 Written Exercises A 1. 0; 25 3. $-5\frac{1}{4}$; 0; $2\frac{3}{4}$ 5. 22 7. 22 9. 41 11. $-5\frac{2}{3}$ 13. 3
15. -14 17. 5.1 19. $1\frac{1}{2}$ B 21. $-\frac{1}{6}$ 23. $\frac{1}{12}$ 25. 2.6 27. $y = z - s$ 29. $d - b = y$ 31. $y = -t - r$
C 33. $z = s + r$ 35. $m - q = z$ 37. $z = -h + g$

Pages 296–297 Written Exercises A 1. 13 3. $-\frac{18}{5}$ or $-3\frac{3}{5}$ 5. 35 7. $\frac{23}{4}$ or $5\frac{3}{4}$ 9. $-\frac{9}{7}$ 11. $d = \dfrac{v}{g}$

13. $\dfrac{a}{l} = w$ 15. $\dfrac{i}{p} = r$ 17. $d = \dfrac{c}{\pi}$ 19. $\dfrac{s}{t} = c$ B 21. -5 23. 14 25. $-17\frac{6}{7}$ 27. 0.4 29. -0.2

31. $\dfrac{c}{2\pi} = r$ 33. $\dfrac{v}{b} = h$ C 35. $n = \dfrac{s}{u} = \dfrac{1}{2}$ 37. $m = -\dfrac{r}{u} = \dfrac{1}{10}$ 39. $p = -\dfrac{t}{t} = -1$

41. $c = -\dfrac{s}{t} = -20$

Pages 300–301 Written Exercises A 1. 4; $\frac{1}{4}$; $\frac{1}{4}$; 3 3. $-\frac{2}{4}$; 0; -10 5. -18; $-\frac{1}{18}$; $-\frac{1}{18}$; -2
7. $z = \dfrac{b}{c + d}$ 9. $\dfrac{g}{v + r} = q$ 11. $h = \dfrac{-3}{3 + j}$ 13. $\frac{13}{7}$ 15. -7 17. 2 19. 0.1 21. $-\frac{1}{9}$ 23. 5 25. 6 27. 1
B 29. 9 31. 3 33. -16 35. 4 C 37. $\frac{5}{2}$ 39. 3 41. $-17\frac{1}{3}$

Page 303 Problems 1. 7 meters 3. 2 years 5. $1500 7. 16 cm 9. 3 cm

Pages 306–307 Written Exercises A 1. E 3. A 5. D 7. -11; -11; -8
9. 4; 4; -2
11. -9; -9; -5 13. {the directed numbers ≥ 10}
15. {the directed numbers > 12} 17. {the directed numbers > -13}
19. {the directed numbers ≥ -5} 21. {the directed numbers < -4}
23. {the directed numbers ≥ 0} B 25. {the directed numbers ≤ -16}
27. {the directed numbers ≤ 4} 29. {the directed numbers ≥ -13}
31. {the directed numbers ≥ -8} 33. {the directed numbers > 12}
C 35. {the directed numbers ≥ 7} 37. {the directed numbers $\leq -\frac{1}{4}$}
39. {the directed numbers < 4}

Pages 309–310 Written Exercises A 1. $-6(-5) > -5(-5)$ 3. $26 \cdot \frac{1}{13} \geq 13 \cdot \frac{1}{13}$
5. $-4 \cdot 4 > -10.4$ 7. $-2(-3) \leq -5(-3)$ 9. $-\dfrac{d}{3}(-3) \leq 7(-3)$

11. $m \geq 3$ -5 -4 -3 -2 -1 0 1 2 3 4 5   **13.** $w \leq 20$ 17 18 19 20 21 22 23 24 25 26 27

15. $d \leq -2$ -5 -4 -3 -2 -1 0 1 2 3 4 5

17. $s > -5$ -5 -4 -3 -2 -1 0 1 2 3 4 5

19. $-3 < b$ -5 -4 -3 -2 -1 0 1 2 3 4 5

21. {the directed numbers $\leq \frac{9}{7}$}   **23.** {the directed numbers $< -\frac{1}{2}$}
25. {the directed numbers ≥ -15}   **27.** {the directed numbers ≤ -5}
B **29.** {the directed numbers < 28}   **31.** {the directed numbers > -32}
33. {the directed numbers > -4}   **35.** {the directed numbers > 1}
37. {the directed numbers > -30}   **39.** {the directed numbers ≥ 4}
C **41.** {the directed numbers $< 21\frac{1}{2}$}

Page 314 Review of Skills   **1.** 10^3 **3.** 10^7 **5.** $6 \cdot 10^3 + 3 \cdot 10^2 + 2 \cdot 10 + 5$
7. $8 \cdot 10^4 + 4 \cdot 10^2 + 9 \cdot 10 + 1$ **9.** $11x$ **11.** $-14q$ **13.** 28   **15.** 310   **17.** 7   **19.** 0   **21.** 85   **23.** 76
25. 29   **27.** 110   **29.** 0   **31.** 10   **33.** $-\frac{1}{5}$ **35.** $\frac{4}{3}$ **37.** 0   **39.** 3   **41.** $\frac{11}{3}$

Chapter 12. Addition and Subtraction of Polynomials

Page 317 Written Exercises A 1. Monomial   **3.** Binomial   **5.** Binomial   **7.** Trinomial   **9.** Trinomial
11–15 Answers may vary. One example is given.   **11.** $3 + c^2 - b^5 + b$ **13.** $b^5 + b + 3$
15. $c^2b - c^2 - 3 + b - b^5 + c^2b^5$

Page 319 Written Exercises A 1. $-7c^5 + c$ **3.** $3k^5 - k^3 + 9k^2 - 5$ **5.** $3n^5 + n^2 - 4n$
7. $-8a^7 + 14a^5 + a^3$ **9.** $r^9 + 5r^4 - 2r^2$ B **11.** $-3x^2 + 4xy$ **13.** $7a^6 - 6a^3b + 2a^2b^2$
15. $2a^2 + ab + 2b^2$ **17.** $m^4n^2 + m^3n^2 + m^2 + n^3$ **19.** $3m^4 + 0m^3 + 0m^2 + 0m - 1$
21. $s^5 + 0s^4 + 0s^3 - 3s^2 + 0s + 2$ **23.** $x^7 + 0x^6 + 0x^5 + x^4 + 0x^3 + 0x^2 + 0x + 0$
25. $k^7 + 0k^6 - 2k^5 + 0k^4 + 0k^3 + 0k^2 + k - 19$

Pages 321–322 Written Exercises A 1. 0   **3.** 0   **5.** 6   **7.** -24 **9.** 7   **11.** 3   **13.** -12 **15.** $(-7, 120)$
17. 6; $(-1, 6)$ **19.** 21; $(4, 21)$ **21.** 105; $(8, 105)$
B **23.** $\{(-6, -13), (-4, -7), (-2, -1), (0, 5), (2, 11), (4, 17), (6, 23)\}$
25. $\{(-3, -17), (-2, 0), (0, 4), (2, 8), (3, 25)\}$ **27.** $\{(-4, -40), (-2, -2), (0, 12), (2, 2), (4, -32)\}$
29. $\{(-2, -9), (-1, -2), (0, -1), (1, 0), (2, 7)\}$ C **31.** $\{(-0.1, 3.31), (0, 3), (0.1, 2.71)\}$

Pages 324–325 Written Exercises A 1. $14w + 9$ **3.** $3a + 9a^3$ **5.** $2n^2$ **7.** $2x^2$ **9.** $-10x^2 + 3xy - 2y^2$
11. $5m^2 - 4n^2 - 4p^2$ **13.** $10c - 8$ **15.** $-a - 2b$ **17.** $4x^2 - 5y - 3$ **19.** $2y^2$ **21.** $9.4a + 0.7b + 8$
23. $3m^2 - 5m - 3$ **25.** $k^4 - 3k^3 - 5k^2 + 6k + 3$ **27.** $4x^2 + 4y^2$ B **29.** $7b - 5$ **31.** $13a - 2b - 15$
33. $-5 - d$ **35.** $4d^3 - 8d^2$ **37.** $1\frac{1}{2}s^3$ **39.** $a - 4b$ **41.** $2.4 + 5ab$ **43.** $20x^4 - 12.2x^2y^2 - 2.2y^4$
45. $13x^3 + 2x^2 + x + 3$

Pages 327–328 Written Exercises A 1. $16 - 2b = 16 - 2b$ **3.** $4b - 10 = 4b - 10$
5. $5z^2 - 2z + 3 = 5z^2 - 2z + 3$ **7.** $2m^2 + 5m - 8 = 2m^2 + 5m - 8$
9. $a^4 - a^3 + 2a^2 - 9a = a^4 - a^3 + 2a^2 - 9a$ **11.** $2y - 2$; $2y - 2$ **13.** $-8x + 6y$; $-8x + 6y$
15. $x^2 + y^2$; $x^2 + y^2$ **17.** -2 **19.** 13   **21.** -3 **23.** 0   **25.** 11.0   **27.** 1   B **29.** $24 = 24$ **31.** $137 = 137$
33. $-22 = -22$ **35.** $1\frac{1}{2} = 1\frac{1}{2}$

Pages 331–332 Written Exercises A **1.** $-2m^2 + 3m - 15$ **3.** $-\dfrac{m^2}{10} - \dfrac{4mn}{2} + \dfrac{2n^2}{7}$

5. $-12x^9 - 6x^7 - 3x^5 - x - 1$ **7.** $-5x^2 - 3x + 7$ **9.** $-k^6 + 4k^3 + 7k^2 + 2k - 1$

11. $-3y^2 - 13y - 5$ **13.** $x^2 - 4x + 10$ **15.** $2x^3 - 25x^2 + 10$ **17.** $-1.8x^5 + 6.2x^3 + 0.7x - 15$

19. $y^7 - 2y^5 + 6y^3 - y + 2$ B **21.** $4x^2 + 4x$ **23.** 0 **25.** 0 **27.** $-8r^2 + 5s - 6t - 8$ **29.** $\frac{5}{8}x^2 + 1$

C **31.** $-s^3 + 5s^2 - 4s - 8$ **33.** $-a^5 + 2a^2 - 6a + 10$ **35.** $2y^3 - 15y^2 + 3y - 7$

37. $2x^4 + x^3 + x^2 + 5x - 8$

Pages 334–336 Written Exercises A **1.** $r + 9s$ **3.** $7p$ **5.** $4x^2 + 5$ **7.** $2b^2 + 13c - 5$

9. $m^2 + 7m + 4$ **11.** $-2k + 9h^2$ **13.** $k - 10 - 2k - 15$ **15.** $-2y^2 - 4y - y^2 + 3y$

17. $2x - y + 8 - 13x - 4y - 9$ **19.** $-w - 12$ **21.** $6t^2 - 4$ **23.** 0 B **25.** w **27.** $3d - 3c$

29. $4d^2 - 10d - 1$ **31.** $4r^2 + 3r + 11$ **33.** $h^2 - 21h - 13$ **35.** $b^2 + 6b - 2$ **37.** $7x^2 - 32x + 7$

39. $k^2 + 5k - 5$ C **41.** $k - 18$ **43.** $x^2 + 2x + 5$

Pages 337–338 Problems **1.** $a^2 + 6a - 10$ **3.** $a^2 + \frac{5}{6}a - 4$ **5.** $z^2 - 3z + 3$ **7.** $-4a^2 - 5a - 3$

9. $\frac{11}{12}x^3 + 2\frac{1}{4}x^2 + x + 5$

Pages 339–340 Written Exercises A **1.** 1 **3.** 6 **5.** 8 **7.** 17 **9.** -1 **11.** 45 **13.** 12 B **15.** -17

17. $-\frac{17}{3}$ or $-5\frac{2}{3}$ **19.** -6 **21.** 4 **23.** -14 **25.** -1 C **27.** -4 **29.** 5

Page 344 Review of Skills **1.** F **3.** C **5.** B **7.** 16 **9.** 216 **11.** 125 **13.** $12 - 3t$ **15.** $rs - rt$

17. $2mn + 2mp$ **19.** Mult. Prop. of 1 **21.** Comm. Prop. **23.** Assoc. Prop. **25.** 6 **27.** 2 **29.** 7 **31.** 9

Chapter 13. Multiplication and Division of Polynomials

Page 347 Written Exercises A **1.** 4 **3.** 72 **5.** 54 **7.** 36 **9.** 36 **11.** 216 **13.** $12x^2$ **15.** $(x + y + z)^2$

17. $-15h^2$ **19.** $-14z^4$ **21.** $-84y^4$ **23.** $-6x^2$ **25.** $24m^6$ **27.** $-42t^2$ **29.** 2^3 **31.** x^7 **33.** p^5 **35.** $-a^5$

37. y^6 B **39.** 4 **41.** 6 **43.** 2

Page 349 Written Exercises A **1.** $28x^2y^3$ **3.** $48b^3$ **5.** $-60m^3n$ **7.** $-30x^2y$ **9.** $-3a^2b^2$ **11.** $-6s^4r$

13. $-36c^2b$ **15.** $3x^3y^2$ B **17.** $21y^8x^6$ **19.** $-0.4b^7a^5$ **21.** $-35c^8b^3a^3$ **23.** $81n^3m^3$ C **25.** $25x^2y$

27. $-14b^4a^3$ **29.** $20m^4n^2$ **31.** $\frac{1}{2}x^2y^2$ **33.** $-\frac{1}{2}m^3n$ **35.** $2.04a^2b^2c^2$

Page 351 Written Exercises A **1.** $81a^4b^4$ **3.** $81x^4y^4z^4$ **5.** $64m^2n^2$ **7.** $100x^4y^4$ **9.** $-a^3b^3$ **11.** r^8s^8

13. n^{16} **15.** 0 **17.** $1{,}000{,}000$ **19.** 256 **21.** 729 **23.** $25{,}000{,}000$ **25.** -125 **27.** 2500 **29.** 400

31. $25x^2y^4$ **33.** $0.04a^2b^4c^6$ **35.** $144a^2b^2c^2$ B **37.** $109a^4b^4$ **39.** $5m^3$ **41.** $-r^7s^9$ C **43.** $1.1a^5t^4$

Pages 353–354 Written Exercises A **1.** $-x^3 - 6x^2 + 7x$ **3.** $5d^2 - 10ad + 5c^2$ **5.** $-d^3 + ad^2 - bd^2$

7. $x^5 + 3x^3 - 7x^2$ **9.** $y^6 - y^5 - y^3$ **11.** $-5x^3 + 10x^2 + 25x$ **13.** $-0.6k^3 + 0.8k^2 + 2k$

15. $4m^4 - m^3n + 2m^2n^2$ **17.** $2x^2 - 3xy$ **19.** $-6m^2 - 8mn$ **21.** $8a^2 + 8ab$ **23.** $18y^2 - 6xy$

25. $-6x^3 + 8x^2w$ **27.** $7x^3y + 2x^2y^3$ B **29.** $60a^2b - 24a^3b^5 + 36a^8b^5 - 6a^2b^{11}$

31. $-\frac{1}{2}a^5b^2 + \frac{1}{2}a^6b^3 - a^4b^3 + a^3b^4$ **33.** $-63x^6y^4 + 84x^5y^5 - 105x^4y^6$ **35.** 11 **37.** -1 **39.** -4 **41.** 16

C **43.** -16 **45.** -13

Pages 354–355 Problems A **1.** $\frac{1}{2}m^2 + m$ **3.** $6x^3 + 15x$ B **5.** $50\pi t + 225\pi$ **7.** $8x^2 + 3x$ **9. a.** y^6 **b.** $6y^4$

Pages 357–358 Written Exercises A **1.** $a^2 + 9a + 20$ **3.** $12m^2 + 86m + 52$ **5.** $x^2 + 3x + 2.25$

7. $x^2 - 81$ **9.** $x^2 - 4$ **11.** $24 + 59n + 36n^2$ **13.** $18 - 36y + 16y^2$ **15.** $9 - 3n - 20n^2$ **17.** $25t^2 - 1$

19. $16x^2 + 44x + 10$ **21.** $2x^2 + 5x + 3$ **23.** $2 - 3b - 2b^2$ **25.** $2s^2 + 11s + 12$ **27.** $8q^2 - 82q + 20$
29. $1.5w^2 - w - 2.5$ B **31.** $8 - 32m - 40m^2$ **33.** $4n^2 - m^2$ **35.** $6m^2 + 3.5mn + 0.5n^2$
37. $x^3 - 9x^2 + 25x - 21$ **39.** $s^3 - t^3$ **41.** $36 + 13x^2 + x^4$ **43.** $a^4b^4 + 2a^2b^2 + 1$
45. $42 - 11r - 15r^2 - 2r^3$ **47.** $2x^2 - 2xy + 0.5xz + 8yz - 4y^2 - 3.75z^2$

Page 359 Problems A **1. a.** $2x^2 + 9x + 4$ **b.** $6x + 10$ **3.** $t^2 + t - 30$ B **5. a.** $c^3 - 4c$ **b.** $c^2 - 4$
C **7.** $w^2 + 4w - 40 = w^2 - 4$

Pages 361–362 Written Exercises A **1.** $b^2 + 8b + 16$ **3.** $x^2 - 2ax + a^2$ **5.** $9x^2 - 12cx + 4c^2$
7. $x^2 - y^2$ **9.** $9k^2 - q^2$ **11.** $a^2c^2 + 2abc^2 + b^2c^2$ B **13.** $k^4 + 12k^2 + 36$ **15.** $9x^2 - 30xwy + 25w^2y^2$
17. $9 + 6xy^2 + x^2y^4$ **19.** $16m^2 + 80m + 100$ **21.** $9b^2c^2 + 60ab^2c + 100a^2b^2$ **23.** $16m^2 + 56bm + 49b^2$
25. $324 + 72ab + 4a^2b^2$ **27.** $s^2t^4 - 4st^2 + 4$ **29.** $0.16m^4 + 0.72m^2n + 0.81n^2$ C **31.** $4y^6 - 4y^3 + 1$
33. $z^6 - 2x^3z^3 + x^6$ **35.** $0.04 + 0.2m^2 + 0.25m^4$ **37.** $16k^2 - k + \frac{1}{64}$ **39.** $6\frac{1}{4}p^2 - p + \frac{1}{25}$

Page 362 Problems **1.** $9x^2 + 48x + 64$ **3.** 15 cm **5.** 20; 21 **7.** 35; 36

Pages 366–367 Written Exercises A **1.** $\dfrac{a-b}{bm}$ **3.** $\dfrac{a-b}{m-b}$ **5.** $\dfrac{(x+1)^2}{x^2+2x+1}$ **7.** $\dfrac{4}{3(a+b^2)}$
9. $a^3b - ab^3 = a^3b - ab^3$ **11.** $40mnx + 32mny = 40mnx + 32mny$
B **13.** $b^2x - 0.25a^2x = b^2x - 0.25a^2x$ **15.** $3x^3 + 2x^2 - 3x - 2 = 3x^3 + 2x^2 - 3x - 2$ C **17.** $4y^7 - 4y^3$
19. $x^3y + x^2y^2 + xy^3$ **21.** $\frac{1}{8}m^3n^2p^3$ **23.** $4.6x^5 + 1.15x^3$ **25.** $\frac{25}{4}n^2 - \frac{1}{9}$ **27.** $3t^3 - 0.2t^2 - 0.05t$
29. $3s^2 - 5s$

Page 370 Written Exercises A **1.** x^4 **3.** $\dfrac{1}{(rs)^5}$ **5.** $-\dfrac{1}{2m^5}$ **7.** $\dfrac{q^2}{5}$ **9.** $\dfrac{4}{3x^2z}$ **11.** $\dfrac{5n}{3m}$ **13.** $\dfrac{2y}{3x}$ **15.** $\dfrac{2a}{3b^2}$
B **17.** $-\dfrac{1}{xyz}$ **19.** $\dfrac{3}{xyz}$ **21.** $\dfrac{3xy^2}{z}$ **23.** $\dfrac{10b}{3d}$ **25.** $21x^3y$ **27.** $-2r^4$ **29.** $\dfrac{2p}{3m^3n}$ C **31.** $\dfrac{a^2b}{0.9}$ or $\dfrac{10a^2b}{9}$
33. $\dfrac{y^4}{0.7x^{10}}$ or $\dfrac{10y^4}{7x^{10}}$ **35.** $\dfrac{-8x^{p-q}}{y^{p-q}}$ **37.** $\dfrac{(-69)^r}{23^r}$ or $(-3)^r$

Pages 372–373 Written Exercises A **1.** $4x - 3$ **3.** $-7b - 4$ **5.** $-6x^2 - 12$ **7.** $-2a^2 - 2a - 3$
9. $y - 2$ **11.** $k + 9a$ **13.** $2q + 9q^2$ **15.** $3 + 4b^2$ **17.** $-3m^3 - 7$ **19.** $2b - 4$ B **21.** $\dfrac{t^2}{s} + st + \dfrac{1}{s}$
23. $y^2 - xy + x^2$ **25.** $2t^2 - 8t - 5$ **27.** $s^2 - s + 27$ **29.** $mx + n + ax^2$ **31.** $\dfrac{4y^3}{x} + \dfrac{8y^2}{x^2}$

Pages 375–376 Written Exercises A **1.** $x + 3$ **3.** $b - 2$ **5.** $y - 2$ **7.** $m + 5$ **9.** $p + 9$ **11.** $3x + 2$
13. $3c - 3$ B **15.** $a - b$ **17.** $2b - c$ **19.** $s + 2$ **21.** $3t + 3$ **23.** $2a + 1$ **25.** $3b - 1$ **27.** $x + a$
29. $5a - 4b$ C **31.** $y^2 + 2$ **33.** $t + 2$ **35.** $3b - 2$ **37.** $x^2 + 1$

Page 380 Review of Skills **1.** 6: 1, 2, 3, 6; 36: 1, 2, 3, 4, 6, 9, 12, 18, 36; GCF: 6
3. 4: 1, 2, 4; 19: 1, 19; GCF: 1 **5.** 10: 1, 2, 5, 10; 15: 1, 3, 5, 15; GCF: 5
7. 14: 1, 2, 7, 14; 35: 1, 5, 7, 35; GCF: 7 **9.** 7: 1, 7; 28: 1, 2, 4, 7, 14, 28; GCF: 7 **11.** Prime
13. Not Prime **15.** Not Prime **17.** 5 **19.** 2 **21.** 3 **23.** $5 \cdot 11$ **25.** $11 \cdot 11$ **27.** $2 \cdot 2 \cdot 5 \cdot 5$
29. $2 \cdot 2 \cdot 2 \cdot 3 \cdot 5$ **31.** Always **33.** Never

Chapter 14. Products and Factoring

Page 383 Written Exercises A **1.** 1, 5; not possible **3.** 1, 2, 4; $2 \cdot 2$ **5.** 1, 3, 7, 21; $3 \cdot 7$

7. 1, 3, 5, 9, 15, 45; $3 \cdot 3 \cdot 5$ **9.** 1, 2, 4, 8, 16; $2 \cdot 2 \cdot 2 \cdot 2$ **11.** 1, 2, 3, 5, 6, 10, 15, 30; $2 \cdot 3 \cdot 5$
13. 5; $2^2 \cdot 3 \cdot 5$ **15.** 5; 5; $2^2 \cdot 5^2$ **17.** 5; $3^2 \cdot 5$ **19.** 47 B **21.** 5^3 **23.** $2^4 \cdot 3^2$ **25.** $2 \cdot 3^3 \cdot 5$ **27.** $2^3 \cdot 5^3$
29. $2^3 \cdot 3 \cdot 5$ **31.** $2 \cdot 97$

Pages 385–386 Written Exercises A **1.** $13y$ **3.** $2mn$ **5.** $3c^5$ **7.** $5r^2s^4$ **9.** $-9ab^2c^3$
11-19. Answers may vary. Two examples are given for each. **11.** $4 \cdot 4t$; $2 \cdot 8t$ **13.** $2m(6n)$; $(3m)(4n)$
15. $-9rs \cdot 2rs$; $-6r \cdot 3rs^2$ **17.** $(-5c)(-d^2)$; $5d \cdot cd$ **19.** $(-x)(3yz)$; $-3z \cdot xy$
21. $5ab$: $5 \cdot a \cdot b$; $10ab$: $2 \cdot 5 \cdot a \cdot b$; GCF: $5ab$ **23.** $14m^2n$: $2 \cdot 7 \cdot m \cdot m \cdot n$; $7mn$: $7 \cdot m \cdot n$; GCF: $7mn$
25. $18w^2$: $2 \cdot 3 \cdot 3 \cdot w \cdot w$; $-3w$: $-1 \cdot 3 \cdot w$; GCF: $3w$ **27.** 15: $3 \cdot 5$; $3mn$: $3 \cdot m \cdot n$; GCF: 3 B **29.** $6y$
31. 5 **33.** b **35.** x C **37.** $7r^2s(1)$; $7r^2s(-4r^2s)$ **39.** $2(2rst)$; $2(-x)$ **41.** $1 \cdot 5s^3$; $1 \cdot rs^2$; $1 \cdot 5r$

Page 388 Written Exercises A **1.** $k^2 + 3$ **3.** $7r + 4$ **5.** $5 - b$ **7.** $2y + 3$ **9.** $5(m - 3)$ **11.** $a(6a + 7)$
13. $ab(b - a)$ **15.** $ab(20a + 3b)$ **17.** $xy(9z - 1)$ **19.** $az(z - 12w)$ **21.** $3y(y - 1)$ **23.** $n(50m - 33n)$
25. $2m^2 + m - 1$ **27.** $t^2 + 1 - t$ **29.** $3 + 6x^2 - 2x$ B **31.** $x(3x^2 - x + 5)$ **33.** $4(t^2 - 7t + 7)$
35. $x(x^2 + x - 1)$ **37.** $2v(3v^2 + 13v + 4)$ **39.** $3m^2(m^2 - 19m + 37)$

Pages 390–391 Written Exercises A **1.** $(3s + t)(r + 4)$ **3.** $(x - w)(y + 3)$ **5.** $(10a + c)(2 + b)$
7. $(5 - a)(b + c)$ **9.** $(x^2 - 5)(y + 7)$ **11.** $(2t + r)(s + 4)$ **13.** $(m + r)(n - 3)$ **15.** $(n - t)(m + 3)$
17. $(r^2 + t^2)(s - 2)$ B **19.** $(2p - 3r)(q + 5s)$ **21.** $(c - 2)(a + d)$ **23.** $(n^2 - 3)(m + 3)$

Page 393 Written Exercises A **1.** 3; 3 **3.** 1; 1 **5.** 2; 2 **7.** $(z + 2)(z - 2)$ **9.** $(a + 6)(a - 6)$
11. $(r + 30)(r - 30)$ **13.** $(3p + 12)(3p - 12)$ **15.** $(t^2 + 10)(t^2 - 10)$ **17.** $(k^2 + 1)(k^2 - 1)$
19. $(y + 4)(y - 4)$ **21.** $(mn - 6)(mn + 6)$ **23.** $(8 + r)(8 - r)$ **25.** $(n + m)(n - m)$ **27.** $(y^2 - x^2)(y^2 + x^2)$
B **29.** $5(n + 2)(n - 2)$ **31.** $3(r + s)(r - s)$ **33.** $4(1 + r)(1 - r)$ **35.** $2(mn + 9)(mn - 9)$
37. $3(x + 5)(x - 5)$ **39.** $\pi(x + y)(x - y)$ C **41.** $\left(\dfrac{x}{3} + 1\right)\left(\dfrac{x}{3} - 1\right)$

Page 395 Written Exercises A **1.** $2x$ **3.** mn **5.** 10 **7.** $6x$ **9.** $2z$ **11.** $6w$ **13.** $4st$ **15.** $c^2 + 4c + 4$
17. $k^2 + 10k + 25$ **19.** $9n^2 + 24mn + 16m^2$ **21.** $(p + 1)^2$ **23.** $(t + 5)^2$ **25.** $(m + 2)^2$ **27.** $(10 + x)^2$
29. $(y + 8)^2$ **31.** $(z + 15)^2$ **33.** $(4c + 3)^2$ B **35.** $(mn + 1)^2$ **37.** $(k + 8t)^2$ **39.** $(10x + y)^2$
41. $(3rs + 5t)^2$ **43.** $(3x + 10y)^2$ C **45.** $(x + \frac{1}{2})^2$

Pages 397–398 Written Exercises A **1.** s^2 **3.** $25y^2$ **5.** 64 **7.** $9x^2$ **9.** $100k^2$ **11.** x^2y^2
13. $z \cdot z$; $-z(-z)$ **15.** $w \cdot w$; $-w(-w)$ **17.** $5y \cdot 5y$; $-5y(-5y)$ **19.** $xy \cdot xy$; $-xy(-xy)$
21. $r^2 - 4r + 4$ **23.** $k^2 - 10k + 25$ **25.** $1 - 2y + y^2$ **27.** $a^2 - 2ab + b^2$ **29.** $4x^2 - 12x + 9$
31. $4s^2 - 4sy + y^2$ **33.** $(t - 4)^2$ **35.** $(s - 5)^2$ **37.** $(c - 3)^2$ **39.** $(r - 3)^2$ **41.** $(3k - 1)^2$ **43.** $(x - 7)^2$
B **45.** $(4a - b)^2$ **47.** $(3m - n)^2$ **49.** $(6y - 3z)^2$ C **51.** $3(n - 1)(n - 1)$ **53.** $2(m - 5)(m - 5)$
55. $(k - \frac{1}{2})(k - \frac{1}{2})$

Page 400 Written Exercises A **1.** 2; 4 **3.** 8; 2 **5.** 12; 2 **7.** 4; 10 **9.** 8; 3 **11.** $(n + 2)(n + 1)$
13. $(k - 3)(k - 2)$ **15.** $(x - 5)(x - 4)$ **17.** $(t - 12)(t - 2)$ **19.** $(z - 4)(z - 3)$ **21.** $(c + 8)(c + 3)$
B **23.** $(r + 5s)(r + 4s)$ **25.** $(m + 5n)(m + 3n)$ **27.** $(8 - x)(2 - x)$ **29.** $(4 + q)(1 + q)$
C **31.** $(b - 10)(b - 2)$ **33.** $(m - 4)(m - 2)$

Page 402 Written Exercises A **1.** 9; 2 **3.** 6; 7 **5.** 9; 7 **7.** 5; 8 **9.** 11; 4 **11.** $(t + 9)(t - 7)$
13. $(r + 11)(r - 2)$ **15.** $(s - 3)(s + 2)$ **17.** $(c - 8)(c + 7)$ **19.** $(n - 15)(n + 3)$ **21.** $(w + 8)(w - 6)$
23. $(b + 9)(b - 4)$ **25.** $(x + 8)(x - 5)$ **27.** $(b + 17)(b - 1)$ B **29.** $(y + 23x)(y - x)$
31. $(r + 50a)(r - 2a)$ **33.** $(yz - 19)(yz + 3)$ **35.** $(a - 18c)(a + 4c)$ **37.** $(x + 13y)(x - 5y)$
39. $(b + 19c)(b - 2c)$ C **41.** $(x + 19y)(x - 3y)$ **43.** $(z + 10)(z - 5)$ **45.** $(b + 12a)(b - 2a)$

Pages 406–409 Cumulative Review **1.** 7 **3.** −13

5. −2

7. > **9.** > **11.** {the directed numbers < 3}

13 {the directed numbers > 0.5} **15.** 0 **17.** −9 **19.** 7

21. −12 **23.** 4; Add. Prop. of Opp. **25.** 5; Comm. Prop. **27.** 18 **29.** 3.4 **31.** 2; 3; 4 **33.** −44
35. −3 **37.** −36 **39.** 8 **41.** $-\frac{1}{2}$ **43.** −8 **45.** $-\frac{1}{10}$ **47.** −12 **49.** 2

51. $w < -3$

53. $s \geq -10$

55. Trinomial **57.** $x^2 + 2x + 2$ **59.** $-x^6 + 4x^4 + 2$ **61.** $3t - 4$ **63.** $2k + 1$ **65.** $2k + 5c$
67. $b^2 + 2b - 1$ **69.** $-6t + 1$ **71.** $11t^2 - 4t - 2$ **73.** $4x^2 - 24$ **75.** −20 **77.** −3 **79.** $12q^3$ **81.** $6t^4x^2$
83. $9y^4$ **85.** $-t^4 + 2t^3 + t$ **87.** $2r^2 + 2r - 12$ **89.** $4y^2 + 8y + 4$ **91.** $t^2 - 9$ **93.** $49a^2 - 1$

95. $3a + 5$ **97.** $\frac{ab}{2} - 1 + \frac{1}{b}$ **99.** $x + z$ **101.** $2^3 \cdot 3^2$ **103.** $2 \cdot 5 \cdot 11$ **105.** $2 \cdot 7 \cdot 3^2$ **107.** $2ab$ **109.** $7x$

111. $13(3k^4 + k - 1)$ **113.** $(2 + r)(t + s)$ **115.** $(5 + b)(5 - b)$ **117.** $(2t - 3)(2t - 3)$
119. $(x + 6)(x + 5)$ **121.** $(x - 11)(x - 2)$ **123.** $(x - 7)(x + 5)$ **125.** $(x - 8)(x + 3)$

Extra Practice Exercises

Page 410 **1.** 22 **3.** 24 **5.** 10 **7.** 8 **9.** 14 **11.** 55 **13.** 5 **15.** 8 **17.** 24 **19.** 72 **21.** 65 **23.** 0 **25.** 48
27. 20 **29.** 2 **31.** 8 **33.** 60 **35.** 6 **37.** 19 **39.** 18 **41.** 12 **43.** 30 **45.** 15 **47.** 32 **49.** 16 **51.** 31 **53.** 0
55. 8 **57.** 35 **59.** 8

Page 411 **1.** 4 **3.** 6 **5.** 7 **7.** 3 **9.** 6 **11.** 13 **13.** 3 **15.** 4 **17.** 10 **19.** 333 **21.** 7 **23.** 14 **25.** 120
27. 45 **29.** $\frac{1}{5}$ **31.** $\frac{1}{2}$ **33.** $\frac{1}{4}$ **35.** $\frac{5}{4}$ or $1\frac{1}{4}$ **37.** $\frac{7}{16}$ **39.** $\frac{3}{4}$ **41.** $\frac{3}{7}$ **43.** 4 **45.** 1 **47.** 1 **49.** 28 **51.** 1

Page 412 **1.** 0.5 **3.** 0.57 **5.** 0.6 **7.** 0.5 **9.** 0.9 **11.** 0.625 **13.** 0.16 **15.** 0.18 **17.** $\frac{1}{5}$ **19.** $1\frac{4}{5}$ **21.** $\frac{3}{8}$
23. $2\frac{4}{25}$ **25.** $3\frac{1}{8}$ **27.** $1\frac{1}{100}$ **29.** 0.45; 45% **31.** 0.3; 30% **33.** 1; 100% **35.** 0.125; 12.5% **37.** 0.50; 50%
39. 0.6; 60% **41.** 0.32; 32% **43.** 0.52; 52% **45.** $\frac{4}{25}$ **47.** $\frac{6}{25}$ **49.** $\frac{9}{25}$ **51.** $\frac{9}{20}$ **53.** $\frac{3}{25}$ **55.** $\frac{63}{200}$ **57.** $1\frac{1}{5}$ **59.** $\frac{81}{125}$
61. $4\frac{1}{5}$

Page 413 **1.** $\frac{5}{7}$ **3.** $\frac{2}{3}$ **5.** $\frac{3}{4}$ **7.** $\frac{2}{3}$ **9.** $\frac{7}{16}$ **11.** $9\frac{1}{3}$ **13.** $\frac{1}{4}$ **15.** 3 **17.** $5\frac{5}{6}$ **19.** $4\frac{1}{2}$ **21.** $\frac{3}{14}$ **23.** $2\frac{7}{18}$ **25.** $7\frac{2}{3}$
27. $\frac{4}{5}$ **29.** $6\frac{16}{21}$ **31.** $1\frac{1}{3}$ **33.** $5\frac{1}{2}$ **35.** $\frac{1}{12}$ **37.** $\frac{1}{2}$ **39.** $\frac{5}{12}$ **41.** $\frac{5}{12}$ **43.** $\frac{5}{9}$ **45.** $\frac{2}{3}$

Page 414 **1.** $\frac{1}{12}$ **3.** $\frac{1}{12}$ **5.** $\frac{1}{5}$ **7.** $\frac{2}{27}$ **9.** 1 **11.** 4 **13.** $\frac{1}{2}$ **15.** $2\frac{3}{4}$ **17.** $1\frac{1}{8}$ **19.** 3 **21.** $7\frac{1}{3}$ **23.** 1 **25.** $\frac{1}{4}$ **27.** $1\frac{1}{2}$
29. $\frac{5}{6}$ **31.** $\frac{23}{42}$ **33.** $\frac{1}{6}$ **35.** $\frac{2}{15}$ **37.** $1\frac{1}{9}$ **39.** $\frac{1}{2}$ **41.** $3\frac{1}{3}$ **43.** $\frac{3}{4}$ **45.** $\frac{2}{3}$

Page 415 **1.** 4.372 **3.** 3.338 **5.** 3.496 **7.** 3.01 **9.** 11.958 **11.** 25.73 **13.** 151.14 **15.** 11.23 **17.** 63.121
19. 5.42 **21.** 0.035 **23.** 0.4424 **25.** 21.19 **27.** 0.43216 **29.** 0.26 **31.** 2.875 **33.** $167.72 **35.** $6.35
37. 0.2736 **39.** 1.5

Page 416 **1.** 96 **3.** 0.052 **5.** 33.61 **7.** 6 **9.** 3 **11.** 250 **13.** $4\frac{2}{3}$ **15.** 86.382 **17.** 13.652 **19.** 8 **21.** {2}
23. {1.28} **25.** {4} **27.** {$\frac{1}{2}$} **29.** {2} **31.** {4}

Page 417   **1.** $11 = 11$ **3.** $5.02 = 5.02$ **5.** $2 = 2$ **7.** $1.5 = 1.5$ **9.** m **11.** s **13.** h **15.** $2 = 2$
17. $100 = 100$ **19.** $45 = 45$ **21.** $28 = 28$ **23.** $2 = 2$ **25.** 9 **27.** 36 **29.** 6 **31.** 128 **33.** 6

Page 418   **1.** 23.56 cm^2 **3.** 196 cm^2 **5.** 154 mm^2 **7.** 24.16 m **9.** 242 cm **11.** 0.2 m^2 **13.** 63 cm
15. 13 cm **17.** 30 mm

Page 419   **1.** 0, true; 1, true; 3, true; 5, false   **3.** 2, false; 4, true; 7, true   **5.** 4, false; 8, false; 30, true
7. 0, true; 4, true; 7, false; 8, false   **9.** 0, false; $\frac{1}{2}$, false; $\frac{1}{5}$, false; 2, true   **11.** 2, false; 4, true; 6, true
13. 7, true; 16, true; 17, false   **15.** 2, false; 7, true; 10, true; 12, true
17. 0, false; 7, false; 8, true; 10, true   **19.** 0, false; $\frac{1}{5}$, false; $\frac{1}{2}$, true; $\frac{4}{5}$, true   **21.** $24r$ **23.** $14 - 12m$
25. $14vw + 28$ **27.** $4s$ **29.** $5x + 10$ **31.** $7a + 27 + 26c$ **33.** $8p$ **35.** $3z + 14y$ **37.** $36qr + 2st$
39. $68w + 34$ **41.** $8f$

Page 420   **1.** Subtract 7; $m = 8$ **3.** Add 4; $s = 20$ **5.** Add 1; $x = 1\frac{7}{8}$ **7.** Subtract $\frac{2}{7}$; $k = \frac{2}{7}$ **9.** 16.9
11. 29   **13.** 579   **15.** 54   **17.** 26   **19.** 5   **21.** 14   **23.** 32   **25.** 0.05   **27.** 12   **29.** 4   **31.** 6   **33.** 5   **35.** 4.1
37. 81   **39.** 0.4   **41.** 2.4   **43.** 2016   **45.** 5.04

Page 421   **1.** 6   **3.** 0.52   **5.** Teresa: 172 cm, Brandon: 165 cm   **7.** Mia: $3\frac{1}{2}$ hours, Jon: $4\frac{1}{2}$ hours
9. 8, 10, 12   **11.** 42   **13.** 3   **15.** Width: 4 cm, length: 17 cm   **17.** 15 dimes

Page 422   **1.** -12 **3.** -22 **5.** -25 **7.** 4   **9.** $-\frac{1}{12}$ **11.** -13 **13.** 5.25   **15.** -19 **17.** $3\frac{3}{8}$ **19.** 3
21. -8 **23.** 7   **25.** 19   **27.** -12 **29.** 12   **31.** -2 **33.** 33   **35.** 28   **37.** 83   **39.** -9 **41.** $1\frac{2}{3}$ **43.** 48
45. 8   **47.** 64   **49.** 2.21

Page 423   **1.** -36 **3.** -54 **5.** -0.495 **7.** -14 **9.** 24   **11.** -19.6 **13.** -3 **15.** -1.97 **17.** 1.05
19. $-2\frac{2}{3}$ **21.** -27 **23.** -21 **25.** -30 **27.** 16   **29.** 0   **31.** $\frac{1}{16}$ **33.** $-\frac{4}{105}$ **35.** -75 **37.** -108 **39.** 0
41. 16   **43.** -10 **45.** -8 **47.** $-\frac{3}{4}$ **49.** -64 **51.** 0

Page 424   **1.** $x - 12y$ **3.** $2ab - 4c$ **5.** $18m + 6p$ **7.** $4w - 3u - 3$ **9.** $6d + 24$ **11.** $4r^2 + 4$
13. $\dfrac{h}{2} - \dfrac{j}{7}$ **15.** $2d + 4f$ **17.** $24 - 19w$ **19.** $a + 4b$ **21.** -8 **23.** -16 **25.** -20 **27.** 4   **29.** -5 **31.** 7
33. -6 **35.** $-\frac{3}{4}$ **37.** $-\frac{5}{3}$ **39.** $-\frac{3}{2}$ **41.** -2 **43.** -4

Page 425   **1.** 5   **3.** -17 **5.** 1   **7.** -4 **9.** $\frac{1}{2}$ **11.** 1.26   **13.** -55 **15.** 9   **17.** 18   **19.** 6.09   **21.** $-\frac{11}{4}$ **23.** 16
25. 2   **27.** 9.1   **29.** 11   **31.** $\frac{1}{3}$ **33.** -9 **35.** 16   **37.** -21 **39.** 8

Page 426   **1.** 1.5 m **3.** \$196   **5.** $1\frac{1}{2}$ years   **7.** 336 cm^3 **9.** 504.7 m^2 **11.** 2.7 cm^2 **13.** 4 cm

Page 427   **1.** $11s + 16$ **3.** $9x^2 + 23z + 4$ **5.** $-2h^3 + 12k^2 - 8$ **7.** $2a + 2c$ **9.** $8.5y - 5.7x + 7.1$
11. $3s^2 + 4s - 2$ **13.** $8f - 6q^2$ **15.** $-25p + 7q^2$ **17.** $40t + w^2 - 27$ **19.** $22 - 8p$ **21.** $11a - 5b$
23. $14f - 19$

Page 428   **1.** 7   **3.** 10   **5.** 2   **7.** 6   **9.** 8   **11.** 7   **13.** -6 **15.** -0.8 **17.** -4 **19.** 16   **21.** $(x + 3)^3$
23. $-98z^5$ **25.** $3f^3$ **27.** b^8 **29.** $144p^3$ **31.** $14x^2y^2z$ **33.** $-6j^3$ **35.** $-56y$ **37.** $4f^2q^6$ **39.** $-8s^6t^3$
41. $-72h^5i^6$ **43.** $\frac{2}{21}p^2r^2$ **45.** $64s^3t^3$ **47.** $18a^4b^6c$ **49.** $-\frac{5}{6}c^4d^6$ **51.** e^{20} **53.** $c^{13}d^8 + 5c^3d^3$

Page 429   **1.** $7ef + 5f^2$ **3.** $40mn^2 + 100m^3n^5 - 120m^4n^7$ **5.** $-\frac{1}{2}u^3v$ **7.** $4c^3d^2 + 56cd$
9. $a^2 + 2ab + b^2$ **11.** $12 + 45n + 42n^2$ **13.** $14.8 - 2.88t - 3.6t^2$ **15.** $25f^2 - 81$ **17.** $f^2 + 12f + 36$
19. $e^2g^2 + 2e^2gh + e^2h^2$ **21.** $81s^2 - \frac{4}{25}t^2$ **23.** $z^2 + 6yz + 9y^2$ **25.** $36 - 4j^2$ **27.** t^2 **29.** $-\dfrac{1}{3m}$

31. $-\dfrac{8y^4}{x}$ **33.** $-\dfrac{8f^2g^2}{e}$ **35.** $-\dfrac{a^5b^4c}{7}$ **37.** $\dfrac{1.2s^3t^2}{r^2}$ **39.** $\dfrac{j^4}{8i^3}$ **41.** $\dfrac{3x^4}{2}$

Extra Practice: Word Problems

Pages 430–431 **1.** $\frac{7}{15}$; $\frac{2}{3}$ **3.** \$2; \$1.80 **5.** $5r$ **7.** $20d$ **9.** 6 **11.** 16 m **13.** 28
15. 10 classical, 6 country-western, 19 rock **17.** $30 + {}^-9 = 21$ **19.** 18 cm **21.** $x^2 - 3x + 2$
23. $6x^5 - 6x^3$ **25.** $6y^2 + 10y - 4$ **27.** 15, 16

Extra Practice: Fractions and Decimals

Pages 432–433 **1.** $\frac{1}{3}$ **3.** $\frac{1}{9}$ **5.** $\frac{5}{12}$ **7.** $\frac{2}{7}$ **9.** $\frac{9}{11}$ **11.** $\frac{3}{10}$ **13.** $\frac{72}{10}$ **15.** $\frac{2}{18}$ **17.** $\frac{18}{24}$ **19.** 0.6 **21.** 1.0
23. 0.4 **25.** $0.\overline{7}$ **27.** $\frac{4}{5}$ **29.** $\frac{13}{20}$ **31.** $\frac{3}{25}$ **33.** $\frac{3}{2}$ **35.** $\frac{5}{4}$ **37.** $0.3 = 30\%$ **39.** $0.9 = 90\%$
41. $0.25 = 25\%$ **43.** $0.52 = 52\%$ **45.** $\frac{7}{10}$ **47.** $\frac{4}{25}$ **49.** $\frac{31}{50}$ **51.** $\frac{11}{20}$ **53.** $\frac{17}{25}$ **55.** $\frac{7}{20}$ **57.** $\frac{1}{3}$ **59.** 1 **61.** 1
63. $11\frac{4}{5}$ **65.** $\frac{7}{12}$ **67.** $\frac{7}{8}$ **69.** 6.74 **71.** 13.36 **73.** 2.88 **75.** 28.82 **77.** $\frac{5}{12}$ **79.** $\frac{2}{5}$ **81.** $\frac{1}{8}$ **83.** $\frac{5}{9}$ **85.** $\frac{15}{32}$
87. 30.6 **89.** 0.00615 **91.** 31.4846 **93.** 29.0745 **95.** $\frac{5}{6}$ **97.** 1 **99.** $\frac{8}{3}$ **101.** 9 **103.** $\frac{16}{3}$ **105.** 3.09
107. 25.8 **109.** 20.5 **111.** 0.235 **113.** 245 **115.** 0.0026

Extra Practice: Formulas

Pages 434–435 **1.** 42 m **3.** 30.5 cm **5.** 8 m **7.** 132 cm **9.** 3.5 cm **11.** 37.21 km² **13.** 270 cm²
15. 125 km² **17.** 2 m **19.** 18.84 cm³ **21.** 706.5 mm³ **23.** 4 cm

Extra Practice: Factoring

Pages 436–437 **1.** $(x + 3)(y + 2)$ **3.** $(7b - 1)(a + 2)$ **5.** $2(m + 1)(n + 3)$ **7.** $(d + 5f)(d - 7)$
9. $(x - 2)(y + 6)$ **11.** $(3k + 4)(2n + 1)$ **13.** $(5n - 3)(3x - 2)$ **15.** $(x + 5)(x - 5)$
17. $(m + 6)(m - 6)$ **19.** $(xy + 1)(xy - 1)$ **21.** $(mn + 4)(mn - 4)$ **23.** $5(c + 3)(c - 3)$
25. $(4y + 5)(4y - 5)$ **27.** $3(4z + 3)(4z - 3)$ **29.** $(x + 5)^2$ **31.** $(y + 9)^2$ **33.** $(b + 13)^2$ **35.** $(xy + 2)^2$
37. $(2ab + 1)^2$ **39.** $(5x + 1)^2$ **41.** $(2c + 3d)^2$ **43.** $(n - 3)^2$ **45.** $(t - 9)^2$ **47.** $(cd - 1)^2$ **49.** $(st - 2)^2$
51. $(5x - 2)^2$ **53.** $(3n - 4)^2$ **55.** $(4x - 3y)^2$ **57.** $(y + 1)(y + 8)$ **59.** $(x + 5)(x + 7)$
61. $(y + 2)(y + 9)$ **63.** $(k + 2)(k + 8)$ **65.** $(z + 3)(z + 5)$ **67.** $(b + 1)(b + 13)$ **69.** $(d + 4)(d + 6)$
71. $(x - 1)(x - 2)$ **73.** $(z - 2)(z - 3)$ **75.** $(w - 2)(w - 5)$ **77.** $(x - 3)(x - 4)$ **79.** $(p - 1)(p - 9)$
81. $(d - 3)(d - 9)$ **83.** $(b - 3)(b - 5)$ **85.** $(x - 3)(x + 5)$ **87.** $(z - 2)(z + 3)$ **89.** $(a - 1)(a + 8)$
91. $(n + 1)(n - 10)$ **93.** $(c - 3)(c + 7)$ **95.** $(d + 2)(d - 9)$ **97.** $(g - 4)(g + 5)$

Mixed Review (pages 447–451)

Chapters 1–4 **1.** 5 **3.** 10 **5.** 22 **7.** 1, 2, 4; 4 **9.** 58, 59, 60 **11.** 30 **13.** 4 **15.** 0, 1, 2 **17.** $48mn$
19. 25.08 **21.** 6.215 **23.** 6 **25.** $\frac{5}{8}$ **27.** > **29.** 22 **31.** 61 **33.** 6, 7, 8 **35.** $\frac{5}{2}$ **37.** 20
39. 0, 1, 2, 3, 4, 5, 6 **41.** 36; 4 **43.** Thirty-two divided by some number

Chapters 5–7 **1.** 8 **3.** 9 **5.** 1 **7.** Assoc. prop. addition **9.** yes, no, yes, no **11.** true, true, false
13. $(0, {}^-1)$, $(\frac{1}{3}, 0)$, $(1, 2)$, $(1\frac{1}{3}, 3)$ **15.** 12, 18, 30 **17.** $\frac{3}{2}$ **19.** 7 **21.** 10 **23.** $x - 2.5 = 28$; $x = 30.5$;
30.5°C **25.** a. false, false, true b. true, true, true, true **27.** $x - 3$

Chapters 8–11 **1.** −15 **3.** 22 **5.** $-\frac{3}{32}$ **7.** 17 **9.** 1.6 **11.** $-3a^2 + 3a - 2$ **13.** 6, −6, −1, 2 **15.** −25
17. −2, 0, 2 **19.** 12 **21.** 122 m **23.** −6 **25.** > **27.** < **29.** $3x^2 - 3x - 6$ **31.** −5 **33.** −9 **35.** −36
37. $3\frac{1}{2}$ **39.** $r \geq -30$; {directed numbers greater than or equal to −30}

Chapters 12–14 **1.** $4m^2 - n^2 - 4p^2$ **3.** 3 **5.** $3d^2 - 9d - 1$ **7.** $24x^7$ **9.** $-18m^2 - 9m + 5$
11. $3x^3 + 11x^2 + 8$ **13.** $(x + y)(3 + x)$ **15.** $-2a^2 - 3a - 1$ **17.** $10ab^2$ **19.** $4x^2 - 20x + 25$
21. $-2x^2 - 10x + 9$ **23.** $3r$ **25.** $5 \cdot 3^3$ **27.** $-24a^3b^4c^7$

Answers to Self-Tests

Chapter 1

Page 13 **1.** 14 **2.** 2 **3.** False **4.** True **5.** 8 **6.** 4 **7.** 20 **8.** 12

Page 25 **1.** The product of three and some number is equal to twenty-one. **2.** $n + 12 = 48$ **3.** 24
4. 11 **5.** 11 **6.** 3 **7.** 4 **8.** 3

Chapter 2

Page 39 **1.** 794, 795, 796, 797 **2.** 0; 5; 1; 9; 7 **3.** $^-1$, $^+1$; $^-4$, $^+4$; $^+2$, $^-2$; $^+4$, $^-4$; $^-2$, $^+2$ **4.** $^-8$, $^-7$, $^-6$
5. $\{^-1, 0, ^+1, ^+2\}$ **6.** {the integers between $^+2$ and $^+7$}
Page 50 **1.** $^-\frac{3}{4}$; $^+3\frac{1}{4}$; $^+2\frac{1}{4}$; $^+1\frac{1}{2}$; $^-2$ **2.** $<$ **3.** $^-5$, $^-2\frac{3}{4}$, $^+\frac{1}{2}$, $^+\frac{9}{10}$, $^+1$
4. **5.** $^-3$, $^+3$; $^+1$, $^-2$; $^+2$, $^+2$; $^-1$, $^-2$ **6.**

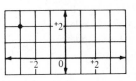

Chapter 3

Page 64 **1.** $r = 7$, $s = 14$, $t = 21$; $\{7, 14, 21, \ldots\}$ **2.** $d = 3$, $e = 6$, $f = 9$; $\{3, 6, 9, \ldots\}$ **3.** 77
4. $\{1, 2, 5, 10\}$ **5.** $\{1, 3\}$; 3 **6.** $18 = 2 \times 9$ **7.** $15 = (2 \times 7) + 1$

Page 77 **1.** $x = 30$ **2.** $\frac{2}{23}$ **3.** $\frac{4}{7}$ **4.** seven eighths; 0.875 **5.** $2.06 = 2.060 = 2.0600$ **6.** $0.4 = 40\%$
7. $0.20 = 20\%$

Page 80 **1.** 5 m 90 cm **2.** 0.87 m **3.** 3 cm 9 mm **4.** 72.0 cm

Chapter 4

Page 99 **1.** $\frac{1}{2}$ **2.** $\frac{9}{10}$ **3.** $\frac{6}{55}$ **4.** $\frac{12}{7}$ **5.** 10.842 **6.** \$7.81 **7.** 9.541 **8.** 34.52 **9.** 6

Page 114 **1.** 15 **2.** $3(14) = 42$; $6(7) = 42$ **3.** 28 **7.**
4. $2s$, s, 2 **5.** $6xy$ **6.** $(1, 8), (2, 9), (3, 10), (4, 11), (5, 12), (6, 13)$

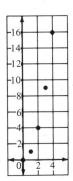

Chapter 5

Page 133 **1.** $k + 2 < 14$ **2.** $r \cdot 7 > 5$
3. Answers may vary. Example: The product of 3 and m is greater than 1.
4. Answers may vary. Example: 4 multiplied by n minus 3 is less than 10. **5.** {8} **6.** {5} **7.** $x = 4.5$
8. $y = 10$ **9.** $r = 12$ **10.** $s = 8$

Page 139 **1.** $4k - 1.3$ **2.** $m - 7$, $m + 7$ **3.** $P = 2l + 2w$; 46 cm **4.** $A = s^2$; 56.25 m

Pages 148–149 **1.** True; True; False **2.** **3.** False; False; True
4. **5.** $8l \le 120$; {the numbers of arithmetic ≤ 15}
6. $7 + 2n < 19$; {the numbers of arithmetic < 6}

Chapter 6

Page 160 **1.** $6 + x$ **2.** $m(n + 3)$ **3.** $9 = 4 + a$ **4.** $z = x^2 \cdot 6$ **5.** $3 - k = m \cdot 5$ **6.** $k + 3 = m$
7. Closed under addition and subtraction; not closed under subtraction and division
8. Closed under multiplication and division; not closed under addition and subtraction **9.** 13; 21; 30
10. $\frac{1}{2}$; $\frac{2}{5}$; $\frac{1}{15}$

Page 172 **1.** Commutative property of multiplication **2.** Commutative property of addition
3. Commutative property of multiplication **4.** Associative property of multiplication
5. Distributive property **6.** Distributive property **7.** 1264 **8.** 534 **9.** 244 **10.** 123

Page 178 **1.** 0 **2.** Every number is a solution. **3.** No solution
4. Every number except 0 is a solution. **5.** $\{(0, 3), (\frac{1}{2}, 3\frac{1}{2}), (1, 4), (1\frac{1}{2}, 4\frac{1}{2})\}$
6. $\{(2, 0), (3, 1), (4, 2)\}$

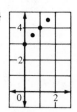

Chapter 7

Page 193 **1.** $7a$ **2.** $r + 2s$ **3.** $4t + 5$ **4.** $4mn + 4m$ **5.** 25 **6.** 8.3 **7.** 8 **8.** 0.9 **9.** 15 **10.** 1.6

Page 203 **1.** 2 **2.** 1.6 **3.** 8 **4.** 8 **5.** 16 **6.** 11 **7.** 37, 39 **8.** 8 **9.** 3 **10.** 8 **11.** 12

Chapter 8

Page 224 **1.** 2 **2.** $^-5$ **3.** 3 **4.** $^-5$

5. 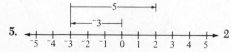 2 **6.** 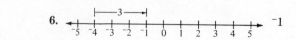 $^-1$

7. Right; $>$ **8.** Right; $>$ **9.** True   **10.** False   **11.** False

Page 231   **1.** {1}   **2.** {⁻1, 0, 1}   **3.** {⁻1, 0, 1}   **4.** {⁻1, 0, 1}
5. {the directed numbers < 2}
6. {the directed numbers > 1}
7. {⁻3 and the directed numbers $> $⁻3}
8. {the directed numbers between ⁻2 and ⁻1}

Chapter 9

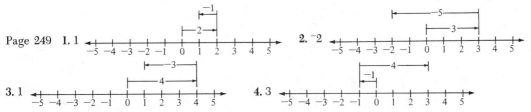

Page 249   **1.** 1   **2.** ⁻2

3. 1   **4.** 3

5. Negative five; ⁻5   **6.** The opposite of negative 6; 6   **7.** 0   **8.** 0   **9.** ⁻7   **10.** 4   **11.** −9   **12.** −7   **13.** 3
14. −5   **15.** Additive property of inverses   **16.** Associative property
17. Property of the opposite of a sum   **18.** Commutative property

Page 258   **1.** 5   **2.** −2   **3.** −9   **4.** 1   **5.** 32   **6.** 9   **7.** 20   **8.** 10   **9.** {(3, 1), (2, 0), (−2, −4), (0, −2)}
10. {(4, 0), (3, −1), (−1, −5), (4, 0)}

Chapter 10

Page 275   **1.** −12   **2.** −40   **3.** 0   **4.** 1   **5.** −10   **6.** −2   **7.** Commutative property
8. Commutative property   **9.** Distributive property   **10.** Positive; 24   **11.** Negative; −120
12. −4(−6) + (−4)2; 16 = 16   **13.** −8(−9) + (−8)(−14); 184 = 184   **14.** −8m + 4n   **15.** −21k + 2

Pages 286–287   **1.** −9   **2.** 3   **3.** −4   **4.** $\frac{7}{3}$ **5.** 1   **6.** 49   **7.** −24   **8.** −$\frac{16}{9}$ **9.** −9   **10.** 8   **11.** 7   **12.** −2
13. {(0, 0), (−4, 2), (4, −2), (−5, $\frac{5}{2}$)}

Chapter 11

Page 304   **1.** 5   **2.** −5   **3.** −5   **4.** −3   **5.** −$\frac{1}{2}$ **6.** 4   **7.** 6$\frac{1}{2}$ meters   **8.** 11 meters

Page 311
1. $x > 3$ **2.** $t \leq -1$
3. $-3 \leq t$ **4.** $m < 3$

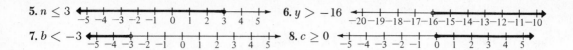

5. $n \leq 3$ **6.** $y > -16$

7. $b < -3$ **8.** $c \geq 0$

Chapter 12

Page 329   **1.** Binomial   **2.** Binomial   **3.** Trinomial   **4.** Monomial   **5.** $9x^2 + 2x + 5$ **6.** $m^3 + m^2 + 2$
7. $x^2y + xy^2 + 1$ or $xy^2 + x^2y + 1$ **8.** $y^5 + 2y + 1$ **9.** 4   **10.** -4 **11.** $12a - 9$ **12.** $5a^2 + ab + b^2$
13. Associative property   **14.** Commutative property

Page 336   **1.** 0   **2.** $7x^2 - 1$ **3.** $2y^3 + 3y^2 - y$ **4.** $-z^3 - 11z$ **5.** $6m - 1$ **6.** $t + 3$ **7.** $-2x^2 - 2x$
8. $n^2 + 13$

Page 340   **1.** $9ab - 1$ **2.** $16x + 8$ **3.** $2x^2 - 9$ **4.** 4   **5.** 6   **6.** 3   **7.** 3

Chapter 13

Page 355   **1.** $-10x^3$ **2.** $-3n^3$ **3.** 4^5 **4.** z^{11} **5.** $-6t^3s^3$ **6.** $-8m^3n$ **7.** $8x^3y^3$ **8.** a^4 **9.** 256   **10.** 729
11. $-3x^3 - x^2 + 5x$ **12.** $4a^2 + 12ab + b^2$

Page 367   **1.** $b^2 - b - 6$ **2.** $n^2 - 4n - 5$ **3.** $6t^2 + 19t - 7$ **4.** $4x^3 + 5x^2 + 3x + 2$ **5.** $n^2 - 8n + 16$
6. $4b^2 + 24b + 36$ **7.** $n^2 - 16$ **8.** $4b^2 - 36$
9. Distributive property; associative property; commutative property; Reciprocal Property

Page 376   **1.** x^4 **2.** $2n^2$ **3.** $-\dfrac{3}{zx^2}$ **4.** $\dfrac{5}{x}$ **5.** $2z + 4$ **6.** $3b - \dfrac{6}{b}$ **7.** $-4n + 2$ **8.** $c + d$ **9.** $m + 3$
10. $2n - 5$

Chapter 14

Page 391   **1.** 2^3 **2.** $2 \cdot 3^2$ **3.** $2 \cdot 3 \cdot 7$ **4.** $5 \cdot 7 \cdot m \cdot n$ **5.** $-2 \cdot 2 \cdot 3 \cdot x \cdot x \cdot y$ **6.** $2 \cdot 5 \cdot a \cdot a \cdot b \cdot b \cdot b$
7. $4ab$ **8.** $3xy$ **9.** $w^2 + 2$ **10.** $4 - x$ **11.** $(x + y)(2 + x)$ **12.** $(4n + m)(m - 2)$

Page 403   **1.** $(b + 3)(b - 3)$ **2.** $(x + 5)(x - 5)$ **3.** $(z + 5)(z + 5)$ **4.** $(y + 2)(y + 2)$ **5.** $(x - 6)(x - 6)$
6. $(m - 9)(m - 9)$ **7.** $(x + 3)(x + 2)$ **8.** $(n - 7)(n - 2)$ **9.** $(c + 4)(c - 2)$ **10.** $(c - 4)(c + 2)$

Additional Answers

Page 15 Written Exercises A 1. The sum of six and nine **2.** The sum of some number and ten **3.** Sixteen divided by some number **4.** The difference of twenty and seven **5.** The difference of some number and six **6.** The product of four and some number **7.** The product of five and eight **8.** The sum of four and some number **9.** The product of some number and ten **10.** Eighteen divided by two **11.** The difference between twenty-five and some number **12.** Forty-five divided by some number **13.** The sum of ten and some number is equal to forty-two. **14.** Twenty-seven is equal to the sum of some number and thirteen. **15.** The difference of thirty-five and some number is equal to nineteen. **16.** Sixty-four is equal to the difference between some number and twenty-six **17.** The sum of thirty-seven and some number is equal to ninety-five. **18.** Forty-eight is equal to the product of three and some number. **19.** Forty-five divided by nine is equal to some number. **20.** Forty divided by some number is equal to five. **21.** The difference of some number and sixty-eight is equal to fourteen. **22.** Some number is equal to the product of one hundred eleven and three. **23.** Thirty-nine is equal to the sum of some number and ten. **24.** The product of some number and forty-one is equal to two hundred forty-six. **25.** Seventy-five is equal to the difference of ninety-two and some number. **26.** Three hundred is equal to the product of some number and twenty-five.

Page 39 Written Exercises 18. {the integers between $^-1$ and $^+4$} **19.** {the integers greater than $^+99$} **20.** {the integers between $^-3$ and $^+3$} **21.** {the integers between $^+9$ and $^+14$} **22.** {the integers between $^-9$ and $^-4$} **23.** {the integers less than $^+1$} **24.** {the integers between $^-1$ and $^+1$} **25.** {the integers between $^-5$ and 0} **26.** Answers may vary. For example, {the integers between 0 and $^+1$} **27.** {the integers between $^-10$ and $^-7$ and their opposites} *or* {the integers between $^+7$ and $^+10$ and their opposites} **28.** {the integers between 0 and $^+3$ and their opposites} *or* {the integers between 0 and $^-3$ and their opposites} **29.** {the even integers between 0 and $^+6$, inclusive} **30.** {the negative integers} **31.** {the integers greater than $^+4$}

Page 142 Written Exercises B 21. {the numbers of arithmetic less than 7} **22.** {the numbers of arithmetic less than 6} **23.** {the numbers of arithmetic greater than 4} **24.** {the numbers of arithmetic greater than 5} **25.** {the numbers of arithmetic greater than 2.5} **26.** {the numbers of arithmetic less than 3} **27.** {the numbers of arithmetic greater than 11} **28.** {the numbers of arithmetic less than 4} **29.** {the numbers of arithmetic} **C 30.** {the numbers of arithmetic less than 10.8} **31.** {the numbers of arithmetic greater than 21} **32.** {the numbers of arithmetic less than 40.4} **33.** {the numbers of arithmetic greater than 0} **34.** {the numbers of arithmetic less than 5} **35.** {the numbers of arithmetic whose squares are greater than 47} **36.** {the numbers of arithmetic less than 4} **37.** {the numbers of arithmetic greater than 10} **38.** {the numbers of arithmetic greater than 1}

Page 145 Written Exercises B 22. {the numbers of arithmetic except $2\frac{2}{3}$} **23.** {the numbers of arithmetic greater than or equal to 1} **24.** {the numbers of arithmetic less than 3} **25.** {the numbers of arithmetic less than or equal to $4\frac{1}{2}$} **26.** {the numbers of arithmetic greater than $3\frac{1}{2}$} **27.** {the numbers of arithmetic less than or equal to 5}

Page 230 Written Exercises 7. {the directed numbers greater than 1} **8.** {the directed numbers less than or equal to $2\frac{1}{3}$} **9.** {the directed numbers greater than $^-2$} **10.** {the directed numbers less than or equal to 0} **11.** {the directed numbers less than or equal to $^-5$} **12.** {the directed numbers between $^-2$ and 2} **13.** {the directed numbers less than or equal to 10} **14.** {the directed numbers between $^-3$ and 0} **C 23.** {$^-8$ and the directed numbers between -8 and -3} **24.** {0 and $2\frac{1}{3}$, and the directed numbers between 0 and $2\frac{1}{3}$} **25.** {$^-5$ and the directed numbers between $^-5$ and $^-10$} **26.** {the directed numbers greater than or equal to 2} **27.** {the directed numbers between $^-6.6$ and $^-4.2$} **28.** {the directed numbers between $^-2.6$ and 1.2} **29.** {the directed numbers, except 5} **30.** {$^-0.0005$ and 0, and the directed numbers between $^-0.0005$ and 0} **31.** {$^-8$ and the directed numbers between $^-8$ and $^-2$} **32.** {the directed numbers less than or equal to $^-15$}

Page 322 Written Exercises B **22.** $\{(-3, 18), (-1, 4), (0, 3), (1, 6), (3, 24)\}$ **23.** $\{(-6, -13), (-4, -7), (-2, -1),$ $(0, 5), (2, 11), (4, 17), (6, 23)\}$ **24.** $\{(-1, 10), (-\frac{1}{2}, 8\frac{1}{4}), (0, 7), (\frac{1}{2}, 6\frac{1}{4}), (1, 6)\}$ **25.** $\{(-3, -17), (-2, 0), (0, 4), (2, 8),$ $(3, 25)\}$ **26.** $\{(-\frac{1}{10}, -13\frac{19}{20}), (-\frac{1}{5}, -12\frac{4}{5}), (0, -15), (\frac{1}{5}, -16\frac{4}{5}), (\frac{1}{10}, -15\frac{19}{20})\}$ **27.** $\{(-4, -40), (-2, -2), (0, 12),$ $(2, 2), (4, -32)\}$ **28.** $\{(-6, 33), (-4, 15), (-2, 5), (0, 3), (2, 9), (4, 23), (6, 45)\}$ **29.** $\{(-2, -9), (-1, -2), (0, -1),$ $(1, 0), (2, 7)\}$ C **30.** $\{(-10, 22), (-5, 17), (0, 12), (5, 7), (10, 2)\}$ **31.** $\{(0.1, 3.31), (0, 3), (0.1, 2.71)\}$ **32.** $\{(-5, -67),$ $(-3, -25), (-1, 1), (0, 8), (1, 11), (3, 5), (5, -17)\}$

Page 362 Written Exercises **19.** $16m^2 + 80m + 100$ **20.** $x^4 + 16x^2y^2 + 64y^4$ **21.** $9b^2c^2 + 60ab^2c + 100a^2b^2$
22. $m^2n^4 - 10mn^2 + 25$ **23.** $16m^2 + 56bm + 49b^2$ **24.** $0.64 - 1.6a + a^2$ **25.** $324 + 72ab + 4a^2b^2$
26. $4x^4 + 4x^2y^2 + y^4$ **27.** $s^2t^4 - 4st^2 + 4$ **28.** $9x^4 + 24x^2z^2 + 16z^4$ **29.** $0.16m^4 + 0.72\,m^2n + 0.81n^2$
30. $16c^2 - 0.8cd + 0.01d^2$ C **31.** $4y^6 - 4y^3 + 1$ **32.** $z^4 - 2y^3z^2 + y^6$ **33.** $z^6 - 2x^3z^3 + x^6$
34. $p^2 + 0.6pq^2 + 0.09q^4$ **35.** $0.04 + 0.2m^2 + 0.25m^4$ **36.** $0.49 + 4.2x + 9x^2$ **37.** $16k^2 - k + \frac{1}{64}$
38. $a^2 + ab + \frac{1}{4}b^2$ **39.** $6.25p^2 - p + \frac{1}{25}$

Pages 366–367 Written Exercises

C **17.** $4y^3(y^4 - 1) = 4y^3 \cdot y^4 - 4y^3 \cdot 1$ distributive prop.
$\qquad\qquad\qquad = 4y^3 \cdot y^4 - 4(y^3 \cdot 1)$ assoc. prop.
$\qquad\qquad\qquad = 4y^3 \cdot y^4 - 4(1 \cdot y^3)$ commutative prop.
$\qquad\qquad\qquad = 4y^3 \cdot y^4 - (4 \cdot 1)y^3$ assoc. prop.
$\qquad\qquad\qquad = 4y^7 - 4y^3$ mult. prop. of 1

18. $(mn)(mn + 1) = mn \cdot mn + mn \cdot 1$ distributive prop.
$\qquad\qquad\qquad = m(n \cdot m)n + mn \cdot 1$ assoc. prop.
$\qquad\qquad\qquad = m(m \cdot n)n + mn \cdot 1$ commutative prop.
$\qquad\qquad\qquad = (m \cdot m)(n \cdot n) + mn \cdot 1$ assoc. prop.
$\qquad\qquad\qquad = (m \cdot m)(n \cdot n) + mn$ mult. prop. of 1
$\qquad\qquad\qquad = m^2n^2 + mn$

19. $(x^2 + xy + y^2)(xy) = x^2(xy) + xy(xy) + y^2(xy)$ distributive prop.
$\qquad\qquad\qquad\quad = (x^2 \cdot x)y + x(y \cdot x)y + (y^2 \cdot x)y$ assoc. prop.
$\qquad\qquad\qquad\quad = (x^2 \cdot x)y + x(x \cdot y)y + (x \cdot y^2)y$ comm. prop.
$\qquad\qquad\qquad\quad = (x^2 \cdot x)y + (x \cdot x)y \cdot y + x(y^2 \cdot y)$ assoc. prop.
$\qquad\qquad\qquad\quad = x^3y + x^2y^2 + xy^3$

20. $(12a^2b)(4ab) = 12(a^2b \cdot 4)ab$ assoc. prop.
$\qquad\qquad\qquad = 12(4 \cdot a^2b)ab$ commutative prop.
$\qquad\qquad\qquad = (12 \cdot 4)[a^2(b \cdot a)b]$ assoc. prop.
$\qquad\qquad\qquad = (12 \cdot 4)[a^2(a \cdot b)b]$ commutative prop.
$\qquad\qquad\qquad = (12 \cdot 4)(a^2 \cdot a)(b \cdot b)$ assoc. prop.
$\qquad\qquad\qquad = 48a^3b^2$

21. $(\frac{1}{2}mnp)(\frac{1}{4}m^2np^2) = \frac{1}{2}(mnp \cdot \frac{1}{4})m^2np^2$ assoc. prop.
$\qquad\qquad\qquad\quad = \frac{1}{2}(\frac{1}{4} \cdot mnp)m^2np^2$ commutative prop.
$\qquad\qquad\qquad\quad = (\frac{1}{2} \cdot \frac{1}{4})(mnp \cdot m^2np^2)$ assoc. prop.
$\qquad\qquad\qquad\quad = \frac{1}{8}m^3n^2p^3$

22. $(0.5a^2 - b)(b + 0.5a) = 0.5a^2 \cdot b + 0.5a^2(0.5a) - b \cdot b - b(0.5a)$ dist. prop.
$\qquad\qquad\qquad\qquad = 0.5a^2 \cdot b + 0.5(a^2 \cdot 0.5)a - b \cdot b - (b \cdot 0.5)a$ assoc. prop.
$\qquad\qquad\qquad\qquad = 0.5a^2 \cdot b + 0.5(0.5 \cdot a^2)a - b \cdot b - (0.5 \cdot b)a$ commutative prop.
$\qquad\qquad\qquad\qquad = 0.5a^2 \cdot b + (0.5 \cdot 0.5)(a^2 \cdot a) - b \cdot b - 0.5(b \cdot a)$ assoc. prop.
$\qquad\qquad\qquad\qquad = 0.5a^2b + 0.25a^3 - b^2 - 0.5ba$

23. $2.3x^2(2x^3 + 0.5x) = (2.3x^2)(2x^3) + (2.3x^2)(0.5x)$ dist. prop.

$\qquad\qquad\qquad\quad = 2.3(x^2 \cdot 2)x^3 + 2.3(x^2 \cdot 0.5)x$ assoc. prop.

$\qquad\qquad\qquad\quad = 2.3(2 \cdot x^2)x^3 + 2.3(0.5 \cdot x^2)x$ comm. prop.

$\qquad\qquad\qquad\quad = (2.3 \cdot 2)(x^2 \cdot x^3) + (2.3 \cdot 0.5)(x^2 \cdot x)$ assoc. prop.

$\qquad\qquad\qquad\quad = 4.6x^5 + 1.15x^3$

24. $\dfrac{2}{3}\,m\left(\dfrac{1}{2}\,mn - \dfrac{n^2}{4}\right) = \dfrac{2}{3}\,m \cdot \dfrac{1}{2}\,mn - \dfrac{2}{3}\,m \cdot \dfrac{n^2}{4}$ dist. prop.

$\qquad\qquad\qquad\qquad = \dfrac{2}{3}\left(m \cdot \dfrac{1}{2}\right)mn - \dfrac{2}{3}\left(m \cdot \dfrac{1}{4}\right)n^2$ assoc. prop.

$\qquad\qquad\qquad\qquad = \dfrac{2}{3}\left(\dfrac{1}{2} \cdot m\right)mn - \dfrac{2}{3}\left(\dfrac{1}{4} \cdot m\right)n^2$ comm. prop.

$\qquad\qquad\qquad\qquad = \left(\dfrac{2}{3} \cdot \dfrac{1}{2}\right)(m \cdot mn) - \left(\dfrac{2}{3} \cdot \dfrac{1}{4}\right)(m \cdot n^2)$ assoc. prop.

$\qquad\qquad\qquad\qquad = \dfrac{1}{3}\,m^2n - \dfrac{1}{6}\,mn^2$

25. $\left(\dfrac{5n}{2} - \dfrac{1}{3}\right)\left(\dfrac{5n}{2} + \dfrac{1}{3}\right) = \dfrac{5n}{2} \cdot \dfrac{5n}{2} + \dfrac{5n}{2} \cdot \dfrac{1}{3} - \dfrac{1}{3} \cdot \dfrac{5n}{2} - \dfrac{1}{3} \cdot \dfrac{1}{3}$ dist. prop.

$\qquad\qquad\qquad\qquad\quad = \dfrac{5}{2}\left(n \cdot \dfrac{5}{2}\right)n + \dfrac{5}{2}\left(n \cdot \dfrac{1}{3}\right) - \dfrac{1}{3} \cdot \dfrac{5}{2}\,n - \dfrac{1}{3} \cdot \dfrac{1}{3}$ assoc. prop.

$\qquad\qquad\qquad\qquad\quad = \dfrac{5}{2}\left(\dfrac{5}{2} \cdot n\right)n + \dfrac{5}{2}\left(\dfrac{1}{3} \cdot n\right) - \dfrac{1}{3} \cdot \dfrac{5}{2}\,n - \dfrac{1}{3} \cdot \dfrac{1}{3}$ comm. prop.

$\qquad\qquad\qquad\qquad\quad = \left(\dfrac{5}{2} \cdot \dfrac{5}{2}\right)(n \cdot n) + \left(\dfrac{5}{2} \cdot \dfrac{1}{3}\right)n - \dfrac{1}{3} \cdot \dfrac{5}{2}\,n - \dfrac{1}{3} \cdot \dfrac{1}{3}$ assoc. prop.

$\qquad\qquad\qquad\qquad\quad = \dfrac{25}{4}\,n^2 + \dfrac{5}{6}\,n - \dfrac{5}{6}\,n - \dfrac{1}{9}$ substitution prin.

$\qquad\qquad\qquad\qquad\quad = \dfrac{25}{4}\,n^2 - \dfrac{1}{9}$

26. $3.7y(2y + 0.3y - 1.2) = 3.7y \cdot 2y + 3.7 \cdot 0.3y - 3.7y \cdot 1.2$ dist. prop.

$\qquad\qquad\qquad\qquad\quad = 3.7(y \cdot 2)y + 3.7(y \cdot 0.3)y - 3.7(y \cdot 1.2)$ assoc. prop.

$\qquad\qquad\qquad\qquad\quad = 3.7(2 \cdot y)y + 3.7(0.3 \cdot y)y - 3.7(1.2 \cdot y)$ comm. prop.

$\qquad\qquad\qquad\qquad\quad = (3.7 \cdot 2)(y \cdot y) + (3.7 \cdot 0.3)(y \cdot y) - (3.7 \cdot 1.2)y$ assoc. prop.

$\qquad\qquad\qquad\qquad\quad = 7.4y^2 + 1.11y^2 - 4.44y$ substitution prin.

$\qquad\qquad\qquad\qquad\quad = 8.51y^2 - 4.44y$

27. $(0.1t + t^2)(3t - 0.5) = 0.1t \cdot 3t - 0.1t \cdot 0.5 + t^2 \cdot 3t - t^2 \cdot 0.5$ dist. prop.

$\qquad\qquad\qquad\qquad\quad = 0.1(t \cdot 3)t - 0.1(t \cdot 0.5) + (t^2 \cdot 3)t - (t^2 \cdot 0.5)$ assoc. prop.

$\qquad\qquad\qquad\qquad\quad = 0.1(3 \cdot t)t - 0.1(0.5 \cdot t) + (3 \cdot t^2)t - (0.5 \cdot t^2)$ comm. prop.

$\qquad\qquad\qquad\qquad\quad = (0.1 \cdot 3)(t \cdot t) - (0.1 \cdot 0.5)t + 3(t^2 \cdot t) - (0.5 \cdot t^2)$ assoc. prop.

$\qquad\qquad\qquad\qquad\quad = 0.3t^2 - 0.05t + 3t^3 - 0.5t^2$ substitution prin.

$\qquad\qquad\qquad\qquad\quad = -0.2t^2 - 0.05t + 3t^3$

28. $\tfrac{1}{4}x(x^3 - \tfrac{1}{3}x^2 + \tfrac{1}{2}) = \tfrac{1}{4}x \cdot x^3 - \tfrac{1}{4}x \cdot \tfrac{1}{3}x^2 + \tfrac{1}{4}x \cdot \tfrac{1}{2}$ dist. prop.

$\qquad\qquad\qquad\qquad\quad = \tfrac{1}{4}x \cdot x^3 - \tfrac{1}{4}x \cdot \tfrac{1}{3}x^2 + \tfrac{1}{4}(x \cdot \tfrac{1}{2})$ assoc. prop.

$\qquad\qquad\qquad\qquad\quad = \tfrac{1}{4}x \cdot x^3 - \tfrac{1}{4}x \cdot \tfrac{1}{3}x^2 + \tfrac{1}{4}(\tfrac{1}{2} \cdot x)$ comm. prop.

$\qquad\qquad\qquad\qquad\quad = \tfrac{1}{4}x \cdot x^3 - \tfrac{1}{4}x \cdot \tfrac{1}{3}x^2 + (\tfrac{1}{4} \cdot \tfrac{1}{2})x$ assoc. prop.

$\qquad\qquad\qquad\qquad\quad = \tfrac{1}{4}x^4 - \tfrac{1}{12}x^3 + \tfrac{1}{8}x$

29. $\left(\dfrac{3}{5}rs - r\right)\left(\dfrac{5s}{r}\right)$ $= \dfrac{3}{5}rs \cdot 5 \cdot \dfrac{s}{r} - r \cdot 5 \cdot \dfrac{s}{r}$ dist. prop.

$ = \dfrac{3}{5}(rs \cdot 5)\dfrac{s}{r} - (r \cdot 5)\dfrac{s}{r}$ assoc. prop.

$ = \dfrac{3}{5}(5 \cdot rs)\dfrac{s}{r} - (5 \cdot r)\dfrac{s}{r}$ comm. prop.

$ = \left(\dfrac{3}{5} \cdot 5\right)\left(rs \cdot \dfrac{s}{r}\right) - 5\left(r \cdot \dfrac{s}{r}\right)$ assoc. prop.

$ = 3s^2 - 5s$

30. $\left(\dfrac{xy}{3}\right)\left(\dfrac{3x}{y}\right)$ $= \dfrac{1}{3}(xy \cdot 3)\dfrac{x}{y}$ assoc. prop.

$ = \dfrac{1}{3}(3 \cdot xy)\dfrac{x}{y}$ comm. prop.

$ = \left(\dfrac{1}{3} \cdot 3\right)\left(xy \cdot \dfrac{x}{y}\right)$ assoc. prop.

$ = 1 \cdot x^2$ substitution prin.

$ = x^2$ mult. prop. of 1